Xth INTERNATIONAL ASTRONAUTICAL CONGRESS
LONDON 1959

X. INTERNATIONALER ASTRONAUTISCHER KONGRESS

Xe CONGRÈS INTERNATIONAL D'ASTRONAUTIQUE

PROCEEDINGS

BERICHT   COMPTES RENDUS

HERAUSGEGEBEN VON   EDITORIAL BOARD   COMITÉ DES RÉDACTEURS

W. von BRAUN · A. EULA · B. FRAEIJS de VEUBEKE · F. HECHT
W. B. KLEMPERER · J. M. J. KOOY · F. I. ORDWAY III · E. SÄNGER
K. SCHÜTTE · L. I. SEDOV · L. R. SHEPHERD · J. STEMMER

SCHRIFTLEITUNG   EDITOR-IN-CHIEF   RÉDACTEUR EN CHEF
F. HECHT

MIT 464 FIGUREN   WITH 464 FIGURES   AVEC 464 FIGURES

I
⟨P. 1—504⟩

SPRINGER-VERLAG WIEN GMBH 1960

ISBN 978-3-662-38961-4   ISBN 978-3-662-39914-9 (eBook)
DOI 10.1007/978-3-662-39914-9
Softcover reprint of the hardcover 1st edition 1960

Vorwort

Der X. Internationale Astronautische Kongreß, eine Jubiläumsveranstaltung der International Astronautical Federation (I.A.F.), wurde in besonders eindrucksvoller Weise in der Woche zwischen dem 30. August und dem 5. September 1959 von der British Interplanetary Society in London durchgeführt, wo schon 1951 der zweite dieser Kongresse stattgefunden hatte. Die Vorträge von 1959 zeigen den fast unglaublichen Fortschritt dieser acht Jahre, den wohl kaum einer der Teilnehmer jenes historischen Kongresses erwartet hätte, der die Gründung der I.A.F. brachte. Die astronautischen Forscher und Techniker dürfen berechtigte Genugtuung über die vielfache Bestätigung eines großen Teiles ihrer damaligen Gedanken und Projekte empfinden.

Die beiden vorliegenden Berichtsbände enthalten die beim X. I.A.F.-Kongreß gehaltenen Vorträge, soweit deren Manuskripte von den Verfassern vor oder nach dem Kongreß zur Verfügung gestellt wurden. Von fast sämtlichen übrigen Vorträgen liegen wenigstens mehrsprachige Zusammenfassungen vor, außerdem Kurzfassungen von vier Arbeiten, die in den „Astronautica Acta" erschienen sind. Der British Interplanetary Society sei für die Mühe, möglichst alles Vortragsmaterial zu beschaffen, wärmstens gedankt.

Wie im Vorjahr wurden die einzelnen Beiträge zum Zweck schnellerer Veröffentlichung nicht in alphabetischer Reihenfolge nach den Namen der Verfasser, sondern je nach der Fertigstellung der Korrekturen und Abbildungen angeordnet. Dem ersten Band wurden wieder ein thematisch geordnetes Inhaltsverzeichnis und ein alphabetisches Verzeichnis aller Mitarbeiter vorangestellt.

Der Herausgeber spricht den Übersetzern der meisten Zusammenfassungen der einzelnen Artikel ins Französische beziehungsweise Englische, den Herren Professor Ing. BAUDOUIN FRAEIJS DE VEUBEKE (Liége), Dr. WOLFGANG B. KLEMPERER (Santa Monica, Calif.) und Mr. FREDERICK I. ORDWAY, III (Washington, D.C.) für ihre überaus wertvolle Hilfe den besten Dank aus. Auch dem Springer-Verlag gebührt besonderer Dank für die gewohnt gute Herstellung und technische Ausstattung der beiden Berichtsbände.

Friedrich Hecht, Wien

Contents — Inhalt — Sommaire

Contents — Inhalt — Sommaire   VII

VI. Artificial Satellites — Künstliche Satelliten — Satellites artificiaux

VII. Space Biology and Medicine — Raumfahrtbiologie und -medizin — Biologie et médecine spatiales

Authors — Autoren — Auteurs

Determination of Air Density and the Earth's Gravitational Field from the Orbits of Artificial Satellites

By

D. G. King-Hele[1]

(With 9 Figures)

(Received July 9, 1959)

Abstract — Zusammenfassung — Résumé

Determination of Air Density and the Earth's Gravitational Field from the Orbits of Artificial Satellites. Methods for evaluating air density and scale height from the changes in the orbits of satellites are presented, and are used to determine air density at heights between 200 and 400 Km., and to trace the variation of density with time. The methods take account of the oblateness of the earth and atmosphere, the tumbling of the satellites, and the rotation of the atmosphere.

The evaluation of the successive terms in the earth's gravitational potential with the aid of satellites is discussed, and results so far obtained are outlined.

Bestimmung der atmosphärischen Dichte und des Erdschwerefeldes aus den Bahnen künstlicher Satelliten. Es werden Methoden zur Auswertung der Luftdichte und der Höhenstufen aus den Bahnänderungen von Satelliten angegeben. Sie werden benützt, um die Luftdichte in Höhen zwischen 200 und 400 km zu bestimmen und die zeitliche Dichtenänderung zu verfolgen. Die angewendeten Verfahren setzen die Abplattung der Erde und der Atmosphäre, die sich überschlagende Bewegung der Satelliten und die Rotation der Atmosphäre in Rechnung.

Die Auswertung der aufeinander folgenden Terme im Gravitationspotential der Erde mit Hilfe von Satelliten wird erörtert; die bis jetzt erzielten Ergebnisse werden skizziert.

Détermination de la masse volumique de l'air et du champ de gravitation par analyse des orbites de satellites artificiels. Des méthodes sont présentées pour évaluer la masse volumique de l'air à partir des altérations observées dans les orbites de satellites. Elles sont appliquées au calcul à des altitudes comprises entre 200 et 400 km. et permettent de retracer la variation de la densité avec le temps. L'aplatissement de la terre et de l'atmosphère et sa rotation ainsi que la rotation transversale du satellite sont pris en considération.

On discute l'évaluation des termes successifs du potentiel de gravitation à l'aide des satellites et les résultats obtenus jusqu'à présent sont esquissés.

I. Introduction

The main perturbations to earth-satellite orbits, those due to the atmosphere and the oblateness of the earth, are of quite different types, and, to a first approximation, can be treated separately. Because of this fortunate separation of the main effects, the properties of the atmosphere and the gravitational field can be

---

[1] Royal Aircraft Establishment, Farnborough, Hants., United Kingdom.

analysed in a surprisingly detailed manner by observing how the orbit of an actual satellite departs from the fixed ellipse which represents the orbit over spherical earth in vacuo. The present paper reviews some of the work on these topics done during the past year at the Royal Aircraft Establishment, Farnborough.

The main effects of the earth's atmosphere and oblateness on satellite orbits of appreciable eccentricity may be summed up as follows. The atmospheric drag reduces the eccentricity and the length of the major axis, (and consequently the period of revolution). For a given satellite, the rate of change of all these quantities depends primarily on the air density near perigee, and the rates of change gradually increase as the perigee height slowly decreases. The main effects of the earth's oblateness are quite different: a rotation of the plane of the orbit, about the earth's axis; a rotation of the major axis of the orbit, in the orbital plane, but with no significant change in its length; and a slight distortion of the orbital ellipse, but with no significant change in eccentricity.

Though the main effects of the atmosphere and the non-spherical components of the earth's gravitational field are distinct, it proves necessary to take account of some secondary effects which do interact: for example, the side-force created by the rotation of the atmosphere slightly alters the rate of rotation of the orbital plane; and the third harmonic in the earth's gravitational field (which expresses asymmetry about the equator) causes a small oscillation in perigee distance, which has to be taken into account in studies of the air density.

There are many minor perturbations to the orbit, the most important being those due to the gravitational attractions of the moon and the sun. Most of the other perturbations, e.g. electromagnetic and relativity effects, have proved too small to be worth considering at present.

II. Determination of Average Air Density, at Heights between 200 and 400 Km.

The action of atmospheric drag makes the orbit of an earth satellite slowly contract, and its period of revolution T decrease. When the eccentricity of the orbit exceeds about 0.02, the drag at perigee is much greater than at apogee, and the rate of decrease of T is determined by the total loss in velocity due to the integrated effect of drag in the region near perigee. If the mass and dimensions of the satellite are known, and estimates are made of its aerodynamic drag, values of air density at heights near perigee can be obtained.

Such estimates of air density have been made in a number of previous papers [1—11]. Since the effective cross-sectional areas and drag coefficients of the satellites are difficult to determine exactly, the various authors have inevitably made different estimates of these quantities. Also most of the previous papers refer to only one or two satellites. In this section of the present paper, which is a revised version of [13], values of density are derived from all the ten satellites launched before the end of 1958 whose orbits are known, the satellites are treated consistently, and a more accurate method of determining density is used.

1. Previous Estimates

The results from the previous papers [1—11], and values [12] obtained from instruments aboard Sputnik 3 (1958 δ 2), are collected in Fig. 1, together with two proposed interim model atmospheres [6, 14] and the A.R.D.C. model atmosphere [15], which was probably the most widely-used standard atmosphere in the year prior to satellite launchings. As is well known, the results from satellites show the density at heights between 200 and 400 Km. to be 5—15 times greater than indicated by the A.R.D.C. model. In Fig. 1, the abscissa is the density rela-

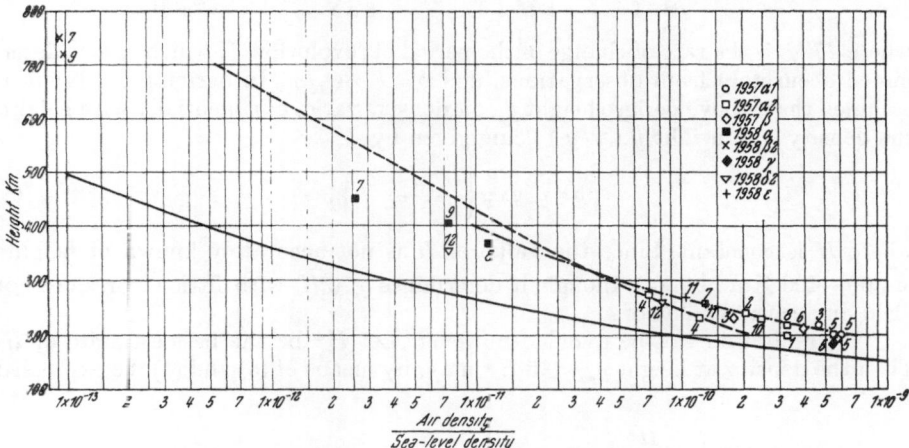

Fig. 1. Values of air density obtained by means of satellites [1—12], with proposed "standard atmospheres" [6, 14, and 15]. Reference numbers are shown beside the plotted points

------- N.R.L. atmosphere (1958) [14]
—·—·—· Smithsonian atmosphere (1957—2) [6]
————— A.R.D.C. atmosphere (1956) [15]

tive to sea level, the sea-level density being taken as 1.23 Kg./m³ or 0.0765 lb./cu. ft. The scatter of the points in Fig. 1 is not significant, since it could be due to the different methods and assumptions of the different investigators.

2. The Present Method of Analysis

The drag D encountered by a satellite moving through air of density ϱ is expressed in the usual way in terms of a drag coefficient C_D, as

$$D = \frac{1}{2} \varrho V^2 S C_D , \tag{1}$$

where S is a reference area, which is taken as the mean cross-section of the satellite perpendicular to the direction of motion, and V is the velocity of the satellite relative to the ambient air. Because of the rotation of the atmosphere, V differs slightly from the velocity v of the satellite relative to the earth's centre, and, if the atmosphere is assumed to rotate with the same angular velocity as the earth, it can be shown [16] that the drag is given with adequate accuracy by

$$D = \frac{1}{2} \varrho v^2 F S C_D , \tag{2}$$

where

$$F = \left(1 - \frac{r_{pc} w}{v_{po}} \cos i_o\right)^2 , \tag{3}$$

r_{po} = initial distance of perigee from earth's centre,
v_{po} = velocity of satellite relative to earth's centre at initial perigee,
w = angular velocity of earth,
i_o = initial orbital inclination.

For most of the satellites so far launched, the value of F lies between 0.90 and 0.95.

1*

If a satellite of mass m is in an orbit with semi-major axis a and eccentricity e between 0.015 and 0.15, the air density at perigee, ϱ_p, can be expressed [5, 17] as

$$\varrho_p = -\frac{dT}{dt}\frac{1}{3\delta}\sqrt{\frac{2e}{\pi a H}}\left\{1-2e-\frac{H}{8\,ae}+0\left(e^2,\frac{H^2}{a^2 e^2}\right)\right\},\qquad(4)$$

where dT/dt is the rate of change of the period of revolution T, which can be determined accurately from observations, and $\delta = F\,SC_D/m$. In deriving eq. (4) it is assumed that, above perigee height y_p, air density varies exponentially with height, the density ϱ at height $y\,(>y_p)$ being given by

$$\varrho = \varrho_p \exp\left\{-\frac{y-y_p}{H}\right\},\qquad(5)$$

where H is constant. Since the value of H is not accurately known at heights of 200—400 Km., eq. (4), though it determines $\varrho_p\sqrt{H}$ with little error, does not yield an accurate value of ϱ_p.

This limitation can be avoided however. Let H^* be the best estimate of H. Then the density at height $y_p + 0.5\,H^*$ may, by means of (5) and (4), be expressed as

$$\varrho_{p+0.5H^*} = \varrho_p \exp\left(-\frac{H^*}{2H}\right) =$$

$$= -\frac{dT}{dt}\frac{1}{3\delta}\sqrt{\frac{2e}{\pi a H^*}}\left[\sqrt{\frac{H^*}{H}}\exp\left(-\frac{H^*}{2H}\right)\right]\left\{1-2e-\frac{H^*}{8\,ae}+0\left(e^2,\frac{H^2}{a^2 e^2}\right)\right\}.\ (6)$$

The function in square brackets is insensitive to the value of H^*/H, and may be taken as 0.593, with error less than 2.5%, if the estimated value H^* does not differ from the true value H by a factor of more than 1.5. Since H is usually known to within a factor of 1.5, and since the likely error in m/SC_D is as a rule appreciably greater than 2.5%, it is legitimate to replace the term in square brackets in (6) by this constant, so that (6) becomes

$$\varrho_{p+0.5H^*} = -\frac{0.158}{\delta}\frac{dT}{dt}\sqrt{\frac{e}{a H^*}}\left\{1-2e-\frac{H^*}{8\,ae}+0\left(e^2,\frac{H^2}{a^2 e^2}\right)\right\}.\qquad(7)$$

The error arising from neglect of terms in e^2 and $H^2/a^2 e^2$ is less than 3.5% if $0.015 < e < 0.15$. The use of eq. (7) to obtain the density at height $0.5\,H^*$ above perigee should not lead to errors of more than 4.5% if H^* is not in error by a factor of more than 1.5. If H^* is in error by a factor of 2, the error in (7) rises to 12%.

3. Evaluation of δ

In applying eq. (7) to actual satellites, the chief difficulty is the evaluation of $\delta = FSC_D/m$. The value of SC_D depends on the shape of the satellite, the manner in which it is rotating, and the mechanism of reflexion of the air molecules. Diffuse reflexion, in which the air molecules are re-emitted in a random manner after striking the surface (or, more strictly, according to the KNUDSEN cosine law), has been chosen as the most likely mechanism, in preference to specular reflexion, in which the molecules are reflected as if from a mirror. Though there is scope for argument about the speed of re-emission of the molecules in diffuse reflexion, the drag coefficients obtained under the various possible assumptions do not differ by more than about 5%. It has been assumed here that the re-emission speed has a Maxwellian distribution about the speed appropriate to the satellite's temperature, and drag from electrical forces has been ignored.

The value of SC_D is most difficult to estimate for the near-cylindrical satellites, Explorers 1, 3 and 4 and Atlas (1958 α, γ, ε and ζ). It appears very probable, both on general dynamical principles and from observations [7], that each of these cylindrical satellites, and the rockets of the Russian satellites [10, 18], have rotated about their axis of maximum moment of inertia, i.e. an axis perpendicular to their length, though the angle between this axis of rotation and the direction of motion has varied. If this picture is correct, the two extreme modes of rotation are (a) travelling exactly like an aeroplane propeller, and (b) tumbling end-over-end: in (a) the axis of spin and the direction of motion are parallel; in (b) they are at right angles; and in practice the angle may be anywhere between these extremes. A recent study [19] has shown that, under the assumptions of diffuse reflexion, a rotating cylinder of length l and diameter d $(d \sim 0.1\ l)$ has a value of SC_D of about 2.2 ld under regime (a), and about 1.5 ld under regime (b). For any motion between these extremes SC_D lies between 1.5 ld and 2.2 ld. The near-cylindrical satellites have been treated here as cylinders, with SC_D taken as 1.85 ld, the mean of the values under regimes (a) and (b). If each satellite has rotated about its axis of maximum of inertia, this value of SC_D will not be in error by more than 19 %.

The spherical satellites Sputnik 1 and Vanguard 1 (1957 α 2 and 1958 β 2) both have cylindrical antennae, and their drag coefficient C_D has been taken as 2.2, with S as the mean cross-section during one rotation. This value should not be in error by more than 5 %, since $C_D \simeq 2.1$ for a sphere and $C_D \simeq 2.2$ for a rotating cylinder. For the conical Sputnik 3 (1958 δ 2), S has been taken as the mean of the cross-sections in modes of rotation (a) and (b), with $C_D = 2.3$. For Sputnik 2 (1957 β), no weights and dimensions are available, but the value of δ can be inferred from Sputnik 1. Since the perigee heights of Sputniks 1 and 2 were virtually the same [10], eq. (4) gives

$$\delta_2 = \delta_1 \sqrt{\frac{a_1 e_2}{a_2 e_1}} \left(\frac{dT}{dt}\right)_2 \bigg/ \left(\frac{dT}{dt}\right)_1, \tag{8}$$

where suffixes 1 and 2 refer to Sputniks 1 and 2 respectively. Since the factors on the right hand side of (8) are known, δ_ε can be evaluated. Values of δ for the rockets of Sputniks 1 and 3 (1957 α 1 and 1958 δ 1) can be found similarly. This indirect method should be reliable if, as is believed [20, 21], each of the satellites 1957 α 1, 1957 β and 1958 δ 1 retained a virtually constant value of δ until the last day of its life.

The values of m/SC_D and δ obtained by these various methods for the satellites with known orbits launched in 1957 and 1958 are listed in Table I. If the assump-

Table I. *Values of Mass m, m/SC$_D$ and δ for Satellites 1957 α—1958 ζ*

Satellite		Mass m Kg.	m/SC_D Kg./sq. m.	δ sq. m./Kg.
Sputnik 1	1957 α 2	83	110	0.0088
Sputnik 1 rocket	1957 α 1	—	62	0.015
Sputnik 2	1957 β	—	58	0.016
Explorer 1	1958 α	14	23	0.039
Vanguard 1	1958 β 2	1.5	23	0.040
Explorer 3	1958 γ	14	23	0.039
Sputnik 3	1958 δ 2	1327	190	0.0049
Sputnik 3 rocket	1958 δ 1	—	59	0.016
Explorer 4	1958 ε	17.5	29	0.032
Atlas	1958 ζ	3960	28	0.032

tions already stated are justified, the error (standard deviation) in the tabulated values of m/SC_D and δ will probably be about 10 %. It is interesting to note that the values of m/SC_D for Sputnik 1 rocket, Sputnik 2 and Sputnik 3 rocket are very similar: this tends to confirm the speculation that a similar rocket was used for all three.

4. Evaluation of Air Density

If values of δ are available, the air density at height 0.5 H^* above perigee can be found from eq. (7). For the Russian satellites, values of dT/dt, a and e have been taken from orbital determinations made in Britain [2, 22—25]. For the U.S. satellites, values have been obtained from the orbital data regularly issued as part of their prediction services by the Smithsonian Astrophysical Observatory, Cambridge, Mass., and Project Space Track, Bedford, Mass. The values of H^* chosen are consistent with values of H given later in this paper.

The resulting values of density, from all the satellites in Table I, are plotted in Fig. 2, and a curve has been drawn through the points to indicate likely average values for density. It is usual when constructing a 'model atmosphere' to make

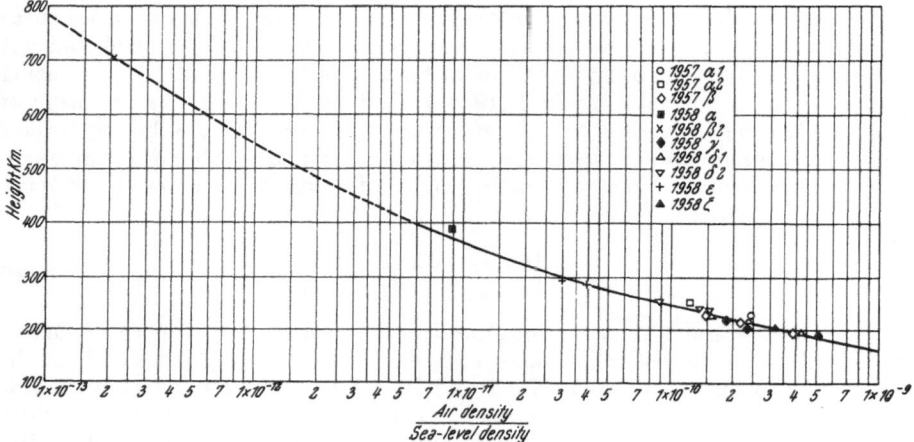

Fig. 2. Values of air density obtained from satellites 1957a—1958ζ, allowing for rotation of the atmosphere

a plausible assumption about the variation of temperature and air composition with height, and thence derive density. Since the primary measurements are of density, however, it seems more logical to take the density as the basic parameter and then to deduce temperature from it. This procedure has the further merit that it shows the large errors in temperature which can arise, even when a good fit is achieved for the density.

Several features of Fig. 2 are worth noting.

(1) The 10 different satellites give surprisingly consistent values of air density. None of the 15 plotted points in the cluster between 190 and 260 Km. differs from the curve by a factor greater than 1.4, and the average factor of difference is about 1.2. Probably, therefore, the assumptions made were not grossly in error, as they would have been if, for example, one of the cylindrical satellites had flown like an arrow instead of rotating.

(2) Even if the method of analysis were perfectly accurate, the points in Fig. 2 would not all lie on the same curve, since the effective air density is known to

vary from day to day and month to month, by up to 30 % at 200—250 Km. height [20, 21] and by a factor of 3 at 700 Km. [26]. The curve in Fig. 2 can therefore only represent average values.

(3) The points in Fig. 2 cover latitudes between 50° N and 35° S; but there is no sign of any regular variation with latitude, and the spread of the points suggests that, between 50° N and 35° S, density does not depart from its average value by a factor of more than about 1.5. This is in contrast with results from rocket experiments [27], which suggest a variation by a factor of up to 5 between 59° N and 35° N on a summer day. The accuracy of initial lifetime-estimates for the Russian satellites, whose perigees moved from an initial latitude near 50° N towards the equator, also strongly suggests that density does not vary greatly within this range of latitude.

(4) The result from Vanguard 1 in Fig. 2, at a height of 700 Km., should be treated with caution, partly because it is unsupported, and partly because the effect of charged drag [28] is likely to be important at this height, where over 5 % of the atoms may be ionized.

Table II lists values of average density from the curve in Fig. 2, for heights between 200 and 400 Km., where the results are most reliable.

Table II. *Average Air Density at Heights between 200 and 400 Km.*

Height Km.	Air Density / Sea-Level Density	Density (gm/c. c.)
200	3.5×10^{-10}	4.3×10^{-13}
220	2.1×10^{-10}	2.5×10^{-13}
240	1.2×10^{-10}	1.5×10^{-13}
260	7.6×10^{-11}	9.3×10^{-14}
280	4.7×10^{-11}	5.8×10^{-14}
300	3.1×10^{-11}	3.8×10^{-14}
320	2.1×10^{-11}	2.6×10^{-14}
340	1.5×10^{-11}	1.8×10^{-14}
360	1.1×10^{-11}	1.3×10^{-14}
380	7.8×10^{-12}	9.6×10^{-15}
400	5.8×10^{-12}	7.1×10^{-15}

III. Variations in Air Density

1. Method

During 1958 the Royal Aircraft Establishment, Farnborough, provided a prediction service for satellites which passed over or near Britain, and simple methods of predicting the times of transit and the geometry of the orbit were developed [29], which relied on a judicious mixture of observation and theory, and required only a desk calculating machine for computation. Sputniks 2 and 3 and the rocket of Sputnik 3 were the satellites of most interest in Britain, since none of the early U.S. satellites (except Explorer 4) reached our latitudes. The prediction service relied almost entirely on simple visual observations relative to the stars, sent in by volunteer observers. Most of these observations were accurate to 1° in direction and 2 seconds in time, and they were used to estimate the time at which the satellites passed through apex, the point of maximum latitude north. The r.m.s. error in determining apex time was usually not more than 3 seconds. From two apex times approximately a day apart, the mean nodal period of revolution T in the interval could be determined with r.m.s. error usually not more than 0.3 sec., or 1 part in 20,000. From this series of values of T, a series of 'observed values' of dT/dt can be obtained with adequate accuracy.

If the air density at a given height and the effective cross-sectional area S of the satellite both remained constant from day to day, the daily change in period would increase smoothly as the satellite's orbit slowly sank lower into the atmosphere: all the terms in eq. (4) would either change slowly and smoothly or remain constant. In reality, for every satellite so far launched, the rate of decrease of T has been irregular, thus implying irregularities in drag, which could be caused by changes in either the effective atmospheric density near perigee or the effective cross-section S. Most of the satellites so far launched have probably rotated about their axis of maximum moment of inertia, and, if so, the mean cross-section during a complete rotation should remain constant as long as the mode of rotation of the satellite is unchanged, though it would not be constant if, for example, tumbling end-over-end changed to spinning like an aeroplane propeller. A clue to the mode of rotation is provided by the fluctuation in brightness of a satellite: if this remains almost the same for many months it is unlikely that the mode of rotation has changed appreciably.

For the rocket of Sputnik 3 (1958 δ 1), the rate of decrease of period was particularly erratic, but the brightness fluctuated regularly with a period which increased slowly from 8.5 seconds in July to 9.5 seconds in November. This strongly suggests that the mean cross-section did not vary significantly and that irregularities in the rate of decrease of period can be ascribed to variations in atmospheric density. Similar conclusions apply [20] to Sputnik 2 (1957 β). For both these satellites the perigee height decreased from 226 Km. initially to about 180 Km. ten days before the end of the satellite's life, and any conclusions about air density relate to this height band.

2. Results

Over 1,000 observations of Sputnik 2 and the rocket of Sputnik 3 have been analysed in the manner described in section III.1, to obtain the rate of change of

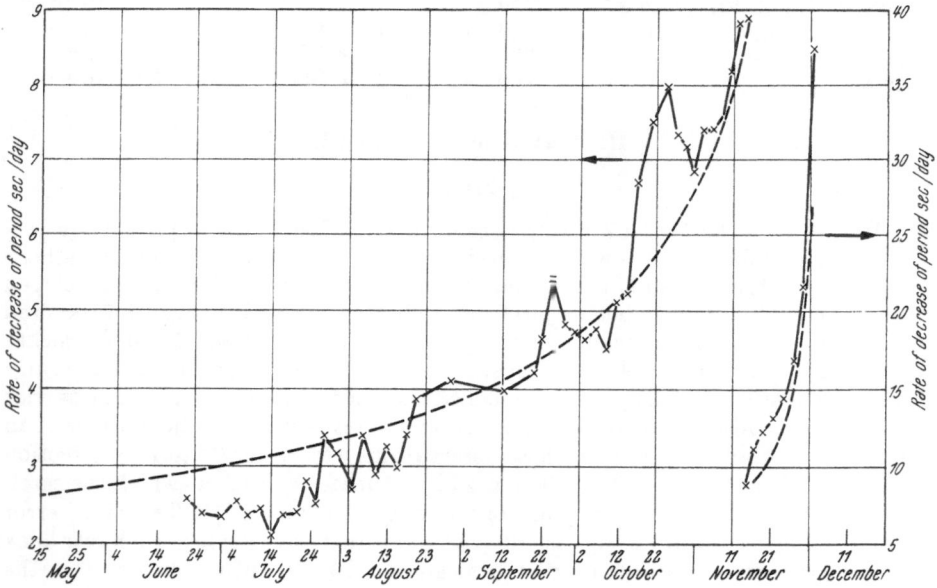

Fig. 3. Rate of change of period for Sputnik 3 rocket (1958 δ 1)
– – – – – – Theoretical curve
—x—x— Observed values

nodal period at intervals of 3 or 4 days. The results for Sputnik 3 rocket are shown in Fig. 3, in which the great majority of the observational values plotted are believed to be in error by less than 0.1 sec./day. For comparison, a theoretical curve, calculated on the assumption that density at a given height is constant, is also shown in Fig. 3. This curve is derived from the simplest theory, which gives [20]

$$\frac{dT}{dt} = -\frac{3\,e_0\,T_0}{4\,t_L\,\sqrt{1-t/t_L}},$$

where t is time after launch, t_L is the total lifetime, e_0 the initial eccentricity, and T_0 the initial period. In assessing irregularities in dT/dt, no advantage is gained by using a subtler theory.

Two main features stand out in Fig. 3, and even more in Fig. 4, where the observed values have been divided by the theoretical to give a curve which represents, in effect, the ratio $\dfrac{\text{air density}}{\text{average air density}}$. First, there is the general impression of irregularity. The value of density on any particular day is a poor guide to the likely value 3 days later: for instance, on 17th October the density

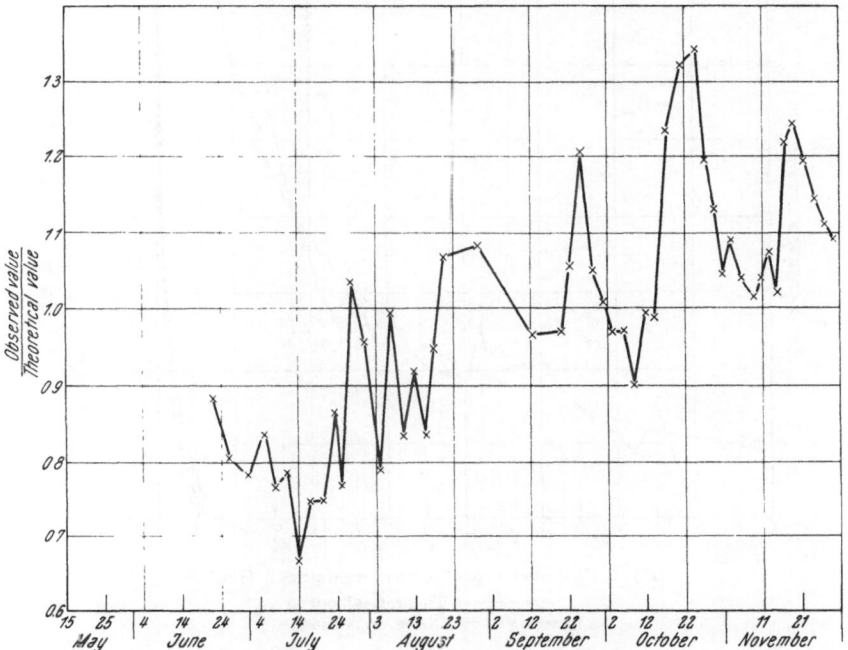

Fig. 4. Rate of change of period of Sputnik 3 rocket (1958 δ 1). Ratio of observed values to theoretical

was nearly 25 % higher than on the 14th. More detailed analysis shows that this irregularity occurs over even shorter time-intervals: on successive days during November, 1958, density differed from the average by +2 %, +11 %, —1 % and +18 %.

The second main feature of Fig. 4 is more interesting and more fruitful: the maximum values of density, and also the minimum values, show a strong tendency to recur at intervals of about 28 days.

Four possible causes of this 28-day periodicity in air density are worth considering. First, is it caused [10, 30] by the movement of the perigee point from daylight into darkness? The answer seems to be 'no', because perigee took about 3 months to perform a complete cycle of movement from light to darkness and back again. The change in air density from day to night therefore seems to have only a minor effect.

A second possible cause of the 28-day oscillation is variation of density with latitude. Between mid-June and mid-November, 1958, the perigee of Sputnik 3 rocket moved from latitude 40° N to latitude 10° S, changing by roughly 10° every 28 days. The 28-day periodicity in dT/dt might therefore seem to indicate a periodic variation of density with latitude, with maxima at intervals of 10°: is this possible? Again the answer is 'no', for two main reasons. First, the effect of drag is spread over a range of latitudes near perigee: if the density at a given altitude were high at perigee latitude and low at latitudes 5° on either side, the effect would be much the same as low density at perigee latitude and high density

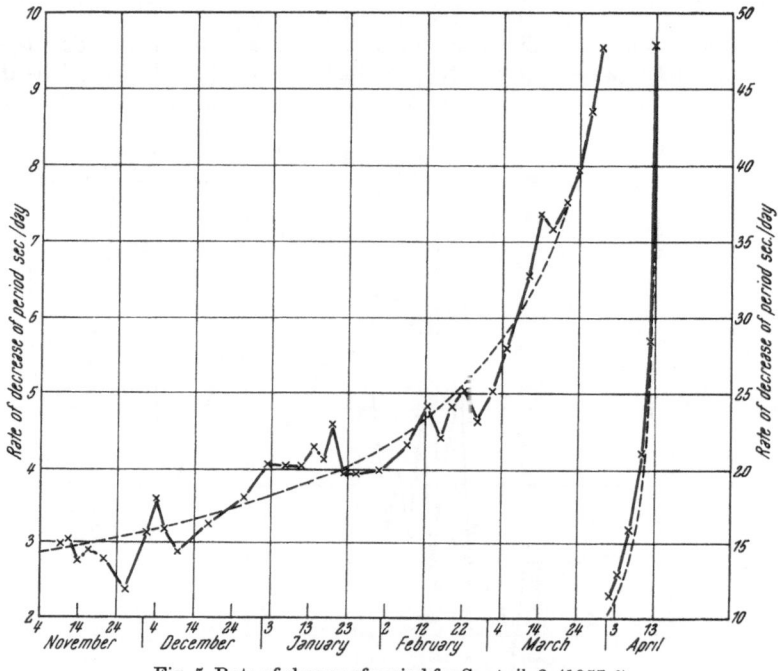

Fig. 5. Rate of change of period for Sputnik 2 (1957 β)
— — — — — Theoretical curve
— x — x — Observed values

at 5° on either side. Second, other satellites, for which perigee latitude changes at quite different rates, show the 28-day oscillation.

A third possible cause of the oscillations is lunar tides in the atmosphere: any 'tidal bulge' caused by the moon should travel round the earth once every 28 days. But, in 28 days, the right ascension of the perigee of Sputnik 3 rocket moved 90° westwards, chiefly because of the westward rotation of the orbital plane caused by the earth's oblateness. Consequently the period of revolution of the moon, relative to perigee, was only about 22 days. Maximum values of density would therefore be expected to occur at intervals of 22 days or, more probably, 11 days. So lunar tides, though they exist, do not seem to have an important effect on the density.

The fourth and most likely cause of the oscillations is solar disturbances. It has long been known that streams of charged particles shoot out radially from the sun, rather like a jet of water from a revolving sprinkler, and since the sun rotates about its axis, relative to the earth, once every 27 or 28 days[1], the earth tends to pass through these streams at intervals of about 27 days. The impact of these streams of particles on the upper atmosphere gives rise to well-known 27-day periodicities in geomagnetic activity, cosmic rays and the aurora. The evidence from Sputnik 3 rocket suggests that air density at heights between 180 and 220 Km. exhibits similar periodicity.

Do the results from other satellites confirm this suggestion? For Sputnik 1 and its rocket (1957 α 2 and 1) the data available to us are not precise enough for any conclusions to be drawn. For Sputnik 3 (1958 δ 2), increases in drag occurred about 80, 110 and 140 days after launch, but detailed results have not yet been published. For the United States satellite Explorer 1 (1958 α), and for Explorer 3

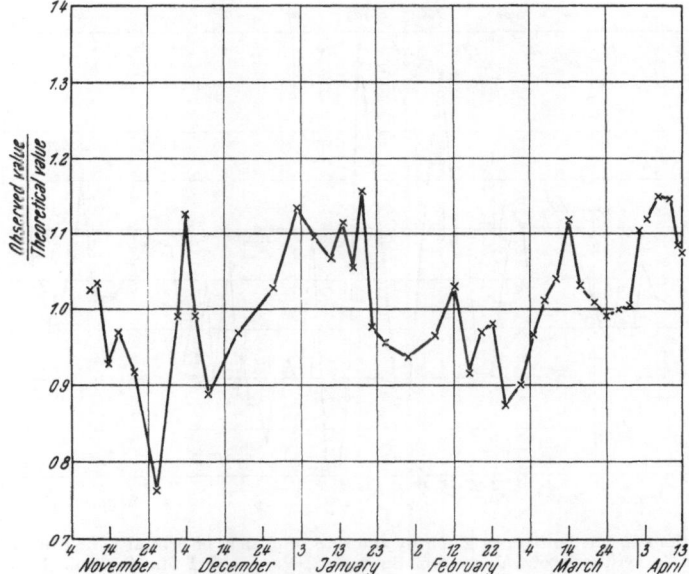

Fig. 6. Rate of change of period of Sputnik 2 (1957 β). Ratio of observed values to theoretical

(1958 γ) except during its last weeks, perigee crossed the equator at intervals of between 24 and 30 days; the oscillations in dT/dt produced by the periodic change in perigee latitude therefore had a period between 24 and 30 days, and, in the present state of knowledge, cannot be reliably distinguished from solar influences of similar period. Our information on Explorer 4 (1958 ε) is limited, but its drag, like that of Sputnik 3 rocket (Fig. 4), increased sharply about 22 August, 1958 and decreased about 8 September. Atlas (1958 ζ) had a lifetime of only one month, too short to yield any conclusive result. For Vanguard 2 (1959 α) and Discoverer 2 (1959 γ) no results are yet available. That leaves Sputnik 2 (1957 β) and Vanguard 1 (1958 β2).

The observations of Sputnik 2 have been analysed in the same way as those of Sputnik 3 rocket to give the results plotted in Figs. 5 and 6. A 28-day oscillation is again discernible, especially during 1958, though it is not so obvious as with Sputnik 3 rocket. The variations have been found by NONWEILER [31] to corre-

---

[1] It is 27 days at the sun's equator, 28 days at latitude 25°.

spond fairly well with solar flares, and by PRIESTER [32] to show excellent correlation with the sun's 20 cm. radiation.

Results for Vanguard 1, as given by JACCHIA [26], show that this satellite exhibited a similar oscillation to Sputnik 3 rocket, with maxima of density occurring at the same times. The amplitude of the oscillations was however much larger for Vanguard; this is to be expected, since the air density is much lower at its perigee height of 650 Km., and changes are therefore likely to be relatively greater. JACCHIA [33] has further compared the variations in dT/dt for Vanguard 1 with the 10.7 cm. solar radiation, and has found the correspondence to be 'little short of perfect'.

Thus it appears probable that the air density in the upper atmosphere, at heights between 200 and 700 Km., and over a wide range of latitudes, is strongly

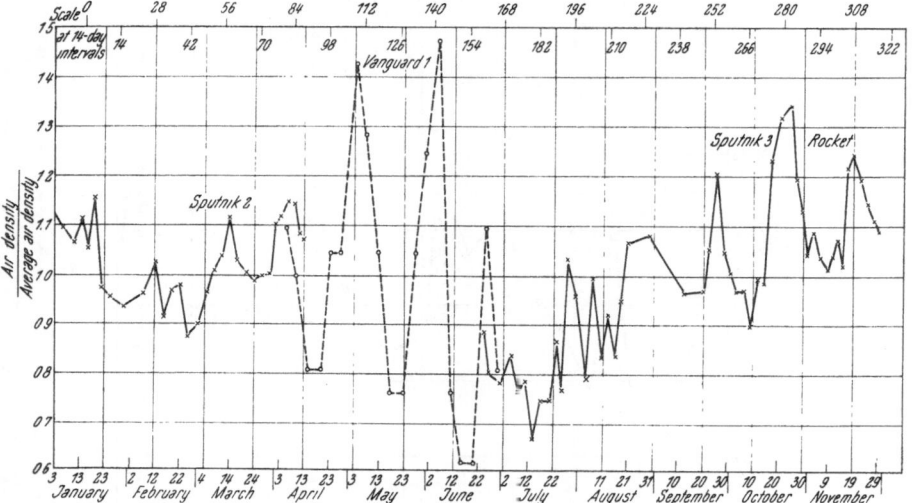

Fig. 7. Variations in air density, as given by the rate of change of orbital period of satellites during 1958: a synthesis of values from Sputnik 2, Vanguard 1 and Sputnik 3 rocket

under solar influence, and exhibits the 27/28-day periodicity which is characteristic of solar effects. Superposed upon the 28-day oscillation, there is an appreciable day-to-day irregularity, which makes exact prediction impossible for satellites which are appreciably affected by air drag.

The persistence of the 28-day oscillation during 1958 is strikingly shown in Fig. 7, in which Figs. 4 and 6 are linked with the results [26] from Vanguard 1: the Vanguard oscillations, though of larger amplitude, fit almost exactly into the pattern, and the 14-day scale at the top of the figure shows how regular the oscillations are.

IV. Scale Height and Air Temperature, at Heights between 200 and 400 Km.

The air temperature in the upper atmosphere may be expressed in terms of the scale height, which is a measure of the rate of change of density with height, and the mean molecular weight. Unfortunately, neither of these quantities is as yet accurately known at heights between 200 and 400 Km., and the values derived are subject to considerable error.

The scale height at any altitude may be defined as the air pressure divided by the rate of decrease of pressure with height. If pressure varies exponentially with altitude, the scale height is equal to the increase in altitude corresponding to a decrease in pressure by a factor e ($=2.718\ldots$). The scale height is closely related to the coefficient H of section II, which is the increase in altitude corresponding to a decrease in density by a factor e. The scale height is given by $H/(1-H')$, where H' is the rate at which H changes with height and is usually small. The air temperature for heights between 200 and 400 Km. may be expressed approximately as

$$T = \frac{1.1\,MH}{1-H'}\ \text{°K},\qquad (9)$$

if H is in Km.

The value of H should in theory be obtainable from the slope of the curve in Fig. 2; in practice, the accuracy is poor. The values derived in this way rise from $H = 37$ Km. at 200 Km. height to $H = 71$ Km. at 400 Km. height. But it is possible to link the points in Fig. 2 with other curves, of different slopes. For example, a straight line between 200 and 400 Km. height, with $H = 50$ Km., fits the points almost as well.

Of the other methods of finding H, the most direct is to utilize the theoretical equation connecting the eccentricity e with the distance r_p of perigee from the earth's centre. The simplest form of this equation, derived on the assumption of spherically symmetrical earth and atmosphere, is [20]

$$H = \frac{r_{po} - r_p}{\frac{1}{2}\ln\dfrac{e_o}{e} + 0\,(e) + 0\left(\dfrac{H}{ae}\right)},\qquad (10)$$

where zero suffix denotes initial values, and e_0 and e lie between 0.02 and 0.2. Eq. (10) is useful as a first approximation, but a more accurate form is needed which gives the terms in $0(e)$ and $0(H/ae)$, and takes into account the effect of the third harmonic in the earth's gravitational field and the oblateness of the atmosphere. If i is the orbital inclination, ω the argument of perigee, and R the earth's equatorial radius, and ae is written as x, the more accurate form of eq. (10) is found [16] to be

$$H = \frac{r_{po} - r_p + 4.3 \sin i\,(\sin \omega_o - \sin \omega)}{\frac{1}{2}\left(1 - \dfrac{3H}{a_o}\right)\ln\dfrac{x_o}{x} + \ln\dfrac{x}{x_o}\left(\dfrac{8x_o - 3H}{8x - 3H}\right) - \dfrac{x_o - x}{a_o} + \varepsilon R\displaystyle\int_{x_o}^{x}\dfrac{\cos 2\,\omega \sin^2 i}{x^2}\,dx},\qquad (11)$$

if distances are measured in Km., $0.02 < e < 0.2$, and the coefficient J_3 of the third harmonic in the earth's gravitational potential (see section II) is taken as -2.2×10^{-6}. The value of ε, which represents the oblateness of the atmosphere, may with adequate accuracy be taken equal to the earth's flattening, 1/298. The last term in the numerator in eq. (11) represents the effect of J_3, and is most important initially when $r_{po} - r_p$ is very small. The last term in the denominator, which depends on the oblateness of the atmosphere, is usually negligible until e falls below 0.04.

Eq. (11) has been applied to all the satellites so far launched for which reasonably good orbital information is available, but the resulting values of H are disappointingly scattered, and it is evident that the observational results are not yet of an accuracy adequate to match the theory. Consequently, it is not worth giving more than a summary of the results, which suggest that H rises from about 45 Km. at 180 Km. height to about 80 Km. at 250 Km. height, with an error factor which might be as large as 1.5. Since Fig. 2 shows that the average value of H between 200 and 400 Km. height is about 50 Km., these figures, if true, would suggest that H, and hence the air temperature, was higher between 200 and 300 Km. height than between 300 and 400 Km., the figures do not, however, appear to be reliable enough to justify this conclusion, which would be in conflict with almost all the 'model atmospheres' proposed for this heightband.

It is probably better to accept the evidence of Fig. 2, which indicates that the average value of H between 200 and 400 Km. height is about 50 Km., and that H tends to increase with height. If a value had to be chosen, H might be assumed to rise from 45 Km. at 200 Km. height to 60 Km. at 400 Km. height.

There is also some difficulty in determining the molecular weight of the air. The oxygen is largely in atomic form above 200 Km. height, but the height-range within which nitrogen changes from molecular to atomic form is still disputed. Above about 300 Km., however, the main constituents of the air are believed to be atomic oxygen and atomic nitrogen, in proportions as yet unknown, [12, 34—36], and the molecular weight may be taken as 15. With the values of H already suggested, the average temperature between 200 and 400 Km. height, as given by (9), would be a little over $1000°$ K. It should be emphasized however that any values of temperature are far less reliable than those of density.

V. Winds in the Upper Atmosphere

Analysis of kinetheodolite observations of Sputnik 2 [24] showed that the inclination of the orbit to the equator changed from $65.32°$ initially to about $65.19°$ at the end of its lifetime. The decrease became much more rapid towards the end of the satellite's life, thus suggesting that the change was due to the atmosphere and increased greatly as perigee came nearer to the earth's surface.

An obvious cause of this change is rotation of the atmosphere, which creates a lateral force on the satellite. This force will tend to reduce the orbital inclination, for a satellite which goes from west to east, and its magnitude will depend on the mean wind speed v_w near perigee height and perigee latitude.

If the angular velocity of the atmosphere in the region near perigee differs from the angular velocity of the earth by a constant factor A, and the orbital eccentricity does not exceed 0.2, the change in the inclination i is found [37] to be

$$\Delta i = \frac{A \sin i}{6} \frac{(2 I_2 - 4 e I_1) \cos^2 \omega + I_o - I_2 + 0 (e^2)}{(1 - T \cos i) I_o + 2 e I_1} \Delta T , \qquad (12)$$

where $T = $ period of revolution of satellite, expressed as a fraction of a sidereal day, $\omega = $ argument of perigee, and the I_n are Bessel functions of the first kind and imaginary argument, of order n, of argument ae/H. In eq. (12) the change in i is directly related to the change in T, and irregularities in air density do not affect the analysis.

If the Bessel functions are replaced by their asymptotic expansions, eq. (12) becomes

$$\Delta i = \frac{A \sin i}{3} \frac{\left(1 - \frac{15}{8} k - 2 e\right) \cos^2 \omega + k}{(1 - T \cos i) \left(1 + \frac{k}{8}\right) + 2 e} \{1 + 0 (k^2, e^2)\} \Delta T , \qquad (13)$$

where $k = H/ae$. This form of expansion is most useful, since e and k are of the same order if $0.05 < e < 0.15$. If the terms in k, e and T are ignored, the simpler form derived by Bosanquet [38] is obtained. Eq. (13) shows that the change in i depends mainly on four factors:

(1) *i* itself—the greater the inclination, the greater the change, because the atmospheric cross-wind is then more nearly perpendicular to the satellite's path;

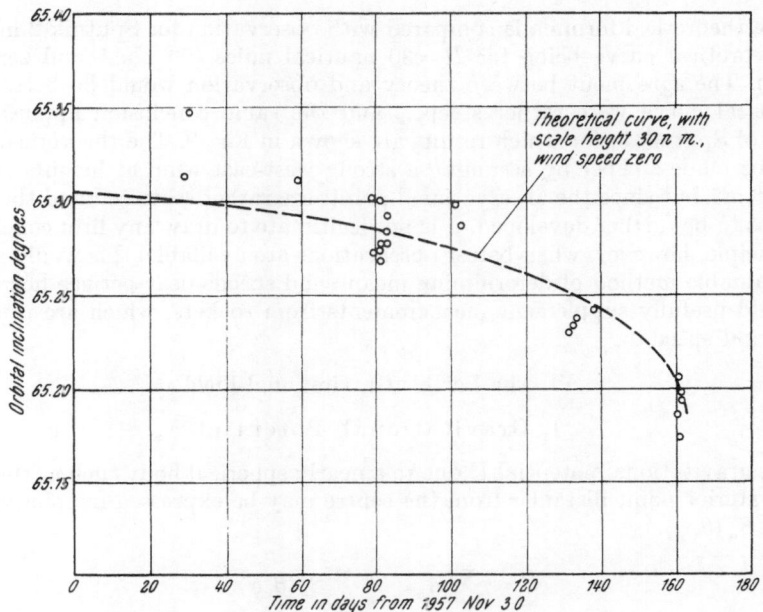

Fig. 8. Mean orbital inclination of Sputnik 2 (1957 β)

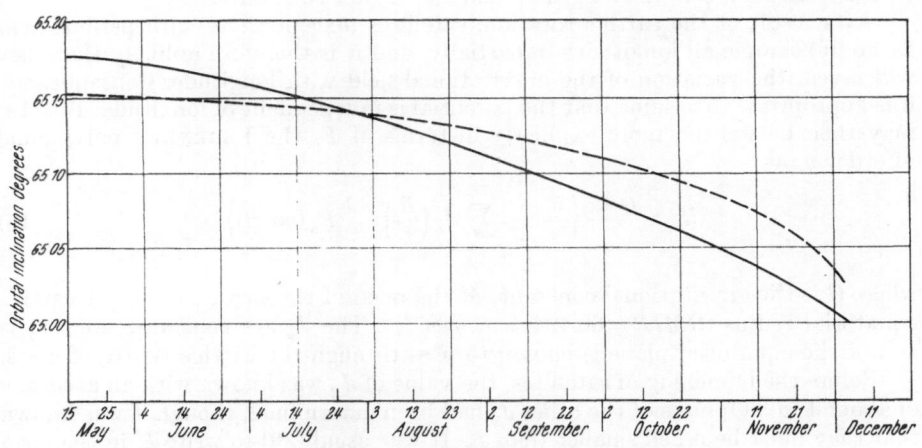

Fig. 9. Orbital inclination of Sputnik 3 rocket (1958 δ 1)
————— Observational values (fitted quadratic)
- - - - - Theoretical curve ($H = 30$ n. m., wind speed zero)

(2) the change in period, which is a measure of how much the atmosphere is affecting the orbit in general;

(3) the angle ω from equator to perigee, the effect being greatest when perigee is at the equator, where the rotational speed is greatest;

and (4) the mean west-east wind speed near perigee, v_w, since the factor A is given by $A = 1 + \dfrac{v_w}{v_r}$, where v_r is the earth's rotational speed near perigee.

The theoretical formula is compared with observation for Sputnik 2 in Fig. 8, the theoretical curve being for $H = 30$ nautical miles (56 Km.) and zero wind ($A = 1$). The agreement between theory and observation would be better if the theoretical curve were rather steeper, and the same conclusion applies to the rocket of Sputnik 3, for which results are shown in Fig. 9. The theoretical curves could be made steeper by assuming a strong west-east wind at heights of 200—-—250 Km.: but since the observational points are rather scattered and the theory requires to be further developed, it is not legitimate to draw any firm conclusions. In principle, however, when better observations are available, this would seem to be a valuable method of determining mean wind speeds near perigee height, and it should usefully supplement measurements from rockets, which are usually of local wind speeds.

VI. The Earth's Gravitational Field

1. Gravitational Potential

The gravitational potential U due to a nearly spherical body such as the earth, at an exterior point distant r from the centre may be expressed in spherical harmonics $S_n(\theta, \varphi)$ as

$$U = \sum_{n=0}^{\infty} A_n r_n^{-n-1} S_n(\theta, \varphi), \tag{14}$$

where θ is taken as the angle between the radius vector and the earth's axis (the co-latitude), φ is the longitude and the A_n are constants.

As a result of the earth's rotation, satellites (except those with periods near 24 hours) sample all longitudes impartially, and it is therefore unlikely that they will reveal the variation of the gravitational field with longitude. Consequently, it is appropriate to assume that the potential is independent of longitude. Eq. (14) may then be written more explicitly in terms of P_n the LEGENDRE polynomial of order n, as

$$U = \frac{GM}{r} \left\{ \frac{R}{r} - \sum_{n=2}^{\infty} J_n \left(\frac{R}{r}\right)^{n+1} P_n(\cos\theta) \right\}, \tag{15}$$

where G is the gravitational constant, M the mass of the earth, and R the earth's equatorial radius $(GM/R = 62.494 \text{ Km.}^2/\text{sec.}^2)$. The J_n are constants, and J_1 is zero if the equatorial plane is chosen to pass through the earth's centre of mass.

Before the launching of satellites, the value of J_2 was known with an accuracy of about 1 in 300; none of the other J_n had been determined, though it was known that they must be much smaller than J_2. It was usual [39] to write U in the form

$$U = \frac{GM}{R} \left\{ \frac{R}{r} + J \frac{R^3}{r^3} \left(\frac{1}{3} - \cos^2\theta \right) + \frac{D}{35} \frac{R^5}{r^5} (35\cos^4\theta - 30\cos^2\theta + 3) + \ldots \right\}, \tag{16}$$

where $J \ (= 1.5 J_2)$ was taken as 1.637×10^{-3}, and $D \left(= -\dfrac{35}{8} J_4 \right)$ was given a conventional value of 10.6×10^{-6}, chosen to make the meridional section of the earth an exact ellipse.

2. Method of Determining the J_n from Orbital Perturbations

The various perturbations to satellite orbits can be expressed by means of orbital theory in terms of the coefficients J_n of eq. (15), and if many satellites with different orbits are accurately observed, it should be possible to evaluate a large number of the coefficients J_n. The perturbations caused by the J_n terms in (15) can conveniently be divided into three groups. First, there are short-term periodic variations, whose period is of the same order as the period of revolution of the satellite: these perturbations are most difficult to measure and have not yet been used in determining the J_n. Second, there are the longer-term periodic perturbations, whose period is of the same order as that of the argument of perigee— between a few weeks and many years. The odd-numbered J_n are particularly active in producing this type of oscillation, and the variation of eccentricity for Vanguard 1 has been used [40, 41] to deduce the value of J_3, as will be described later. Third, there are the long-term secular effects, which build up steadily as time goes on and can be accurately measured. The even-numbered J_n are chiefly responsible for these secular effects, and their measurement provides a powerful method of evaluating the even J_n, which will now be described.

The two orbital elements which suffer a secular change under the influence of the earth's gravitational field are ω, the argument of perigee (i.e. the angle from northward crossing of the equator to perigee, measured along the orbit) and Ω, the right ascension of the ascending node (i.e. the longitude at which the satellite crosses the equator, going north, measured relative to the stars). The value of Ω can be measured much more accurately than ω, and the ensuing discussion will, for brevity, refer only to Ω. Similar methods apply to ω, but they are more difficult to execute.

The rate of change of Ω, which expresses the rate of rotation of the orbital plane of the satellite about the earth's axis, could in principle be expressed in the form

$$\dot{\Omega} = \sum_{n=2}^{\infty} J_n F_n (a, e, i, \omega) + \sum_{n=2}^{\infty} \sum_{m=2}^{\infty} J_n J_m F_{nm} (a, e, i, \omega) + \ldots , \qquad (17)$$

where the F_n and F_{nm} are functions of the orbital elements. In practice, only the terms in the first series, together with the J_2^2 term, need be considered at present. Explicitly, $\dot{\Omega}$ may be written [42] as

$$\dot{\Omega} = - \sqrt{\frac{GM}{a^3}} \left(\frac{R}{p}\right)^2 \cos i \left\{ \frac{3}{2} J_2 + \frac{9}{4} J_2^2 \left(\frac{R}{p}\right)^2 \left(\frac{19}{12} \sin^2 i - 1\right) + \right.$$
$$\left. + \frac{15}{16} J_4 \left(\frac{R}{p}\right)^2 (7 \sin^2 i - 4) + 0 (J_3, J_5, J_6 \ldots) \right\}, \qquad (18)$$

where $p = a (1 - e^2)$. Since p changes by only about 10 % during the life of a typica satellite, and i changes scarcely at all, the factors multiplying J_2, J_2^2 and J_4 in (18) do not vary much during a satellite's life; and the same applies to the factors multiplying J_6, $J_8 \ldots$ But the term in J_3 on the right-hand side of (18), which is [43]

$$-\frac{3}{8} \sqrt{\frac{GM}{a^3}} \left(\frac{R}{p}\right)^3 J_3 e \sin \omega \cot i (15 \sin^2 i - 4), \qquad (19)$$

contains the factor $e \sin \omega$, which is small and periodic. So the term in J_3, and those in J_5, $J_7 \ldots$, are likely to be negligible and, if not, can usually be averaged out to zero by choosing an appropriately long time-interval. Eq. (18) thus essentially provides a means of determining J_2, J_4, $J_6 \ldots$, leaving J_3, $J_5 \ldots$, to be found by other methods. Since the factors multiplying J_2, J_4..., change little during a satellite's life, a single satellite generally yields only one equation between J_2, J_4, J_6... Attempts to extract two equations usually lead to ill-conditioned simultaneous equations. Measuring $\dot{\Omega}$ for one satellite can thus be expected to give only one relationship between J_2, J_4, J_6...; this can be solved for J_2 only if we assume conventional values (e.g. zero) for J_4, J_6,.... Measurement of $\dot{\Omega}$ for two satellites of different inclinations give two distinct relations between J_2, J_4, J_6..., which can be solved for J_2 and J_4, assuming $J_6 = J_8 ... = 0$. And so on. Since the J_n quickly become small as n increases, the process converges rapidly.

3. Evaluation of J_2, J_4, J_6

Values of $\dot{\Omega}$ for Sputnik 2 (1957 β, inclination 65°) were determined from a series of kinetheodolite observations made at trials ranges of the Royal Aircraft Establishment, and the results showed [44] that the previously accepted value of J_2 was in need of revision. It was found that $J_2 = 1082 \times 10^{-6}$ if J_4 was taken as zero, or $J_2 = 1084 \times 10^{-6}$ if J_4 had its conventional value of -2.4×10^{-6}. (The previously accepted value of J_2 was 1091×10^{-6}.) When accurate values of $\dot{\Omega}$ from Vanguard 1 (1958 β 32, inclination 34°) became available [45] it was possible to derive J_2 and J_4 independently, assuming $J_6 = J_8 = ... = 0$. The values obtained [46] were $J_2 = (1083.1 \pm 0.2) \times 10^{-6}$ and $J_4 = (-1.4 \pm 0.2) \times 10^{-6}$. The errors quoted here are the standard deviations stemming from observational errors, and it is possible that somewhat larger errors may arise through neglecting higher-order harmonics. The coefficient most likely to cause such errors is J_6; but a preliminary estimate of J_6, made [46] with the help of the orbital information so far available on Explorer 4 (1958 ε, inclination 50°) gives $J_6 = (0.1 \pm 1.5) \times 10^{-6}$. This causes no significant change in the values of J_2 and J_4 already quoted.

In the results so far given, no allowance has been made for contributions to $\dot{\Omega}$ from sources other than J_2, J_4 and J_6. Three such sources certainly provide appreciable contributions. First, there is the effect of the small J_3 term, as given by the expression (19), with $J_3 = -2.2 \times 10^{-6}$. Second, the side-force on the satellite caused by the rotation of the atmosphere changes Ω by a small amount $\Delta\Omega$, which is given by [47]

$$\Delta\Omega = \frac{A \sin 2\omega}{6} \frac{I_2 - 2e I_1}{(1 - T \cos i) I_0 + 2e I_1} \Delta T, \qquad (20)$$

where the symbols have the same meaning as in eq. (12). For Sputnik 2, this contribution increases the value of $\dot{\Omega}$ by 1 part in 10,000; for Vanguard 1, the effect is negligible. Third, there are lunar and solar perturbations, for which formulae have been given by KOZAI [41]. Their combined effect is to increase $\dot{\Omega}$ by 1 part in 7,000 for Vanguard 1 and 1 part in 20,000 for Sputnik 2. The moon contributes about $\frac{2}{3}$ of the total and the sun $\frac{1}{3}$.

When the three extra contributions described in the last paragraph are taken into account, the values obtained for J_2, J_4 and J_6, from Sputnik 2, Vanguard 1 and Explorer 4, are $J_2 = (1083.0 \pm 0.2) \times 10^{-6}$, $J_4 = (-1.3 \pm 0.2) \times 10^{-6}$, and $J_6 = (-0.1 \pm 1.5) \times 10^{-6}$, if J_5, J_7, J_8... are assumed zero.

4. Evaluation of J_3

The main perturbation to the orbital elements caused by the third harmonic is an oscillation in the distance of perigee from the earth's centre, with the same period as ω, and amplitude [43] $\dfrac{J_3 R \sin i}{2 J_2} + 0\,(J_5)$. The third harmonic also gives rise to an oscillation in the inclination i, of the same period as ω and of amplitude [41, 43] $\dfrac{e\,J_3 \cos i}{2 J_2} + 0(J_3\ e^2,\ J_5)$. By comparing these theoretical expressions with the observed oscillations in the Vanguard 1 orbit, O'KEEFE, ECKELS and SQUIRES [40] and KOZAI [41] have deduced that $J_3 = (-\,2.2 \pm 0.1) \times 10^{-6}$, assuming $J_5 = 0$.

5. Conclusions

At present, therefore, the most likely values of J_2, J_3, J_4 and J_6 appear to be

$$\left.\begin{array}{l} J_2 = (1083.0 \pm 0.2) \times 10^{-6} \\ J_3 = \quad (-2.2 \pm 0.1) \times 10^{-6} \\ J_4 = \quad (-1.3 \pm 0.2) \times 10^{-6} \\ J_6 = \quad (-0.1 \pm 1.5) \times 10^{-6} \end{array}\right\} . \tag{21}$$

If the sea-level surface of the earth were an exact spheroid, the flattening, f, of the earth, defined as

$$\frac{\text{equatorial diameter---polar diameter}}{\text{equatorial diameter}} ,$$

could be expressed in terms of J_2 as

$$f = \frac{1}{2}\,(3\ J_2 + m)\left(1 + \frac{3}{4}\,J_2 + \frac{3}{28}\,m\right) + 0\,(J_2{}^3), \tag{22}$$

where $m = R^3\ w^2\ (1-f)/GM = 0.0034498$ is the ratio of centrifugal to gravitational acceleration at the equator, multiplied by $(1-f)$, and w is the earth's angular velocity. Now that the sea-level surface is known to depart appreciably from a spheroid, it may be that (22) is not the best possible definition of f. For the moment, however, it is still usual to quote the value of f given by (22), which, with the value of J_2 from (21), is $1/f = 298.21 \pm 0.03$, and differs by over 1 part in 300 from the pre-satellite figure of 297.1. The difference between the equatorial diameter and the polar diameter is 500 ft. less than was previously believed.

Acknowledgements

I wish to thank Mrs. D. M. C. WALKER for her help in the preparation of this paper. I am also grateful to Messrs. R. H. MERSON, R. N. A. PLIMMER and G. E. COOK, who have all made contributions to the work described here. Crown Copyright is reserved.

References

1. Staff of Mullard Observatory, Cambridge. Nature 180, 879 (1957).
2. Staff of R.A.E., Farnborough. Nature 180, 937 (1957).
3. T. E. STERNE and G. F. SCHILLING, Smithsonian Astrophysical Observatory Special Report No. 3 (1957).
4. I. HARRIS and R. JASTROW, Science 127, 471 (1958).
5. G. V. GROVES, Nature 181, 1055 (1958).
6. G. F. SCHILLING and T. E. STERNE, J. Geophys. Res. 64, 1 (1959).
7. J. W. SIRY, Paper presented at 5th C.S.A.G.I. Assembly, Moscow, July, 1958.
8. T. E. STERNE, Science 128, 420 (1958).
9. I. HARRIS and R. JASTROW, Science 128, 420 (1958).
10. L. I. SEDOV, Proceedings of the IXth International Astronautical Congress, Amsterdam 1958, p. 456. Wien: Springer, 1959.

11. G. F. SCHILLING and C. A. WHITNEY, Smithsonian Astrophysical Observatory Special Report No. 18, p. 20 (1958).
12. V. I. KRASSOVSKY, Proceedings of the IXth International Astronautical Congress, Amsterdam 1958, p. 614. Wien: Springer, 1959.
13. D. G. KING-HELE, Nature 183, 1224 (1959).
14. I. HARRIS and R. JASTROW, Planet. Space Sci. 1, 20 (1959).
15. R. A. MINZNER and W. S. RIPLEY, The A.R.D.C. Model Atmosphere. U.S. Air Force Surveys in Geophysics No. 86 (1956).
16. D. G. KING-HELE and G. E. COOK, Ministry of Supply Report (1959). (To be published.)
17. T. E. STERNE, Science 127, 1245 (1958).
18. J. G. DAVIES, J. V. EVANS, S. EVANS, J. S. GREENHOW, J. E. HALL, E. L. NEUFELD and J. H. THOMSON, Proc. Roy. Soc. A 250, 367 (1959).
19. G. E. COOK, Ministry of Supply Report (1959). (To be published.)
20. D. G. KING-HELE and D. C. M. LESLIE, Nature 181, 1761 (1958).
21. D. G. KING-HELE and D. M. C. WALKER, Nature 183, 527 (1959).
22. D. G. KING-HELE, Nature 181, 738 (1958).
23. E. C. CORNFORD, Paper presented at 5th C.S.A.G.I. Assembly. Moscow, July, 1958.
24. D. G. KING-HELE and R. H. MERSON, J. Brit. Interplan. Soc. 16, 446 (1958).
25. D. G. KING-HELE, Nature 182, 1409 (1958).
26. L. G. JACCHIA, Nature 183, 526 (1959).
27. H. E. LAGOW, R. HOROWITZ and J. AINSWORTH, Paper presented at 5th C.S.A.G.I. Assembly, Moscow, July, 1958.
28. R. JASTROW and C. A. PEARSE, J. Geophys. Res. 62, 413 (1957).
29. D. G. KING-HELE and D. M. C. WALKER, J. Brit. Interplan. Soc. 17, 2 (1959).
30. G. V. GROVES, Nature 182, 1533 (1958).
31. T. R. NONWEILER, Nature 182, 468 (1958)
32. W. PRIESTER, Naturwiss. 46, 197 (1959).
33. L. G. JACCHIA, Nature 183, 1662 (1959).
34. V. G. ISTOMIN, Paper presented at 5th C.S.A.G.I. Assembly, Moscow, July, 1958.
35. J. W. TOWNSEND, Science 129, 80 (1959).
36. W. W. KELLOGG, Planet. Space Sci. 1, 71 (1959).
37. R. N. A. PLIMMER, Unpublished Ministry of Supply Report (1959).
38. C. H. BOSANQUET, Nature 182, 1533 (1958).
39. H. JEFFREYS, The Earth, 3rd edition, pp. 129, 184. Cambridge: University Press, 1952.
40. J. A. O'KEEFE, A. ECKELS and R. K. SQUIRES, Science 129, 565 (1959).
41. Y. KOZAI, Smithsonian Astrophysical Observatory Special Report No. 22 (1959).
42. D. G. KING-HELE, Proc. Roy. Soc. A 247, 49 (1958).
43. R. H. MERSON and R. N. A. PLIMMER, Ministry of Supply Report (1959). (To be published.)
44. R. H. MERSON and D. G. KING-HELE, Nature 182, 640 (1958).
45. L. G. JACCHIA, Smithsonian Astrophysical Observatory Special Report No. 19 (1958).
46. D. G. KING-HELE and R. H. MERSON, Nature 183, 881 (1959).
47. G. E. COOK, Unpublished Ministry of Supply Report (1959).

A Practical Investigation of Spaceship Control Problems

By

C. A. Cross[1]

(With 5 Figures)

(Received June 18, 1959)

Abstract — Zusammenfassung — Résumé

A Practical Investigation of Spaceship Control Problems. A Spaceship flight simulator has been designed, built, and successfully operated. It consists of a control panel, an electro-mechanical computer, and a planetarium type projector. A person sitting at the control panel and watching the projected display has the experience of flying a rocket-ship in interplanetary space.

This relatively simple quipment has been used to investigate the techniques of manned spaceflight. The computer is an exact physical analogue in which a pen duplicates the motion of the spaceship on a scale of one inch to seventy miles. The pen controls a reference sphere projector and simultaneously plots the spaceship's position on a chart ten inches square. Flight characteristics have been assigned to the spaceship from a consideration of first principles. The spaceship can be turned at rates of up to 2 R.P.M. by firing torque jets, and aiming trials have shown that the standard deviation of the main propulsion rocket motor is 1.65°.

A series of twenty flights have been made to determine the human pilots ability to carry out a simple circumnavigation of a luminous reference sphere 22 miles in diameter in the middle of the 700 mile square navigable area. The success of each flight was judged by the accuracy with which the pilot returned to his starting point, and by the economy of propulsive effort used during the journey. These trials have shown that the spaceship cannot be flown succesfully by direct instinctive interpretation of the projected display. It must be navigated from starting to finishing point by deducing its position in space from the observations, plotting this on a chart, and taking the control action needed to complete the desired journey.

Improvements in navigating equipment and technique show up as a steady reduction in the final position errors, from hundreds of miles at the start to tens of miles at the end of the series. The minimum time needed is also reduced from over an hour to about half an hour. Propulsion economy was not significantly improved, and remained in the region of 50 percent throughout the series.

The simulator described is limited to two dimensions and does not include the effects of a gravitational field. An analysis is presented which suggests that the electro-mechanical computer used cannot be extended to three dimensions, although this should not be difficult with an electronic analogue. Further developments of the present equipment are discussed which include an inverse square law field about the central reference sphere, and the provision of a much more elaborate reference sphere. This would show phase changes, surface details, and would occult the stars like a real solid body. With these developments one could simulate many of the effects of flying a spaceship in a close orbit round the Earth.

Praktische Untersuchung von Raumschiff-Kontrollproblemen. Ein Raumflug-simulator wurde entworfen, gebaut und erfolgreich betrieben. Er besteht aus einem

[1] 284 London Road, Northwich, Cheshire, United Kingdom.

Kontrolltisch, einer elektro-mechanischen Rechenmaschine und einem Projektor, wie er in einem Planetarium verwendet wird. Eine am Kontrolltisch sitzende Person, die das projizierte Bild beobachtet, hat das Gefühl, ein Raketenschiff im interplanetarischen Raum zu fliegen.

Diese verhältnismäßig einfache Einrichtung wurde benützt, um die Technik des bemannten Raumfluges zu untersuchen. Die Rechenmaschine ist eine genaue physikalische Analogiemaschine, in der eine Feder die Bewegung des Raumschiffes im Verhältnis von 1 Zoll zu 70 Meilen wiedergibt. Die Feder kontrolliert einen sphärischen Bezugsprojektor und trägt gleichzeitig die Position des Raumschiffes auf einem Diagramm von 10 Quadratzoll auf. Aus einer Betrachtung der Grundprinzipien wurden dem Raumschiff Flugcharakteristika zugeordnet. Das Raumschiff kann mit Hilfe eines Quer-Strahlantriebes auf Umdrehungsgeschwindigkeiten von 2 Umdrehungen je Minute gebracht werden. Zielversuche haben gezeigt, daß die Standardabweichung des Raketenhauptantriebmotors 1,65° ist.

Eine Serie von 20 Flügen wurde ausgeführt, um zu zeigen, inwieweit menschliche Piloten imstande sind, eine einfache Umfahrung einer leuchtenden Richtkugel auszuführen, die 22 Meilen Durchmesser in der Mitte eines befahrbaren Gebietes von 700 Quadratmeilen hat. Der Erfolg jedes Fluges wurde nach der Genauigkeit beurteilt, mit der der Pilot zu seinem Standpunkt zurückkehrte, und ebenso nach der Ökonomie der Antriebsleistung während der Fahrt. Diese Versuche zeigten, daß das Raumschiff nicht erfolgreich mittels direkter instinktiver Interpretation des projizierten Umweltbildes geflogen werden kann. Vielmehr muß es vom Start bis zum Endpunkt gelenkt werden, indem seine Position im Raum aus den Beobachtungen, die auf einem Diagramm aufgetragen werden, abgeleitet und die Kontrollaktion eingeleitet wird, welche die gewünschte Fahrt vollenden soll.

Verbesserungen in der Navigationsausrüstung und -technik erscheinen als ständige Reduktion der Fehler in der endgültigen Position, von hunderten Meilen beim Start bis zu Zehnern von Meilen am Ende der Reihe. Die erforderliche Mindestzeit läßt sich auch von mehr als einer Stunde auf etwa eine halbe Stunde verringern. Die Ökonomie des Antriebes wurde nicht wesentlich verbessert und verharrte im Bereich von 50% während der Versuchsserie.

Der beschriebene Simulator ist auf zwei Dimensionen begrenzt und schließt nicht die Wirkungen eines Schwerefeldes ein. Es wird eine Analyse angegeben, die vermuten läßt, daß die elektro-mechanische Rechenmaschine nicht auf drei Dimensionen ausgedehnt werden kann, obwohl dies mit einer elektronischen Analogiemaschine nicht schwierig sein sollte. Es werden weitere Entwicklungen der gegenwärtigen Ausstattung diskutiert, die ein quadratisch reziprokes Feld um die zentrale Bezugssphäre zum Inhalt haben; auch die Herstellung einer viel besser ausgearbeiteten Bezugssphäre wird erörtert. Diese würde Phasenwechsel und Oberflächendetails zeigen und die Sterne wie ein wirklicher fester Körper verdunkeln. Mit Hilfe dieser Fortschritte könnte man viele der Effekte simulieren, die beim Flug eines Raumschiffes in einer engen Umkreisungsbahn um die Erde auftreten.

Une enquête pratique sur les problèmes de contrôle des astronefs. Un simulateur de vol spatial a été conçu, réalisé et opéré avec succès. Il consiste en un panneau de contrôle, un calculateur électro-mécanique et un projecteur du genre planetarium. Une personne assise au panneau de contrôle et regardant la projection a l'impression de piloter un astronef dans l'espace interplanétaire. Cet équipement relativement simple a été utilisé pour évaluer les techniques du vol spatial. Dans le calculateur analogique exact, une plume reproduit la trajectoire à l'échelle d'un pouce par 70 miles. La plume contrôle un projecteur sphérique de référence et simultanément marque la position de l'astronef sur une carte de dix pouces carrés. Les caractéristiques de vol de l'astronef ont été choisies sur la base de principes fondamentaux. Des rotations peuvent être effectuées au taux de 2 tours/min. en actionnant des jets transversaux et des essais de visée ont montré que la déviation standard du moteur principal était de 1.65°.

Dans une série de 20 vols l'aptitude des pilotes humains à contourner une sphère lumineuse de référence de 22 miles de diamètre dans une aire navigable de 700 miles carrés a été évaluée. Le succès de chaque vol était jugé d'après la précision du retour au point de départ et l'économie d'effort propulsif dépensé. Les essais ont montré qu'une interprétation instinctive directe de la projection était insuffisante pour un pilotage adéquat. Du départ à l'arrivée il faut déduire la position à partir des observations, l'inscrire sur une carte et prendre les mesures de guidage nécessaires.

L'amélioration dans les équipements de navigation et dans la technique de celle-ci se montre par une réduction constante des erreurs dans la position finale. Celle-ci descend de centaines de miles au départ à des dizaines de miles au cours des derniers essais. Le temps requis est aussi descendu de plus d'une heure à une demi-heure. L'économie de propulsion est restée dans la région des 50% tout au long des essais.

Le simulateur décrit est à deux dimensions et n'inclut pas les effets du champ de gravitation. Une analyse suggère que le calculateur électro-mécanique utilisé ne peut être étendu à trois dimensions; quoique ceci ne doive pas offrir de difficultés avec un calculateur analogique électronique. Un développement de l'équipement actuel est discuté. Il comporte notamment un champ de gravitation quadratique inverse autour d'une sphère de référence centrale et une sphère de référence beaucoup plus détaillée. Elle présenterait des changements de phase, des détails de surface et occulterait les étoiles comme un véritable corps solide. Avec ces améliorations on pourrait simuler un grand nombre des impressions reçues en pilotant un astronef sur une orbite fermée autour de la terre.

Introduction

Flight simulators are used in the aeronautical field to train pilots without incurring the hazards and expense of flying the real aeroplanes. The trainee sits at a set of aircraft controls connected to a computer which simulates the aircraft's responses. These responses are appropriately displayed so that the pilot can see the results of his control efforts and take continuing action to maintain control. In most of these simulators [1] the display is only an instrument flight panel, and the pilot must fly "blind", without seeing his external environment. The SHORT and HARLAND helicopter flight simulator however provides a complete landscape outside a helicopter cabin mock up [2], and the pilot here encounters a close approximation to flying a real helicopter.

This paper describes the construction, calibration and preliminary trials of a spaceship flight simulator in which the external environment of stars and planets is correctly represented. As no-one has yet built a spaceship the computer cannot be set up to imitate the responses of a known vehicle, but has characteristics which are derived from first principles. It follows that this machine is not intended to train space pilots for an established duty, but instead allows us to define and explore some of the problems of controlling a rocket ship in free space. Whilst the use of spaceflight simulators has been discussed in the literature [3], this new application was not envisaged. The work presented here appears to be the first attempt to construct such a machine.

Definitions and Restrictions

With unlimited resources it would in theory be possible to build a simulator of quite general application, which would compute and display correctly the motion of a spaceship throughout the whole solar system. By limiting its application the apparatus can however be greatly simplified whilst still accurately representing the flight of the spaceship within the limits imposed. Thus the first simulator which I built [4] was an extremely simple one which dealt only with a rocket making a vertical landing on the moon.

The present simulator is of much more general application, and only two restrictions are accepted. The motion of the rocket is restricted to a plane surface of limited extent, and there is no sensible gravitational field. Thus the rocket always moves in a straight line at constant velocity until the motor is fired, and when it has finished firing the rocket moves with a new velocity and direction determined by vector addition of the original and the injected velocities. The limitation to two dimensions is not so restrictive as might at first be thought. It is likely that in all local navigation in free space the first problem will be the selection of a plane in which the manoeuvers will be carried out. Thus only this first selection has been made, leaving the rest of the problem open.

With the spaceship well removed from massive bodies, and limited to relatively small movements, the Sun, Planets, and the Stars show no relative motion. To make the spaceships movements apparent a reference body is supplied in the form of a luminous sphere at the centre of the plane surface over which the ship can move. The computer causes the size and position of the sphere to vary appropriately as the spaceship moves around it.

Design Details

The simulator consists of a separate control and instrument panel which is connected by a multicore cable to a combined computer and projector unit.

Fig. 1. View of computer and projector unit, with author seated at the control panel

Fig. 1 shows the author seated at the control panel with the computer projector unit in the background. In use the control panel is arranged inside a small "cabin" with a navigation port which looks out upon the projected display. Thus the pilot sits within a motionless mock up, and when he turns his spaceship one way the computer moves the projected stars the equivalent amount the other way. Whilst changes in the attitude of the ship are at once evident to the pilot because he sees the stars moving, these changes do not have any effect on the ship's velocity or direction of motion, which are quite independant of its attitude. The velocity and direction can only be changed by firing the rocket motor which is the ship's only means of propulsion.

The computer used is an electro-mechanical analogue working in real time, and requires three integrating units to solve the equations of motion and attitude of the ship in two dimensional space. The integrating units are of the ball and disc type, and were obtained from a government surplus air position indicator, where they perform a somewhat similar function.

The computer projector unit is shown diagramatically in Fig. 2. The greater part of the mechanism is mounted on a turntable (1) which revolves about the main yaw axis (2) at a rate determined by the integrating unit (3). The setting of this integrator is controlled by the yaw controls on the panel, and it also drives an impulse transmitter which keeps a repeater compass on the panel synchronised with the yaw attitude of the spaceship.

The turntable rotates under a fixed bridge *(4)* on which the main propulsion drive motor is mounted *(5)*. Its rubber tyred drive wheel bears on the resolving

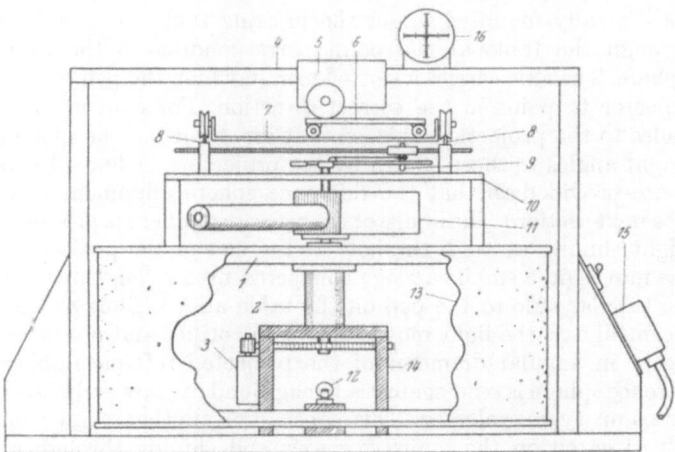

Fig. 2. Elevation of computer-projector

table *(6)* at the point of intersection with the yaw axis. The drive wheel rotation is transmitted to the control panel by a dessyn type position indicator, and is displayed there as an integrating accelerometer reading (marked VELOCITY MPH/

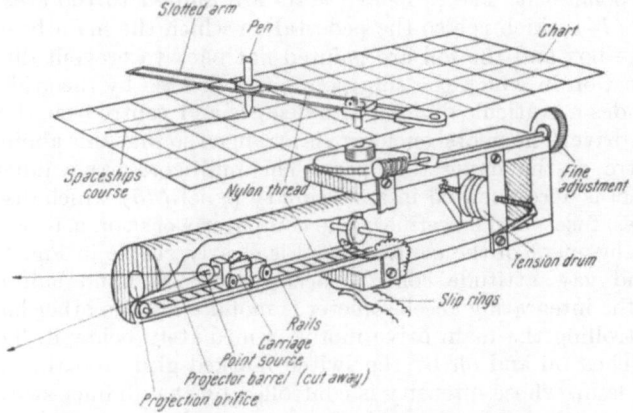

Fig. 3. Cut-away sketch of reference-sphere projector

100, Fig. 1). The resolving table runs on rails that are mounted on a second carriage *(7)*, which itself runs on rails *(8)* mounted at right angles to the first set. Thus the resolving table is free to move in any horizontal direction in which it may be driven by the main drive wheel. Moreover its displacement is unaffected by subsequent rotation on the yaw axis. This displacement is then proportional to the vehicles velocity, and is resolved along the direction of the two sets of rails to control the settings of two integrating units (omitted from the diagram for clarity) which in turn drive the pen *(9)* on the plotting table *(10)* in the same direction as the resolving table has been displaced, and at a speed proportional to its displacement. Thus the pen (representing the spaceship) moves at uniform velocity in a straight line, only suffering changes in velocity and direction when the main drive motor is fired and moves the resolving table to a new position.

The pen is mechanically coupled to the reference sphere projector *(11)* so that the size and position of the reference sphere are correctly reproduced. The projector is centrally mounted under the plotting table (see Fig. 3) and its axle extends through the table at the point corresponding to the location of the reference sphere. The axle carries a slotted arm in which the pen moves, constraining the projector to point in the correct direction. For convenience the arm is at right angles to the projection axis, so that directions on the plotting table are always at right angles to those shown by the projectors. A fine adjustment worm and wheel are provided so that the reference sphere alignment can be exactly adjusted (see next section). The image of the reference sphere is produced by a point source of light shining through the hole at the end of the projector barrel. The point source moves on a small carriage connected by a nylon thread which passes through the tubular axle to the pen on the table above. Thus as the pen moves nearer the central axis the light moves nearer the orifice and produces the appropriate increase in angular diameter of the projected reference sphere.

The reference sphere is seen against a background of stars and planets produced by a planetarium type projector. This consists of another point source of light *(12)*, Fig. 2, situated on the main yaw axis and shining through a cylindrical screen of aluminium foil *(13)*. The foil is 8″ wide by 96″ long representing a complete strip of the heavens 360° by 30° wide. It is pierced with some 2,000 holes varying in diameter from 0.018″ for 6th magnitude to 0.18″ for first magnitude stars, all correctly located to represent actual stars. The strip of foil is readily changed, and so far two strips have been punched to represent the equatorial belt and a polar belt. The projected stars are limited to the area of the screen by the box *(14)* which is also the pedestal on which the main bearing rests. The inside of the box and the foil are painted flat back to prevent diffuse reflection.

The direction in which the ship points is indicated by the grid projector *(16)* which provides a graticule divided into degrees and centred on the aiming point of the main drive. The pilots' enclosure is arranged so that this aiming grid appears in the centre of the navigating port. The multicore cable junction from the control panel is accomodated in a subsidiary panel *(15)* which also houses isolation switches, fuses, and a rectifier. The computer works on a 14 volt A.C. supply.

The arrangement of the control panel is clearly visible in Fig. 1. The compass repeater and yaw attitude controls occupy the left hand half of the control panel, and the integrating accelerometer is mounted in the other half with a press button controlling the main drive motor immediately below it. The aiming grid can be switched on and off by the switch marked grid, and the control panel is lit by a red lamp whose intensity is controlled by the dimmer switch at the right hand corner. A clock completes the instrumentation.

Although the simulator includes a great many modifications made to improve its accuracy and reliability to acceptable levels, it does not differ in principle from the original design which was worked out during the month of September 1957. Treated as a spare time project the initial construction was completed by June 1958. Thereafter a series of modifications was made whilst at the same time flight trials were going on. The changes included the replacement of the home made integrators used at first by those dismantled from an air position indicator, the installation of an instantaneous velocity indicator, and the fitting of the repeater compass on the control panel. These were all completed and the simulator attained its present form by December 1958.

Adjustment and Calibration

Choice of Scale

The simulator provides a chart area 10″ by 10″ over which the spaceship can move at a maximum speed of 0.6″ per minute. This speed was chosen so that major effects could be obtained in a reasonably short time. Common sense suggests that no-one would be prepared to sit in the simulator for hours on end waiting for something to happen. The size of the reference sphere is chosen so that from the chart edge it subtends an angle of 2.5°, exhibiting a disc of a diameter that can be estimated reasonably accurately by comparison with the aiming graticule. On the chart scale this corresponds to a diameter of 0.3125″, and this is in fact the diameter of the hole at the end of the reference sphere projector. The absolute scale is then a matter of free choice. Quite arbitrarily I have assigned a scale of 1″ to 70 miles. This makes the navigable area 700 miles square, with the central reference sphere 22 miles in diameter, and the maximum relative velocity 2,500 miles per hour. (This velocity can be exceeded along the diagonals of the resolving table, but it is the maximum which can be safely applied in any direction.)

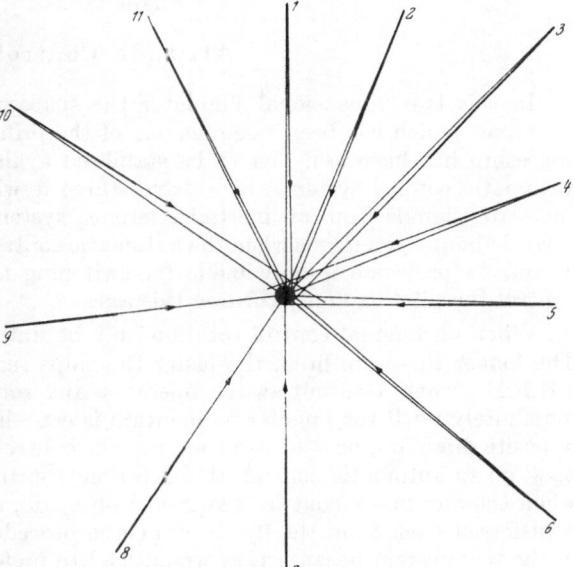

Angular Diameter of the Reference Sphere

In order that the apparent size of the reference sphere shall vary in the proper manner the length of the nylon thread must be adjusted so that the point source is at exactly the same distance from the projection orifice as the pen is from the centre of the chart. This adjustment cannot be made very accurately by direct measurement, so an indirect calibration is employed. With the approximate setting the reference sphere diameter (a) is observed for several different positions of the pen on the chart (x). The observed values of $(\sin a/2)^{-1}$ plotted against x

Fig. 4. Aiming tests

Test Number	Error degrees
1	—1.0
2	—1.3
3	—1.4
4	+1.4
5	—1.1
6	+0.8
7	0
8	0
9	0
10	—3.7
11	—0.5

give a straight line which can be extrapolated to obtain the residual value of x. The length of the nylon thread can then be altered by this amount to give a straight line which now passes through the origin.

Alignment of the Reference Sphere

In order to set up the simulator correctly the aiming point marked by the grid must be made to coincide with the real aiming point. This is best done by adjusting the resolving table so that the rocket is at rest relative to the reference sphere, then turning the rocket until the graticule and reference sphere coincide and firing the main drive motor. If the alignment is correct the rocket will then move straight towards the reference sphere on the chart, and the reference sphere image will grow larger without changing its position amongst the stars. Any error can be measured on the chart and then corrected by means of the fine adjustment on the projector. Fig. 4 shows the result of a series of these aiming trials made after correct setting up and without intermediate adjustments. These tests show that there is no marked systematic error, and the standard deviation of the rocket about its aiming point is 1.65°, which compares reasonably well with the accuracy of current missile rockets.

Flight Trials

Attitude Control

In this two dimensional simulator the spaceship can only rotate about the yaw axis, which has been taken as one of the principal axes. To lend reality the spaceship has been assumed to be stabilised against changes in attitude by an automatic control system which drives three flywheels at speeds determined by the error signals from an inertial reference system. The ship can then only be turned about a given axis when the automatic control on this axis has been switched off. As provision is only made for switching the yaw attitude on to manual control the pilot can only turn on this axis.

When on manual control rotation may be initiated by firing the torque jets. The longer these are fired, the faster the ships turns, up to a maximum rate of 2 R.P.M., when a cutout switch operates. Any rotation, once started, continues indefinitely until the angular momentum is cancelled by firing torque jets in the opposite direction, or is transferred into the control system flywheel by switching back on to automatic control. If this is done rotation is resumed at the same rate when the automatic control is switched off again, and the angular momentum is transferred back from the flywheel. (These procedures would not be encouraged in the real system because they would lead to undesirably large momentum storage capacities for the flywheels. They are undesirable for rather different reasons in the simulator as well, and the aim should be to start and stop all turns with the torque jets.)

The judgement required to effect changes in attitude smoothly and rapidly is soon acquired, and I do not think that attitude control in three dimensions will be especially difficult, although free rotation about an arbitrary axis is a problem of some mathematical complexity [5].

Movement in Space

When the simple art of attitude control has been learnt one can attempt to move about in space. A simple exercise which suggested itself was to circumnavigate the reference sphere, starting at rest from a distant point and returning to this point. During this manoeuver the pilot sees the reference sphere growing in size as he approaches it, moving round the heavens as he circles it, and finally shrinking back to its original size and position as he returns to his starting point.

There are of course a great many ways of effecting this manoeuver, but a simple and economical method is to aim off to one side of the sphere, and coast along right past it. As soon as one is beyond it the motor is used again to halt the rocket and give it a slow drift at right angles to its original motion. When the rocket has drifted far enough across "behind" the reference sphere a further propulsive effort sets it off back towards its starting point, where it may be halted by a final burst. A typical successful run of this kind is shown in Fig. 5, which

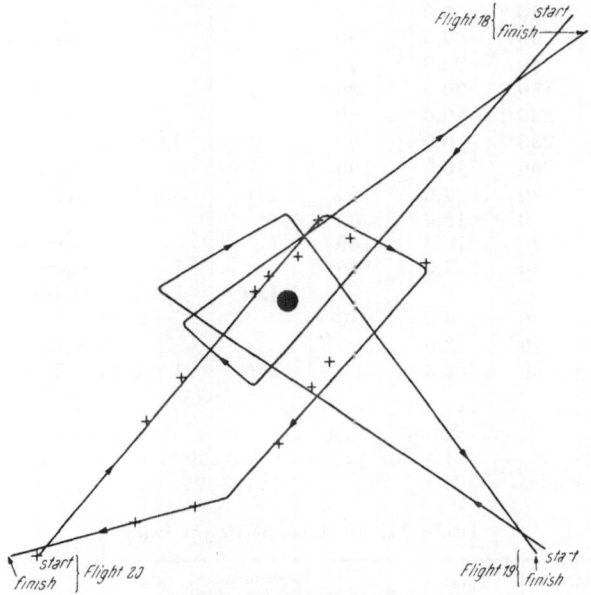

Fig. 5. Chart for flights 18, 19 and 20

reproduces the computer chart on which runs 18, 19, and 20 were carried out. The navigational fixes taken during the course of flight 20 (Table I) have been plotted directly on the chart for comparison with the actual record of the motion traced by the computer. The run includes one course correction, and the steps in its computation and execution are seen in Table I. Fig. 6 also shows the two earlier runs in which the flight programme was carried through without sufficient deviation developing to need correction.

Table II gives critical data on all the runs carried out. The criteria of successful accomplishment of this simple circumnavigation are quite complex. They include the errors in position and velocity when the manoeuver is complete, together with the time taken and the total impulse used from the main drive. The residuals of position and velocity should be as small as possible, and a propulsion efficiency can be calculated by comparing the impulse used (plus the residual velocity) with the absolute minimum (eight times the initial distance divided by the total time taken).

To some extent the significance of the results in Table II is obscured by the changes in the computer which were made during this period. The propulsive efficiency shows no significant improvement over the whole series, but the ear-

Table I. *Flight Number 20*
Reference sphere at 260° and 400 miles.

Flight Plan: 2,400 mph at 255°
3,000 mph at 40°
3,000 mph at 120°
2,400 mph at 265°

Time	Bearing		Distance		Drive		Notes
	Sphere	Self	Angle	Miles	Bearing	DV	
0.00	260	80	2.5	400	255	2,400	
0.05	265	85	5.0	200			
0.07	270	90	7.0	150			On course
0.09	320	140	25.0	40			
0.10	0	180	30.0	30			
0.11	50	230	20.0	50			
0.12	55	235	10.0	100	40	3,000	
0.14	80	260	10.0	100			On course
0.18	110	290	7.0	160	110	3,000	10° error in firing
0.21	180	0	12.0	80			
0.22	200	20	10.0	100			Seriously off course
0.24	220	40	7.0	160			Correction computed
0.26	230				190	1,200	Correction applied
0.27	240	60	4.0	250			
0.29	250	70	3.0	300			New final DV comp.
0.33	260	80	2.5	400	285	2,700	E.T.A. 0.34

Flight Duration .... 33 minutes
Position Error .... 35 miles
Residual Velocity .... 300 mph
Number of Corrections .... one

Table II. *Analysis of Flight Data*

Flight No.	Durn. Mins.	Min. Calc. DV.	Total DV. Used	Eff. Calc. Used	Residuals Vel. mph	Residuals Dist. Miles	No of Corr.	Comments
1	90	2,100	4,900	0.43	—	120	3	
2	77	2,500	6,100	0.41	—	70	2	
3	54	3,700	4,400	0.84	—	30	1	
4	70	2,700	4,550	0.60	—	105	2	
5	60	3,200	6,800	0.47	—	210	2	
6	45	4,300	6,700	0.64	—	25	1	Vector adding device
7	60	3,200	5,200	0.62	—	40	1	brought into use
8	50	3,800	5,000	0.76	—	29	1	
9	47	4,100	5,400	0.74	—	15	1	
10	34	5,700	12,300	0.40	200	20	1	New Integrators and
11	39	4,900	9,900	0.48	250	35	1	residual velocity
12	50	3,800	7,900	0.49	0	21	2	indicator fitted
13	—	—	—	—	—	—	—	Collision with
14	51	3,750	9,900	0.33	150	58	1	reference sphere
15	41	4,700	11,200	0.42	90	40	2	Compass brought
16	47	4,100	9,400	0.43	300	65	1	into use
17	52	3,700	11,300	0.30	900	95	2	
18	33	5,800	10,800	0.54	200	24	0	
19	32	6,000	10,800	0.55	360	14	0	
20	33	5,800	12,300	0.46	300	35	1	

lier values (1 to 9) are not corrected for residual velocity, which was not measured in these runs. Two significant effects can be seen however:

(a) After run 5 a simple computing device was used. It consisted of two celluloid protractors bolted on a perspex strip, and may be seen in the authors hand, Fig. 1. With its help course corrections could be computed in time to apply them, and in subsequent runs the residual error of position is much reduced.

(b) After run 14 the attitude of the ship was read directly from the repeater compass dial. This is much quicker than finding the attitude by comparing the starfield observed round the graticule with a star chart, which was the method used in the earlier runs before the compass was fitted. It might be objected that a compass does not work in space, but in fact the instrument reading would be very simply provided from the inertial reference system against which the ship is stabilised. The compass north bearing is set quite arbitrarily, but when the operating plane includes the pole star this is the natural choice. It will be seen that after run 14 the flight duration is reduced, because the ship could be flown at higher speeds without losing one's bearings.

My experience in flying this simulator suggests that it is quite impossible to fly by visual cues alone. The changes in position and size of the reference sphere cannot be interpreted instinctively, in the way we judge velocity and direction when moving in more familiar environments. The only way I have found to make this interpretation is by measuring quantitatively the changes, and then computing velocity and direction. Thus even the simplest spaceflight turns into a complex problem in navigation. The spaceman who dons a spacesuit, steps out of an airlock and jets casually across the void is due for a nasty shock unless he stays inside those science fiction magazines.

It would seem desirable to check the results I have obtained by training others to operate the simulator. Whilst there has never been any shortage of people prepared to endure a short demonstration of what the space pilot would see and do, I have not yet found anyone prepared to devote 12 to 24 hours to being properly trained and tested in this device.

Future Developments

The first question which must arise is the possibility of extending the simulator to cope with the unrestricted three dimensional problem. I believe that this is impossible with an electro-mechanical computer such as the one described in this paper. This machine centres around an exact physical analogue of the spaceship moving in two dimensions over the chart, whilst the third dimension is used to feed control information to the pen and to extract information from the pen to control the projector. If the analogue is extended into three dimensions there ceases to be any route by which this information can be conveyed.

This does not of course mean that a three dimensional space simulator cannot be built. By substituting an electronic analogue for the mechanical one the dimensional restriction is avoided, and the transition in the computing section from two to three dimensions is accomplished simply by adding a third series of electrical components to represent the third dimension. There would still be some interesting problems to solve however in arranging the output of these computers to control a conventional type planetarium projector which-would provide the complete celestial sphere instead of the strip used in the present apparatus.

Whilst it appears impossible to simulate three dimensions electro-mechanically it does not seem at all difficult to include a central gravitational field. In place of the main rocket motor drive wheel we must provide a drive wheel mounted on the central axle of the reference sphere projector and arranged so that it always pushes the resolving table away from the pen. If this wheel is driven continuously at a rate inversely proportional to the square of distance of the pen from the centre of the chart then the pen is effectively moving in a central inverse square law field. Thus the pen will describe an elliptical, parabolic, or hyperbolic orbit about the reference sphere according to its initial velocity. In order to allow the spaceship to use its rocket motor one would have to transfer the resolving table from the gravity drive to the rocket drive whenever the rocket was fired. Thus the gravitational field would not operate when the rocket motor was in use. In most circumstances this would not cause serious error.

Another fairly simple improvement would be the provision of a more elaborate reference sphere. The present one must be regarded as a self luminous cloud of gas, for it does not show any phases and the stars are visible through it (this is particularly noticeable when the rocket is close to the sphere so that it subtends a large angle). I am now developing a reference sphere projector which will give an image showing phase changes as the rocket moves round it, and with surface detail visible on the illuminated hemisphere. This projector could be mechanically linked to an occulting disc in the star projector which would prevent star images from being formed on the image of the sphere. This solid planetary reference sphere could be combined with gravity simulation to give a demonstration of many of the effects to be found when flying a rocket in a planetary orbit.

Having completed and successfully operated the simulator a number of more frivolous developments suggest themselves, in general aimed at improving the realism by introducing extra "effects". Thus one might provide a more elaborate cabin mock up, including a much more elaborate instrument panel, with the ship environment controls and the communication section etc. which would all be needed in a real spaceship. One could provide proper sound effects, principally rocket motor noise for when the main drive is in use. The whole mock up, with the pilot inside, could be arranged to tip up whenever the main drive was used. This would give a very convincing simulation of the sudden application of thrust. There is no limit to the ingenuity and effort which could go into these ancillary effects, and I incline to the view that an all out effort on these lines could produce an illusion of startling and frightening reality in all except the simulation of zero gravity.

Devices of this kind may well be under development for the training of the Russian and American spacemen of the future, but they are clearly beyond the resources of a single individual. The only space simulator which has been publicly announced so far is that used for training the pilot of the X15 rocket in orientation control during the part of the flight when the ship is in free fall virtually outside the atmosphere. This simulator presumably deals with the full three dimensional case, but I think it unlikely that the external environment has been imitated. The display will probably be limited to an instrument panel dealing only with changes in attitude.

References

1. Advanced Aircrew Trainer with Mechanical Computer. Engineering **13**, 214 (1954).
2. Simulating Helicopter Flight. Engineering **18**, 500 (1957).
3. W. Ley and W. von Braun, The Exploration of Mars, Chapter 7, p. 138. London: Sidgwick and Jackson Ltd., 1956.
4. C. A. Cross, An Analogue Computer for the Vertical Rocket Landing and Take Off Problems. J. Brit. Interplan. Soc. **15**, 7 (1956).
5. R. N. Bracewell and O. K. Garriott, Rotation of Artificial Earth Satellites. Nature **182**, 760 (1958).

Magnetohydrodynamics and its Application to Propulsion and Re-Entry

By

Rudolf X. Meyer[1]

(With 4 Figures)

(Received June 29, 1959)

Abstract — Zusammenfassung — Résumé

Magnetohydrodynamics and its Application to Propulsion and Re-Entry. The first part of this paper is largely a review of some of the basic concepts of magnetohydrodynamics in the regime of continuum fluid mechanics. The acceleration of plasmas to high velocities by means of magnetic fields is considered. The use of magnetic fields for the purpose of reducing aerodynamic heating at re-entry is discussed, and some theoretical and experimental data which have been previously reported, are reviewed.

In a second part of this paper, the theory of the Newtonian approximation to magnetohydrodynamic flow is developed. Results are presented, concerning the magnetohydrodynamic flow in the shock layer of a re-entry body. A similarity solution of the resulting equations is obtained for a circular cone in the case of finite and variable electrical conductivity.

Magnetohydrodynamik und ihre Anwendung für Antrieb und Rückkehr. Der erste Teil der vorliegenden Arbeit ist großenteils eine Übersicht über einige der grundlegenden Auffassungen der Magnetohydrodynamik im Bereich der Kontinuum-Strömungsmechanik. Es wird die Beschleunigung von Plasmen auf hohe Geschwindigkeiten mittels magnetischer Felder betrachtet. Die Benützung magnetischer Felder zur Verringerung der aerodynamischen Erhitzung beim Wiedereintritt in die Atmosphäre wird erörtert; einige theoretische und experimentelle Daten, über die früher berichtet worden ist, werden besprochen.

Im zweiten Teil der Arbeit wird die Theorie der Newtonschen Annäherung an die magnetohydrodynamische Strömung entwickelt. Hinsichtlich der magnetohydrodynamischen Strömung in der Stoß-Schicht eines in die Atmosphäre zurückkehrenden Körpers werden Ergebnisse gebracht. Eine Ähnlichkeitslösung der sich ergebenden Gleichungen wird für einen Kreiskegel im Fall begrenzter und variabler elektrischer Leitfähigkeit erhalten.

La magnétohydrodynamique et ses applications à la propulsion et au problème de la rentrée. La première partie est principalement une revue des concepts de base de la magnétohydrodynamique en régime d'écoulement continu. On considère l'accélération de plasmas à de très hautes vitesses par champs magnétiques et on discute l'utilité de ces champs pour réduire l'échauffement cinétique à la rentrée sur la base de données théoriques et expérimentales publiées.

Dans une seconde partie, l'approximation Newtonienne à l'écoulement magnétohydrodynamique est développée. Des résultats sont présentés pour l'écoulement

---

[1] Member of Senior Staff, Physical Research Laboratory, Space Technology Laboratories, Inc., Los Angeles 45, California, U.S.A.

derrière l'onde de choc à la rentrée. Un principe de similitude des équations résultantes est obtenu pour le cas d'un cone circulaire quand la conductivité électrique est finie et variable.

Introduction

The study of the interaction of magnetic fields with electrically conducting gases is of increasing importance to astronautics. In particular, applications to plasma propulsion and to problems associated with re-entry into a planetary atmosphere have been given attention [1] to [9]. A bibliography of recent date and covering all fields of magneto-hydrodynamics has been compiled by BANKS [10].

In general, a very high temperature is a prerequisite for appreciable magneto-hydrodynamic effects in gases. In the case of hydrogen, in the range of densities of interest in plasma propulsion, a lower limit for this temperature appears to be about 10,000° K. Actually, much higher temperatures of the working fluid have often been postulated in various proposed schemes for magnetohydrodynamic propulsion. In air, at conditions prevailing during re-entry, this temperature limit tends to be somewhat lower (approximately 6,000° K), mainly due to the relatively low ionization potential of nitric oxide, which is formed at elevated temperatures. Seeding of the gas by means of small amounts of alkali vapors or other materials of low ionization potential may be feasible, particularly in the case of rocket propellants [1]. In this case, useful effects may be achieved at temperatures as low as 2,500° K.

The presence of an applied magnetic field can radically change the hydro-dynamic flow of an ionized gas (plasma). Due to the motion through the magnetic lines of force, currents are induced in the gas which produce a body force, modifying therefore the motion. The plasma currents give rise to a magnetic field, which adds to the field produced by external currents. In applications to plasma propulsion and re-entry, the density is usually sufficiently large for the description of the fluid as a continuum to be valid.

The theoretical analysis of these problems requires in general a solution of MAXWELL's equations for the electromagnetic field, coupled with the equations of fluid mechanics containing the additional force term (ponderomotive force). In the case of steady flow of an inviscous gas, for which radiation and heat conduction can be neglected, the equations of magnetohydrodynamics are

$$\nabla \times \vec{H} = \vec{j} \tag{1}$$

$$\nabla \times \vec{E} = 0 \tag{2}$$

$$\nabla \cdot \vec{B} = 0 \tag{3}$$

$$\nabla \cdot (\varrho \vec{u}) = 0 \tag{4}$$

$$\varrho \frac{D\vec{u}}{Dt} = -\nabla p - \vec{j} \times \vec{B} \tag{5}$$

$$\varrho \frac{D}{Dt} \left(h + \frac{u^2}{2} \right) = \vec{j} \cdot \vec{E} . \tag{6}$$

They are supplemented by constitutive equations which usually assume the form

$$\vec{B} = \mu \vec{H} \tag{7}$$

$$\vec{j} = \sigma (\vec{E} + \vec{u} \times \vec{B}) \tag{8}$$

together·with the caloric equation of state $h = h$ (ϱ, p) and an expression for the conductivity $\sigma = \sigma$ (h, ϱ). In these equations, the electrical quantities $\vec{B}, \vec{H}, \vec{E}$ and μ have their usual meaning, $\vec{j}$ is the current density, σ the (scalar) conductivity, $\vec{u}$ the velocity, ϱ the density, p the pressure and h the enthalpy. The electric field strength $\vec{E}$ in these equations is the one observed in the stationary frame, rather than in the rest frame of the gas.

A recent discussion of the energy equation in magnetohydrodynamics [eq.(6)] has been given by CHU [11]. In formulating it, the Joule heating j^2/σ must be taken into account. In many problems which are time-independent and which exhibit certain symmetries, $\vec{E}$ can be shown to vanish identically. In particular, this is the case for rotationally symmetric problems, such as the magnetohydrodynamic flow about a symmetric body at zero angle of attack, provided that no electric potential difference is impressed on the boundaries. In these cases, it follows from eq. (6) that the total enthalpy is conserved along streamlines, just as in the ordinary gasdynamics of an inviscous fluid. The Joule heating is then just equal to the work done against the magnetic field.

The permeability μ can be replaced in all cases of interest here by the permeability of vacuum. The conductivity generally depends very strongly on the temperature, and — to a lesser degree — on the density. The strong temperature dependence is particularly pronounced in cases in which the gas is only weakly ionized [12]. If the degree of ionization is of the order of one per cent or higher, the conductivity very nearly satisfies SPITZER's equation [13] derived for a fully ionized gas and increases approximately as the 3/2 power of the temperature. A somewhat more complicated temperature dependence must be expected, however, in the case of an incomplete ionization of a multiply ionizable gas.

The simple form of OHM's law represented by eq. (8) is valid only in the limit of weak fields. In particular, if the magnetic field strength is such that the LARMOR frequency of the electrons becomes comparable or larger than their collision frequency, this equation does not apply. A more generally valid expression has been derived by SCHLÜTER [14]. For the case of weakly ionized gases, KEMP and PETSCHEK [5] indicate that two new effects arise, namely a HALL current and an "ion slip". In applications to plasma propulsion and re-entry, useful results can often be obtained with OHM's law in the form of eq. (8), although it is necessary in each case to check its validity.

The set of equations given above is mathematically complete. It is not necessary to include here the expression for the net charge density as the divergence of the dielectric displacement, because the net charge can be shown to have a negligible effect in situations described by magnetohydrodynamics [15].

Applications to Re-Entry and Plasma Propulsion

Re-Entry

The possibility of reducing the aerodynamic heating during entry into a planetary atmosphere by means of magnetic fields, has been recognized for some time now. Proposals of this kind usually involve the use of solenoids located in the vicinity of the stagnation point, where the heat transfer to the vehicle is most critical.

Theoretical analyses of the magnetohydrodynamic flow near the stagnation point of blunt bodies at hypersonic speeds have been given by a number of authors, e.g., [6], [7], [8], [16]. The magnetic field modifies the flow in the inviscous portion of the shock layer as well as the flow in the viscous boundary layer, and a number of results have been obtained for both regions. In the case of a body of revolution at zero angle of attack, the plasma current flows on circular paths in planes perpendicular to the axis of symmetry. The direction of the ponderomotive force is such as to impede the flow of the gas through the magnetic lines of force. The fact that the velocity in the shock layer is diminished, tends to reduce the convective part of the heat transfer to the surface of the body.

The qualitative description of the magnetic effects on the boundary layer is more complicated, but computations indicate that an additional reduction of the heat transfer rate is achieved. In spite of the considerable number of theoretical analyses, the problem is far from being completely solved.

The fact that the velocity in the shock layer is reduced, has been demonstrated experimentally by showing that the thickness of the shock layer is increased, if a magnetic field is applied [17]. More experimental work in this field is required, and the adverse effect of the presumably increased radiative transfer resulting from the increased shock-layer thickness needs to be examined.

In order to determine the magnitude of magnetohydrodynamic effects, an estimate of the electric conductivity of the gas is required. In Fig. 1, the conductivity σ of atmospheric air behind a normal shock is plotted for different flight velocities V and altitudes H above sea level [16]. For comparison, the velocity of a low-altitude satellite and the earth's escape velocity are indicated. The seemingly irregular features of the curves of constant conductivity are caused by the effects of dissociation, and formation of NO.

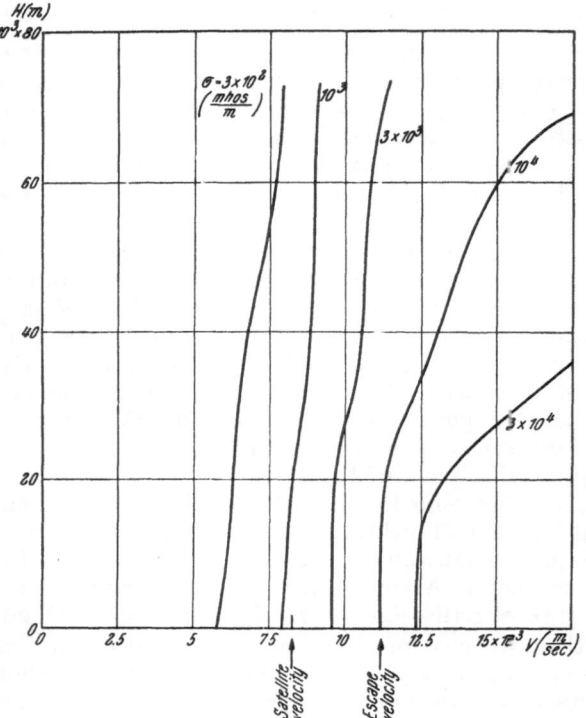

Fig. 1. Electrical conductivity σ of air behind a normal shock, as a function of flight velocity V and altitude H (from [16])

Another important question concerns the magnitude of the magnetic fields which can be achieved in practice. Joule heating of the coils is very appreciable for the time intervals of interest in re-entry problems, limiting the flux density of the applied magnetic field usually to a few thousand Gauss. In Fig. 2, the ratio of the convective heat transfer rate at the stagnation point and of the corresponding peak heating rate in the absence of a magnetic field is plotted for the case of a re-entry of a space vehicle into the earth's atmosphere[1]. The parameter B_0 is the

---

[1] From an unpublished report by W. B. BUSH, R. X. MEYER, A. B. SCHAFFER, and R. W. ZIEMER.

magnetic flux density at the stagnation point. A very appreciable reduction of the peak heating rate and of the total heat input can be achieved at field strengths below 10,000 Gauss.

Propulsion

There is little doubt that magnetohydrodynamics will have to play a vital role, if thermonuclear propulsion can be achieved [9]. The present discussion will be confined, however, to the acceleration of plasmas by means of magnetic fields, without considering the problem of power generation.

In principle, it is possible to accelerate ionized propellants to very high velocities, corresponding to specific impulses comparable to those envisaged for ion propulsion. Assuming that the required electric power must be generated by means of a thermodynamic cycle in an essentially conventional manner, the large weight of such power plants would restrict the thrust of magnetohydrodynamic propulsion devices to low values, similarly as in the case of ion propulsion, and suitable only for transfer between orbits.

Conceptually the simplest configuration consists of a device, in which the velocity of the propellant, the electric current passed through it, and the externally applied magnetic field are perpendicular to each

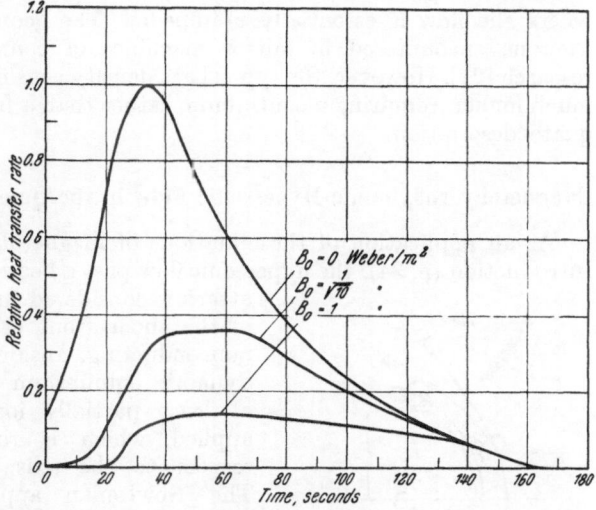

Fig. 2. Convective heat transfer rate at the stagnation point of a space vehicle entering the earth's atmosphere. B_0 = magnetic flux density at the stagnation point

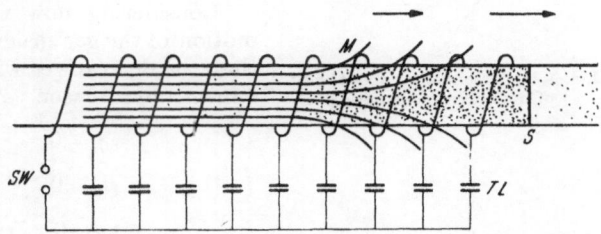

Fig. 3. Acceleration of plasma by means of a magnetic field. S = shock front, M = magnetic field, $T.L$ = transmission line, SW = switch

other. The plasma is accelerated by the ponderomotive force. The current is admitted to the gas through electrodes.

The use of electrodes, which probably would be subject to rapid erosion at the high recovery temperatures considered, can be avoided, if a pulsed operation is adopted. Such devices would have some resemblance to the various forms of electromagnetic shock tubes which have been developed recently [3], [18], [19], [20]. One of these configurations is schematically represented in Fig. 3, where the propellant is ionized by means of a strong shock (S) propagating towards the right. A traveling magnetic field (M) acts in effect as a piston, driving the gas ahead of it. The magnetic field is due to the current in a solenoid would around the tube. This solenoid together with distributed capacitors form a transmission line

$(T.L)$, which — if its capacitors are discharged through the switch (SW) — will propagate a current pulse towards the right. The velocity u_p of the magnetic piston is roughly given by $u_p=(LC)^{-1/2}$ where L and C are the inductance, and capacitance per unit length. Velocities of the order of 10^5 m/sec and larger appear feasible.

To some extent, the gas diffuses through the magnetic field, an effect which must be minimized. In the case of the field configuration shown in Fig. 3, an additional loss of gas is caused by the escape of plasma through the central portion, where the flow is essentially unimpeded. The geometry of this field resembles the one encountered in mirror machines of controlled thermonuclear fusion research [21]. However, the typical gas density considered for propulsion is usually much higher, requiring a continuum, rather than a free particle model for an adequate description.

Magnetohydrodynamic-Hypersonic Flow in the Quasi-Newtonian Approximation

As an application of the equations of magnetohydrodynamics stated in the Introduction (p. 34), the hypersonic flow past a body of revolution at zero angle of attack is considered in some detail. The gas ahead of the shock front (Fig. 4) is cold, and therefore nonconducting. Inside the shock layer, thermodynamic equilibrium is assumed, and the gas is at least partially ionized. A magnetic field is applied, which is rotationally symmetric with respect to the axis of symmetry of the body. The Newtonian approximation, or rather the "Newtonian-plus-centrifugal" approximation (VAN DYKE [22]) familiar from hypersonic aerodynamics is introduced to simplify the problem.

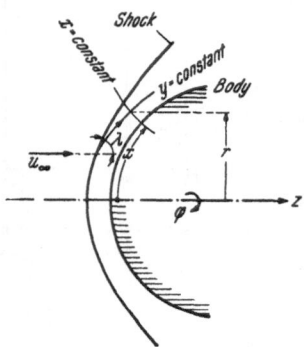

Fig. 4. Hypersonic flow about body of revolution, schematic

Considering now the equations governing the motion of the gas inside the shock layer, it is noted that the derivatives with respect to the azimuth φ vanish by reason of symmetry. Furthermore,

$$j_r=j_z=0 \tag{9a}$$

$$B_\varphi=0 \tag{9b}$$

$$u_\varphi=0 \tag{9c}$$

$$E_r=E_z=E_\varphi=0 \tag{9d}$$

where the subscripts indicate the components in cylindrical coordinates. The postulate of symmetry expressed by these equations is justified by showing that they satisfy eqs. (1) to (8) and the boundary conditions of the problem. Eq. (9a) means that the current flows in circular loops about the z-axis. Eq. (9b) follows from it by an application of STOKES theorem to eqs. (1) and (7). Furthermore, it follows from eq. (9a) that the ponderomotive force $\vec{j}\times\vec{B}$ has no φ — component and that therefore the gas, which ahead of the shock has no azimuthal velocity component, will never acquire one. Finally, since $\vec{j}$ and $\vec{u}\times\vec{B}$ are purely azimuthal, it follows that consistent with eq. (8), $\vec{E}$ can have at most an azimuthal component. But even this component must vanish, as follows from applying eq. (2).

It is convenient to introduce here a slight change in nomenclature, by putting $\vec{B} B_0$ for the magnetic flux density, $\vec{j} \, \varrho_\infty u_\infty{}^2/B_0 R_0$ for the current density, $\vec{u} \, u_\infty$ for the velocity, $\varrho \varrho_0$ for the density, $p \, \varrho_\infty u_\infty{}^2/2$ for the pressure, $h \, u_\infty{}^2/2$ for the enthalpy, and $\sigma \sigma_0$ for the conductivity, where $\vec{B}, \vec{j}, \vec{u}, \varrho, p, h$ and σ are dimensionless ratios. ϱ_0 and σ_0 are the density and conductivity at some fixed point inside the shock layer. Similarly, B_0 and R_0 are a characteristic flux density and a characteristic length of the problem. The free-stream velocity is designated by u_∞, and the free-stream density by ϱ_∞. Similarly, the coordinates are made dimensionless by dividing through R_0.

As a preparation for the limiting process implied in the Newtonian approximation, curvilinear coordinates x and y are introduced. The quantity y is taken proportional to the STOKES streamfunction and is defined such that the mass flow rate through the inside of a circle concentric with the z-axis and passing through the point (r, z) equals $2\pi y \, \varrho_\infty u_\infty R_0{}^2$. Ahead of the shock, y is constant on the surface of circular cylinders of radius $(2y)^{1/2}$. The quantity x is taken constant on orthogonals to the surfaces of constant y. The orthogonals are parameterized by means of the (dimensionless) arc length measured along the body (Fig. 4).

LAME's scale factors k_x, k_y, k_φ are introduced for the curvilinear coordinate system, defined in such a manner that $k_x \, dx$ is the element of (non-dimensional) arc length perpendicular to the surfaces of constant x, and similarly for k_y and k_φ. Clearly

$$k_\varphi = r$$

and, from continuity (eq. [4]),

$$k_y = \frac{\varrho_\infty}{\varrho_0} (r \, \varrho \, u)^{-1} .$$

An expression for k_x is not required.

From eqs. (1) and (7) follows

$$\frac{1}{k_x k_y} \left\{ \frac{\partial}{\partial x} (k_y \, B_y) - \frac{\partial}{\partial y} (k_x B_x) \right\} = j_\varphi \frac{\varrho_\infty u_\infty{}^2 \mu}{B_0{}^2}$$

where B_x and B_y are the components of $\vec{B}$ in the direction of the streamlines and perpendicular thereto. It is convenient to introduce

$$i = j_\varphi (r \, \varrho u)^{-1} \frac{\varrho_\infty}{\varrho_0}$$

which is, except for a constant factor, the current flowing through a rectangular area, with one side being of unit length along the streamline and the other side corresponding to a unit increment of mass flow rate. We also define the parameter

$$Q = \frac{\sigma_0 \, B_0{}^2 \, R_0}{\varrho_0 \, u_\infty}$$

which is a measure for the ratio of ponderomotive to inertial forces inside the shock layer[1], and the magnetic REYNOLDS number

$$R_m = u_\infty R_0 \, \sigma_0 \mu .$$

---

[1] Since the velocity in the shock layer behind an oblique shock in general is of the order of u_∞.

Consequently,

$$\frac{\partial}{\partial x}(k_y B_y) - \frac{\partial}{\partial y}(k_x B_x) = i k_z \frac{\varrho_\infty}{\varrho_0} \frac{R_m}{Q}. \tag{10}$$

Eq. (3), if expressed in the curvilinear coordinates, yields

$$\frac{\partial}{\partial x}(r k_y B_z) + \frac{\partial}{\partial y}(r k_x B_y) = 0. \tag{11}$$

From eq. (5) follows

$$\varrho u \frac{\partial u}{\partial x} + \frac{1}{2} \frac{\partial p}{\partial x} \frac{\varrho_\infty}{\varrho_0} + k_z i B_y \varrho u r = 0 \tag{12}$$

for the direction of increasing x, and

$$\frac{u}{R} - \frac{1}{2} \frac{\partial p}{\partial y} r + i B_z r = 0 \tag{13}$$

for the direction of increasing y, where $R = R(x, y)$ is the nondimensional radius of curvature of the streamlines. The sign of R is chosen such that R taken on the surface of the body is positive, if the body is convex.

For finite conductivity, the RANKINE-HUGONIOT equations of conservation of mass, momentum and energy across the shock are not affected by the presence of the magnetic field. In particular, the total enthalpy is conserved, and therefore from eq. (6),

$$h + u^2 = +1 \tag{14}$$

where use has been made of the hypersonic approximation, namely that the enthalpy of the free-stream is negligible compared with its kinetic energy.

Finally, eq. (8) needs to be considered, resulting in the relation

$$i \varrho r = Q \sigma B_y. \tag{15}$$

Quasi-Newtonian Approximation

This approximation consists in assuming that the density at any point inside the shock layer is very large compared with the free-stream density [22], [23], [24]. In the case of air, and assuming that the component of the free-stream velocity perpendicular to the shock is hypersonic, this ratio actually is roughly between ten and eighteen for points directly behind the shock. The Newtonian approximation can be regarded for instance as the limit in which the dissociation energy per molecule is infinite compared with the product of the temperature and BOLTZMANN's constant[1], but of the same order as the kinetic energy of the free-stream.

As a consequence, the shock layer thickness tends to zero and $R(x, y) \to R(x)$ which is the radius of curvature of the body[2]. With $\varrho_\infty/\varrho_0 \to 0$, $k_z \to 1$, $r(x, y) \to r(x)$, it follows from eq. (10) that B_z is independent of y, $B_z = B_z(x)$. Similarly, from eq. (11), $B_y = B_y(x)$. The magnetic field is seen to be continuous across the shock layer, in this approximation. Eliminating the current density, eqs. (12), (13) and (15) result in

$$\frac{\partial u}{\partial x} + Q B_y^2(x) \frac{c}{\varrho} = 0 \tag{16}$$

---

[1] The latter ratio is large because of the much greater number of translational states of the dissociated gas, compared with the molecular species.

[2] In the experiments reported in [17], Q is very large, and the theory cannot be expected to apply.

and

$$\frac{u}{R\,(x)} - \frac{r\,(x)}{2} \frac{\partial p}{\partial y} + Q B_x\,(x)\, B_y\,(x)\, \frac{\sigma}{\varrho} = 0 \,. \tag{17}$$

Together with eq. (14), which is unchanged, and the caloric equation of state, the last two equations form a system of equations sufficient for the determination of u, p, and ϱ. B_x and B_y are determined by the currents external to the flow field.

The equations apply in the region between the body ($y=0$), and the shock ($y = r^2\,(x)/2$). There are two boundary conditions, namely

$$u = \cos \lambda \quad \text{at} \quad y = \frac{r^2\,(x)}{2}$$

from conservation of the tangential component of the momentum across the shock, and

$$p = 2 \sin^2 \lambda \quad \text{at} \quad y = \frac{r^2\,(x)}{2}$$

from the hypersonic approximation to the conservation of the normal component of momentum. The angle included between the z-axis and the shock (which in the Newtonian approximation is replaced by the angle between z-axis and body) is designated by $\lambda = \lambda\,(x)$.

The application of these results is particularly simple in the case of a cone at zero angle of attack. If the magnetic field on the surface is given by

$$B_x = x^{-1/2} \cos \beta; \quad B_y = x^{-1/2} \sin \beta \tag{18}$$

(where β is a given constant characterizing the direction of the magnetic field on the surface of the cone), a similarity solution exists, since the expressions

$$u = u \left(\frac{2\,y}{x^2} \operatorname{cosec}^2 \lambda \right); \qquad p = p \left(\frac{2\,y}{x^2} \operatorname{cosec}^2 \lambda \right); \qquad \varrho = \varrho \left(\frac{2\,y}{x^2} \operatorname{cosec}^2 \lambda \right) \tag{19}$$

satisfy all equations identically. This is the case even if the conductivity and enthalpy are arbitrary functions of the pressure and density.

Acknowledgment

The author is indebted to Dr. H. M. Lieberstein for a discussion of the energy equation of magnetohydrodynamics.

References

1. E. L. Resler, Jr. and W. R. Sears, The Prospects for Magnetoaerodynamics. J. Aeronaut. Sci. **25**, 235 (1958).
2. J. W. Bond, Jr., Plasma Physics and Hypersonic Flight. Jet Propulsion **28**, 228 (1958).
3. R. M. Patrick, Magnetohydrodynamics of Compressible Fluids. Ph. D. Thesis, Cornell University, Ithaca, N.Y., 1956.
4. V. J. Rossow, On Flow of Electrically Conducting Fluids Over a Flat Plate in the Presence of a Transverse Magnetic Field. NACA TN 3971, 1957.
5. N. H. Kemp and H. E. Petschek, Two-Dimensional Incompressible Magnetohydrodynamic Flow Across an Elliptic Solenoid. J. Fluid Mechanics 4, 553 (1958).
6. J. L. Neuringer and W. McIlroy, Incompressible Two-Dimensional Stagnation Point Flow of an Electrically Conducting Viscous Fluid in the Presence of a Magnetic Field. J. Aero/Space Sci. **25**, 194 (1958).
7. W. B. Bush, Magnetohydrodynamic-Hypersonic Flow Past a Blunt Body. J. Aero/Space Sci. **25**, 685 (1958).

8. N. H. Kemp, On Hypersonic Blunt-Body Flow With a Magnetic Field. J. Aero/ Space Sci. Readers' Forum, **25**, no. 6 (1958).
9. M. U. Clauser, The Feasibility of Thermonuclear Propulsion. Conference on Extremely High Temperatures, H. Fischer and L. C. Mansur (ed.), pp. 209-219. New York: J. Wiley, 1958.
10. R. B. Banks, A Bibliography on Magnetohydrodynamics. Techn. Inst., Northwestern University, Unpublished Report, 1959.
11. B. T. Chu, Thermodynamics of Electrically Conducting Fluids and Its Application to Magneto-Hydromechanics. TN 57-350, Wright Air Dev. Center, 1957.
12. L. Lamb and Shao-Chi-Lin, Electrical Conductivity of Thermally Ionized Air Produced in a Shocktube. J. Appl. Physics **28**, No. 7, 754 (1957).
13. L. Spitzer and R. Härm, Transport Phenomena in a Completely Ionized Gas. Physic. Rev. **89**, 977 (1953).
14. A. Schlüter, Dynamik des Plasmas. Z. Naturforsch. **5a**, 72, and **6a**, 73 (1950-51).
15. W. M. Elsasser, Hydromagnetic Dynamo Theory. Rev. Mod. Physics **28**, 135 (1956).
16. R. X. Meyer, Magnetohydrodynamics and Aerodynamic Heating. J. Amer. Rocket Soc. **29**, 187 (1959).
17. R. W. Ziemer and W. B. Bush, Magnetic Field Effects on Bow Shock Stand-off Distance. Physic. Rev. Letters **1**, 58 (1958).
18. R. G. Fowler, J. S. Goldstein and B. E. Clotfelter, Luminous Fronts in Pulsed Gas Discharges. Physic. Rev. **82**, 879 (1951).
19. V. Josephson, Production of High-Velocity Shocks. J. Appl. Physics **29**, 30 (1958).
20. J. Marshall, Acceleration of Plasma into Vacuum. 2nd U.N. International Conference on the Peaceful Uses of Atomic Energy, Geneva, 1958.
21. R. F. Post, Controlled Fusion Research—An Application of the Physics of High Temperature Plasmas. Rev. Mod. Physics **28**, 338 (1956).
22. M. D. Van Dyke, A Study of Hypersonic Small-Disturbance Theory. NACA TN 3173, 1954.
23 M. J. Lighthill, Dynamics of a Dissociating Gas, Part I, Equilibrium Flow. J. Fluid Mechanics **2**, 1 (1957).
24. W. D. Hayes and R. F. Probstein, Hypersonic Flow Theory. New York: Academic Press, 1959.

Accuracy Limits in Electronic Tracking of Space Vehicles

By

Paul F. von Handel[1] and Fritz Hoehndorf[1]

(With 9 Figures)

(Received June 29, 1959)

Abstract — Zusammenfassung — Résumé

Accuracy Limits in Electronic Tracking of Space Vehicles. The accuracy of advanced electronic tracking of objects moving in the atmosphere is limited by propagation anomalies. The inherent high precision of modern tracking systems cannot be utilized because of unpredictable propagation errors.

This situation does not prevail in tracking of vehicles moving beyond the atmosphere. In this case the refractive index can be measured at the site of observation. It equals unity at the site of the target.

It is the purpose of this paper to show that individual and unpredictable refractive index profiles have only minute effects on range and angular errors as long as the index is known at both ends of the profile, i.e., at the measuring site and at the target.

Therefore, the accuracies to be achieved in space-tracking are considerably higher, they can be determined numerically, and the system's inherent precision can be fully utilized.

Genauigkeitsgrenzen bei der elektronischen Bahnverfolgung von Raumfahrzeugen. Die Genauigkeit fortgeschrittener elektronischer Bahnverfolgungsmethoden von Objekten, die sich in der Atmosphäre bewegen, ist durch die Fortbewegungsanomalien begrenzt. Die den modernen Beobachtungssystemen eigene hohe Präzision ist wegen nicht voraussehbarer Abweichungen in der Fortbewegung nicht ausnützbar.

Diese Situation besteht aber nicht bei der Bahnverfolgung von Fahrzeugen, die sich jenseits der Atmosphäre bewegen. In diesem Fall kann der Refraktionsindex am Beobachtungsort gemessen werden. Er wird am Ort des Zielobjektes gleich eins.

Die vorliegende Arbeit soll zeigen, daß individuelle und nicht voraussehbare Brechungsindexprofile nur kleine Auswirkungen auf die Reichweite und Winkelfehler haben, solange der Index an beiden Enden des Profils bekannt ist, d. h. also am Meßort und am Ort des Zielobjektes.

Deshalb sind die bei Bahnverfolgung im Weltraum erreichbaren Genauigkeiten beträchtlich größer; sie können numerisch bestimmt und die dem System eigene Präzision voll ausgenützt werden.

Limites de précision dans la poursuite électronique des astronefs. La précision des dispositifs électroniques avancés de détection-poursuite d'objets se mouvant dans l'atmosphère est limitée par des anomalies de propagation. Leur caractère imprévisible tient en échec la haute précision inhérente à ces instruments.

Pour les objets se mouvant au délà de l'atmosphère la situation est différente. L'objet de l'article est de monter combien peu les profils particuliers et imprévisibles de l'indice de réfraction influencent les mesures d'angle et de distance, du moment

---

[1] Air Force Missile Development Center, Air Research and Development Command, United States Air Force, Holloman Air Force Base, New Mexico, U.S.A.

que l'indice soit connu au site d'observation, où il peut être mesuré, et à l'endroit
de la cible, où il est supposé être égal à l'unité. Il en résulte que la précision des
dispositifs peut être pleinement utilisée.

I. The Problem Situation

Since the advent of space vehicles, problems of propagation anomalies in electronic tracking and guidance are confronted with a new situation. It consists of

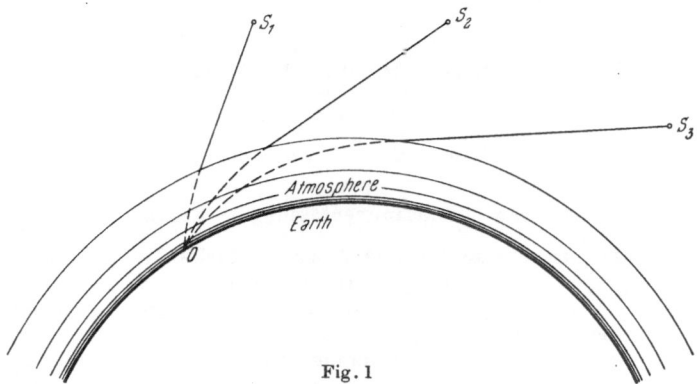

Fig. 1

the fact that for space vehicles we know the refractive index at the target site is
unity, whereas it can be measured, as heretofore, at the site of the tracking system
on the surface of the earth. This is true only, of course, if we neglect the refractive

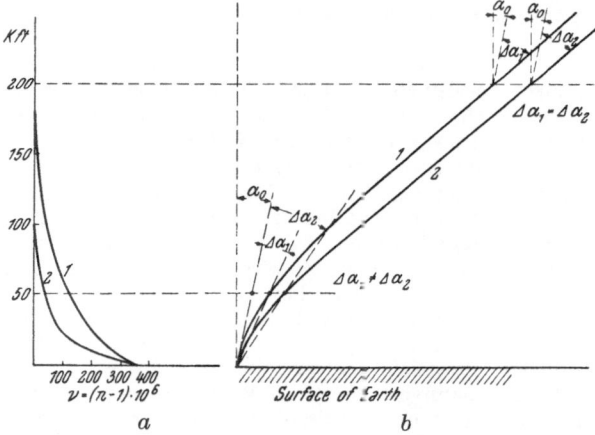

Fig. 2. *a* Refractive index profiles, *b* wave paths

effects of the Ionosphere. But it can be shown that these effects are negligible if
frequencies in the kilo megacycle range are used. We, therefore, shall confine
our considerations to these high frequencies, which appear to be quite permissible,
because modern high-accuracy tracking systems will exclusively use these frequencies, for other reasons.

The only other assumption we make is that of a spherically stratified atmosphere concentric with the earth, neglecting horizontal gradients of the refractive
index. This assumption must be made, because such gradients along the path of a

wave cannot be determined. The azimuthal error limits caused by such gradients will have to be determined by experiment. They are the same as those common to optical measurements in astronomy and not likely to exceed a few seconds of arc.

The new situation, then, consists of the fact that we now know the refractive index on both ends of the wave-path, at its origin on the earth's surface and at its end on the site of the target. It will be shown that we now can define a family of standard refractive index profiles, each member of the family starting at a certain index, n_0, on the earth's surface and approaching unity at 200,000 feet altitude. The errors in angle and range produced by these standard profiles will be computed. Proper correction terms, therefore, can be applied to the corresponding numbers as measured by the tracking radar and we shall call these errors, consequently, correctable errors. They do not impair the accuracy of the system; the corrections are based upon measurable data and precise calculations.

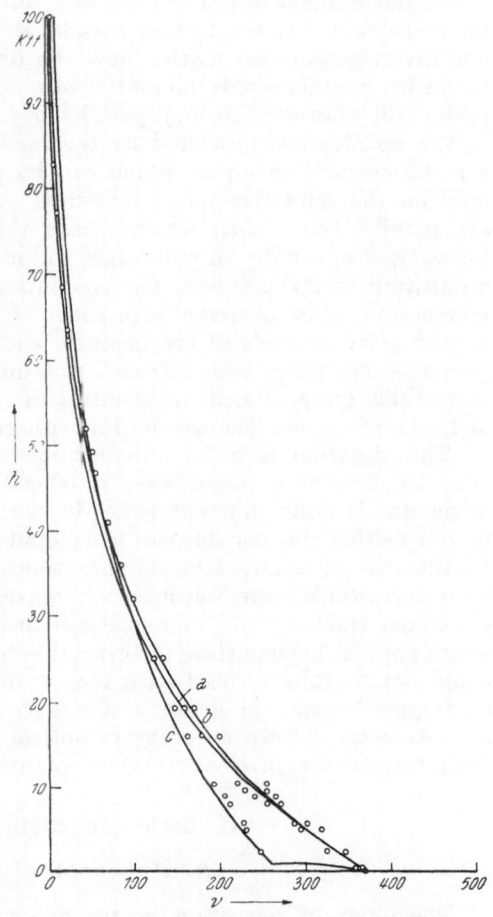

Fig. 3. Refractive index profiles versus altitude

The crucial question now arises concerning the magnitude of the non-correctable errors which originate in the individual, unpredictable, and unmeasurable profiles between the known data at the beginning and at the end of the line. In Fig. 1 we see the site of the radar and three positions of a satellite. The scale of this picture, of course, is strongly distorted in order to make clear what we like to show. If we want to base our calculations upon individual profiles, we would have to gather meteorological data along all the dashed lines where the waves penetrate the atmosphere. Three of these lines are indicated. It is quite clear that this is impossible, so much the more if we realize that these data are subject to continuous change in time.

Let us now have a look at Fig. 2. In Fig. 2a. two refractive index profiles have been drawn schematically, representing two extreme weather conditions. The index on the surface of the earth is the same in both cases and the index always approaches unity at 200,000 feet altitude[1]. At profile 1, the air is almost saturated all the way up to higher altitudes whereas at profile 2, the air is dry from high altitudes down to the lower atmosphere and the humidity increases rapidly from there to the surface. A similar profile (2) will occur with a tem-

---

[1] At 200,000 feet $\nu \leq 10^{-6}$.

perature inversion in the lower atmosphere. In Fig. 2b we see the two schematic
wave paths corresponding to the two profiles. Path 1 has a slight curvature
corresponding to an overall low gradient of index versus altitude (1) whereas
path 2 has a greater curvature in the lower part of the atmosphere corresponding
to a high index gradient and a very low curvature in the upper atmosphere
where the gradient of profile (2) is low. But note that both paths leave the atmosphere beyond 200,000 feet as parallel straight lines. This is the vital point in
our investigation. No matter how the distribution of a stratified atmosphere
might be, in other words, no matter how an individual profile might be, the wave
paths will always come out parallel beyond the atmosphere. We denominate
a_0 the zenith-angle measured by the radar. Then Δa_1 and Δa_2 at 200,000 feet
are the correctable errors which can be precisely computed from information
only on the refractive index measured on the surface (n_0). Both wave paths
are parallel $(\Delta a_1 = \Delta a_2)$ which means that the uncorrectable errors are zero
for a target at infinity. A real target will not be at infinity but relatively far from
an altitude of 200,000 feet. The computation, therefore, will show that the uncorrectable errors decrease with range. It will be seen later that they do not
exceed a few seconds of arc in angle and a few feet in range for all practical
purposes. For target ranges from 1 to 5 million feet, as plotted in Fig. 8, the uncorrectable error, $\Delta^2 a$, is independent of range to a fraction of a second of arc
and, therefore, the change of $\Delta^2 a$ with range is not visible in this figure.

The situation is quite different for targets inside the denser atmosphere
(Fig. 2). Assume a target, say, at 50,000 feet altitude. We see that the error
angle Δa_1 is quite different from Δa_2 for the two profile cases. But the point
is that neither Δa_1 nor Δa_2 can be computed because the profiles are not known.
In the case of a target inside the atmosphere, therefore, Δa_1 and Δa_2 are no
more correctable errors but incorrectable ones of considerable magnitude. Modern
electronic tracking and guidance systems reach an inherent system precision
which goes far beyond these incorrectable errors. The system's capacity, therefore,
could not be fully utilized with targets in the atmosphere. But it can be used
on targets beyond the denser atmosphere and there is no reason why it should
not reach the known accuracy of optical instruments without being biased by
their restrictions on clear weather conditions.

II. Basic Computation and Analysis

A. Meteorological Profiles

The index of refraction, n, for microwaves penetrating the atmosphere is
a function of temperature, air pressure, and partial pressure of water vapor,
according to the equation

$$\nu = (n-1) \times 10^6 \frac{c}{T}\left(p + \frac{be}{T}\right) \tag{1}$$

where $c = 79$ [°K/mb]
$T = $ temp [°K]
$p = $ air pressure [mb]
$b = 4800$ [°K]
$e = $ partial pressure of water vapor [mb]
$n = $ refractive index.

Fig. 3 shows profiles of refractive index versus altitude (MSL). The meteorological
soundings were furnished by the U.S. Weather Service. From measured data
of pressure, temperature, water vapor content, and altitude the profile was

calculated according to eq. (1) and plotted in curve a, Fig. 3. In order to obtain profiles corresponding to extreme weather conditions at the same ground index,

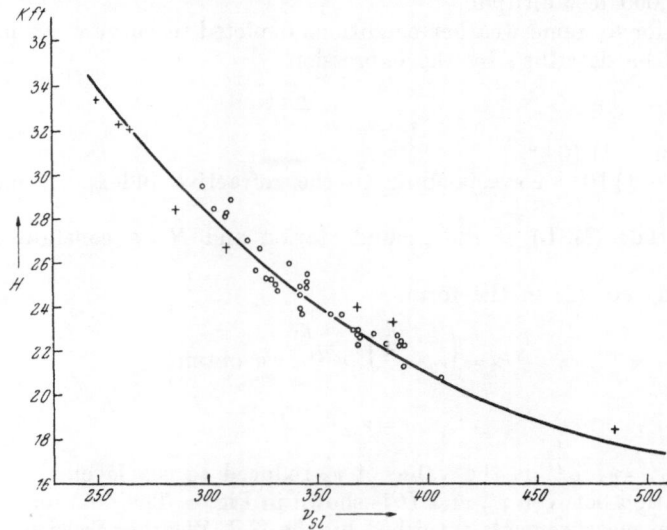

Fig. 4. Relation between v_{SL} and H. + from constructed curves, ○ from measured curves

n_0, we assume that at curve b the air is saturated at altitudes above 1,000 feet while the pressure and temperature distribution versus altitude remains the same.

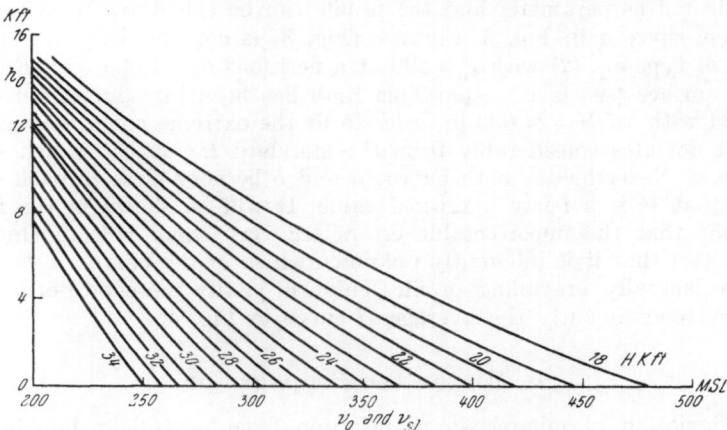

Fig. 5. Graph for the determination of v_{SL} and H for standardized curves if h_0 and v_0 are given.

Illustration:

Given: $h_0 = 5000$ ft

$n_0 - 1 = 280 \times 10^{-6}$

Then: $H = 24\,650$ ft

$n_{SL} - 1 = 344 \times 10^{-6}$

The formula of the approximation will be: $n = 1 + 344e^{-\dfrac{h}{24\,650}} \times 10^{-6}$ where h is the altitude above mean sea level

In curve c, to the contrary, the air is assumed to be extremely dry from high altitudes down to about 1,000 feet and to become rapidly wet in a small transition layer from here down, reaching 75 % relative humidity at the surface. The same conditions will also occur in case of a temperature inversion between zero and 1,000 feet altitude.

Except for extreme weather conditions depicted in curve c, an index profile can always be described by the expression

$$\nu = \nu_0 \cdot e^{-\frac{h - h_0}{H}} \tag{2}$$

where $\nu = (n - 1) \, 10^{+6}$,

$\nu_0 = (n_0 - 1) \, 10^{+6}$ corresponding to the refractive index, n_0, measured on the ground,

$h_0 =$ altitude (MSL) of the ground station and $H =$ a constant for any h_0 and ν_0.

Rewriting eq. (2) in the form

$$\nu = \left(\nu_0 \cdot e^{\frac{h_0}{H}}\right) \cdot e^{\frac{-h}{H}} \quad \text{we obtain}$$

$$\nu = \nu_{SL} \cdot e^{-\frac{1}{H}} \tag{3}$$

where $\nu_{SL} = \nu_0 \cdot e^{\frac{h_0}{H}}$ is the value of ν_0 reduced to sea level.

The relation between ν_{SL} and H is shown in Fig. 4. The dots represent values derived from measurements furnished by the U.S. Weather Service. The crosses were calculated from meteorologically meaningful assumptions. An averaging curve (Fig. 4) was then drawn through these points.

From the relation ν_{SL} versus H (Fig. 4) a family of standard profiles has been drawn in Fig. 5. For each ν_0 measured at a station of altitude h_0 (MSL) the proper H is indicated as parameter and the profile can be calculated from eq. (2). An example is curve a in Fig. 3. Curve c (Fig. 3) is composed of an exponential function of type eq. (2) with $\nu_0 = 260$ at a fictitious $h_0 = 1,000$ feet. From there down to surface (sea level) a parabola limb has been introduced. Curve b was calculated with an $H = 24,000$ in order to fit the extreme conditions mentioned above. It deviates considerably from the standard $H = 22,635$ which was used for curve a. Nevertheless, both curves a and b lie very close to each other indicating that H is a fairly uncritical value. It will be shown in the following paragraphs that the uncorrectable errors are very small indeed which result from the fact that it is inherently unknown which of the curves, a, b, or c, represent the actually prevailing profile. This will justify the choice of a standard profile and consequently the average H-curve in Fig. 4.

B. Microwave Propagation

The deviation of microwave propagation from a straight line in passing through the atmosphere can be calculated by Snell's law of refraction. In its original form it holds only for the special case of plane surfaces

$$\frac{\sin a}{\sin a_0} = \frac{n_0}{n}. \tag{4}$$

For our purposes it has to be generalized to hold for media with curved surfaces. This derivation has been made by H. Knothe [1] and independently by H. M. Dixon [2] who started from Fermat's law and arrived at the same results:

$$\frac{\sin a}{\sin a_0} = \frac{R + h_0}{R + h} \cdot \frac{n_0}{n} \quad \text{see Fig. 6).} \tag{5}$$

It can be seen immediately that the general law [eq. (5)] reduces to eq. (4) if the radii of curvature $R+h_0$ and $R+h$ approach infinity, their difference remaining finite, which means the plane case.

Any electronic range measurement between points O and S (Fig. 6) basically consists of the measurement of a phase; it may be an RF-phase, an IF-phase, or a PRF-phase. In conventional pulse-radars, for instance, the electronic phase angle in the PRF (Pulse Repetition Frequency) is measured which occurs between the illuminating transmitter pulses and the pulses reflected from a target. At a known frequency the phase angle can be expressed by a correspond-

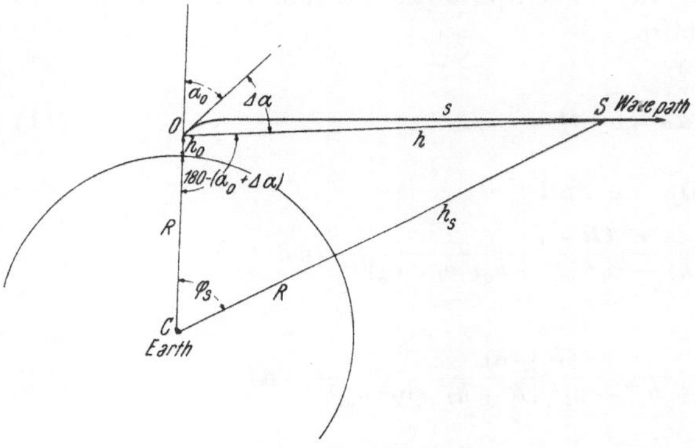

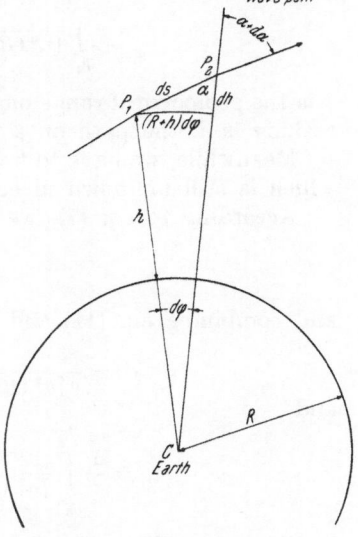

ing time interval, τ, which is the time for a roundtrip of a wavefront from the radar to the target and back. The one-way trip, therefore, needs $\dfrac{\tau}{2}$. The distance to be measured is obtained by multiplying $\dfrac{\tau}{2}$ by the mean signal propagation velocity, $\bar{v}$. The propagation velocity v is not known exactly; it is a function of the refractive index along the path and subject to our calculations. Each electronic system, therefore, has a certain apparent velocity, v_a, which is calibrated to the instrument as a constant. The measured value of range, indicated by the radar system, then is an apparent range, s_a. It is a fictitious range which a wavefront would transgress if propagated with constant velocity, v_a, along a path s (Fig. 6).

$$s_a = \frac{\tau}{2} \cdot v_a. \tag{6}$$

With s_a measured and v_a known ($v_a = 299\ 773.1$ Km/sec), we have $\dfrac{\tau}{2}$.

Next, we have to determine the range s along the curved path. We note the well-known expression for the propagation velocity in a medium with refractive index, n.

$$v = \frac{c}{\sqrt{\varepsilon\,\mu}} = \frac{c}{n}. \tag{7}$$

It follows from Fig. 7 that

$$ds = \frac{dh}{\cos a} \tag{8}$$

Fig. 6

Fig. 7

and we obtain from eq. (5)

$$\cos a = \left[1 - \sin^2 a_0 \left(\frac{R+h_0}{R+h}\right)^2 \left(\frac{n_0}{n}\right)^2\right]^{1/2} .$$

(9)

Substituting eq. (9) in eq. (8) we obtain

$$s = \int_{h_0}^{h_s} \frac{n(R+h)}{[n^2(R+h)^2 - n_0^2(R+h_0)^2 \sin^2 a_0]^{1/2}} \cdot dh .$$

(10)

s is the geometrical range on the curved path (see Fig. 6). The real range we are seeking is d, the path on a straight line. We shall calculate it later.

Meanwhile, we have to find a way to compute the altitude of the satellite h_s, which is still unknown in eq. (10).

According to eq. (7), we have

$$ds = v \cdot dt = \frac{c}{n} dt$$

(11)

and combining eq. (11) and (10), we find

$$dt = \frac{n^2(R+h)}{c\,[n^2(R+h)^2 - n_0^2(R+h_0)^2 \sin^2 a_0]^{1/2}} \cdot dh$$

and

$$\frac{\tau}{2} = \frac{1}{c} \int_{h_0}^{h_s} \frac{n^2(R+h)}{[n^2(R+h)^2 - n_0^2(R+h)^2 \sin^2 a_0]^{1/2}} \cdot dh .$$

(12)

From eq. (12) we find h_s by methods of numerical integration on an electronic computer[1]. All the other values are known: $\frac{\tau}{2}$ is measured [see eq. (6)] and n can be found from eq. (2) and (3)

$$n = 1 + (n_0 - 1)\, e^{-\frac{h - h_0}{H}}$$

where H is determined by Fig. 5 for a measured n_0. and a known h_0. With h_s now known, we find s from eq. (10).

Next, we have to determine the geometrical distance, d, (Fig. 6) that is the real range to the target.

Applying the law of cosines of plane trigonometry, we find

$$d = [(R+h_0)^2 + (R+h_s)^2 - 2(R+h_0)(R+h_s)\cos \Phi_s]^{1/2} .$$

(13)

In this equation Φ_s is still unknown. In order to find Φ as a function of any altitude, h, we proceed as follows. It can be seen from Fig. 7 that

$$d\Phi = \frac{\tan a \cdot dh}{R+h} .$$

(14)

Eliminating a from eq. (14) and (5), we obtain

$$|\Phi_s| = \int_{h_0}^{h_s} \frac{n_0(R+h_0)\sin a_0}{(R+h)\,[n^2(R+h)^2 - n_0^2(R+h_0)^2 \sin^2 a_0]^{1/2}} \cdot dh ,$$

(15)

d can now be calculated from eq. (13).

---

[1] All integrals have been calculated on a Remington Rand 1103A Univac Scientific.

Finally, we want to determine the error in range, Δs_a, which has to be added to the apparent range, s_a, measured on the Radar in order to obtain the real range d. From eqs. (6) and (13), we find

$$\Delta s_a = d - s_a = [(R+h_0)^2 + (R+h_s)^2 - 2(R+h_0)(R+h_s)\cos\Phi_s]^{1/2} - \frac{\tau}{2}v_a. \tag{16}$$

The error in angle, Δa, can readily be found in applying the law of sines (see Fig. 6):

$$\sin(a+\Delta a) = \frac{R+h_s}{d}\sin\Phi_s$$

and

$$\Delta a = \sin^{-1}\left(\frac{R+h_s}{d}\sin\Phi_s\right) - a_0. \tag{17}$$

III. Error Analysis

We now proceed to the crucial point of our problem, that is, to determine the computed, i.e., the correctable errors and the non-correctable ones, the latter setting the accuracy limits in electronic tracking of space vehicles.

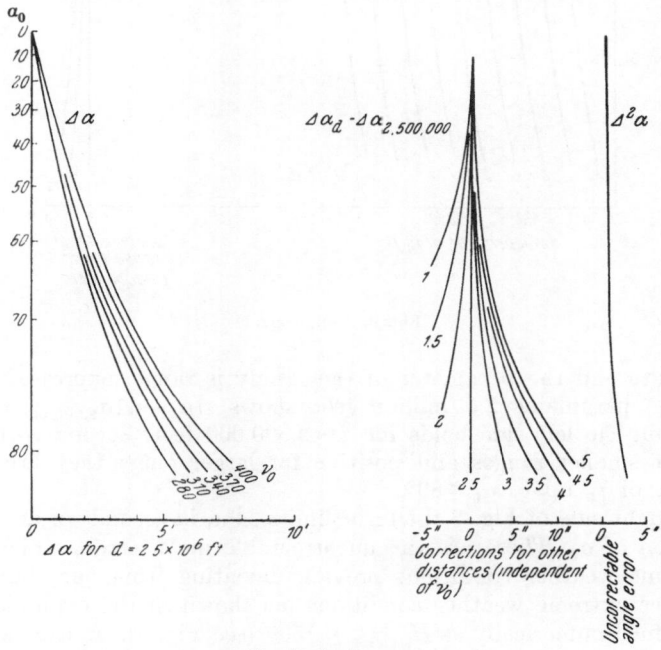

Fig. 8. Angle error

All integrals shown in Chapter II have been solved to an accuracy of 11 decimals by methods of numerical integration. Instead of presenting the results in extensive tables, it will be more convenient to show them in graphical form.

Fig. 8 shows the errors in angle. The apparent zenith-angle, a_0, measured on the radar, is plotted as ordinate.

The family of curves on the left side represents the correctable errors Δa versus a_0. Δa is plotted as abscissa in minutes of arc. The parameters are v_0, the refractive indices, as measured at the site of the radar. The whole family is valid for a target range, d of 2.5×10^6 feet. We see that the correctable errors at high zenith-angles (high a_0) are considerable. But even at lower zenith-angles, say $40°$, they are around one minute of arc which means a lateral deviation of 750 feet at a distance of 2.5×10^6 feet. These errors, $\varDelta a$, therefore, have to be allowed for and added to the reading of a_0 if the results are expected to be precise.

The error angle, $\varDelta a$, also changes slightly with range. Therefore, a second correction term, $\varDelta a_d - \varDelta a_{2,500,000}$, has been plotted in the middle of Fig. 8, taking care of ranges $d \leqq 2,500,000$. The abscissa for this term is plotted in

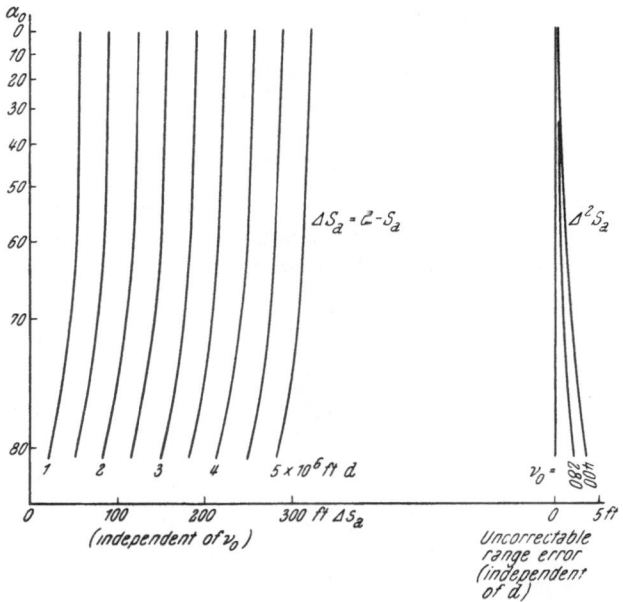

Fig. 9. Range error

seconds of arc and the parameter of the family is range, expressed in millions of feet. The parameter 2.5 million feet shows $\varDelta a_d - \varDelta a_{2,500,000} = 0$ because the family on the left side holds for $d = 2,500,000$ feet. Second corrections are negative for smaller ranges and positive for larger ones; they are practically independent of v_0 up to $a_0 \leqq 80°$.

On the right side of Fig. 8 the error limits, $\varDelta^2 a$, in seconds of arc are plotted as a function of a_0. These are the uncorrectable mean errors which can occur when extreme weather conditions prevail, deviating from our standard index profile. These extreme weather conditions (as shown in the example of Fig. 3) are now defined numerically as $H_{standard} \pm 1500$ (see Fig. 4). It can be seen there that all the points lie between a margin of ± 1500 from the standard H-curve. The plotting of $\varDelta^2 a$ shows that the accuracy limits in electronic angle measurement of space vehicles are in the order of magnitude of a few seconds of arc reaching, therefore, optical tracking accuracy. It is interesting to note that for angles a_0 of $80°$ and smaller $\varDelta^2 a$ is practically independent of d and v_0.

Fig 9 shows a corresponding plotting for errors in range, $\varDelta s_a = d - s_a$. The parameter of the family now is the range, d; the curves are practically independent of v_0 up to $a_0 = 80°$. At a first glance it may be surprising that, contrary to the family of $\varDelta a$ (Fig. 8), the error $\varDelta s_a$ increases with decreasing a_0 and that it also increases with the distance parameter, d. This is because the apparent propagation velocity, v_a, calibrated to the radar as a constant, was 299 773.1 Km/sec in our case and all calculations in this paper have been based upon this figure. If another constant v_a had been calibrated into a specific radar, the respective number has to be changed in eq. (6) and (16). If, for instance, light velocity in vacuum ($c = 299\ 792.9$ Km/sec) had been used for v_a, then the family of curves on the left side of Fig. 9 would show increasing errors Δs_a with increasing a_0 and decreasing d.

Here again, it can be seen that the correctable errors Δs_a are in the order of a few hundred feet and must be allowed for if accurate data are expected.

On the other hand, the uncorrectable errors, $\Delta^2 s_a$, are only a few feet. In this respect, electronic will be more convenient than optical tracking because range can be obtained only by triangulation in optics, using several stations simultaneously.

It is to be expected, therefore, that the general trend of further development in tracking space vehicles will shift toward electronic procedures which promise to achieve the same accuracy without restrictions on weather condition and time of the day.

References

1. P. von HANDEL and H. KNOTHE, The Curvature of Microwave Propagation Through the Troposphere and the Ionospheric Layers. Holloman Air Development Center, January 1956, T.R. No. 3.
2. H. M. DIXON, Study of Refraction Errors in Radar Propagation. White Sands Signal Corps Agency, April 1956, T.R. No. 18A.

The Biological Satellite

By

R. P. Haviland[1], AAS, ARS, BIS

(With 8 Figures)

(Received June 15, 1959)

Abstract — Zusammenfassung — Résumé

The Biological Satellite. Need for Biological Satellite is reviewed. It is concluded that the long term problems of weightlessness and radiation exposure cannot be investigated except by satellite flight. Subjects and measuring techniques are discussed. Critical design limitation are reviewed both from the standpoint of the animal subject and from the design of life support and re-entry equipment. Artist's conceptions of three different vehicles with experiments of varying complexity are shown as suggestions for a suitable research program.

Der biologische Satellit. Es wird die Notwendigkeit, einen biologischen Satelliten zu schaffen, geprüft. Der Verfasser kommt zu dem Schluß, daß die auf lange Sicht bedeutungsvollen Probleme der Gewichtslosigkeit und des Strahlungseinflusses nur auf Satellitenflügen untersucht werden können. In der vorliegenden Arbeit werden die einzelnen Themen und Meßverfahren erörtert. Die kritischen Konstruktionsgrenzen werden sowohl vom Gesichtspunkt des lebenden Passagiers als auch hinsichtlich der Vorkehrungen für die Aufrechterhaltung des Lebens und der Ausrüstung für die Rückkehr besprochen. Als Vorschläge für ein geeignetes Forschungsprogramm werden drei verschiedene Fahrzeuge für Experimente wechselnder Kompliziertheit in künstlerischer Auffassung dargestellt.

Le satellite biologique. En investigant l'utilité d'un satellite biologique on est amené à conclure à sa nécessité pour résoudre à longue échéance les problèmes d'absence de pesanteur et d'exposition aux radiations. Sujets d'expérience et techniques de mesure sont discutés. L'équipement de rentrée et la réalisation d'une ambiance adéquate pour l'animal posent aux conceptions des limitations critiques, qui sont passées en revue. Une conception préliminaire de trois véhicules pour expériences de complexité croissante est présentée comme suggestion pour un programme de recherches.

Introduction

The biological satellite is defined as the satellite vehicle used for carrying living objects into satellite orbits for biological studies. These studies may be divided into two families. One of these is the area of biological research, where the interest lies in the life process itself. The second is more nearly an engineering field and consists of studies directed to the preparation for and conduct of manned satellite operation and of space travel. Since this paper is an engineering paper, it considers only the second of these fields.

[1] Missile and Space Vehicle Department, General Electric Company, Philadelphia 4, Pennsylvania, U.S.A.

Reasons for Engineering Biological Studies

Space is a completely new environment to men and the ability to enter space, to live in it and to return from it safely demands an understanding of the nature of the environment, of the equipment to be used and of the effects of the environment and the equipment on man. It is reasonable to assume that many environmental studies will have been conducted by unmanned instrumented probe vehicles. However, it must always be realized that there may be undiscovered environmental effects which affect man. While the introduction of animal life into space will not uncover all of these, it should uncover all major ones and give a high probability that there are no minor ones. Since some of these effects may be cumulative, it appears that the biological satellite program must include both initial short term flights and later long term flights which check for the presence of cumulative effects.

To a very considerable extent, the interrelation between living objects and the equipment used to place them in space may be checked by ground simulation. However, there are a number of things which are difficult or impossible to simulate and there is always the possibility of error and oversight. Therefore, common sense would indicate that any equipment techniques intended for man's use will be previously checked by animal experiments. This would extend to animal tests of the early ships designed for manned operation.

Specific Biological Studies

The most obvious item of engineering importance is simply proof that an animal can exist in space for a period of time. This step has already been accomplished. The second step is the measurement of response of biological organisms to flight stimuli. This also has been accomplished in part, although at the time of writing the results have not been made available for detail study. Additional programs of measurement are under way and it is expected that by the end of this year, a reasonable amount of this information will be available.

One of the major questions in space operation is that of adaption to the zero g condition. This includes a group of factors which are both short term and long term in nature. The first of these is the efficiency of the animal when subjected to the zero g condition and arises from the changed operation conditions of the various body sensory elements. For example, there may be digestive disturbances, racing heart beat, vision difficulties, and other effects which reduce the efficiency of the subject appreciably. These factors are undoubtedly influenced by training and by mental attitude, so that a number of tests of various subjects will be needed before complete understanding of the problem is secured.

Associated with this problem is the learning of locomotion under zero g conditions. This will probably take an appreciable period of time since it will be necessary to develop a set of reflexes quite different from those normally used. Studies of this are not simple since the subject must be free of all restraint for the study to be meaningful. Further, if a considerable amount of room is provided, it may be possible for the subject to float freely out of reach of any aids such as handgrasps, etc. Possible solutions of this would include use of several subjects to give mutual assistance and the use of small thrust units to produce a small gravity field at intervals which would allow the subject to reach a wall or floor.

One of the problems which cannot be checked with biological subjects is that of free maneuvering in open space. The complexity of a space suit for the subject and of preparatory training appear to be insurmountable, so that this experiment would be conducted as a part of a manned program.

One of the effects of the zero g condition is that the subject is under a lower acceleration stress than is normal. This can cause changes in muscle tone, although the free maneuvering will probably use some of the normal muscles and may cause some to be used greater than a normal amount. The deterioration can probably be prevented by a suitable set of exercises and the restoration can probably be speeded by a retraining program on return. This retraining program is probably also necessary since normal reflexes may have been degraded to the point that the subject must be considered to be hazardous to himself until they are relearned. It seems necessary to check both the deterioration which would occur through disuse and then the restoration which would occur on return to the earth's surface.

Another problem which must be studied is the effect of exposure to energetic particles. This includes investigation of effects on the organism itself and also the effects on future generations through genetic mutation. Since these effects have a high statistical fluctuation it appears that these must also be conducted for a number of subjects and for appreciable periods of time. These effects are very difficult to measure during the course of the experiment so that recovery of the subject for detail examination appears necessary. This is needed to permit design of structure and shielding, and for selection of trajectories and flight durations.

Another of the questions to be investigated is the effect of photon radiation on the organism. Since the earth's atmosphere is a very efficient screen at most photon energies, we do not have good information as the tolerance level of organisms for unusual conditions. A certain amount of this information can be secured from laboratory tests using simulated sources but a final check appears necessary before we can undertake a design of minimum protection such as would be needed for space suits.

A proof test of equipment designs intended for manned operation will not always be easy. This is particularly true of those items which are designed for manipulation by the man. For these it appears that the only approach is a thorough study during design mock-up which includes operating tests under unusual environmental conditions. Such conditions might include upside down operation, underwater tests of the subject and equipment and short term tests during parabolic air flight.

Automatic equipment is much easier to check by using biological subjects. For example, the load on an oxygen regeneration system can be simulated quite closely with a number of animals selected for equivalent oxygen consumption. Such tests will, of course, have been run on the ground prior to flight so that the use of the biological subject is in the nature of a final proof rather than a detailed investigation.

Once a biological satellite is available, it becomes a convenient tool for investigations of new equipment principles. For example, it appears that a water immersion system would be a good method of supporting the human body for high-g re-entry. Simulation of a particular design condition on the ground is not too difficult so that again use in flight would be for final proof. However, there are a number of other designs which could not be checked on the ground, specifically those in which zero g conditions are necessary for proper operations. These will have to be checked in detail by biological satellite test programs.

Subjects for Biological Experiments

Considering only the engineering problems, it appears that only two classes of biological specimens are needed to satisfy desirable experimental programs. These may be called simple subjects and higher subjects.

Because of their short reproduction cycle and the opportunity of investigating long term effects over many generations, the simple subjects appear to be best suited for investigation of the fundamentals of long term exposure to corpuscular and photon radiation. Two types of subjects which have been used in the past for rocket flights are yeasts and the fruit fly. The fruit fly is of particular interest since its chromosome structure is well mapped and the effects of many types of radiation already determined from earth's surface tests.

Because of their short life cycle, the simple subjects offer opportunity to investigate the effects where both sexes have been exposed and also where radiation occurs over a number of generations. These are not really important in the immediate future, but will become more important as space travel and extraterrestial colonialization develops.

The fruit fly also offers an opportunity for investigation of ability to maneuver under free flight conditions although the flight maneuvers which they use are more of academic than engineering interest.

For the other engineering investigations it appears that it is necessary to use selected higher subjects. These might include mice, rats, dogs and primates. Mice and rats are attractive for a number of purposes since their life cycle is relatively short and since they have been thoroughly studied. While they do have intelligence, their manipulative power is low so that they are best used for passive experiments. These include exposure to radiation and adaptability to zero g condition with the limitation that their maneuvering under zero g is probably confined to feeding and similar operations.

Because both mice and rats are available in carefully purified strains, they are also useful for radiation investigation. However, because their life cycle is much longer than, say the fruit fly, there will be some problem in securing exposure for a number of generations.

The dog was chosen for the first injection of life into space. Dogs are very adaptable to training and are quite intelligent. However, their manipulation ability is still relatively low and their lack of grasping limb terminations indicates that they are poorly adapted to maneuvering under zero g. Because of this, it does not seem desirable to conduct any of the long term experiments with dogs.

The primates are available over a wide range of weight and size with an appreciable variation in characteristics. The tailed monkeys are tree dwellers and probably could adapt readily to zero g condition. The tailless monkeys and the larger primates are adapted to earth's surface condition and are generally more intelligent and have greater curiosity and initiative. Since the tailless monkeys are closer to man in characteristics, it appears that the investigations are best conducted with them.

Because the monkey possesses limbs adaptable for manipulation, and can be trained, a variety of tasks and adaptation maneuvers can be investigated. However, since their intelligence is still low, approximately that of a three to four year old human infant, they cannot be trusted to initiate ship control functions. On the other hand their natural curiosity is extremely high and a measure of their adaptation to space conditions can be secured by providing a number of puzzle-like tests in addition to the usual provisions for feeding, exercise, and sleeping.

It appears from these factors that a reasonable biological satellite program will include three classes of living subjects. The part of the program would be conducted with simple subjects, probably the fruit fly, and would be directed to the investigating of long term exposure to radiation. Another part of the program would be conducted with mice or rats. This part would include investigation of effect of radiation on the organisms and to some extent on future generations.

These subjects would also provide some information on long term adaptability to zero g. The remaining part of the program would be conducted with primates of the tailless type, quite probably the Rhesus monkey and Chimpanzee. These subjects would be used somewhat to investigate the effects of radiation on the organism, but would be used to a much greater extent to investigate the adaptability to zero g condition and the long term effects of this condition on muscle tone, motor response, etc. The extent and duration of relearning process on return to normal g conditions would also be investigated. These larger animals would also be used for proof test of equipment designed for manned operation.

Flight Profiles

It is convenient to divide space flight duration into periods on the order of days, on the order of weeks and on the order of months. For these, the ascent and descent profiles are essentially independent of the duration and are not very dependent on the exact altitude of operation.

Flights on the order of days do not appear to be very useful for biological investigations. The basic reason for this is that the time constant of biological response of interest tends to be long. There are two areas in which this type of flight appear to be useful. The first of these is in the proof test of some equipment design, such as the test of restraint seats or water immersion re-entry devices. The second would be in the investigating of the high intensity radiation belts where large exposures are obtainable in very short time periods.

At the other extreme, it does not seem to be necessary to conduct too many experiments for periods of the order of months. The possible exceptions to this are an investigation of the very long exposure to radiation, and for a few tests of the influence of the zero g condition on muscle and motor tone and on the relearning process on return to the earth's surface.

It appears that it will be satisfactory to design a biological satellite of operation for the range of weeks, say over the range of two to eight weeks of time. The smaller animals would probably be used over the longer periods so that the total weight of the subjects and their life support would tend to remain constant. This makes a single design of boost vehicles feasible. There would need to be variations in the payload to accomodate the different balance between subject size and weight and the duration of the life support. This can be probably handled with about three cabin designs, one for maximum duration, a second for maximum subject size and a third for average size and average duration.

Measurement Techniques

A common technique of biological research involves the placement of sensory elements on the subject for recording or telemetry. Such techniques furnish large amounts of valuable information but suffer fundamental drawbacks. The chief of these is the fact that the instrumentation tends to require immobilization of the subject or at best, partial restraint. Additional undesirable features are the facts that many measurements require surgery on the subject and others involve local deterioration when used for appreciable periods of time. For example, the terminals used for measuring galvanic skin response or for electrocardiograph measurements produce skin ulceration if left in place an appreciable time.

Because of these factors, it appears that extensive instrumentation of the subject can only be used for flights of short duration where it is permissible to immobilize the subject. Since most of the studies to be conducted with the biological satellite involve freedom of motion or long duration or both, subject measurements will have to be conducted by indirect measurement techniques. Some of the possible measurements are outlined but there is a fertile field for the exercise of ingenuity in planning such programs.

In addition to sensing elements directed toward the subject, there is need for a set of sensors to measure the environmental conditions that the subject is exposed to. Sensors will also be needed for control and for monitoring the performance of the control system.

One important measurement is the oxygen consumption of the subject or subjects. This can be done on a long term basis by measuring the flow or demand in an open cycle system or by measuring the regenerated waste product accumulation in a closed or partly closed system. With some pains as to instrumentation resolution, time variations in demand can be observed by measuring the rate. For example, suppose that the oxygen supply system is of the open cycle type which maintains constant oxygen partial pressure by a demand regulator. If the interval of opening and closing of this regulator is measured, a good measurement of the flow rate is secured. A cross check on this can be secured by measuring the pressure in the oxygen supply system and differentiating this for rate of decrease.

A second group of measurements would be based on the acoustic noise generated by the subject. The factors which would be measured includes the environmental noise level, noises produced by motion, intentional sounds produced by the subject, noises incident to the digestive process, and even the heartbeat of the subject. The ability to obtain useful data would be determined by the ability to separate these into their components.

There does not seem to be any reason why the background noise cannot be reduced below the limits of detectability for short periods of time. For longer periods, however, it appears that there must be a background level due to the operation of equipment such as air circulators and attitude control elements. Additionally, it appears likely that complete sustained silence would be undesirable because of its depressing effect on the subject.

The separation of the various sounds produced by the subject can be accomplished by correlation with other measurements of subject activity and with measurements made during preflights studies. To avoid saturation of the measurement system, it appears that the sonic spectrum should be divided into bands and that each band should be measured by an instrument having an automatic gain control element. This allows maximum resolution of whatever noise is present. By stopping control and air circulation systems at intervals the background noise can be reduced to essentially zero so that the measured noise is all produced by the subject. Correlation with subject motion, oxygen consumption and such factors will allow determination as to whether the noise is due to say, breathing or heartbeat.

In flights where the subject is allowed free motion, a measure of this motion is the reaction force on the cabin supports. This measuring system is essentially the same as the ballistocardiograph. It appears that the use of automatic gain techniques will allow the equipment sensitivity to be raised to the level of the cardiograph. In addition to the general cabin support reaction it appears that measurements of some individual reaction forces are also desirable. For example, if hand holds are provided around the cabin, the total force on these should be separately measured. This would provide an index of motion habits and to some extent of muscle tone of the subject.

Since it appears that much of the motion of the subject will be by hand-over-hand "climbing" between hand-holds, an additional measurement can be secured by insulating each of these and by measuring the contact potential developed at subject contact. This would give a direct measurement of galvanic skin response and could possibly provide electrocardiograph data if the hand-hold were designed to give contact low resistance.

Temperature measurements of the subject by remote means would not be easy to accomplish but some information could be obtained by burying thermocouples in the hand-holds. An additional possibility is the use of optical techniques in the infrared zone to measure the skin temperature of the subject.

The total energy output of the subject can probably be obtained best from the oxygen consumption. However, a further check on the metabolism could be secured by measuring the fine structure in the performance of the thermal control system of the vehicle. This would not be an easy measurement because of the large influence of solar radiation but should be possible with careful design.

In addition to specific items, general information would be provided by photography, and or by television. The major advantage of visual instrumentation is its enormous data capacity. This would provide information on subject activity, on physical conditions, and to some extent, on mental condition. The trials and failures of the subject during the learning and adaptation process can be observed and used later to establish preflight training programs for subsequent subjects. Difficulties in maneuvering, in feeding and sleeping, etc. could be observed and the information gathered used to improve design of the cabin furnishings. The "tone" of the subject can be observed through motor action and also by observation of skin, hair, eye and nose condition, etc. This evaluation of tone would be aided if the photography were in colour instead of in black and white.

One of the most valuable features of the optical instrument is the fact that it serves as a useful instrument for unexpected events. For example, if a fire should occur, this would be detected immediately through the optical system. Quite probably, the events which lead to the initiation of the fire, its source, and development would also be shown by the photographs. Because of this great range, it seems that either television or photographic coverage should be included in every flight with living subjects.

Limitations on Designs

Since biological payloads are usually small, most of the biological satellite effort will be conducted with a relatively small booster vehicle. Typically, these might use a military missile of the intermediate range class for the first booster and an assembly of research rockets for subsequent stages. Because of their simplicity and high reliability, it is probable that most of the last stages would use solid propellent rocket motors. This may impose some problems in maximum accelerations and vibration.

The acceleration problem is probably tolerable since most of the subjects discussed above are quite rugged. However, if it proves necessary to limit the peak acceleration, it appears that the most reasonable way of doing this is to use additional rocket stages, readjusting the weight distribution so that peak acceleration is kept low.

The vibration environment will be somewhat more difficult to control. It appears that this must be handled as an individual design problem and that it is best solved by placing low frequency filters between life cell and thrust unit. This can be done by using an adapter section which is a pressure stiffened structure or by other more complex means involving structural elements which are filters.

Power will be continuously needed in orbit for instrumentation and also for air circulation. Because of the long duration flights needed for biological studies it appears that batteries will not be suitable due to their weight. Nuclear sources are also unattractive due to stray radiation so that solar energy appears to be the best power source. In the view of the present state of development, this limits the average power demand to the range of about 10 to 50 watts.

Since zero g studies are needed, it appears that the biological satellite must be positively stabilized. Probably the simplest method of accomplishing this is to use the earth as a reference, maintaining one face of the satellite constantly toward the earth by means of a horizon sensing system or equivalent. This also has the advantage of giving automatic antenna orientation which helps keep power demand low. The alternate possibility of stabilization with respect to the sun would simplify the solar power system but would introduce antenna orientation problems.

Since most of the studies involve recovery of subjects, re-entry must be considered in the design. Two points of importance here are the influence of the return kick magnitude and the angle. The curves of [1] show the magnitude of these effects.

It is usually considered best to initiate return at the so called optimum angle since this reduces the re-entry dispersion and requires a minimum impulse for a given range to impact. However, in the present case, it appears better to operate the system with a vertically directed return impulse since this orientation is already provided by the attitude control system. The magnitude of the impulse should be chosen so that re-entry deceleration is kept reasonably low. The exact value would be a compromise between this and the dispersion at impact.

It appears that it is necessary to design the recovery system for water impact. This arises from several considerations, the chief being the much larger percentage of water area on the globe. With flights of great duration, and assuming simple guidance system it would not appear possible to make a sufficiently accurate prediction of impact point prior to attainment of the orbit.

Life Support

Table I summarizes the requirements for maintenance of life for several of the animals described above. It is evident from these figures that the total consumption per animal is relatively small even for periods as long as 8 weeks.

Table I. *Characteristics of Some Biological Subjects*

	Mouse	Rhesus Monkey
Weight	0.045 lb	7.26 lb.
Water intake	0.0077 lb/day	0.66 lb/day
Oxygen intake	0.0063 lb/day	0.50 lb/day
Food intake	0.044 lb/day	0.22 lb/day
Water output	0.0077 lb/day	0.64 lb/day
Carbon dioxide output	0.074 lb/day	0.58 lb/day
Nitrogen output	1.25 cc/day	small
Combustible gas output	3.0 cc/day	small
Heat generated	3.4 Kcal/day	330 Kcal day

Because of this it does not seem to be worthwhile to attempt to design biological satellites for closed cycle or even semi-closed cycle operation. Accordingly, the design problem in the life support system is to provide for the storage of food, water and oxygen and for the removal of the waste products.

Gaseous oxygen stored in pressure vessels appear to be entirely satisfactory for the suggested flights. Oxygen metering to the life cell can be accomplished by maintaining the total cell pressure constant. For longer flights this may not be sufficient and it is probably necessary to permit adjustment of the composition of the cabin atmosphere. Rather than introduce automatic controls, it seems more desirable to perform this

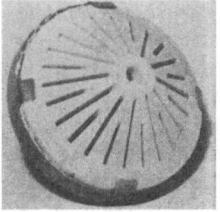

a   *b*

Fig. 1. Hydrogen catalyst. *b* Exploded view

adjustment on an intermittent basis by means of ground signals. The need for the adjustment can be determined from cabin gas analysis information which

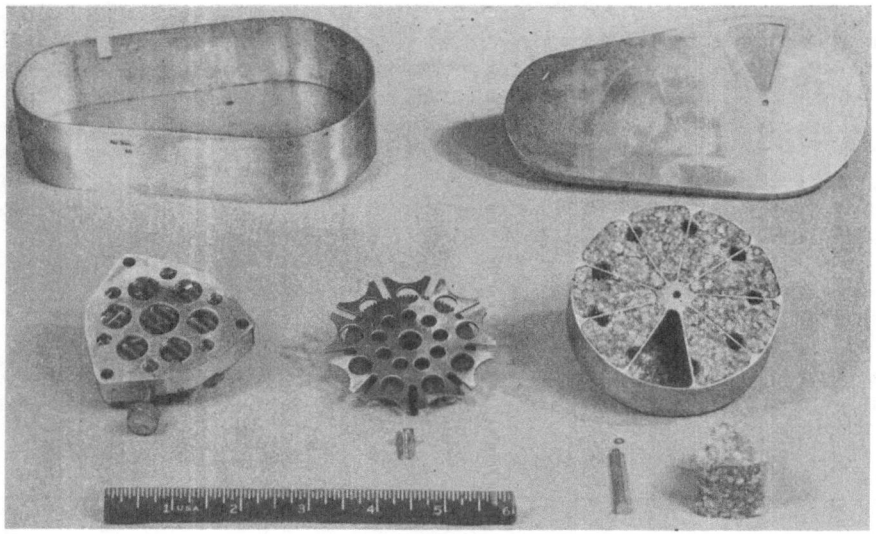

Fig. 2. Animal feeder

is telemetered to the ground. This conclusion is based on belief that the cabin atmosphere should approximate closely normal air composition. The reasons for this have been examined in a preceeding paper [2].

For the removal of carbon dioxide from the atmosphere it appears that absorption by lithium hydroxide is the most suitable method, with sufficient absorber for the entire flight. For longer flights the weight of the chemical becomes unattractive and a regeneration system would be desirable. However, at this time a suitable cycle is not available.

Water removal is closely associated with carbon dioxide removal since the absorption system for carbon dioxide requires that the air previously be dried. Several factors indicate that the cabin should be operated at rather a low relative humidity. The habits of animals are such that it is not possible to have a definite location for evacuation. Further, in the weightless state the waste products are difficult to confine and will tend to float free. The only method of removal which seems practical under these conditions is desiccation and collection of the solids on air filters.

Two possible methods of water removal are absorption and freezing. Absorption units can be regenerated by exposure to vacuum and appear to be simplest for this application. Silicagel appears best for the active element in view of its capacity and stability.

In long term confinement it appears necessary to provide for the removal of trace gases. Many of these are removable by catalytic combustion. For exam-

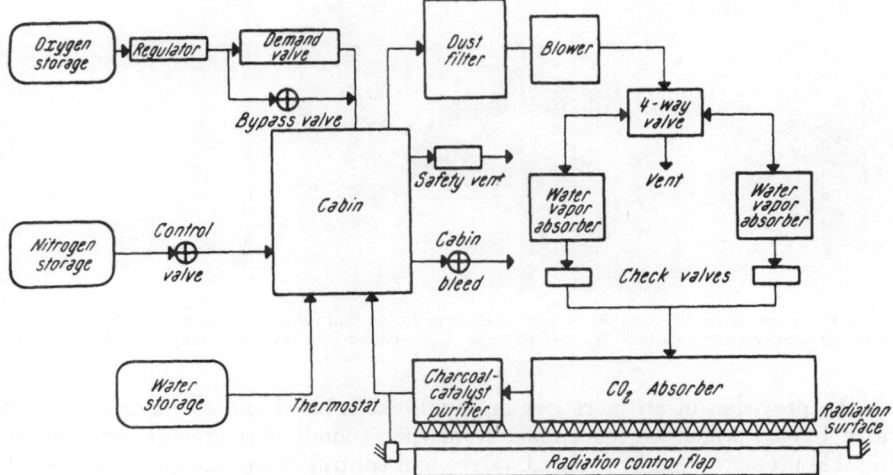

Fig. 3. Life support system block diagram. Control and instrument circuits not shown

ple, a commercially available hydrogen removal unit is shown in Fig. 1. This burns hydrogen and oxygen by catalytic combination. This unit also contains a separate absorber to prevent poisoning of the catalyst by stibine since this unit was designed to use in connection with removal of hydrogen generated by battery charging.

In addition to these catalytic units it also appears necessary to provide further absorption by activated charcoal. This has a very broad absorption spectrum and is highly effective so that only a small quantity of the charcoal needs to be provided. While the quantity of charcoal could be reduced by vacuum regeneration, this does not seem justified in this application.

Animal foods have received an appreciable amount of attention, and it appears that suitable feeders have already been designed. For example, Fig. 2 shows a simple and rugged feeder suitable for use with monkeys. This is intended for use with a gelatin-like preparation which contains all the required nutrients. It could also be used with dry food of the dog bisquit type. These are probably better for the long term flights since the low relative humidity to be maintained in the cabin will tend to dry out the gelatin foods.

A water source will be needed for the larger animals in any event, and for the small ones if dry foods are used. In view of the zero g conditions of operation, nipple supply of the water is used. To prevent cavitation, the water will need to be contained within a collapse bag.

A block diagram of a suitable cabin supply system is shown in Fig. 3. The oxygen supply except for size is similar to that in [3]. The major additional features

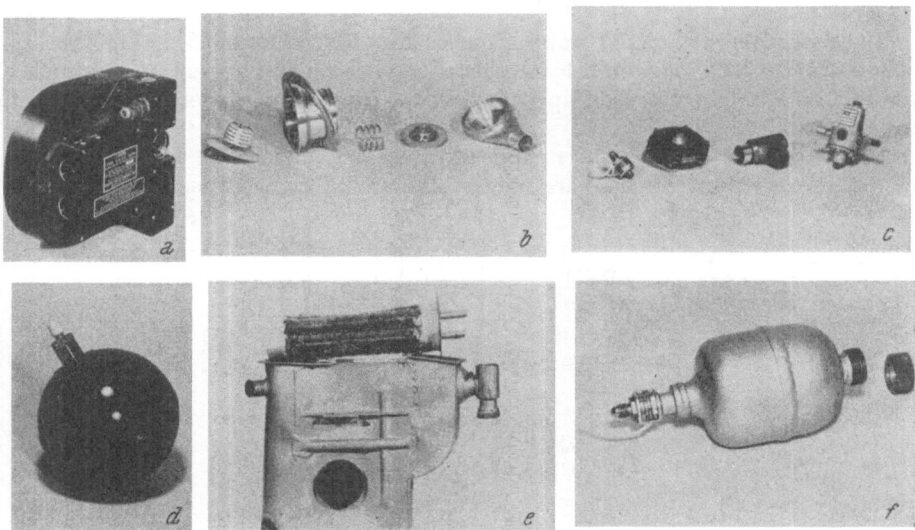

Fig. 4. Components for life support system. *a* Camera for animal movement studies. *b* Demand valve. *c* Oxygen supply components. *d* N_2 storage flask. *e* Air cooler and filter assembly. *f* O_2 storage flask

are the provision of an inert gas supply, a cabin bleed and an oxygen demand valve bypass which are controlled from the ground, plus greater purification. Suitable pressure vessels, demand valves and control valves are shown in Fig. 4.

Take Off and Landing Protection

The animal will be subjected to relatively high *g*'s during take off and landing maneuvers. Most of the proposals for protection under these conditions involve the use of a formed seat. However, this does not seem to be feasible for long term biological subject flights since free motion while in orbit is needed, and since it would be difficult to train the subjects to occupy the seat on signal.

A method of supporting the animal which appears to satisfy all of the requirements of protection during acceleration and of freedom in space is a combination of a foam plastic bed and a restraining net. The bed would be orientated perpendicular to the direction of acceleration and the foam density would be controlled so that the "upper" surface would mold itself around the animal. The net would be used to hold the animal lightly in place against the foam prior to the start of acceleration. During zero *g* flight the net would be moved by a linkage to the side of the cabin away from the foam plastic. This would allow space for in-orbit motion.

In addition to providing approximately optimum orientation of the animal during acceleration, this net would also serve to return the animal to a wall in case the animal were "floating" out of reach of any of the hand-holds provided around the cabin walls. This operation would be initiated by ground command if needed. Similar initiation of net closing would be made before take off or re-entry.

Since the animals tend to be destructive, the net material must be quite strong. It appears that fibreglass cords molded in nylon would give a usable combination of strength and flexibility. In addition, small electrodes could be placed along the net at intervals for contact of potential measurements. Net supports could include strain gauges for reaction force measurement.

Re-Entry Devices Available for Design

Fig. 5 shows a group of re-entry devices which have been constructed for various purposes and which are suitable for re-entry from satellite orbits when provided with the proper heat shield for the operating conditions of the specific test.

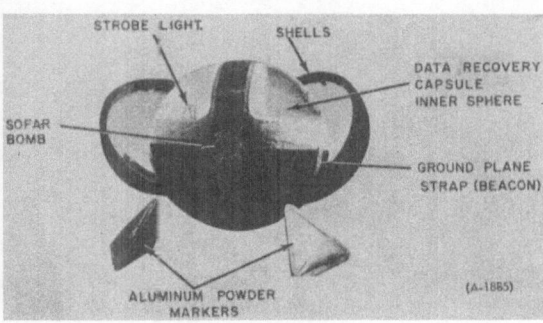

The smaller of these devices is called a data capsule and was originally designed for the recovery of telemetry tape or photographic records. This unit is a sphere and is normally used with an 18″ diameter heat shield. The payload capacity is reasonably

a

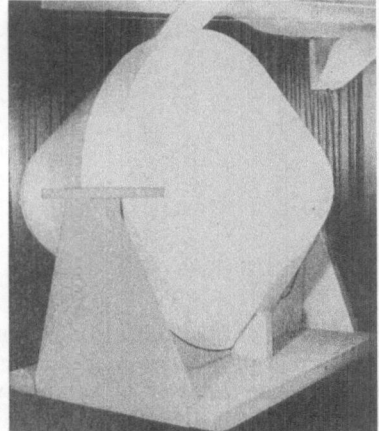

b

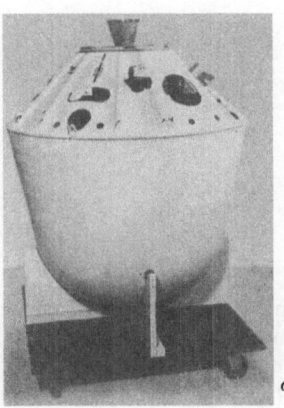

c

Fig. 5. Re-entry devices. *a* Data capsule. *b* Large re-entry body. *c* Quasi-parabolic re-entry body

large, around 18 pounds, but the space available for equipment installation is less than $^1/_4$ of the total volume. This is due to the fact that this unit is designed to impact at terminal velocity. It does not use a parachute.

The next unit is a quasi-parabolic shape of revolution intended for use with parachutes for low velocity impact, and is available in two sizes. The smaller has a base diameter of 20″ and the larger just over 30″. The payload capacity of the smaller unit is about 40 pounds and with a volume of $^3/_4$ of a cubic foot while that of the larger unit is around 75 pounds with a volume of about 2 cubic feet. Both of these units are provided with parachutes to give impact velocity of about 50 ft/sec at sea level.

The largest unit is an early design re-entry device intended to carry large payloads. It has a base diameter of 55″ and a total payload capacity of around 2,000 lbs. The total volume available for installation is 60 cubic feet less the internal structure which depends on the specific application. While this unit was originally designed for terminal velocity impact, the internal structure has been modified for use with a parachute to give an impact velocity of 50 feet per second. Designs and mock-ups of this unit are available for still larger sizes up to 8 feet base diameter.

All of these units use ablating heat protection. The exact material, method of fabrication and distribution over the re-entry surface depends on the con-

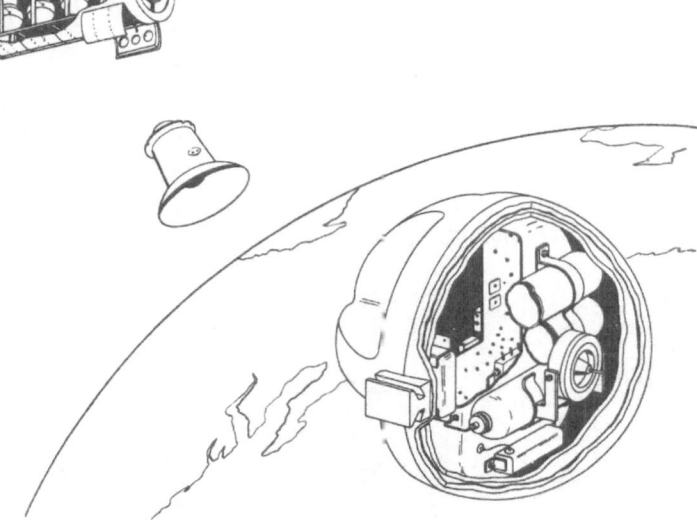

Fig. 6. Multiple capsule biological satellite for radiation effect studies

ditions of re-entry and may be computed using published data [4]. Designs are available for low angle satellite vehicle re-entry for all of these bodies. For some, designs have been prepared for velocities up to the velocity of escape and for re-entry angles up to 20° above the horizontal.

Because of weight and space limitation, the spherical design is suitable only for experiments with small subjects. The period of operation may be extended by providing a separate air system so that only the air supply for the re-entry and recovery phase need be installed within the sphere. Because of the relatively low weight of these units, it is possible to carry several of these on a single rocket, and return them to earth in sequence.

The quasi-parabolic shapes can be used for larger animals for short periods or for small animals for extended times. Again, separate air systems may be provided to extend the period of operation. However, because volume for experimental equipment is still small, these shapes are not desirable for larger animals and long times.

The largest shape appear to be completely suitable for any of the subjects discussed above and for any period of time. There is ample volume for the subject and its life cell and in the necessary air purification and other life support equipment. (Larger versions have sufficient capacity to carry one or two men for short periods and so may serve as a basis for some of the early manned investigations.)

An "artist's concept" of ways of using the above re-entry shapes for biological experiments are shown in Figs. 6, 7 and 8. These should be regarded as schematic drawings which show major design features and should not be considered to be complete.

Fig. 6 uses four of the spherical data capsules in a single satellite body for a radiation experiment. The subjects shown are fruit fly colonies, one in each of the data capsules. The satellite would be launched to penetrate the inner VAN

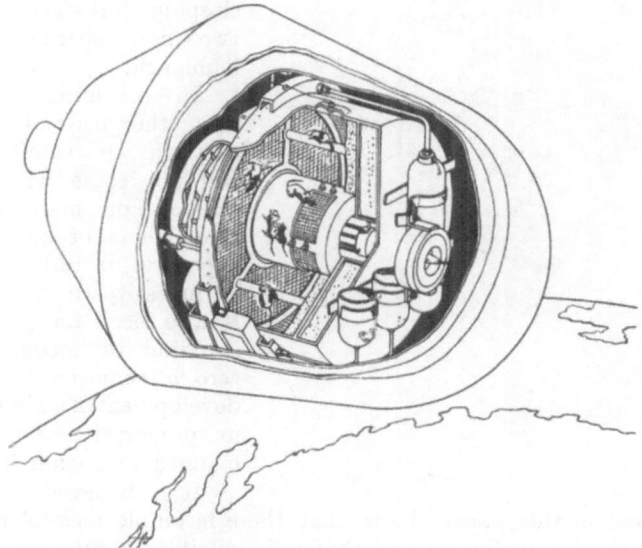

Fig. 7. Biological satellite cell for mouse experiments

ALLEN radiation belt. One capsule would be ejected after the first two penetrations of the belt, the second after about one day, the third after about one week and the last after approximately a month of exposure. In this design no provisions for acceleration protection are made, and the cabin is basically just an enclosed volume.

The second design shown in Fig. 7 would use the larger of the parabolic shapes to carry mice (or rats) for extended exposure to zero g condition. This unit would use an exterior air supply carried in the satellite body to allow the test to be conducted for a number of weeks with a colony of 6 to 8 mice. Attitude stabilization would be used to give zero g for this period. The foam bed-net combination is used to give acceleration protection. A single cabin is shown, with a central feeder, although it is possible that a sectioned cabin would be better. Mesh is attached to the walls and a number of "climbing posts" are included to allow for motion.

If sufficient payload capacity were available for the booster, flight conditions could be chosen so that simultaneous exposures to radiation were obtained. However, it appears that the experiment would still be of sufficient value even if the operating altitude were low.

The largest design of Fig. 8 is intended to carry two Rhesus Monkeys for long period tests. This cabin also uses the bed-net system of acceleration system. The cabin is made large and is kept free of internal structure to allow maximum opportunity for learning of locomotion in free fall. Hand-holds are placed around the cabin to aid this. Operating condition for this would be similar to the mouse capsule.

Provisions for sleeping have been purposely omitted from these animal designs, to allow investigation of sleep during free fall. If this should prove to be impossible under zero g, it is always possible to use the net to provide a restraint. In this case later designs would include special sleeping provisions, such as two nets spaced to form a hammock.

In conclusion, it is noted that the need for biological research in satellite vehicles will not cease with the introduction of man into space. The effects of zero g's and of exposure to radiation need to be investigated at length. There is also need for greater understanding of locomotion under zero g condition, and for the development of training programs in preparation for long term manned operation in space.

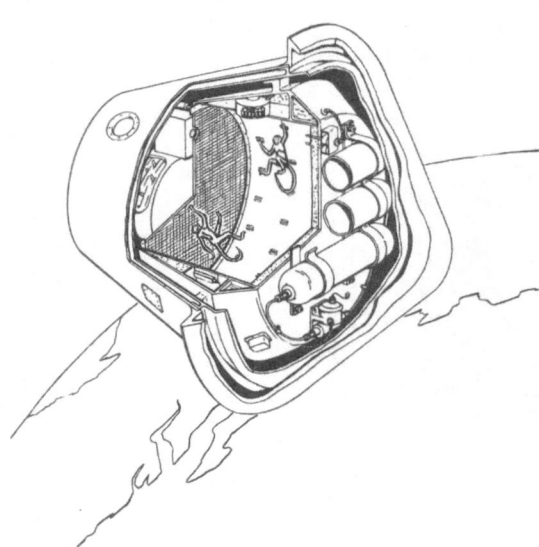

Fig. 8. Biological satellite cell for monkey flight test

It is believed that the material presented in this paper shows that there is ample technology available for the conduct of experiments and that it is possible to commence studies of the type outlined immediately.

Acknowledgements

The approach to the problems discussed here is based on discussion with Messrs. D. G. Simons, S. Gottlieb, M. F. Clarke, R. W. Lawton, E. S. Miller, J. H. Hoffnagle, Jr., F. Bernt and others. Solutions shown to re-entry, recovery and life support problems are the results of work at the Missile and Space Vehicle Department, General Electric Company.

References

1. C. V. Carter and W. W. Huff, Jr., The Problem of Escape from Satellite Vehicles. IAS Report 59-41.
2. R. P. Haviland, Air for the Space Ship. J. Astronautics 3, No. 2, Summer 1956.
3. D. G. Simons and D. P. Parks, Climatization of Animal Capsules During Upper Stratosphere Balloon Flights. ARS Preprint 241—55.
4. J. H. Quillinan, L. London and B. L. Aston, Configuration Selection of Reentry Vehicles. To be presented IAS Summer Meeting, Los Angeles, June 16, 1959.

Les aériens à grand gain

Par

G. Callede[1]

(Avec 4 Figures)

(Reçu le 18 juin 1959)

Résumé — Zusammenfassung — Abstract

Les aériens à grand gain. Les transmissions spatiales à très grande distance supposent l'utilisation d'antennes de gain très élevé. La surface d'un aérien (ou, pour un aérien à rayonnement longitudinal, sa longueur) étant proportionnelle au gain, on ne peut obtenir de très grandes valeurs de gain que du côté terrestre de la transmission. Ce sont ces aériens terrestres qui font l'objet de la communication.

Après avoir rappelé les expressions qui relient le gain à la largeur angulaire du diagramme, nous donnons, pour les aériens à rayonnement longitudinal et pour les aériens à ouverture rayonnante, les relations entre gain et dimensions, qui font apparaître l'intérêt de ce dernier type. En particulier, l'antenne à réflecteur parabolique circulaire, par sa simplicité et son indépendance — relative — vis à vis de la fréquence, semble devoir être retenue.

Nous rappelons les relations fondamentales simples de ce type d'aérien, pour lequel nous calculons le gain maximum possible, en fonction des tolérances de fabrication acceptables. Un gain de 65 dB environ paraît, en conséquence, être à la limite des possibilités, pour les ondes centimétriques. L'exigence relative à la précision de fabrication (le dixième de la longueur d'onde) correspond d'ailleurs à la règle de Lord RAYLEIGH pour les miroirs de télescopes.

Une dernière question est rapidement évoquée, il s'agit de la température de bruit de l'aérien. Les amplificateurs modernes à très faible facteur de bruit mettent à l'ordre du jour les recherches relatives aux antennes à basse température de bruit. Cette caractéristique est liée au niveau des lobes secondaires et du rayonnement diffus, qui fait l'objet de recherches auxquelles on commence seulement à faire allusion.

Enfin un tableau récapitule quelques réalisations actuelles, dont plusieurs sont utilisées conjointement pour la radioastronomie et les liaisons spatiales.

Antennen mit hohem Gewinnfaktor. Räumliche Signalübertragungen auf sehr große Entfernungen setzen Antennen mit einem sehr hohen Gewinnfaktor voraus. Da der Gewinnfaktor einer Antenne deren Oberfläche (oder bei Antennen mit Abstrahlung in Längsrichtung deren Länge) proportional ist, kann man einen hohen Antennengewinn nur auf der Bodenseite eines räumlichen Übertragungssystems erzielen. Solche bodenseitigen Antennen bilden den Gegenstand der vorliegenden Mitteilung.

Nach einem kurzen Überblick über die Beziehungen, welche den Antennengewinn mit der Winkelöffnung des Antennendiagramms verknüpfen, werden für Antennen mit longitudinaler Abstrahlung und für solche mit strahlender Fläche die Zusammenhänge zwischen Abmessungen und Antennengewinn wiedergegeben. Aus diesen geht die große Bedeutung des letzteren Antennentyps hervor. Insbesondere ist die Antenne

---

[1] Laboratoires de Physique Appliquée, Sud-Aviation, France (actuellement: Laboratoire des Hautes Energies, Ecole Normale Supérieure, Boîte Postale n° 2 Orsay, S.-et-O., France).

mit kreisrundem parabolischem Reflektor infolge ihrer Einfachheit und ihrer relativ großen Frequenzunabhängigkeit beachtenswert.

Es werden die einfachen grundsätzlichen Beziehungen für diesen Antennentyp wiedergegeben, mit welchen dann der höchstmögliche Gewinn in Abhängigkeit von tragbaren Fertigungstoleranzen berechnet wird. Ein Gewinnfaktor von etwa 65 dB erscheint für cm-Wellen an der Grenze des Möglichen zu liegen. Die Anforderung an die Genauigkeit der Fertigung (zulässige Toleranz etwa $^1/_{10}$ der Wellenlänge) entspricht im übrigen etwa der von Lord Rayleigh für Spiegelteleskope gegebenen Regel.

Eine weitere Frage wird gestreift, nämlich die dem Antennenrauschen entsprechende Temperatur. Moderne Verstärker mit ihrem sehr niedrigen Rauschpegel haben die Erforschung von Antennen mit niedriger Rausch-Temperatur in den Vordergrund gerückt. Die Rauscheigenschaften einer Antenne sind aufs engste mit der Größe unerwünschter Nebenkeulen im Strahlungsdiagramm und dem Anteil einer diffusen Abstrahlung verknüpft. Dem Einfluß solcher Rauschquellen beginnt man jetzt erhöhte Aufmerksamkeit zu schenken.

Eine Tafel am Schluß stellt die Daten verschiedener ausgeführter Antennen zusammen, von welchen einige sowohl für die Radioastronomie als auch für räumliche Nachrichtenverbindungen benützt werden.

High Gain Antennas. The use of very high gain antennas is assumed for long distance space transmissions. Its surface (or, for an end-fire antenna, its length) being proportional to gain, one can only obtain very high gain values from the terrestrial end of the transmission. These are terrestrial antennas which are the object of the communication.

After having recalled the expressions that connect the gain to the angular width of the diagram, we give (for antennas of the end-fire type and for radiating aperture antennas), the relations between gain and dimensions that generate interest in the latter. In particular, the circular parabolic reflector type antenna, by its simplicity and its relative independence vis à vis frequency, apparently should be retained.

We recall the simple fundamental relations of this type of antenna, for which we calculate the maximum possible gain as a function of acceptable manufacturing tolerances. A gain of approximately 65 dB appears consequently to be at the limit of possibilities for wavelengths measured in centimeters. The relative demands on manufacturing precision (tenth of a wavelength) corresponds moreover to Lord Rayleigh's rule for telescopic mirrors.

A last question is quickly evoked, concerning the antenna noise temperature. Modern low noise factor amplifiers bring to the forefront research concerning low noise temperature antennas. This characteristic is tied to the level of secondary diffuse radiating lobes which are the subject of research about which mention is beginning to be made.

Finally, a table recapitulates some current results, some of which are utilized jointly by radio astronomy and space communications.

I. Introduction

La liaison radioélectrique entre un astronef et le sol suppose l'emploi d'aériens de gain élevé, comme le montre la relation classique entre puissances émise et reçue:

$$P_r = P_e \, G_e \, G_r \left(\frac{\lambda}{4\,\pi\,d}\right)^2 \tag{1}$$

P_r et P_e étant les puissances émise et reçue
G_e et G_r étant les gains des aériens à l'émission et à la réception
λ la longueur d'onde
d la distance.

Si on suppose la longueur d'onde, la distance et la puissance émise imposées, il faudra rendre G_e et G_r aussi grands que possible. Le problème se pose évidemment de façon différente pour G_e et G_r.

Comme nous allons le préciser, un gain élevé suppose de grandes dimensions d'aériens, et un très grand gain ne pourra être envisagé que pour l'aérien terrestre.

On devra se contenter pour l'aérien spatial de performances plus modestes, non seulement à cause des dimensions prohibitives, mais aussi à cause des difficultés de pointage de l'aérien spatial vers la Terre.

Dans la suite de cet exposé, nous traiterons, sauf indication contraire, des aériens à grand gain du côté terrestre de la transmission.

II. Gain d'un aérien

Rappelons que le gain d'un aérien s'exprime par comparaison avec un aérien isotrope. Nous pouvons raisonner à l'émission sans restriction de généralité, en vertu du théorème de réciprocité.

Soit P la puissance totale émise par l'aérien considéré.

Soit $p(D)$ la puissance par unité d'angle solide qu'il émet dans la direction D.

Supposons que l'aérien isotrope de référence émette la même puissance totale P. Il émet donc, dans toute direction, une puissance par unité d'angle solide égale à $p_0 = \dfrac{P}{4\pi}$.

Par définition, le *gain de l'aérien considéré dans la direction D* est égal au rapport:

$$g(D) = \frac{p(D)}{p_0} = \frac{4\pi p(D)}{P}. \tag{2}$$

On a par ailleurs

$$P = \int_0^{4\pi} p(D)\, d\Omega \tag{3}$$

($d\Omega$ = angle solide élémentaire autour de D).
Soit

$$g(D) = 4\pi \frac{p(D)}{\displaystyle\int_0^{4\pi} p(D)\, d\Omega}. \tag{4}$$

Nous venons de définir le gain dans la direction D. Il existe en général une direction D_0 dans laquelle le gain est maximum. Le gain $g(D_0)$ dans cette direction privilégiée est appelé *par définition* le *gain de l'aérien*:

$$G = g(D_0) = 4\pi \frac{p(D_0)}{\displaystyle\int_0^{4\pi} p(D)\, d\Omega}. \tag{5}$$

Notons enfin qu'on appelle parfois *fonction gain* la quantité

$$h(D) = \frac{p(D)}{p(D_0)} = \frac{g(D)}{g(D_0)} = \frac{g(D)}{G}, \tag{6}$$

dont le maximum est égal à 1. D'où:

$$G = \frac{4\,\pi}{\displaystyle\int_0^{4\,\pi} \frac{p\,(D)}{p\,(D_0)}\,d\,\Omega} = \frac{4\,\pi}{\displaystyle\int_0^{4\,\pi} h\,(D)\,d\,\Omega}\,. \tag{7}$$

III. Diagramme et lobe de rayonnement. Pencil beams

Rappelons que la connaissance de la directivité d'un aérien définie par $h\,(D)$, permet de tracer une surface (S) dont le rayon polaire ϱ est proportionnel à $h\,(D)$. On définit aussi le diagramme dans un plan (P): c'est une courbe (C) dont le rayon polaire ϱ est proportionnel à $h\,(\theta)$, θ définissant les directions D du plan (P).

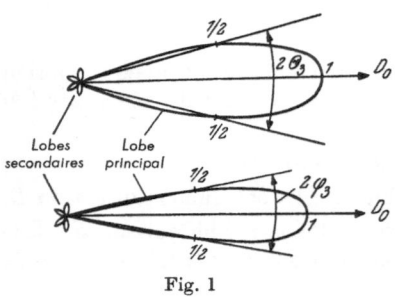

A moins que des formes particulières de diagramme ne soient recherchées (par exemple le rayonnement en "cosécante carrée" pour certains aériens de radar), les aériens à grand gain présentent en général un lobe de rayonnement étroit, en forme de faisceau (en anglais pencil beam).

Ces lobes sont descriptibles par leurs diagrammes dans 2 plans principaux (Fig. 1), qui sont les plans pour lesquels le diagramme a une ouverture maximum et minimum. L'ouverture angulaire du diagramme peut être caractérisée par l'angle qui sépare les directions à demi puissance: $2\,\theta_3$ et $2\,\varphi_3$ sur Fig. 1 (l'indice 3 indique 3 dB).

C'est ce type de lobe en faisceau que nous allons considérer par la suite. Il peut être intéressant de donner de leurs diagrammes une équation approchée:

$$\begin{cases} h\,(\theta) = \mathrm{e}^{-\alpha\,\theta^2} \\ h\,(\varphi) = \mathrm{e}^{-\beta\,\varphi^2} \end{cases} \tag{8}$$

avec

$$\alpha = \frac{L2}{\theta_3{}^2} \quad \text{et} \quad \beta = \frac{L2}{\varphi_3{}^2}$$

$$L2 = \mathrm{Log}_e\,2 = 0{,}693\,.$$

Ces valeurs approchées ne tiennent pas compte des lobes secondaires, mais elles rendent bien compte de la forme du lobe principal.

IV. Relation entre gain d'un aérien et largeurs à demi puissance du diagramme

L'expression (7):

$$G = \frac{4\,\pi}{\displaystyle\int_0^{4\,\pi} h\,(D)\,d\,\Omega} \tag{9}$$

permet d'établir une relation simple entre G et les largeurs à demi puissance $2\,\theta_3$ et $2\,\varphi_3$. Nous ferons le calcul dans le cas où:

a) le lobe est de révolution autour de l'axe de rayonnement maximum:

$$2\,\theta_3 = 2\,\varphi_3$$

Figure caption:
Fig. 1

(Labels in figure: $\frac{1}{2}$, $2\,\theta_3$, 1, D_0, $\frac{1}{2}$, Lobes secondaires, Lobe principal, $\frac{1}{2}$, $2\,\varphi_3$, 1, D_0, $\frac{1}{2}$)

b) l'ouverture du diagramme est faible, de sorte que pour tous les angles considérés :

$$\sin \theta \sim \theta$$

c) la fonction gain peut être approximée par :

$$h(D) = h(\theta) = e^{-a\theta^2} \quad a = \frac{L2}{\theta_3{}^2}$$

$$\Omega = 2\pi(1 - \cos\theta)$$

$$d\Omega = 2\pi \sin\theta \, d\theta$$

$$G = \frac{4\pi}{\displaystyle\int_0^{} h(\theta)\, d\Omega} = \frac{2}{\displaystyle\int_0^{} h(\theta) \sin\theta \, d\theta} = \frac{2}{\displaystyle\int_0^{} e^{-a\theta^2} \sin\theta \, d\theta}.$$

On peut écrire $\sin\theta \sim \theta$ car, bien que l'intégration soit faite en principe entre o et π, la zone effective d'intégration est limitée à une valeur θ_0, faible (Fig. 2), la fonction $e^{-a\theta^2}$ étant pratiquement nulle entre θ_0 et π. D'où :

$$G = \frac{2}{\displaystyle\int_0^{} e^{-a\theta^2} \theta \, d\theta} = \frac{-4a}{\left| e^{-a\theta^2} \right|_0^\pi} = 4a = \frac{4L2}{\theta_3{}^2} = \frac{16L2}{(2\theta_3)^2} \qquad (10)$$

avec θ_3 exprimé en radians
et

$$G = \frac{16L2\left(\dfrac{180}{\pi}\right)^2}{(2\theta_3)^2} = \frac{36\,000}{(2\theta_3)^2} \quad (11)$$

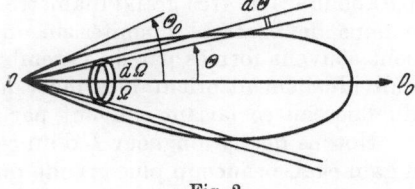

Fig. 2

avec θ_3 exprimé en degrés.

Cette valeur est une valeur théorique limite, qui suppose que le diagramme n'a pas de lobes secondaires. En pratique, il n'en est pas ainsi et la puissance diffusée dans les lobes secondaires est à peu près égale à la moitié de la puissance totale, ou encore égale à toute la puissance rayonnée dans le lobe principal. C'est dire que l'intégrale précédente doit être multipliée par 2 :

$$G = \frac{18\,000}{(2\theta_3)^2} \quad (2\,\theta_3 \text{ en degrés}). \qquad (12)$$

Cette nouvelle relation est une très bonne approximation des valeurs rencontrées en pratique.

Elle exprime un résultat bien connu : le gain varie en raison inverse du carré de l'ouverture du faisceau.

Nous avons tiré les valeurs du tableau suivant, qui nous permettront de fixer les ordres de grandeur :

2 θ_3 (degrés)	Ouverture à demi-puissance
	Gain G
10°	180 = 22,6 dB
5°	720 = 28,6 dB
2°	4.500 = 36,5 dB
1°	18.000 = 42,6 dB
0,5°	72.000 = 48,6 dB

Pour augmenter le gain de 6 dB, il faut diviser les ouvertures à demi-puissance par 2.

Une première limitation apparaît donc, dans la voie à suivre pour l'accroissement du gain: l'ouverture diminue, et la précision du pointage de l'antenne dans la direction intéressante doit augmenter.

V. Types d'antennes appropriés aux transmissions spatiales

Il existe un certain nombre de types d'antennes susceptibles de fournir un faisceau étroit, donc un gain élevé. On peut citer les réseaux à rayonnement longitudinal ou transversal, et les aériens à ouverture.

Ces aériens ont été étudiés en vue d'applications au radar et à la radioastronomie. La technique radar requiert des formes particulières de lobes, les diagrammes dans les 2 plans pouvant différer par la valeur de l'ouverture (antennes de site, antennes de gisement), ou même par la forme (antenne en cosécante carrée). Dans la presque totalité des cas, ces antennes de radar sont constituées par un réflecteur "éclairé" par une source primaire (cornet).

En radioastronomie on a utilisé soit de très grands réseaux, soit des réflecteurs paraboliques de très grand diamètre. Les réflecteurs sont, dans la plupart des cas connus, des ensembles montés sur un support orientable. Au contraire, les réseaux sont souvent formés par un ensemble d'antennes fixées au sol, qui peuvent être individuellement orientables, mais dont l'ensemble n'est pas mobile, l'orientation du faisceau se faisant souvent par variation de phase.

Notons que la longueur L d'un réseau à rayonnement axial, par exemple, est, à gain égal, beaucoup plus grande que le diamètre D d'un réflecteur parabolique. En effet L varie comme G, et D comme $\sqrt{G}$ (voir ci-dessous). Il n'est donc pas possible de monter sur un support orientable un grand réseau longitudinal.

Le réflecteur parabolique semble donc présenter un grand avantage sous ce rapport.

Un autre avantage du paraboloïde est de pouvoir être utilisé à toute fréquence, sous réserve que les mailles du grillage qui constitue le réflecteur soient de dimensions inférieures à $\lambda/10$ environ (et aussi sous réserve que les tolérances de fabrication soient de l'ordre de $\lambda/10$, comme nous le montrons plus loin). Il suffit de changer la source primaire d'excitation pour changer la fréquence de fonctionnement, ce qui peut permettre d'affecter le même aérien à des utilisations variées. Cette propriété est vraisemblablement précieuse en radioastronomie, où les aériens peuvent être des constructions énormes (exemple: Jodrell Bank — paraboloïde de 115 m de diamètre) de coût très élevé.

Notons enfin que les liaisons astronautiques, pour lesquelles des aériens spéciaux ont déjà été construits, utilisent aussi ces grands aériens de radioastronomie: la station de Jodrell Bank a été utilisée pour du "tracking" de satellites et de sondes spatiales.

Nous pouvons donc conclure que le réflecteur parabolique paraît être le type même des aériens terrestres pour liaisons spatiales. Nous donnons ci-dessous ses principales caractéristiques.

VI. Caractéristiques des réflecteurs paraboliques

La surface du réflecteur est une portion de paraboloïde de révolution, limitée à un cercle (C) de section droite, de diamètre D. Le rapport du diamètre D à la distance focale f est de l'ordre de 3: $D/f \sim 3$. Cette surface réflectrice est matérialisée par un grillage métallique dont la maille doit être de dimensions inférieures à $\lambda/10$.

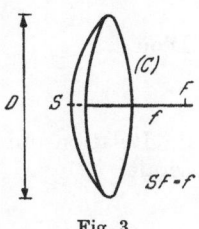

Fig. 3

Le réflecteur est "éclairé" par une source primaire placée au foyer F. Cette source doit être suffisamment directive pour ne pas rayonner à côté du paraboloïde ("spill over"). Elle peut être constituée par un cornet une hélice, ou un Yagi.

Si A est la surface du plan d'ouverture, $A = \dfrac{\pi D^2}{4}$, on sait que le gain théorique maximum G_t s'exprime en fonction de A par:

$$G_t = \frac{4\pi A}{\lambda^2} = \pi^2 \left(\frac{D}{\lambda}\right)^2 \# 10 \left(\frac{D}{\lambda}\right)^2 . \tag{13}$$

Le diagramme est de révolution, son ouverture à demi-puissance est

$$2\,\theta_3 \# 60\,\frac{\lambda}{D} \text{ (degrés)} \tag{14}$$

d'où

$$(2\,\theta_3)^2 = 3.600 \left(\frac{\lambda}{D}\right)^2 = 3.600\,\pi^2\,G_t \# 36.000\,G_t . \tag{15}$$

On retrouve la relation (11) ce qui montre que les relations (13) et (14) sont compatibles.

Le gain pratiquement atteint est

$$G = k\,G_t \sim 5 \left(\frac{D}{\lambda}\right)^2 \tag{16}$$

$k \sim 0,5$ est appelé facteur de gain.

On voit que le diamètre D varie comme la racine du gain G.

VII. Antennes a rayonnement longitudinal. Antenne hélicoidale

Revenons rapidement sur ce type d'aérien. Le gain est proportionnel à la longueur L de l'antenne:

$$G = K\,L .$$

Nous allons donner comme exemple l'antenne en hélice en mode axial, étudiée en particulier par John D. Kraus [1].

Soient D le diamètre de l'hélice

$C = D/\pi$ sa circonférence

S le pas

n le nombre de spires

$L = n\,s$ la longueur

Le rayonnement en mode axial a lieu pour $\dfrac{C}{\lambda} \sim 1$.

Le gain et la largeur à demi-puissance s'écrivent, d'après Kraus:

$$G_t = 15 \left(\dfrac{C}{\lambda}\right)^2 \dfrac{L}{\lambda} \sim 15 \dfrac{L}{\lambda} \tag{17}$$

$$2\varphi_3 = 2\theta_3 = \dfrac{52}{\dfrac{C}{\lambda}\sqrt{\dfrac{L}{\lambda}}} \sim \dfrac{52}{\sqrt{\dfrac{L}{\lambda}}} \text{ (degrés)} . \tag{18}$$

D'où

$$G_t = \dfrac{40.500}{(2\,\theta_3)^2} .$$

Relation qui montre bien qu'il s'agit d'un gain théorique. Le gain pratique serait:

$$G = 7{,}5 \dfrac{L}{\lambda} . \tag{19}$$

Notons enfin que le rayonnement émis est polarisé à peu près circulairement, et que la source isotrope de référence est supposée aussi à polarisation circulaire.

Il semble que ce type d'antennes ne présente pas d'intérêt pour réaliser au sol de très grands gains: sa longueur varie en effet trop vite avec le gain, ce qui lui fait préférer le paraboloïde mais, même à bord d'un véhicule spatial, les gains qui seront prochainement requis sont trop élevés: en effet, d'après E. Rechtin (Guidance Research Division 7 J.P.L.) le gain des antennes de véhicule évoluerait comme suit [4].

Année	1958	1960	1962
Gain	6 dB	16 dB	36 dB
	= 4	= 10	= 4 000

Le tableau suivant donne les dimensions correspondantes de l'hélice et du paraboloïde:

		hélice		paraboloïde	
année	gain G	$\dfrac{L}{\lambda} = \dfrac{G}{7{,}5}$	L à $\lambda = 10$ cm	$\dfrac{D}{\lambda} = \sqrt{\dfrac{G}{5}}$	D à $\lambda = 10$ cm
1958	4	0,53	5 cm		
1960	40	5,3	53 cm	2,8	28 cm
1962	4000	530	53 m	28	2,8 m

Il semble donc que, même pour les aériens du véhicule, le paraboloïde soit préférable.

VIII. Limitation supérieure du gain par la précision de la surface du reflecteur

Revenons donc au réflecteur parabolique. Avec le gain, augmentant à la fois la finesse du faisceau et le diamètre du réflecteur, et par conséquent les difficultés de pointage et de construction, qui ne sont d'ailleurs pas indépendantes. Mais une autre difficulté va apparaître, qui impose au gain une limite supérieure: il s'agit de la précision de la surface du réflecteur.

Des recherches effectuées à la C.S.F. (Compagnie Générale de T.S.F. Paris) par J. Robieux [2] ont permis à cet auteur d'apporter les conclusions suivantes, que nous lui empruntons:

Une antenne dont la précision de fabrication[1] est ε rayonne dans la direction principale de rayonnement un champ dont la valeur moyenne est N dB au-dessous de la valeur théorique, N étant donné par:

$$N \# + 107 \left(\frac{\varepsilon}{\lambda}\right)^2 . \tag{20}$$

Cette diminution ne dépend que de la tolérance de fabrication. Les fluctuations autour de la valeur moyenne sont faibles.

Le gain théorique est (paraboloïde à ouverture circulaire):

$$G_t = \left(\frac{\pi D}{\lambda}\right) \# 10 \left(\frac{D}{\lambda}\right)^2 = 10\, h^2$$

en posant:

$$\frac{D}{\lambda} = h$$

soit, en décibels:

$$N_t = 10 \log G_t = 10 + 10 \log h^2 \tag{21}$$

avec une diminution du gain théorique égale à:

$$N = 107 \left(\frac{\varepsilon}{\lambda}\right)^2 = 107\, (h \cdot 10^{-m})^2 \# (h \cdot 10^{1-m})^2 \tag{22}$$

d'où le gain réel:

$$N_r = N_t - N = 10 + 10 \log h^2 - (h \cdot 10^{1-m})^2 . \tag{23}$$

En fonction de h, N_r a un maximum: dérivons N_r par rapport à h:

$$\frac{d N_r}{d h^2} = \frac{4,3}{h^2} - 10^{2(1-m)} . \tag{24}$$

N_r est max pour $h^2_{max} = 4,3 \cdot 10^{2(m-1)}$

$$h_{max} \# 2 \cdot 10^{m-1} \tag{25}$$

d'où:

$$N_{max} = -8 + 20\, m . \tag{26}$$

On tire de cette équation le tableau:

m	Tolérance relative 10^{-m}	Gain max. (dB) N_{max}
2	10^{-2}	32
3	10^{-3}	52
4	10^{-4}	72

La relation $h_{max} \# 2.10^{m-1}$ fixe, pour m donné, une limite max. à h, donc à la dimension de l'antenne:

m	$h_{max} = \dfrac{D}{\lambda}_{max}$
2	20
3	200
4	2.000

---

[1] Il s'agit de la distance entre surface réelle et surface théorique, normalement à cette dernière.

Si on augmentait les dimensions au-delà de ces valeurs, le gain n'augmenterait pas, mais cette valeur est supérieure aux valeurs qu'il est raisonnable d'adopter pour obtenir un gain important, à m donné.

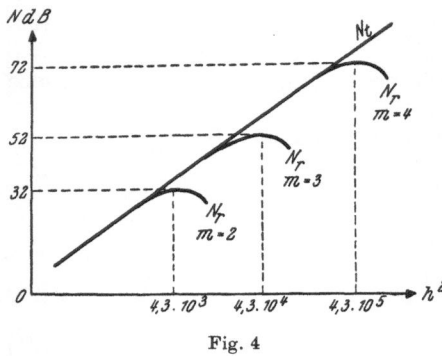

Fig. 4

En effet, la courbe du gain en fonction de h^2 a l'allure ci-contre (Fig. 4). Pour de petites valeurs de h^2, la courbe de N_r est pratiquement confondue avec celle de N_t.

Pour $h = 10^{m-1}$, la différence est de 1 dB. On a alors :

$$h = 10^{m-1}$$

$$N_r = -11 + 20\,m\;.$$

Pour une dimension double, qui est celle du maximum, on gagne 3 dB seulement, en ayant quadruplé la surface.

Nous admettrons donc, avec J. Robieux qu'il est raisonnable de se limiter à ces 2 valeurs :

$$h = \frac{D}{\lambda} = 10^{m-1} \qquad N_r = -11 + 20\,m$$

soit

$$m = \frac{N_r + 11}{20} \tag{27}$$

$$\frac{D}{\lambda} = 10^{m-1} \tag{28}$$

$$\frac{\varepsilon}{D} = 10^{-m} \tag{29}$$

$$\text{et } \frac{\varepsilon}{\lambda} = 10^{-1} \text{ (indépendant de } N_r). \tag{30}$$

Les équations précédentes déterminent le diamètre D de l'aérien et sa précision de fabrication ε en fonction de son gain N_r (en dB). On en tire le tableau suivant :

Gain N_r (dB)	m	$\dfrac{D}{\lambda} = 10^{m-1}$	$\dfrac{\varepsilon}{D} = 10^{-m}$
29	2	10	10^{-2}
49	3	10^2	10^{-3}
69	4	10^3	10^{-4}

Conclusion

Une fabrication à $\frac{\varepsilon}{D} = 10^{-4}$, associée à un diamètre D de 1.000 λ donnant un gain théorique de 69 dB, paraît être à la limite des possibilités actuelles. Notons que le gain pratique correspondant est $69 - 3 = 66$ dB.

La précision ε englobe non seulement la précision de fabrication, mais aussi les tolérances de déformations mécaniques et thermiques. A ce sujet, l'antenne de guidage de la sonde spatiale soviétique Mechta est un exemple de corrections locales de la déformation du réflecteur (voir ci-dessous). De telles corrections ont d'ailleurs été introduites aussi en optique sur les miroirs de télescopes [7].

IX. Comparaison avec l'optique

Les résultats précédents méritent d'être rapprochés de ceux de l'optique des miroirs de téléscope, avec laquelle ils présentent une remarquable concordance.

La condition qui vient d'être établie (J. ROBIEUX):

$$\frac{\varepsilon}{\lambda} = \frac{1}{10} \tag{30}$$

est très voisine de celle que la règle de Lord RAYLEIGH (GEORGES BRUHAT, [3, § 142]) impose aux miroirs paraboliques de télescopes:

$$\frac{\varepsilon}{\lambda} < \frac{1}{4} \, . \tag{31}$$

Dans la pratique, grâce à la mise en œuvre de remarquables techniques de retouche de la surface, on arrive effectivement à respecter les tolérances de $\lambda/10$.

Citons l'exemple limite du plus grand miroir actuel, celui du grand télescope du mont Palomar (U.S.A.):

$$D = 5 \, m \, \frac{D}{\lambda} = 10^7 = 10^{\,m-1} \qquad m = 8$$

la condition $\frac{\varepsilon}{\lambda} = \frac{1}{10}$ correspond à

$$\frac{\varepsilon}{D} = 10^{-\,m} = 10^{-8} \, .$$

X. Puissance de bruit des aériens
(voir P. MAGNE [5])

Une antenne dirigée vers le cosmos reçoit une certaine puissance de bruit P_b. On appelle température de bruit de l'antenne la température T de la source qui donnerait la même puissance de bruit:

$$P_b = K \, T \, df \tag{32}$$

$K =$ constante de BOLTZMANN $= 1{,}38 \cdot 10^{-23}$ Joule/degré

T en $^\circ K$

$df =$ bande passante dans laquelle on reçoit P_b.

D'autre part, l'amplificateur associé est équivalent à une source P_b' ramenée à son entrée, dont la température serait T', cette valeur T', qui est la température de bruit de l'ampli, est liée à son facteur de bruit par

$$(F - 1) \, T_0 = T' \quad \text{d'où} \quad P_b' = K \, T' \, df \tag{33}$$

($T_0 = 290^\circ \, K$ d'après les standards I.R.E. de définition de F).

La puissance de bruit est $P_b'' = P_b + P_t' = K \, (T + T') \, df$ à l'entrée et $W P_b'' = W K \, (T + T') \, df$ à la sortie

(W étant le gain de l'ampli).

Pour les communications spatiales envisagées, T et T' auraient les ordres de grandeurs suivants:

a) Pour un aérien dirigé vers le ciel, à 1 000 MHz, la température de bruit est de l'ordre de 50° K [6].

$$T \sim 50^\circ \, \text{K}$$

Réalisations d'aériens à grand gain

Source	Type de l'aérien	Dimensions	Gain, largeur de lobe, fréquence	Emplacement	Observations
— Aviation Week 2 Février 1959 (Couverture)	Paraboloïde ouverture circulaire	Diamètre: 26 m Hauteur: 34 m 200 tonnes		Gladstone Camp Irwin Californie (U.S.A.)	Voir ci-dessous Note 1
— Control Engineering, Juin 1958, p. 84 et — Interavia Décembre 1958, p. 1264	— idem —	Diamètre: 18 m	Gain: 30 dB d'où: $f = 230$ MHz $\lambda = 1,3$ m	Cape Canaveral — Floride (U.S.A.) et Kanae, Hawai (U.S.A.)	Aérien de télémesures utilisé avec Pioneer
— Interavia Décembre 1958, p. 1264	Réseau de 4 hélices		Gain: 10 dB à $f = 230$ MHz ?	Kanae Hawai (U.S.A.)	Aérien de télémesures A été remplacé par l'aérien précédent
— idem —	Paraboloïde ouverture circulaire	Diamètre: 115 m		Jodrell Bank — Université de Manchester (Grande Bretagne)	Radioastronomie Télémesures de satellites et astéroïdes
— Révue Française d'Astronautique N° 1	— idem —	Diamètre: 25 m		Stackert — Münstereifel (Allemagne)	Radiotélescope Voir aussi Interavia Octobre 1956
	— idem —	Diamètre: 85 m		Macclesfield (Grande Bretagne)	Radiotélescope
— Révue Française d'Astronautique N° 2	— idem —	Diamètre: 28 m			
— Aviation Week 26 Janvier 1959	— idem —	Diamètre: 26 m	$\lambda = 12$ cm Gain $= 54$ dB $\theta_3 = 0,3°$	U.R.S.S.	Guidage Mechta Voir ci-dessous Note 2

Note 1. On espère assurer avec cet aérien le "tracking" de "sondes spatiales" jusqu'à une distance de 64 Millions de km. en 1960 et de 6,4 Milliards de km. en 1962

Note 2. La distorsion de la surface du réflecteur due aux variations de température était mesurée par thermocouples, l'information envoyée à un calculateur qui commandait des vérins hydrauliques qui corrigeaient les déformations de la surface.

b) Pour un ampli paramétrique ou un maser:

$$T \sim 25^{\circ}\,\mathrm{K}$$

ce qui correspond à

$$F = 1 + \frac{T}{T} = 1 + \frac{25}{290} \# 1{,}09\,.$$

Le bruit en provenance de l'aérien sera donc prépondérant.

On peut espérer le diminuer, car une partie de ce bruit provient non du lobe principal, mais des lobes secondaires et du rayonnement diffus, qui, comme nous l'avons signalé, rayonnent autant d'énergie que le faisceau principal (en adoptant le langage à l'émission) et sont intéressés par des sources terrestres chaudes.

HAL GETTINGS [4] citant E. RECHTIN, signale que des aériens à faible température de bruit (par abaissement des lobes secondaires) sont à l'étude.

Une autre méthode, signalée par le même auteur, consisterait à blinder l'antenne de la terre en la plaçant dans une vallée et en recouvrant le creux de la vallée par une surface métallique réflectrice.

XI. Quelques réalisations d'aériens à grand gain

Nous allons donner maintenant quelques exemples de réalisations. Les renseignements que l'on peut trouver dans la littérature sont assez sommaires: photos, indications des dimensions, quelquefois du gain.

Nous avons rassemblé dans le tableau suivant un certain nombre de renseignements qui peuvent donner une idée des réalisations actuelles.

Bibliographie

1. J. D. KRAUS et CLAUDE WILLIAMSON, Characteristics of Helical Antennas Radiating in the Axial Mode. J. Appl. Physics 19, 87 (1948).
2. J. ROBIEUX, Influence de la précision de fabrication d'une antenne sur ses performances. Annal. Radioéléctricité (1956).
3. G. BRUHAT, Optique. Paris: Masson, 1954.
4. HAL GETTINGS, To the Edge of Interstellar Space. Missiles and Rockets 5, No. 15, 19 (1959). (Analyse d'une étude de EBERHARD RECHTIN intitulée: Communication Technique for Space Exploration.)
5. P. MAGNE, Article sur le bruit à paraître dans le n° 3 ou 4 de la Rév. Franç. Astronaut.
6. Antenna Noise Temperature — P.I.R.E. 4A May (1958).
7. COUDER, Correction des déformations thermiques des miroirs de télescopes. C. R. Acad. Sci., Paris, 231, 1290 (1950); séance 6. 12. 1950.

Some Remarks on the Optimum Operation of a Nuclear Rocket

By

G. Leitmann[1], BIS

(With 2 Figures)

(Received June 26, 1959)

Abstract — Zusammenfassung — Résumé

Some Remarks on the Optimum Operation of a Nuclear Rocket. Recent papers have dealt with performance optimization of nuclear rockets, that is, rockets in which energy source and working fluid are separate. Their results are modified to include the constraint arising from the energy-limited nature of such rockets. The exhaust-speed program leading to maximum characteristic speed is derived and shown to be independent of mass flow rate. The characteristic speed corresponding to the optimum exhaust program is found to be a function of total working fluid mass only, increasing with increasing working fluid mass. Thus, maximum characteristic speed of energy-limited rockets is obtained at lowest feasible power operation.

Einige Bemerkungen über die optimale Wirksamkeit einer Kernrakete. In der letzten Zeit erschienene Arbeiten haben sich mit dem Erreichen der optimalen Ausführung von Kernraketen befaßt, d. h. von Raketen, in denen die Energiequelle und das Arbeitsmedium voneinander getrennt sind. Die Ergebnisse dieser Arbeiten werden abgeändert, um den Zwang, der sich aus der energiebeschränkten Natur solcher Raketen ergibt, zu berücksichtigen. Es wird das Ausströmgeschwindigkeitsprogramm, das zu einer charakteristischen Maximalgeschwindigkeit führt, abgeleitet und gezeigt, daß es unabhängig von der Masserflußrate ist. Die dem optimalen Ausströmprogramm entsprechende charakteristische Geschwindigkeit wird als bloße Funktion der gesamten Masse des Arbeitsmediums erkannt; sie nimmt mit zunehmender Masse des Arbeitsmediums zu. Auf diese Weise wird die charakteristische Maximalgeschwindigkeit von energiebeschränkten Raketen bei niedrigst erreichbarer Triebwerkstätigkeit erhalten.

Quelques remarques sur le fonctionnement optimum des fusées nucléaires. L'optimisation des performances de fusées nucléaires, où les sources d'énergie et le fluide moteur sont séparés, a fait l'objet de travaux récents. Les résultats sont modifiés pour inclure la contrainte provenant des limitations d'énergie dans de telles fusées. La programmation de la vitesse d'éjection conduisant à la vitesse caractéristique maximum est dérivée et démontrée être indépendante du débit massique. La vitesse caractéristique correspondant au programme optimum est uniquement fonction croissante de la masse totale de fluide moteur. Par conséquent la vitesse caractéristique maximum en présence d'une source d'énergie limitée est obtenue avec une puissance opérationnelle aussi faible que réalisable.

---

[1] Associate Professor of Engineering Science, Department of Engineering, University of California, Berkeley, California, U.S.A., and Research Scientist, Lockheed Missiles and Space Division.

List of Symbols

E total available energy
K constant
K_1 specific mass of power plant, m_P/P
K_2 fraction of working fluid assigned to tankage
P power output
V speed of rocket (relative to inertial space)
V_c characteristic speed of rocket (in field-free vacuo)
V_e exhaust speed (relative to rocket)
V_{e0} initial exhaust speed
d m_D/m_0
e_0 E/m_0
e_1 E/m_1
f m_F/m_0
m instantaneous rocket mass

m_0 initial rocket mass
m_1 final rocket mass
m_D mass of payload and structure
m_F total mass of working fluid
m_P mass of power plant (exclusive of energy source)
m_S total mass of energy source (fissionable mass)
m_T mass of tankage
p m_P/m_0
s m_S/m_0
t time
t_b operating time
u $(1-f)^{-1}$
y m/m_0
λ undetermined constant multiplier

I. Introduction

Recently published work has dealt with the derivation of criteria for the optimum operation of rockets in which energy source and working fluid are independent—in contrast to chemically powered rockets in which source of energy and the working fluid are one and the same.

One note [1] was devoted to the determination of the exhaust speed program for an *energy-limited* rocket in order to attain maximum characteristic speed. The effect of constant gravitational acceleration was taken into account. However, the optimum exhaust speed program was obtained under the assumption of *constant mass flow rate* and no optimization with respect to internal mass distribution was made. In this note[1] it will be shown that 1) the optimum exhaust speed program is independent of mass flow rate, and 2) for constant power operation a certain nonconstant flow rate is optimum.

Other recent papers [2, 3] contained a discussion of the optimum distribution between power plant and working fluid masses for nuclear-powered ion rockets operating at constant power output. In the later paper the question of the optimum exhaust speed program was raised but left unanswered. Mass distribution criteria were derived for the case of *constant exhaust speed* and for the case of *constant acceleration*; however, the fact that *nuclear-powered rockets possess a limited amount of available energy* was not introduced. In the present note, in addition to items 1) and 2) above, it will be shown that 3) the optimum exhaust speed program under conditions of constant power output is indeed the one resulting in constant acceleration, and 4) there exists no finite extremum of characteristic speed with respect to internal mass distribution, i.e. between power plant and working fluid masses, *except* in the case of constant exhaust speed and described available energy per unit *final* rocket mass.

The total, initial rocket mass is divided here as follows

$$m_0 = m_S + m_P + m_D + m_F + m_T \tag{1.1}$$

or equivalently

$$1 = s + p + d + f + K_2 f, \tag{1.2}$$

---

[1] It has been brought to the author's attention that some considerations of this note are also contained in a paper by M. P. Blanc, Mém. Artillerie Française (1954).

where it is assumed that the tankage mass increases linearly with working fluid mass.

It is also assumed that the mass of the energy source (fissionable material) is small compared with the remaining rocket mass, i.e.

$$s \ll 1 \tag{1.3}$$

so that the final rocket mass can be approximated by

$$m_1 = m_0 - m_{\text{,}} \tag{1.4}$$

or

$$\frac{m_1}{m_0} = 1 - f. \tag{1.5}$$

II. Constant Exhaust Speed

In recently reported work [2, 3] the optimum distribution of power plant and working fluid masses was derived for the case of constant exhaust speed. In the context of that investigation the criterion of optimization was maximum V_e^2/t_b.

We shall derive the optimum mass distribution which yields maximum characteristic speed for constant exhaust speed—taking into account the energy-limited character of a nuclear rocket.

For constant exhaust speed

$$V_c = -V_e \ln (1 - f). \tag{2.1}$$

The specific power is given by

$$P/m_0 = p/K_1 = -\frac{1}{2} V_e^2 \frac{dy}{dt} \tag{2.2}$$

so that for constant power the mass flow rate is constant and

$$V_e^2 = 2 \frac{p \, t_b}{K_1 f}. \tag{2.3}$$

However,

$$\frac{E}{m_0} = e_0 = \frac{p}{K_1} t_b \tag{2.4}$$

so that

$$V_c = -\sqrt{2 e_0} \, \frac{\ln (1 - f)}{\sqrt{f}}. \tag{2.5}$$

Thus, if the available energy per unit *initial* rocket mass is prescribed, the characteristic speed increases with increasing working fluid mass—and there exists no optimum distribution between power plant and working fluid masses. The characteristic speed possesses a maximum when

$$\frac{1}{2} \ln (1 - f) = -\frac{f}{1 - f}, \tag{2.6}$$

that is $f = 1$ and $V_c = \infty$.

If, on the other hand, the available energy per unit *final* rocket mass is specified, then

$$V_c = \sqrt{2 e_1} \, \frac{\ln u}{\sqrt{u - 1}} \tag{2.7}$$

where

$$\frac{E}{m_1} = e_1 = u \, e_0.$$

In that case there exists an optimum mass distribution, since V_c is maximum when

$$\frac{1}{2} \ln u = \frac{u-1}{u} \tag{2.8}$$

corresponding to

$$u = 4.9,$$
$$f = 0.797 ,$$

and

$$V_c = 0.805 \sqrt{2e_1},$$
$$= 1.59 \, V_e .$$

It should be noted that this result is the same as that obtained previously [3] for the case of maximum V_c^2/t_b, variable working fluid mass, and $K_2 = 0$.

III. Optimum Exhaust Speed

If the exhaust speed can be varied during the period of operation, it is of interest to find the exhaust speed program resulting in maximum characteristic speed. Thus, it is desired to extremize

$$V_c = -\int_1^{1-f} V_e \, y^{-1} \, dy \tag{3.1}$$

subject to the constraint that the specific energy is prescribed, that is

$$e_0 = -\int_1^{1-f} \frac{1}{2} V_e^2 \, dy . \tag{3.2}$$

This is an isoperimetric variational problem [4]. The functional to be extremized is then

$$J = \int_1^{1-f} \left[V_e \, y^{-1} - \frac{1}{2} \lambda V_e^2 \right] dy \tag{3.3}$$

where λ is a constant undetermined multiplier. The resulting EULER-LAGRANGE equation is

$$y \, V_e = \lambda^{-1} = \text{constant.} \tag{3.4}$$

The constant in eq. (3.4) can be found from eq. (3.2) whence

$$y \, V_e = V_{e0} = \sqrt{2e_0 \frac{1-f}{f}} . \tag{3.5}$$

Since

$$V = -\int_1^{y} V_e \, y^{-1} \, dy \tag{3.6}$$

it follows that

$$V_e - V = V_{e0} \tag{3.7}$$

and

$$V_e = \sqrt{2e_0 \frac{f}{1-f}} . \tag{3.8}$$

If the available energy per unit *final* rather than per unit *initial* mass is given

$$V_e = \sqrt{2 e_1 \frac{u-1}{u}}$$

$$= \sqrt{2 e_1 f}.$$

(3.9)

Eqs. (2.5), (3.7), and (3.8) agree with those obtained previously [1] for the case of *constant mass flow rate*. It is clear that the optimum exhaust speed program and the resulting characteristics speed do not depend on the mass flow rate. However, we shall see that for *constant power operation* mass must be expended at a certain non-constant rate.

From eq. (3.8) and (3.9) it follows that if either e_0 or e_1, is specified, V_e depends only on f and increases with increasing f. Hence, *no optimum distribution between power plant and working fluid masses exists in either case*.

The specific power corresponding to the optimum exhaust speed program is

$$\frac{P}{m_0} = \frac{p}{K_1} = - \frac{V_{e0}{}^2}{2 y^2} \frac{dy}{dt}$$

(3.10)

so that for constant power operation the corresponding mass flow program is given by

$$y = \left(1 + \frac{2 e_0}{V_{e0}{}^2 t_b} \, t\right)^{-1}.$$

(3.11)

Furthermore, then the acceleration is also constant and given by

$$\frac{dV}{dt} = \frac{2 e_0}{V_{e0} t_b} = \frac{2 p}{K_1 V_{e0}}.$$

(3.12)

Substitution of eq. (2.4), which applies for constant power operation, in eq. (3.8) yields

$$V_e = \sqrt{\frac{2 p \, t_b}{K_1} \frac{f}{1-f}},$$

(3.13)

which is precisely the result obtained recently under the assumption of constant acceleration [3]. Thus the condition of constant acceleration, used previously [3], as an example of a possible program, is in fact the one resulting from the employment of optimum exhaust speed program.

However, it is clear now that once the available energy is prescribed and the optimum exhaust speed program is used, the characteristic speed can be increased only by increasing the working fluid mass at the expense of power plant mass (payload, structure, energy source mass, and initial mass being fixed).

IV. Alternate Derivation of Optimum Exhaust Speed

The results of III can also be deduced from the following arguments. From the expression for specific power one has

$$V_e{}^2 = \frac{2 p}{K_1} \left(-\frac{dy}{dt}\right)^{-1}$$

(4.1)

so that for constant power operation

$$V_e = \sqrt{\frac{2 p}{K_1}} \int_0^{t_b} \left(-\frac{dy}{dt}\right)^{1/2} y^{-1} \, dt.$$

(4.2)

Thus, it is desired to maximize

$$I = \int_0^{t_b} \left(-\frac{dy}{dt}\right)^{1/2} y^{-1} \, dt . \tag{4.3}$$

For energy-limited operation t_b is fixed for given level of power. The EULER-LAGRANGE equation is then

$$0 = \frac{d}{dt} [y \sqrt{-dy/dt}]^{-1} - 2y^{-2} \sqrt{-dy/dt} . \tag{4.4}$$

Eq. (4.4) possesses a first integral

$$y^{-2} \, dy/dt = K = \text{constant}. \tag{4.5}$$

To evaluate K eq. (4.5) can be integrated to yield

$$K t_b = -\frac{f}{1-f} . \tag{4.6}$$

Upon combining eq. (4.1), (4.5), (4.6) and

$$\frac{dV}{dt} = -V_e y^{-1} \, dy/dt \tag{4.7}$$

one finds

$$\frac{dV}{dt} = \sqrt{\frac{2p}{K_1 t_b} \frac{f}{1-f}} = \text{constant} \tag{4.8}$$

and thence

$$V_c = \sqrt{\frac{2p \, t_b}{K_1} \frac{f}{1-f}} . \tag{4.9}$$

Thus, we have again the results of III.

V. Maximum Acceleration

In the preceeding sections it was found that the optimum exhaust speed program, i.e. the one yielding maximum characteristic speed per unit energy converted, leads to constant acceleration. One may now ask for that mass distribution which yields maximum constant acceleration — or equivalently — minimum operating time for prescribed characteristic speed. Whereas before energy had to be considered as a given and the corresponding maximum attainable speed as a derived quantity, this is not so in the case of optimizing mass distribution[1]. In a real situation the requirements of the mission will determine the characteristic speed and, whatever program is used, sufficient energy will have to be supplied. In that sense as much energy as required is available and energy cannot be considered as an a priori fixed quantity.

From eq. (4.9)

$$t_b = \frac{1}{2} K_1 V_c^2 \frac{1-f}{fp} . \tag{5.1}$$

If initial mass, payload, and structure are prescribed

$$p + (1 + K_2) f = \text{constant} = 1 - d . \tag{5.2}$$

It is desired to find that distribution between p and f, for given V_c, which leads to minimum t_b. That is, t_b as given by eq. (5.1) is to be minimized as a function of p and f subject to eq. (5.2). As previously shown [2] the resulting relation is

[1] Clarifying discussion of this point with Dr. PRESTON-THOMAS is gratefully acknowledged.

$$p = f(1-f)(1+K_2).$$ (5.3)

The minimum operating time is then

$$t_b = \frac{K_1 V_c^2}{2(1+K_2)f^2}$$ (5.4)

where

$$f = 1 - \sqrt{1 - \frac{1-d}{1+K_2}}.$$ (5.5)

VI. Conclusions

For the sake of clarity the main results of the foregoing sections are reiterated here:

For $V_e = $ constant, the maximum attainable characteristic speed is

$$V_c = -\sqrt{2e_0}\,\frac{\ln(1-f)}{\sqrt{f}}$$

or

$$V_c = 0.805\sqrt{2e_1}.$$

For variable V_e, according to eq. (3.5), the maximum attainable characteristic speed is

$$V_c = \sqrt{2e_0\,\frac{f}{1-f}}$$

or

$$V_c = \sqrt{2e_1 f}.$$

Fig. 1 shows a plot of $V_c/\sqrt{2e_0}$ as a function of f for both V_e constant and V_e variable. It is easily seen that little advantage derives from using the optimum variable exhaust speed program rather than the best constant exhaust speed. However, the advantage of the variable exhaust speed program becomes more pronounced

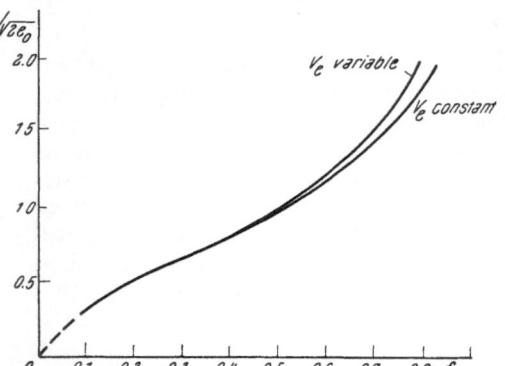

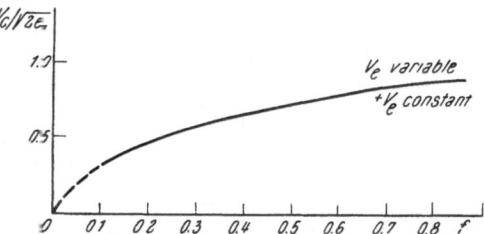

Fig. 1. Characteristic speed—energy ratio, initial mass prescribed

Fig. 2. Characteristic speed—energy ratio, final mass prescribed

as f increases, i.e., a larger and larger part of the rocket mass is allotted to working fluid.

Fig. 2 shows a graph of $V_c/\sqrt{2e_1}$, as a function of f. Of course, here $V_c/\sqrt{2e_1}$ for V_e constant has one value only, corresponding to $f = 0.797$. Again, the variable exhaust speed program becomes greatly preferable only for values of f approaching unity.

The following general conclusions may be drawn concerning the optimum operation of energy-limited rockets with separate energy source and working fluid:

a) The optimum exhaust speed program, that is, the one leading to maximum characteristic speed, does not depend on the mass flow rate. It is such that the product of exhaust speed and instantaneous rocket mass remains constant, or to put it differently, the difference of exhaust speed and rocket speed, i.e., the exhaust speed relative to inertial space, stays constant.

b) The characteristic speed resulting from operation with the optimum exhaust speed program depends only on the total mass of the working fluid and increases with it. For specified payload, structure, energy source, and initial mass an increase in working fluid mass leads to a decrease in power plant mass, and hence a decrease in operating power, with a corresponding increase in operating time. It is reasonable to expect that a point will be reached when decrease in power no longer results in decrease of power plant mass. Past that point the characteristic speed cannot be increased within the context of the assumptions made. Of course, other considerations such as limitations on the operating time may dictate operation at a lower characteristic speed—energy ratio.

References

1. R. H. OLDS, Jet Propulsion 28, 405 (1958).
2. H. PRESTON-THOMAS, J. Brit. Interplan. Soc. 16, 508 (1958).
3. H. PRESTON-THOMAS, Realities of Space Travel, edited by L. J. CARTER, p. 174. New York: McGraw-Hill, 1957.
4. R. COURANT and D. HILBERT, Methoden der mathematischen Physik. Berlin: Springer, 1931.

Note

After this paper had been submitted for publication two contributions dealing with the problem under study appeared. The first [5] of these concerns the determination of the mass ratio u for constant V_e, which maximizes the "efficiency"

$$\frac{\frac{1}{2} m_1 V_c^2}{E}$$

i.e., the ratio of the kinetic energy provided to the burn-out mass over the energy demanded from the energy source. It is seen that such a requirement is equivalent to finding that value of u which maximizes V_c for given e_1. The results are the same, of course.

The second [6] of these paper deals with the details of the analysis which led to a previously published paper [2]. As mentioned in Section II the results, for V_e constant, of maximizing V_c for given e_1, are the same as those deduced previously [2] for maximizing V_c^2/t_b and $K_2 = 0$, but e_1, unconstrained and only m_F variable. This is obviously so since, for $K_2 = 0$

$$V_c^2/t_b = \frac{2 m_P}{K_1 m_F} \ln^2 \frac{m_F + m_P + m_D}{m_P + m_D}$$

$$= \frac{2 m_P}{K_1 (m_P + m_D)} \frac{\ln^2 u}{u - 1} ,$$

or

$$V_c^2 = 2 e_1, \ln^2 u/u - 1 .$$

Added References

5. R. R. NEWTON, Jet Propulsion 28, 752 (1958).
6. H. PRESTON-THOMAS, J. Brit. Interplan. Soc. 16, 575 (1958).

Les limites de la survie en cabine étanche au cours des voyages interplanétaires

F. Violette[1], H. Boiteau[1] et S. Bernard[1]

(Reçu le 8 juin 1959)

Résumé — Zusammenfassung — Abstract

Les limites de la survie en cabine étanche au cours des voyages interplanétaires. L'homme est un hétérotrophe obligatoire, et notre survie est étroitement solidaire du milieu qui nous entoure. Un voyage de l'homme dans l'espace n'est concevable que si certains éléments constitutifs de ce milieu l'accompagnent. L'épuisement des éléments indispensables entraine la cessation de la survie.

Pour les longs voyages, il est nécessaire de chercher à récupérer les éléments indispensables et à les intégrer dans un système cyclique où ils resserviraient indéfiniment. Ce problème peut être considéré comme résolu dans le cas de la fourniture d'oxygène et de l'élimination du CO_2. Mais la réutilisation de l'azote, du phosphore, du soufre et des métaux indispensables sont autant de problèmes qui restent encore à résoudre.

Dans l'état actuel de nos connaissances, on peut conclure que l'homme peut survivre en cabine étanche en symbiose avec un organisme végétal pendant un assez long temps. La limite extrême d'une telle symbiose sera inéluctablement fixée par l'épuisement des stocks de matière organique azotée, soufrée et phosphorée indispensable et dont une nouvelle synthèse ne pourra être réalisée.

Die Grenzen der Überlebensfähigkeit in einer abgeschlossenen Kabine bei interplanetarischen Reisen. Der Mensch ist ein zwangsläufig heterotropher Organismus und unsere Überlebensfähigkeit ist direkt abhängig von dem Milieu, das uns umgibt. Eine Reise des Menschen in den Raum ist nur denkbar, wenn ihn bestimmte grundlegende Elemente dieser Umgebung begleiten. Die Erschöpfung der unerläßlichen Grundstoffe hat das Aufhören der Lebensfähigkeit zur Folge.

Auf Reisen von langer Dauer ist es notwendig zu versuchen, die unentbehrlichen Grundstoffe zurückzugewinnen und sie in ein zyklisches System einzufügen, wo sie unbegrenzt neu verwendet werden können. Dieses Problem kann im Fall der Sauerstoffanlieferung und der Beseitigung des Kohlendioxyds als gelöst betrachtet werden. Hingegen sind die Wiedergewinnung des Stickstoffes, Phosphors, Schwefels und der unentbehrlichen Metalle ebensosehr Probleme, die noch zu lösen sind.

Nach dem gegenwärtigen Stand unserer Kenntnisse kann man schließen, daß der Mensch in einer abgeschlossenen Kabine in Symbiose mit einem pflanzlichen Organismus während ziemlich langer Zeit überleben kann. Die äußerste Grenze einer solchen Symbiose wird durch die Erschöpfung der Bestände an stickstoff-, schwefel- und phosphorhaltiger organischer Substanz unabwendbar bestimmt, auf die nicht verzichtet werden kann, zumal deren Neuherstellung nicht verwirklicht werden kann.

Survival Limits in Sealed Cabins during Interplanetary Travel. Man is a forced heterotroph, and our survival is rigidly influenced by the medium that surrounds us.

---

[1] Centre d'Enseignement et de Recherches de Médecine Aéronautique, 5 bis, Avenue de la Porte de Sèvres, Paris 15e, France.

A manned voyage into space is only conceivable if certain elements characteristic of this medium accompany him. The exhaustion of the indispensable elements leads to his demise.

For long trips it is necessary to look towards the recuperation of the indispensable elements and to integrate them into a cyclic system where they would be used indefinitely. This problem can be considered resolved in the case of furnishing of oxygen and the elimination of carbon dioxide. But the regeneration of nitrogen, of phosphorus, of sulfur, and of the indispensible metals represent problems that remain to be resolved.

In the present state of our knowledge one can conclude that man can survive in a sealed cabin in symbiosis with a vegetable organism during a fairly long period of time. The extreme limit of such a symbiosis will be inevitably fixed by the emptying of indispensable stocks of nitrogenized, sulfurized and phosphorized organic material and of which a renewed synthesis will not be realisable.

On sait qu'il est nécessaire de distinguer deux sortes de voyages en Astronautique suivant que leur durée sera courte (quelques jours) ou prolongée (au-delà d'une ou deux semaines).

Les premiers représentent une brève incursion dans l'Espace. On emporte à bord du véhicule, des rations de survie compatibles avec l'activité physique et intellectuelle qui sera imposée. Une marge de sécurité sera prévue pour que le retour s'effectue avant l'épuisement des réserves nutritives. Le problème consiste à établir le minimum indispensable et à le conditionner sous la forme la plus légère et la moins encombrante possible. Dans ce cadre peuvent entrer les opérations à bord d'un satellite terrestre, une expédition autour de la Lune et à la rigueur un séjour pas trop long sur cette planète.

Tout à fait différents sont les problèmes qui se posent lorsqu'on cherche à dépasser l'orbite lunaire. On doit alors tenter de créer un cycle fermé de transformations reproduisant les conditions de Survie Terrestre. Les produits de déchet du métabolisme humain seront repris par d'autres organismes qui les transformeront en substances nutritives à nouveau assimilables.

Tout revient donc à réaliser la transposition du cycle de survie terrestre à une portion d'espace aussi restreinte qu'une cabine de quelques mètres cubes. Ceci nous amène à étudier le déroulement du cycle de Survie sur la Terre, à envisager les rapports que l'homme contracte avec le milieu où il vit et les autres êtres vivants avec lesquels il cohabite. C'est un vaste problème de Biologie Générale.

Les auteurs américains Brockmann [5], Myers [29] et Taylor [40] nous annoncent comme bientôt réalisable le programme de cycles écologiques fermés — que convient-il d'en penser ? C'est ce que nous allons maintenant tenter d'examiner avec quelque détail, nous attachant particulièrement à l'aspect biologique général des diverses questions à résoudre.

L'ensemble des différents problèmes peut se résumer ainsi: on emporte à bord des quantités limitées de tous les facteurs indispensables à la survie de l'espèce humaine. On dispose d'une source d'énergie suffisante. En prenant exemple sur les phénomènes naturels observés, dont on pense qu'ils seront reproductibles et contrôlables, on cherche à mettre au point des systèmes écologiques fermés, tels que les aliments métabolisés par l'homme normal, après épuisement, soient repris par d'autres organismes qui les ramènent à leur point de départ, sous une forme à nouveau utilisable par l'être humain, en vue de sa survie. Le cycle se perpétuerait identique à lui-même, jusqu'à la fin du voyage, des mois ou des années plus tard.

Précisons tout de suite qu'à notre avis, un pareil programme relève, à l'heure présente, de la pure utopie. Un bref rappel des principaux facteurs qui participent au maintien du système de survie sur notre planète, suffira pour nous convaincre:

l'ambition qui consiste à les vouloir reproduire en vase clos de petites dimensions, en escomptant des résultats analogues, dépasse de très loin, et beaucoup plus encore, nos capacités techniques actuelles.

A. La Terre, vaisseau spatial

Il est permis de considérer notre planète, à l'échelle astronomique, comme un gigantesque vaisseau spatial en mouvement sur son propre orbite, et satellite du Soleil, à la vitesse de 30 Kms par seconde — ce qui représente trois fois et demi la vitesse du satellite terrestre. Un champ de gravitation maintient dans le proche environnement de ce vaisseau, tous les éléments indispensables à la survie des êtres vivants: ce sont le carbone, l'hydrogène et l'oxygène, ces deux derniers intervenant très souvent sous forme d'eau, chargée en électrolytes divers, milieu naturel et obligatoire des réactions du métabolisme. L'azote, le phosphore et le soufre ne sont pas moins essentiels, encore que moins importants quantitativement. Enfin, il faut ajouter à cette liste, certains métaux à doses catalytiques; au premier rang, le fer et le magnésium, d'autres encores dont nous reparlerons.

Ces éléments critiques — qu'un seul vienne à faire défaut, et la vie s'arrêterait — sont en perpétuel remaniement, et font partie intégrante de cycles écologiques fermés.

Au cours de ces cycles, les atomes font tout d'abord partie de molécules inorganiques. Des mécanismes variés — c'est tout le problème de l'autotrophie, sous son double aspect phototrophe et chimiotrophe — incorporent les atomes dans des molécules de matière vivante, d'un niveau énergétique plus élevé. L'énergie indispensable est prélevée, en dernière analyse, aux dépens de la lumière solaire. Sitôt édifiée, la substance vivante se dégrade, au cours des multiples chaines métaboliques dont les grandes lignes nous sont connues, et dans lesquelles les phénomènes d'oxydo-réduction jouent un rôle de premier plan; ainsi l'énergie accumulée se libère progressivement, permettant la survie temporaire des organismes intéressés. Mais les déchets du métabolisme font retour au monde minéral, où ils sont repris par les autotrophes en vue d'un nouveau cycle. Ainsi, se crée dans une zone privilégiée — notre biosphère — un système fermé de transformations qui comporte, à un certain stade, un état transitoire compatible avec notre survie. Au delà de la biosphère s'étend l'Espace, que l'homme se propose d'explorer.

Cette image — fort prétentieuse et d'un goût discutable — est devenue classique: la Terre, ramenée aux dimensions d'un navire de l'Espace, qui n'existe encore que dans notre imagination, et qui serait construit de main d'homme. Si nous reprenons à notre compte cette comparaison, c'est qu'elle paraît utile pour amorcer la critique de certains travaux, qui envisagent avec un optimisme à notre avis exagéré, la réalisation prochaine des grands voyages interplanétaires. Est-il raisonnable, dans l'état actuel de nos connaissances, de prétendre créer de toutes pièces — sans sortir des étroites limites de poids et de volume que nous imposent les techniciens de l'astronautique — un monde en miniature qui reproduise intégralement toutes les conditions réalisées par notre biosphère, assurant la survie de notre espèce pour une durée quasi-illimitée.

Certains auteurs américains, l'enthousiasme aidant, sont très affirmatifs: "La possibilité technique de réaliser un tel cycle qui régénère les éléments, n'est pas en cause: il existe bien dans la nature. Point n'est besoin de recherche, au sens strict du terme; il suffit d'améliorer les techniques, en les orientant vers les procédés qui encombrent et pèsent le minimum, tout en assurant des résultats satisfaisants." (Taylor [40].)

Nous serons beaucoup moins formels, et chercherons à serrer de près le problème pour en mesurer l'ampleur et les difficultés. Nous voudrions, dans ce chapitre, évoquer la complexité des phénomènes qui interfèrent dans la biosphère, pour maintenir la permanence de notre survie. Ensuite confronter ces exigences, si nombreuses et variées, avec les données expérimentales actuellement acquises, qui sont encore éparses et disparates. Enfin, faire la critique du système proposé, et tenter, si possible, de tracer la limite entre des espoirs raisonnablement permis, et ce qui relève de la pure imagination.

B. Rappel des données actuelles sur le métabolisme des éléments indispensables à la survie de l'Organisme animal, au sein de la biosphère terrestre

I. Présentation de la biosphère

Le milieu naturel dans lequel se déroulent les phénomènes biologiques s'étend en altitude jusqu'à 20.000 mètres environ. A ce niveau, déjà stratosphérique, s'étend une couche diffuse d'ozone, absolument indispensable à toute survie, car elle filtre le rayonnement solaire, et retient la majeure partie des radiations ultra-violettes, qui, sans elle, exerceraient sur les organismes une action léthale. Cet ozone provient de l'action des rayons U.V. de courtes longueurs d'onde sur l'oxygène.

Au dessous, c'est la biosphère, où la vie peut s'exercer, au sein d'un milieu inorganique à qui elle empruntera sa matière; celle-ci se présente sous trois états physiques différents: l'un est solide, la lithosphère, c'est l'écorce terrestre sur une dizaine de mètres d'épaisseur; l'autre est surtout liquide (il comporte aussi les neiges et les glaces), c'est l'hydrosphère, dont la masse de l'eau de mer représente environ 98%. Le troisième est gazeux, c'est l'atmosphère elle même, siège de courants de convection qui lui assurent une composition moyenne à peu près stable. Il va sans dire que ces trois portions de la biosphère réalisent entre elles des échanges continuels qui ne sont pas sans influencer grandement les modalités de la survie des organismes qui ont élu domicile dans chacune d'elles.

II. Les organismes en présence

Tel est le décor de la biosphère. Avant d'aborder l'étude des cycles biologiques, il convient de présenter les personnages appelés à y jouer un rôle, animaux, végétaux et micro-organismes. Ce dernier terme, assez vague, et qui recouvre encore quantité d'inconnues, concerne principalement le monde des Bactéries. Là encore, impossible de dissocier ces trois aspects des manifestations vitales, dans une étude d'ensemble de la biosphère. Ils sont solidaires, à un point tel, que la disparition de quelques espèces entrainerait rapidement la cessation de la survie des autres. Ils diffèrent entre eux, tant du point de vue de leur anatomie que de leurs aptitudes. Chacun joue son rôle propre, et constitue un maillon plus ou moins important du cycle fermé de transformations de la matière qui assure la survie de l'ensemble. C'est ainsi que certains seront spécialisés dans la décomposition de la substance organique après sa mort, ramenant ses éléments constitutifs à l'état minéral. D'autres seront chargés — dans certains cas seulement — de transformer ces produits résultant de l'activité des premiers, pour les faire aboutir à un nouvel état, toujours inorganique, mais utilisable par les suivants. D'autres encore — ce sont les producteurs autotrophes — seront capables de réintégrer cette matière minérale au sein de molécules vivantes, et le cycle se ferme.

D'autres enfin (ce sont tous les hétérotrophes, et l'homme est de ceux là) n'apportent au cycle de survie qu'une bien médiocre contribution. Certes, ils dégradent certaines substances organiques, que les autotrophes obligatoires ne peuvent assimiler sans minéralisation préalable. Mais ils sont essentiellement les consommateurs purs et simples de la matière vivante élaborée par ces derniers. Est-il besoin de préciser que la survie des hétérotrophes, dont les exigences sont maximales, est totalement tributaire de l'activité biologique de tous les précédents.

III. Étude des cycles particuliers des éléments fondamentaux

Cette vue très générale du cycle de survie dans la biosphère, mérite d'être analysée plus intimement; il convient de reprendre chacun des éléments fondamentaux cités ci-dessus, qui participent à l'édification et au métabolisme de la matière vivante. Nous apprécierons la part respective qui revient à chacun, tant du point de vue structural que fonctionnel, avant de décrire schématiquement les grandes lignes du cycle qui l'amène à traverser le monde vivant, avant de faire retour à une forme inorganique.

1. Le carbone

Il est bien inutile d'insister sur l'importance du carbone. Cet élément constitue l'unité structurale de base: l'enchainement des atomes de carbone les uns aux autres forme le squelette fondamental, à partir duquel s'organisent, autour des valences restées libres, les divers groupements fonctionnels qui confèrent son originalité à la molécule de matière vivante.

La présence du 0,03% de CO_2 dans l'atmosphère terrestre suffit à entretenir la permanence du cycle. Sous cette forme, oxydée au maximum, le carbone est prélevé, puis réduit; enfin il réapparait sous forme de glucides qui s'incorporent immédiatement au métabolisme de la cellule responsable de ces transformations. Ainsi prennent naissance les chaines carbonées biologiques. La description de ce phénomène, abolument fondamental, est connue depuis longtemps; mais son analyse biochimique, à coup sûr fort complexe, est loin d'être achevée, et soulève encore maintes discussions dans lesquelles nous n'entrerons pas: de notre point de vue, il suffit de préciser que le phénomène se présente sous deux aspects:

— le premier, de beaucoup le plus important, utilise l'énergie électromagnétique de la lumière solaire [26]; l'hydrogène de l'eau comme agent réducteur, et les chlorophylles comme médiateurs chimiques; c'est la photosynthèse des végétaux verts et des Algues vertes.

— le second, de rendement beaucoup moindre, n'est pas directement tributaire de la lumière solaire; il couple la réduction de CO_2 qui nécessite de l'énergie, avec une réaction chimique exergonique, en général une oxydation; c'est la chimiosynthèse, apanage de certaines bactéries.

Déjà, au cours de la vie des organismes, du CO_2 est rejeté par les cellules, vers le monde minéral; c'est le terme ultime du métabolisme aérobie le produit de déchet de la respiration, par l'intermédiaire du cycle tricarboxylique.

Mais c'est à leur mort que les chaines carbonées constitutives de toutes les structures protoplasmiques, vont subir la minéralisation bactérienne, et reviendront ainsi à leur point de départ: le CO_2. Cette oxydation s'opère à des vitesses variables, et peut comporter nombre de produits intermédiaires, suivant la molécule considérée; si elle est simple et rapide pour les sucres solubles, elle sera lente et fort compliquée pour la celullose des végétaux. Ainsi le CO_2 rendu à l'atmosphère, équilibrera quantitativement celui que la photosynthèse surtout, incorpore au monde biologique.

2. L'azote

Pas plus que pour le carbone, il n'est nécessaire d'insister sur l'importance de l'azote, élément constitutif obligatoire des protéines, lesquelles — comme leur nom l'indique — représentent le matériel premier de toute structure protoplasmique. L'azote y apparait sous forme de liaison peptidique ou de groupement aminé.

Le cycle est facile à décrire, encore que, là aussi, certains mécanismes biochimiques demeurent obscurs. Une grande réserve naturelle: l'azote moléculaire atmosphérique. Seuls dans toute la biosphère, certains micro-organismes du sol sont capables de l'utiliser en vue d'une synthèse protéique. Ce sont des Algues bleu-vert, et surtout des bactéries qui peuvent être libres (aérobies ou non, parfois photosynthétiques) ou symbiotiques, telles les Rhizobium associés aux Légumineuses: dans ce dernier cas, il est certain que la plante-hôte contribue à l'élaboration d'un ou plusieurs métabolites indispensables à la fixation par la bactérie de l'azote moléculaire; leur nature exacte reste à découvrir.

Cet azote protéique, qu'il soit d'origine bactérienne, animale ou végétale, subit une minéralisation dont le terme ultime est l'ammoniaque: c'est l'œuvre de micro-organismes dits ammonifiants, beaucoup plus variés que les précédents.

Troisième chainon: l'oxydation de l'azote ammoniacal, et la formation de nitrites, puis de nitrates, par action de bactéries autotrophes et chimiosynthétiques en ce sens que l'énergie libérée au cours de la réaction oxydative, leur sert pour réduire le CO_2 et l'incorporer dans leur structure. Les nitrates sont assimilables par le système radiculaire des Végétaux, où ils retournent à l'état protéique: c'est par l'intermédiaire des plantes que l'azote bactérien atteint enfin le monde animal.

Puis d'autres micro-organismes se chargent de réduire les nitrates: c'est la dénitrification, qui régénère l'azote moléculaire et ferme ainsi le cycle.

Le siège de toutes ces transformations se situe préférentiellement dans le sol; l'océan, pauvre en organismes fixateurs d'azote, reçoit ce dernier surtout par l'eau de ses affluents.

3. Le phosphore

Non moins fondamental que les précédents, le phosphore est essentiel à toutes les manifestations du dynamisme vital. Il entre dans la constitution des phospholipides et des acides nucléiques; comme tel, il préside à toute division cellulaire et à la transmission du patrimoine génétique. Plus généralement, il s'avère indispensable à la mise en réserve de l'énergie nécessaire à la survie de toute cellule vivante; les phénomènes respiratoires et fermentaires, tout au long des multiples chaines métaboliques où abondent les phosphorylations, ont pour but principal le stockage des liaisons riches en énergie, sous la forme de cet accumulateur universellement répandu; l'adénosine-triphosphate. Ainsi, grâce à l'intervention du phosphore, l'énergie globale contenue dans les aliments se trouve-t-elle fractionnée en petites portions, immédiatement disponibles à la demande, suivant les besoins variables du métabolisme cellulaire.

Ce sont encore les micro-organismes qui assurent la transformation du phosphore minéral en orthophosphates assimilables par les végétaux: le mécanisme est mal connu, on n'a pas déterminé de bactéries spécifiques. Un problème consiste à solubiliser les phosphates insolubles; les carbonates jouent un rôle essentiel.

Lors de sa traversée du monde vivant, le phosphore restera sous sa forme la plus oxydée et la plus hydratée: les orthophosphates, sous forme de sels ou d'esters. Lors de la minéralisation de la matière organique en général, les ions PO_4 seront libérés par l'entremise une fois de plus des micro-organismes.

Contrairement à ce qui se passait pour l'azote, c'est l'océan qui s'enrichit en phosphore, et ce au détriment de la lithosphère. C'est dire l'importance du retour à la terre des phosphates océaniques, le rôle capital joué par les oiseaux de mer, les pêcheries, qui nous ramènent cet élément fondamental dont les réserves terrestres vont en s'appauvrissant.

4. Le soufre

Il intervient parfois dans le métabolisme sous une forme très oxydée, le radical sulfate SO_4, notamment pour solubiliser certains produits de déchet, en vue de leur détoxication. Mais surtout il se rencontre à l'état plus réduit, dans des vitamines (thiamine, biotine), et dans plusieurs acides aminés indispensables couramment répandus dans les protéines: méthionine, cystéine et cystine. Des ponts disulfures relient très souvent entre elles les chaines polypeptidiques; quant au radical thiol, on connait son extrème importance en tant que groupe actif des protéines enzymatiques; il n'est que de se souvenir de l'aspect clinique si bruyant et trop souvent fatal, de l'intoxication par l'arsenic: c'est l'expression du blocage des radicaux — S.H., peu de temps compatible avec la survie, si on ne fournit pas très vite de nouveaux thiols actifs, sous forme de dimercaptopropanol (B.A.L.).

Par bien des côtés, le cycle du soufre présente des analogies avec celui de l'azote. Toutefois, les stades en sont moins bien délimités, il est certainement plus complexe. Les micro-organismes revendiquent encore une place prépondérante: ils minéralisent le soufre organique des débris végétaux et animaux, en formant des sulfates directement assimilables par les racines des plantes supérieures, par le canal desquelles les produits soufrés, ramenés à l'état organique, rejoignent le monde animal.

Carbone, azote, phosphore et soufre, en milieu aqueux, sous des formes plus ou moins oxydées ou réduites, tels sont les éléments de base qui viennent du monde inorganique, et traversent la biosphère avant d'y faire retour. Il en est beaucoup d'autres. Mais du point de vue qui nous préoccupe, il importe de faire certaines remarques, concernant le cycle de survie dans son ensemble.

IV. Quelques réflexions sur ces cycles: mise en évidence de l'extrême complexité du phénomène global

1. Les interférences entre les cycles

Nous venons de décrire comment se comportait chacun des éléments fondamentaux dont l'assemblage, organisé en macromolécules, constitue la structure des protoplasmes. C'est dire à quel point les cycles particuliers vont interférer sans cesse: ce n'est que pour la commodité de l'exposé qu'il est classique de les séparer. Telle bactérie, métabolisant le soufre oxydera SH_2 en S, et utilisera à la fois l'énergie et l'hydrogène dégagés pour réduire CO_2 et incorporer le carbone à ses glucides. Telle autre, pour faire de même, empruntera l'énergie à l'oxydation des nitrites en nitrates; autant d'exemples — que l'on pourrait multiplier — des interrelations constantes entre le cycle du carbone, celui de l'azote, celui du soufre, etc...

2. La place de l'homme dans le cycle de survie

Tous les êtres vivants participent aux transformations subies par les éléments fondamentaux, mais nous avons déjà précisé que la part prise par chaque type d'organisme était très variable: l'homme n'est qu'un consommateur pas du tout indispensable au bon fonctionnement de l'ensemble (son absence au cours des ères géologiques précédant le quaternaire, n'empêchait pas la vie, surtout végétale, d'être florissante). Son action cependant est loin d'être négligeable: l'agronomie consiste à favoriser au maximum l'activité des autres organismes, en vue d'obtenir le meilleur rendement en aliments consommables par le monde animal.

3. La place des micro-organismes

A l'opposé, le monde des micro-organismes — Bactéries, Algues, Levures, Champignons inférieurs, moisissures — toutes formes élémentaires de la vie sur le plan structural, mais doué d'étonnantes aptitudes, dont nous saisissons les effets — les effets seulement, car l'analyse biochimique des mécanismes est à peine entamée. Ce monde encore fort mystérieux si on le considère globalement, assume par ses seules activités la majeure partie du cycle général de survie: il n'est, pour s'en convaincre, que de jeter un coup d'œil sur n'importe lequel des cycles particuliers décrits ci-dessus. A lui seul il survivrait, alors que les animaux supérieurs et l'homme dépendent étroitement de lui.

Pour qui cherche à se faire une idée du mécanisme général de survie de l'homme sur sa planète, dans le but — ne l'oublions pas — de reproduire ce cycle en miniature, les micro-organismes occupent une place cruciale, dont il sera impossible de les frustrer, à bord du vaisseau spatial. Sommes nous donc si bien renseignés sur leurs milieux de culture, sur leurs exigences métaboliques, sur les facteurs externes et internes qui conditionnent leurs aptitudes, dont notre propre survie dépend.

Un des lieux de prédilection de l'activité des micro-organismes est la lithosphère. Dans un très récent Traité de Microbiologie du sol (Paris: Dunod, 1958), J. POCHON et H. DE BARJAC, de l'Institut Pasteur, cherchent à mesurer l'importance considérable, pour l'équilibre des cycles biologiques, des interrelations entre les divers organismes vivant dans le sol. Les auteurs insistent sur le fait qu'une colonie bactérienne, étudiée isolément au laboratoire, aura un comportement entièrement différent de celui de la même colonie, considérée au sein de son milieu naturel, et conservant ses rapports normaux avec les autres espèces vivantes qui l'environnent. Dans presque tous les cas, nous l'avons vu, pour parvenir à l'homme à l'état d'aliments assimilables, les produits biologiques de l'activité des micro-organismes doivent passer par l'intermediaire des Végétaux: autour des racines de ces derniers existe une zone baptisée rhizosphère, au sein de laquelle pullulent des bactéries dont l'action, couplée avec celle de la plante, contribue à créer un certain état biochimique favorable aux échanges. Il va sans dire que le degré d'hydratation, les variations du pH, l'équilibre d'oxydo-réduction, et bien d'autres facteurs, influent grandement sur cet état.

On ne peut en dire plus: l'étape biochimique n'est pas franchie en ce qui concerne ces phénomènes. On connait certes, l'énuméré de nombreux facteurs: la carence en un seul suffit à assurer le déséquilibre du système dans son ensemble. Outre le fer et le magnésium — dont le rôle est connu et dont l'importance vitale n'est plus à souligner — le sodium et le potassium, le manganèse, le bore, le molybdène, le cuivre et le zinc, sans oublier le cobalt, au centre de la molécule de vitamine B_{12}. Certains servent à équilibrer la balance ionique; d'autres de plus, intervenant à très petites doses, mais rigoureusement indispensables — ce

sont les oligoéléments — font partie intégrante, d'une manière spécifique, des
protéines enzymatiques, où ils sont impliqués dans les liaisons chélates.

Au sein des micro-organismes, une place tout à fait à part (et qui aura dans
notre discussion l'importance que l'on devine) doit être réservée aux bactéries
pathogènes pour l'homme et les animaux supérieurs. Nous trouvons là une illustration excellente de l'intrication extrême des rapports entre organismes vivants:
dans le sol, par exemple, des équilibres vont se créer entre espèces microbiennes
différentes, équilibres qui résultent d'un double courant d'activités biologiques,
l'un synergique et l'autre antagoniste. Ce que nous saisissons à l'occasion du
travail expérimental, n'est que la résultante de toutes ces actions biochimiques
dont l'existence est prouvée mais l'analyse encore très fragmentaire. Nous savons
toutefois, à propos de l'antagonisme qui oppose entre elles les populations microbiennes, qu'un facteur essentiel intervient (outre la compétition nutritive); c'est
la production de substances inhibitrices ou antibiotiques. Certaines de ces substances ont pu être isolées et on sait quelle extraordinaire fortune fut la leur
en médecine clinique. Mais si nous sommes capables de les utiliser en vue de la
destruction élective de tel ou tel pathogène, nous ne pouvons encore préciser la
part exacte que prennent les antibiotiques dans l'équilibre dynamique qui se
crée avec d'autres substances (encore inconnues), pour réaliser l'état global d'une
population de micro-organismes telle que nous l'observons. Cet équilibre chimique,
il faudrait pouvoir en faire l'étude cinétique et thermodynamique, préciser les
mécanismes enzymatiques qui le régulent: trop de facteurs inconnus, voire in-
soupçonnés, y sont impliqués pour que nous puissions y prétendre. C'est dire que
si la réalité de cet équilibre ne fait pas de doute, son contrôle nous échappe entièrement. Un fait est certain: il se déplace sans cesse, puisque des résistances se
créent, qui obligent le biochimiste à renouveler périodiquement l'arsenal théra-
peutique qu'il tient à la disposition du clinicien.

4. Le mécanisme général du cycle de survie nous échappe

Nous venons de noter la complexité des interrelations existant entre espèces
différentes de micro-organismes, au sein de la biosphère. Nous avons signalé le
caractère constamment évolutif de ces interférences. Si on cherche à généraliser
le phénomène, on est amené à évoquer, par exemple, les relations qui existent
entre vie en milieu marin et vie en milieu terrestre, et, au maximum, les échanges
constants parfaitement équilibrés dans notre biosphère, entre monde inorganique
et monde biologique. Au cours de notre étude systématique de chaque cycle,
nous avons souligné quelques uns des aspects de cet état de choses. Dans le maintien de cet équilibre global — qui conditionne la survie des espèces vivantes — le
rôle est évident des saisons, des climats, des phénomènes géologiques, donc de
mille facteurs encore incontrôlables. Est-il besoin d'ajouter que dans l'état
actuel de nos connaissances, une vue d'ensemble du cycle de survie et des facteurs qui le régissent, est impossible. Il faudrait, pour en savoir davantage, avoir
recours à des hypothèses finalistes qui n'ont pas leur place dans cette discussion.

Il est probable que le volume relativement considérable occupé par la biosphère,
constitue une des conditions essentielles de son bon fonctionnement. Toutefois,
si parfaitement agencés que soient les mécanismes du cycle de survie, ils comportent des points faibles. Nous en retiendrons un exemple particulièrement démon-
stratif, et bien connu des cliniciens: le goître endémique par carence iodée. Si les
riverains du bord de la mer en sont protégés, il n'en est pas de même pour certaines populations montagnardes très éloignées des océans. L'iode entre dans la
constitution des hormones thyroïdiennes; sa présence en quantité suffisante est

liée à une relative proximité de l'océan; sa diffusion dans certaines zones de la biosphère n'est point assurée de façon satisfaisante. Les doses exigées par l'organisme humain sont pourtant infinitésimales: la carence, cependant, suffit à provoquer le ralentissement métabolique et la déchéance intellectuelle que l'on sait.

V. Conclusions

A l'issue de cette revue, très succincte et fort incomplète, de quelques uns des phénomènes cycliques qui interfèrent sur notre planète pour garantir notre survie à long terme, il est utile de faire le point: nous avons cherché à mettre en évidence, outre certains aspects bien élucidés, les multiples incertitudes qui persistent encore sur ce vaste problème. Il convient en effet de ne pas se leurrer et de conserver en mémoire le fait fondamental: tout blocage métabolique d'un seul des éléments de base du cycle, entraine à brève échéance la mort des organismes carencés. La méthode dite du "milieu minimum" compatible avec la survie et l'équilibre physiologique, a depuis longtemps fait ses preuves; c'est à elle que nous devons la mise en évidence de nombreux oligoélements, dont la liste, d'ailleurs n'est peut-être pas close. (Quelques suppléances sont possibles; dans l'ensemble cependant, la specificité reste très étroite.)

Nous avons insisté sur l'étroite solidarité qui liait entre eux les êtres vivants, sur le rôle primordial des micro-organismes, ceux-là précisément dont les échanges métaboliques sont bien loin d'être tous éclaircis.

Toutes ces difficultés n'ont pas empêché les auteurs américains de s'attaquer au problème de la reproduction durable du cycle de survie en vase clos. Les résultats de ces travaux nous sont parvenus un peu moins d'un an après le lancement du premier satellite: voyons ce que cela donne.

C. Les tentatives actuelles de réalisation pratique d'un cycle biologique fermé

Elles sont encore extrêmement modestes, et le dossier qui les résume est fort mince. Cependant, le réalisateur des fusées Redstone et Jupiter, a établi le plan d'un véhicule pesant 1.700 tonnes, dont 35 de charge utile, comportant un équipage de 12 hommes, en vue d'une expédition sur Mars qui durerait deux ans et demi. "La probabilité de leur retour sains et saufs sera plus grande que celle qui concernait plus d'une des courageuses équipes qui prirent autrefois le départ, à la recherche de continents nouveaux... [5]". On voit que les ambitions ne sont pas petites.

1. La régénération de l'oxygène à partir du CO_2 dégagé

On s'adresse aux agents biologiques naturels capables de réaliser cette transformation, à savoir les organismes photosynthétiques [12, 13, 32] (on admet qu'au cours du voyage, la lumière solaire ne fera pas défaut: est-ce prouvé?). Le choix du végétal devra satisfaire plusieurs critères qui apparaissent comme contradictoires:

— il devra avoir une très haute activité photosynthétique, pour assurer un rendement maximum, et couvrir — sous faible volume, en pesant peu — la totalité des échanges respiratoires de l'être humain, tant en absorbant le CO_2 qu'en fournissant l'oxygène.

— il devra lui-même avoir peu d'exigences métaboliques, pour présenter des conditions de culture faciles qui couvriront ses propres besoins.

Si on compare l'activité photosynthétique de divers Végétaux (en milligrammes de CO_2 fixés par cm² et par heure) on constate que les Algues type *Chlorella*, et certaines plantes adaptées à la vie en haute altitude (*Gentiana algicola*, *Veronica bellidioides*) sont les plus efficaces. Certes, le rendement des plantes supérieures est beaucoup plus élevé que celui des Algues: ainsi *Gentiana* est environ 50 fois plus efficace que *Chlorella*. Mais si on préfère sacrifier un peu du rendement à la facilité des conditions de culture, il faut s'adresser aux Mousses, aux Hépatiques, aux Lichens.

Les Algues toutefois, ont la faveur générale: on sait que *Chlorella*, depuis des années, constitue le materiel de choix à partir duquel des hypothèses diverses ont été émises sur le mécanisme biochimique de la photosynthèse. De plus, la considérable surface que présentent les Algues par rapport à leur volume, fait qu'on peut obtenir avec elles un rendement assez bon pour un poids relativement faible.

Chlorella pyrenoïdosa fixerait approximativement 13,4 gr de CO_2 pour 100 gr de poids sec et par heure [29]. Dans ces conditions, 1,1 Kg de poids frais de cette Algue suffirait à absorber tout le CO_2 émis en une heure par un homme d'équipage.

2. *Quelques exemples de transformations chimiques réalisées par les micro-organismes*

On se borne à indiquer quelques bactéries autotrophes, dont nous avons étudié le rôle essentiel, en précisant la transformation qu'elles sont capables d'accomplir. Nous reconnaissons un maillon, tantôt du cycle du carbone tantôt du cycle de l'azote ou du soufre, mais rien de plus. Aucune précision sur un éventuel contrôle de l'activité de ces bactéries, ni sur la façon dont elles supporteront les conditions très défavorables que l'Espace créera pour elles comme pour l'homme.

Rhodopseudomonas palustris: CO_2 + matière organique + lumière = (CHOH), élément d'une chaine glucidique + matière organique oxydée.

Thiobacillus thiooxydans: $S + 1^1/_2 O_2 + H_2O = SO_4H_2$

Nitrosomonas europea: $NH_3 + 1^1/_2 O_2 = NO_2H + H_2O$

Beggiatoa alba: $H_2S + {}^1/_2 O_2 = H_2O + S$

Clostridium aceticum: $4 H_2 + 2 CO_2 = CH_3 \cdot COOH + 2 H_2O$.

Thiorhodaceae: $CO_2 + 2 H_2S + \text{lumière} = (CHOH) + H_2O + 2 S$

Athiorhodaceae: $CO_2 + 2 H_2 + \text{lumière} = (CHOH) + H_2O$

3. *Les Algues, source possible d'aliments*

Non seulement *Chlorella* pourra jouer le rôle d'échangeur de gaz et contribuer à assurer les fonctions respiratoires des passagers, mais on songe à la faire figurer à leur menu pour une part essentielle. En effet, l'analyse chimique a montré que *Chlorella* est un aliment presque complet, couvrant à peu près toutes les exigences de la ration alimentaire de l'homme: acides aminés indispensables (sauf les soufrés), lipides, glucides, vitamines et sels minéraux. Cependant, un calcul simple montre que, pour satisfaire ses besoins en hydrates de carbone, un seul des voyageurs de l'Espace devra consommer journellement environ 4,250 Kg d'Algues — ce qui, bien sûr, est difficile à concevoir [5]. Faudra-t-il intercaler dans le cycle un nouveau maillon entre *Chlorella* et l'homme, sous la forme d'un Herbivore — des Lapins, des Chèvres — dont le tube digestif transformera partiellement les aliments végétaux en substances animales, plus facilement assimilables par l'espèce humaine? La survie propre de cet animal soulèverait de nouvelles difficultés; ses exigences alimentaires et respiratoires, l'odeur qu'il dégagera seront autant de facteurs qui contribuent à rendre cette éventuelle solution tout à fait inacceptable pour l'instant.

Quoi qu'il en soit, à l'heure présente, la recherche s'oriente vers la production massive de *Chlorella*, et sur l'étude approfondie de la physiologie de ces algues [9, 27]. Quelles sont leurs exigences personnelles en vue d'obtenir le rendement photosynthétique maximum [39]; dans quelle mesure peuvent elles réutiliser les produits de déchet de l'organisme humain pour leur nutrition; quelle est l'action des conditions escomptées en vol spatial sur leur croissance, leurs aptitudes, leur valeur nutritive; comment varie leur composition chimique en fonction de certains facteurs, tels, par exemple, l'intensité lumineuse, la méthode d'illumination et la longueur d'onde utilisée [26, 27]. la composition du milieu nutritif, l'épaisseur, la densité et le pH de la solution, sa turbulence, son degré d'aération, sa concentration en CO_2. Autant de problèmes qui font actuellement partie du plan de travail, autant de réponses précises qu'il faudra obtenir avant de concrétiser ce projet de cycle biologique fermé, qui, sans cela, risque de demeurer à l'état de simple ébauche théorique.

4. Traitement des déchets

Il s'agit cette fois, non plus de s'en débarrasser, mais de les transformer en produits à nouveau assimilables qui puissent réintégrer le cycle, sous forme soit d'aliments directement consommables par l'homme, soit de substances utilisables par le végétal.

La récupération de l'eau et des sels minéraux à partir de l'urine, serait ici particulièrement précieuse. Les ions pourraient servir à équilibrer le milieu de culture de la plante.

On espérait pouvoir, par un traitement approprié, récupérer la cellulose à partir des féces. Des expériences récentes ont montré que contrairement aux notions classiques, la cellulose, en principe matière de lest mal digestible, est en réalité dégradée jusqu'à un stade avancé dans le tube digestif de l'homme, par l'action des bactéries intestinales.

Le problème de l'élimination des substances toxiques d'origine respiratoire, urinaire, sudorale ou fécale, se pose ici avec plus d'acuité encore que lors des expéditions de courte durée. Il y a lieu cependant d'établir une distinction entre substances toxiques: certaines, qui le sont pour les Végétaux, ne le sont pas pour l'homme et vice-versa. Encore un problème qu'il convient d'inscrire à l'actif de la recherche future.

D. Critique du système proposé

Si on confronte les exigences du cycle de survie tel qu'il existe dans la biosphère terrestre, avec les moyens dont on dispose aujourd'hui pour le reproduire en circuit fermé à très petite échelle, on se persuade sans peine que l'heure est bien loin d'être sonnée, où les passagers du vaisseau spatial s'embarqueront avec espoir de retour pour des voyages qui dureraient des mois ou des années. Nous sélectionnerons parmi les critiques que l'on peut adresser aux tentatives de réalisation actuelles, celles qui nous paraissent les plus convaincantes.

1 — Les conditions expérimentales dans lesquelles sont étudiés les organismes dont les activités physiologiques tendent à s'équilibrer, ne sont pas encore au point, principalement en ce qui concerne le choix du végétal. Les résultats sont encore très fragmentaires et souvent contradictoires, ils portent sur des plantes qui ne sont pas cultivées dans des conditions analogues. Impossible dès lors, d'étudier l'influence d'un seul facteur que l'on ferait varier systématiquement, toutes choses égales par ailleurs. Enfin, le nombre des organismes étudiés en ce sens, est encore trop faible, on ne peut appliquer les méthodes statistiques, et les conclusions que l'on cherche déjà à tirer de ces quelques expériences ne peuvent pas avoir une portée générale. Bref, le recul manque, et les impératifs auxquels doit répondre une méthode expérimentale rigoureuse, ne sont pas encore satisfaits dans ce domaine.

2 — Rappelons toutes les inconnues que nous avons cherché à mettre en évidence en brossant le tableau du cycle de chacun des éléments indispensables à

notre survie. Tant que la biochimie et la physiologie des micro-organismes qui
jouent un rôle essentiel dans ces phénomènes, ne seront pas davantage appro-
fondies, toute tentative de mise en équilibre et de contrôle permanent d'un cycle
artificiel est immanquablement vouée à l'échec dans de très courts délais.

3 — Si cependant nous supposons le problème résolu, tous ces écueils ayant
été surmontés, le cycle de survie équilibré à l'intérieur du vaisseau spatial et
fonctionnant, d'autres objections persistent; la permanence du cycle est-elle
assurée, pour qu'on puisse garantir la survie à long terme de l'équipage ?

a) Il est a peu près certain que le rendement d'un pareil cycle diminuera très
vite à mesure qu'il tournera. Si, par exemple on cherche à débarrasser l'urine
de ses ions pour en faire le milieu du culture des Algues *Chlorella*, dans le cas
très probable où le rendement n'atteindrait pas 100 % d'efficacité, l'eau obtenue
après quelques cycles sera de plus en plus saumâtre, et deviendra vite de ce
fait impropre à cet usage.

Les mécanismes artificiels mis en place devront avoir une très haute efficacité,
beaucoup plus élevée que ceux qui fonctionnent au sein de la biosphère. Celle-ci
dispose d'un volume d'extension considérable alors qu'à l'intérieur du navire
spatial, tous les phénomènes devront avoir lieu au dedans d'une enceinte qui ne
dépassera pas, tout au moins au début, quelques mètres cubes. Si nous osons
risquer cette comparaison, il semble que le vaste volume qu'occupe la biosphère,
lui confère en quelque sorte un état "tamponné" par rapport à un système analogue en miniature. Ainsi, dans la biosphère, une brusque variation d'un des
facteurs de survie en un point, ne déséquilibre pas le fonctionnement global du
cycle; au contraire, en vase clos de petites dimensions, toute baisse brutale du
rendement d'un seul des mécanismes artificiellement mis en place, suffit à assurer
la non-viabilité de l'ensemble. Est-il besoin de préciser que cette notion, purement
intuitive, demande à être vérifiée ?

b) si, comme on l'annonce, la production de *Chlorella* est intensifiée dans
l'intention de faire de cette Algue l'aliment de base des premiers pionniers, il
convient de noter que, quelles que soient ses qualités nutritives, elle ne saurait
en aucun cas couvrir tous leurs besoins à long terme. L'homme ne peut se contenter
de protéines végétales présentées à chaque repas sous forme d'une pâte verte,
surtout si on exige de lui une activité physique et intellectuelle relativement
soutenue. D'une manière plus générale d'ailleurs, au sein de ce tout petit univers
où l'homme devra voisiner sans trêve et de tout près avec tous les organismes
vivants chargés d'assurer le cycle de survie, on conçoit sans difficulté quelle
nauséeuse intimité se créera rapidement entre les passagers et le système; nous
pensons qu'il est inutile d'insister davantage sur ce point.

c) Enfin et surtout, l'infection de l'ensemble des organismes intéressés par les
bactéries pathogènes, est une quasi-certitude, qu'il s'agisse de celles qui s'atta-
quent à l'homme ou de celles qui visent préférentiellement les végétaux. Nous
savons qu'un contrôle efficace des interactions entre micro-organismes est actuellement chose impossible. Certes, des antibiotiques judicieusement utilisés pourraient
juguler des infections passagères; mais, dépassant le but à atteindre, ils neutrali-
seront en même temps d'autres systèmes bactériens indispensables au fonctionnement du cycle de survie. Tant que le mécanisme d'action des antibiotiques ne
sera pas mieux analysé, permettant un contrôle et une adaptation de leur efficacité
(ni trop, ni trop peu), tout progrès dans ce domaine nous apparait comme illusoire.

Le problème reste entier du comportement en milieu spatial de tous ces organismes. Comment, par exemple, réagiront-ils à l'absence de pesanteur ? Avant de
répondre, il conviendrait d'être plus amplement informé sur leur physiologie en
milieu terrestre.

E. Conclusions

Chacune des ces objections, à elle seule, suffit, nous semble-t-il, pour annihiler des espoirs trop rapidement conçus. Le cycle de survie, tel qu'il existe dans la biosphère, doit être considéré comme un tout; il est autorégulé dans la mesure où les cycles de chacun des éléments qui le constituent, s'articulent entre eux de façon adéquate. A quoi sert-il de prévoir la mise en place du cycle du carbone, en ajustant le métabolisme d'un autotrophe avec celui d'un hétérotrophe, si on ne cherche pas dans le même temps à perpétuer le cycle de l'azote et celui du soufre? Certes, des Souris, convenablement alimentées par ailleurs, ont vécu 12 jours en circuit fermé, sans fourniture d'oxygène, en présence d'un végétal (*Anacystis nidulans*): le taux d'oxygène s'est même élevé de 38% [12]. L'expérience en soi ne manque pas d'intérêt. Mais on ne peut la retenir dans le cas qui nous préoccupe; certes, le problème de l'absorption du CO_2 et de la régénération de l'oxygène semble résolu; mais à long terme les Souris ne pourraient pas survivre, du fait de l'épuisement progressif dans l'enceinte close, du matériel biologique azoté, phosphoré ou soufré. Tous ces aspects du phénomène global, répétons le, sont intimement liés; le fait qu'un espoir de solution apparait pour l'un d'eux ne fait rien pour résoudre le problème du cycle de survie dans son ensemble.

On peut donc conclure que, contrairement aux opinions citées ci-dessus, une somme considérable de travaux doit être encore effectuée dans le domaine de la recherche fondamentale, avant qu'on puisse affirmer — sans parler de la réaliser pratiquement — qu'un pareil système est simplement possible. Les voyages interplanétaires de longue durée, ceux qui se prolongeraient des mois ou des années et qui viseraient Vénus ou Mars, en sont encore au stade d'un débat purement académique. Du point de vue biologique, dans l'état actuel de nos connaissances, il ne semble pas raisonnable d'y croire, tout au moins dans un proche avenir.

Bibliographie

1. A. ANANOFF, L'astronautique. Paris: Arthème Fayard, 1950.
2. J. F. BARCROFT, Significance of hemoglobin. Physiol. Rev. **4**, 329 (1924).
3. N. J. BOWMANN, The Food and Atmosphere Control Problem on Space Vessels. I: Chemical Purification of Air—II: The Use of Algae for Food and Atmosphere Control. J. Brit. Interplan. Soc. **12**, 118, 123, 159 (1953).
4. N. J. BOWMANN, Nutritional and Other Physiological Problems on Extended Space Flights. Rocket News Letter, Chicago 1949, **2**, 6—11 (No. 7 Feb.), 8—16 (No. 8 March).
5. M. C. BROCKMANN, A. S. HENICK, G. W. KURTZ et G. TISCHER, Closed Cycle Biological Systems for Space Feeding. I: Systems 1 and 2, Food Technology **12**, 449 (1958); II: System 3, Food Technology **12**, 454 (1958).
6. H. G. CLAMANN, Continuous Recording of Oxygen, Carbon Dioxide and Other Gases in Sealed Cabines. J. Aviat. Med. **23**, 330 (1952).
7. C. A. DEMPSEY et Coll., Long Term Human Confinement in Space Equivalent Vehicles. J. Astronautics **4**, 52, 59 (1957).
8. C. E. DRYDEN et Coll., Artificial Cabin Atmosphere Systems for High Altitude Aircraft. W.A.D.C. Tech. Report 55—353, 1956.
9. C. P. EDWARDS, The Culture of Algae. U.S.A.F.O.S.R. Rapp. 57—378, 1957.
10. R. M. FENNO, Man's Milieu in Space. A Summary of the Physiologic Requirements of Man in a Sealed Cabin. J. Aviat. Med. **25**, 612 (1954).
11. B. FINKELSTEIN, Feeding Crews in Air Vehicles of the Future. Food Technology **12**, 445 (1958).
12. J. G. GAUME, Nutrition in Space Operations. Food Technology **12**, 433 (1958).
13. J. G. GAUME, Plants as a Means of Balancing a Closed Ecological System. J. Astronautics **4**, 72 (1957).

14. A. Gibert et A. Adeline, L'utilisation des plantes pour la régénération de l'oxygène à partir du gaz carbonique. Fusées 2, 69 (1957).
15. A. Gibert, H. Boiteau, C. Jacquemin, J. Fabre et A. Adeline, Etat actuel de l'experimentation animale et humaine dans le vol en gravité nulle. Méd. Aéronaut. 13, 177 (1958).
16. J. Gustavson, Synthetic Atmospheres for Space Ships. Astronautics 2, 48, 62 (1957).
17. J. B. S. Haldane, The Purification of Air During Space Travel. J. Brit. Interplan. Soc. 14, 87 (1955).
18. S. W. F. Hanson, Some Observations on the Problems of Space Feeding. Food Technology 12, 430 (1958).
19. R. P. Haviland, Air for the Space Ship. J. Aeronautics 3, 31, 48 (1956).
20. W. T. Ingram, Environmental Problems Connected with Space Ship Occupancy. In: American Astronautical Society, 3rd Annual Meeting, Proceedings. New York: The Society, 1957.
21. E. B. Konecci, Human Factors and Space Cabins. Astronautics 3, 42, 71 (1958).
22. M. F. Lee et Coll., Basic Requirements for Survival of Mice in a Sealed Atmosphere. J. Aviat. Med. 25, 399, 432 (1954).
23. C. Lewis, U.S.A.F. School Simulates Living in Space. Aviat. Week 68, 49, 51, 53, 55, 57, 58, 59, 61 (1958).
24. H. J. Masson, Study of Methods for Obtaining Oxygen from Carbon Dioxide. U.S.A.F.O.S.R. Rapp. T.N. 57—379, June 1957.
25. A. Moyse, La photosynthèse. Année biol. 28, 217 (1952); 29, 165 (1953).
26. A. Moyse, L'utilisation de l'énergie solaire. Année biol. 32, 25 (1956).
27. A. Moyse, Influence de divers facteurs sur la croissance accélérée des cellules de Chlorelles. Rev. Gén. Bot. 63, 167 (1956).
28. W. D. Murray, A Gondola for Physiological Research. J. Aviat. Med. 25, 354 (1954).
29. J. Myers, Basic Remarks on the Use of Plants as Biological Gas Exchangers in a Closed System. J. Aviat. Med. 25, 407 (1954).
30. N. R. Nicoll, Design of Life Compartment Necessary for Space Travel. J. Brit. Interplan. Soc. 13, 277 (1954).
31. H. E. Ross, Air Conditioning Problems of Space Travels. Bull. Brit. Interplan. Soc. 1, 49 (1946).
32. H. E. Ross, Green Plants as Atmosphere Regenerators. Bull. Brit. Interplan. Soc. 2, 7 (1947).
33. D. G. Simons, Nutrition in Space Flight. Balloon Flight Experience. Food Technology 12, 436 (1958).
34. D. G. Simons, Surface Temperatures of Animal Capsules Floating above 80,000 Feet. Holloman Air Development Center, Technical Report 56—6, May 1956.
35. D. G. Simons et D. P. Parks, Climatization of Animal Capsules During Upper Stratosphere Balloon Flights. Jet Propulsion 26, 565 (1956).
36. M. Specht, Toxicology of Travel in the Aeropause. In: Physics and Medicine of the Upper Atmosphere, p. 171—181. Albuquerque, N.M.: University of New Mexico Press, 1952.
37. H. Strughold, The U.S. Air Force Experimental Sealed Cabin. J. Aviat. Med. 27, 50 (1956).
38. M. V. Strumza, Les conditions respiratoires dans les fusées intercontinentales. Fusées 2, 241 (1957).
39. H. Tamiya, T. Iwamura, Shibata, Hase et T. Nihei, Correlation Between Photosynthesis and Light Independant Metabolism in the Growth of Chlorella. Biochim. Biophys. Acta 12, 23 (1953).
40. A. A. Taylor, Present Capabilities and Future Needs for Space Feeding. Food Technology 12, 442 (1958).

On trouvera de plus une importante bibliographie d'ordre général concernant la médecine spatiale dans:
"Bibliography of Space Medicine" édité en 1958 par la "National Library of Medicine", Reference Division (Washington, D.C.).

Rocket Postal Service
Part II: Intercontinental Connection

By

Glauco Partel[1] and Antonio Angeloni[2]

(With 3 Figures)

(Received July 6, 1959)

Abstract — Zusammenfassung — Résumé

Rocket Postal Service. Part II: Intercontinental Connection. The first part of this study, which considers shipping of mail and lightweight valuable goods by means of unmanned vehicles, has been delivered in previous international congresses this year both in Paris and in Rome. Part I was concerned with the continental trajectories of the service, for which a ramjet-propelled vehicle was shown as the best solution. Collection mail centers, both in Europe and in America, for instance Paris and New York, were envisaged, to and from which all the peripheral towns had to be connected by the ramjet vehicles.

In this IInd Part, we discuss the intercontinental trajectories (taking as an example an arc of one radiant, corresponding approximately to the distance Paris—New York). The best vehicle for such a mission is a rocket glider. Velocities and flight times of the trajectory are analyzed, and weights of the vehicles are given for a mail payload of 100 kilograms per vehicle. It is shown that the flight frequency should be of at least one vehicle per hour. Operating costs as well as postage costs to cover the expenses are derived.

The postal system described fits in today's economical plane. Moreover, missile technology has already developed most of the required components, and therefore there is no need of substantial new inventions. Apart from allowing the transmittal of communications in the same period of time as a cable, it appears that the rocket postal service in question could repay the capital investments over a relatively short period of time.

Raketenpostdienst. Teil II: Interkontinentale Verbindung. Der erste Teil dieser Studie, welche die Verschiffung von Poststücken und wertvollen Leichtgütern mit Hilfe unbemannter Raketenfahrzeuge behandelt, wurde in diesem Jahr (1959) sowohl in Paris als auch in Rom bei internationalen Kongressen vorgelegt. Teil I befaßte sich mit den kontinentalen Flugbahnen dieses Dienstes, für welche, wie gezeigt werden konnte, Flugkörper mit Staustrahltriebwerken sich als beste Lösung erwiesen hatten. Es wurden Postsammelzentren sowohl in Europa wie auch in Amerika, z. B. in Paris und New York, ins Auge gefaßt, wohin alle peripher liegenden Städte durch Staustrahl-antrieb-Flugkörper Verbindung haben sollten.

Im vorliegenden Teil II erörtern wir die Interkontinentalbahnen; dabei wird als Beispiel ein Kreisbogen von einem Ausstrahlungspunkt aus genommen, der ungefähr der Distanz Paris—New York entspricht. Das beste Fahrzeug für einen solchen Zweck ist ein Raketengleiter. Geschwindigkeiten und Flugzeiten der Bahn werden analysiert und Gewichte der Fahrzeuge für eine angenommene Nutzlast von 100 kg je Fahrzeug

---

[1] Director of Planning, Missile Systems Consulting Company, Rome, Italy.

[2] Preliminary Design, SISPRE, Rome, Italy.

berechnet. Es wird gezeigt, daß die Flugfrequenz mindestens 1 Flugzeug je Stunde sein sollte. Betriebs- und Frachtkosten zur Deckung der Ausgaben werden berechnet.

Das beschriebene Postsystem eignet sich gut für das heutige ökonomische System. Überdies hat die Technologie der Flugkörper bereits die meisten der erforderlichen Komponenten entwickelt, so daß keine wesentlichen neuen Erfindungen dafür mehr nötig sind. Der Transport von Mitteilungen kann in der gleichen Zeit wie ein Kabeltelegramm erfolgen; daneben scheint es, daß der hier vorgeschlagene Raketenpostdienst in einem verhältnismäßig kurzen Zeitraum die Kapitalinvestitionskosten hereinbringen könnte.

Service postal par fusées. 2ème partie: Connexions intercontinentales. La première partie de cette étude dévolue à l'expédition de lettres et de colis précieux par fusées sans pilotes, a été exposée à l'occasion de congrès antérieurs à Paris et Rome. Elle concernait les trajectoires continentales pour lesquelles la propulsion par statoréacteur apparaissait comme la meilleure solution. Des centres de triage étaient envisagés, tels New-York et Paris, reliés par stato-fusées aux villes périphériques.

Cette seconde partie analyse les trajectoires intercontinentales, prenant comme exemple un rayon d'action d'un radian qui correspond approximativement à la distance Paris-New York. Le planeur-fusée semble être le meilleur véhicule pour cette mission. La fréquence du service devrait être d'au moins un véhicule à l'heure pour une charge payante de 100 kilos. Les frais d'opération et l'affranchissement nécessaire à les couvrir sont établis.

Le système postal décrit s'intègre dans l'économie actuelle. La technologie a mis à notre disposition la plupart des équipements requis, si bien qu'il n'y a pas nécessité d'inventions nouvelles importantes. Outre qu'il assure la transmission d'informations dans un temps comparable à celui du cable, le service postal par fusées pourrait être amorti sur une période relativement courte.

List of Symbols

c	gas discharge velocity	v_f	velocity at destination (end of gliding flight)
D	aerodynamic drag		
E	efficiency	v_b	velocity at burn-out
g	acceleration of gravity	P_s	payload of step s
h_0	altitude at gliding-flight start	P_n	payload of last step
L	lift	$W_s i$	initial weight of step s
m	mass of gliding vehicle	$W_s b$	burn-out weight of step s
n	total number of steps; number of flights per day	w_s	propellant weight of step s
		β	structural coefficient
r	radius of flight path	φ, ψ	arcs of gliding flight path and of its trace on the earth's surface
R	Earth's radius (6350 kms)		
s	generic distance covered on the flight path; generic step of n steps	σ, γ	distance of the arc centers φ and ψ, and angle between σ and the initial radius
s_0	space at time zero ($=0$)		
S	total distance in gliding-flight	θ	generic angle between velocity vector and the horizontal
t	generic time		
T	total time of gliding flight	θ_i	angle θ at gliding-flight start
t_0	time at gliding-flight start	θ_f	angle θ at destination (end of gliding flight)
v	generic velocity		
v_0	velocity at gliding-flight start	ϱ	mass ratio

I. Foreword

In a first part of a report on rocket postal services, a European mail net has been dealt with, based on an economic and commercial base. This first report has been presented at two different rocket and electronic congresses, one hold in Paris and another one in Rome. In that report the vehicles suitable for the Euro-

pean connections were analyzed and the different times involved in the calculation
of the total time en route between the collecting post office of the take-off site
and that of the landing site were discussed, pointing out that the rapidity of mail
shipment is mainly dependent upon the flight velocity and the flight frequency.
In illustrating the rocket postal net, connections between Paris, Rome, Vienna,
Madrid, Stockholm and London have been considered, with a service of 96 single
trips and 96 return trips per day on each connection, that is, a vehicle take-off
every 15 minutes. The postal vehicle considered for the European service was
propelled by ramjets, capable of developing a speed of 750 meters per second at
an altitude of 20,000 meters, having a payload of 10 kilos per vehicle. After an
analysis of the operating costs of such a mail net, it has been pointed out that the
rocket postal service allows the transmittal of communications in the same period
of time as a cable, with the advantage—with respect to the latter—to be able of
sending many more words, as well as authentic documents, drawings, etc. It
appeared that the rocket postal service could repay the capital investments over
a relatively short period of time.

In this second report we discuss the intercontinental connection of
a rocket postal service. The character of the work is more that of an
operations research than of a feasibility study.

Without entering details about vehicles,
propulsion units, auxiliary equipment, we limit
ourselves to general considerations on dimensions. weights, energies required and we try to
derive the order of magnitude of the relative
costs.

II. Flight Path

As typical flight path, we assume an arc of
one radiant, corresponding to about 6350 kilometers on the earth's surface.

We make reference essentially to work
developed by A. J. EGGERS, H. J. ALLEN, and
S. E. NEICE [1], who have compared three
types of vehicles:
  —the ballistic missile.
  —the glide vehicle,
  —the skip vehicle.

From such a comparison, the glide vehicle
results far superior, both with respect to the
burn-out velocity required, and to the effects
of the aerodynamic heating.

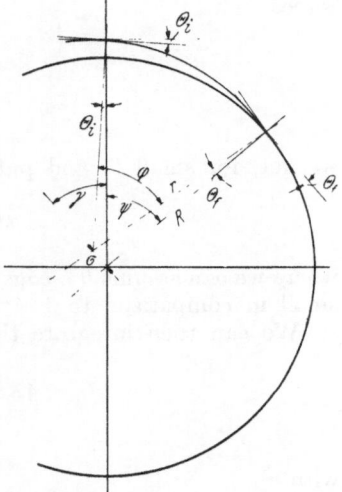

Fig. 1. Illustration of notation. Flight
path assumed as an arc of a circle
with radius r equal to the Earth's
radius R

From the said work of EGGERS and others, results are also presented for the
velocities required at the beginning of the unpowered flight to cover different
ranges.

In our case, we can realize in a direct and simple way what initial velocity is
required, assuming as a first approximation that the glide vehicle follows a flight
path constituted by an arc of a circle having a radius r, equal to the earth's
radius R.

Then, we have from Fig. 1 that the circle will have its center displaced from
the earth's center for the small length necessary to have the circular trajectory
passing between the gliding-flight start and the landing sites.

We deem the initial altitude should have an order of $h_0 = 30$ kms, and, since the angle ψ (Fig. 1) is one radiant, the angle γ is about $32° 20'$, since $\gamma = 90° - \frac{1}{2} \theta_i - \psi$, assuming $\theta_i = 20'$. Since we also have

$$\sigma \cos \gamma \approx h_0$$

it is apparent that for $h_0 = 30$ kms, σ is of an order of 36.5 kms. Since

$$\sigma \sin \gamma = R \sin \theta_i$$

where θ is the angle between the tangent to the trajectory and the horizontal, $\theta_i = 0° 11'$. On the other hand,

$$\frac{\sigma}{R} \approx \theta_i$$

hence, θ_i should be $0° 20'$.

That is, the difference between the angles subtended to the effective trajectory and to its projection on the earth's surface is small, and is

$$\varphi - \psi = \varDelta \theta .$$

Expressing now the usual equilibrium conditions of the normal and tangential forces

$$m \frac{v^2}{R} = m g \cos \theta - L$$

$$m \frac{dv}{dt} = m g \sin \theta - D$$

we get, for small θ, and putting $L/D = E$

$$g (E \theta - 1) + \frac{v^2}{R} = E \frac{dv}{dt}$$

where we can assume $\theta = $ const., since it has been shown that it varies little and is small in comparison to 1.

We can then integrate the expression, for $E = $ const., arriving at

$$\varDelta v = \frac{A}{\sqrt{\dfrac{A}{R}}} \cdot \tan \left(\frac{1}{E} \sqrt{\frac{A}{R}} \varDelta t \right)$$

with

$$A = g (\theta E - 1) .$$

Since the term

$$\frac{1}{E} \sqrt{\frac{A}{R}} \varDelta t$$

for the case in question is small enough, we assume

$$\tan \approx \text{arc}.$$

This conclusion remains valid also for A negative, since

$$(1/i) \tan i x = \tanh x \approx x.$$

That is

$$\varDelta v = \frac{A}{E} \varDelta t$$

or

$$\frac{\varDelta v}{\varDelta t} = \frac{dv}{dt} = \text{const.}$$

We then have

$$s - s_0 = \frac{1}{2} \frac{A}{E} t^2 + v_0 t$$

that is, for $t_0 = 0$, $t_f = T$

$$\left. \begin{array}{l} v_f = \dfrac{g\,(\theta\,E - 1)}{E}\,T + v_0 \\[2mm] S = \dfrac{1}{2} \dfrac{g\,(\theta\,E - 1)}{E}\,T^2 + v_0\,T\,. \end{array} \right\}$$

In the assumed simplification, the motion occurs at a constant acceleration, which is positive, negative or zero, according to the magnitude of θ and of the efficiency factor. These equations allow us, even if in an approximate way, an evaluation of the required velocities and times.

In the case we are taking into consideration,

$$S \approx R = 6,350,000 \text{ meters.}$$

Therefore, we have

$$v_0 = \frac{6,350,000}{T} = \frac{1}{2E} g\,(\theta\,E - 1)\,T\,.$$

In this report θ is given by the geometrical conditions of the flight path, that is, by the range, the radius, and the initial altitude.

Since range, initial altitude, and radius are considered steady, the different velocities and times are referred to the same θ.

In the specific case integrating the equations for $\theta = \text{const.} = 0° 15'$, we get for the minimum v_0 ($v_f = 0$):

$E =$	2	3	4
$v_0 =$	7900	6350	5370 mt/sec
$U =$	27	33	38.3 minutes.

Therefore, we can take into consideration, with an initial altitude of 30 kms, and an average efficiency factor of the vehicle from 3.5 to 4, a v_0 of about 5,500 mt/sec, and a time of about 38 minutes.

Let us keep the above data for the operating calculations.

We neglect the effect of the aerodynamic heating, this being a particular problem we can overcome using suitable materials.

III. Vehicles

Let us now see which propulsion units are necessary to reach the altitude and the velocity determined above, carrying out the first analysis in vacuum and with no gravity.

We designate

$\varrho =$ mass ratio (take-off weight/burn-out weight);
$\beta =$ structural factor

$$\left(\beta = \frac{\text{vehicle weight without payload and propellant}}{\text{propellant weight}} \right).$$

The final velocity is of course

$$v_b = c \ln (\varrho_1 \cdot \varrho_2 \cdot \varrho_3 \cdots \varrho_n)$$

where the subscripts $1, 2, 3, \ldots n$ represent the number of the respective steps, and c is the discharge velocity. We take ϱ and β as having the same value for all steps. We then have

$$v_b = c \ln \varrho^n = c \cdot n \cdot \ln \varrho \, .$$

Since we aim at reaching ideal velocities of 6250 mt/sec (in order to get effective velocities of 5500 mt/sec), and assuming to use a propellant with $c = 2500$ mt/sec, we have

$$\ln \varrho = \frac{2.5}{n}$$

that is

$$\varrho = e^{\frac{2.5}{n}} \, .$$

Then, for vehicles of 2, 3, 4 steps we should have $\varrho = 3.5$, 2.3, 1.9, respectively.

We are interested in determining, for the above ϱ and for the practical structural factors β, the total weights, the weights of the different steps and of the propellant, referring these to the payload.

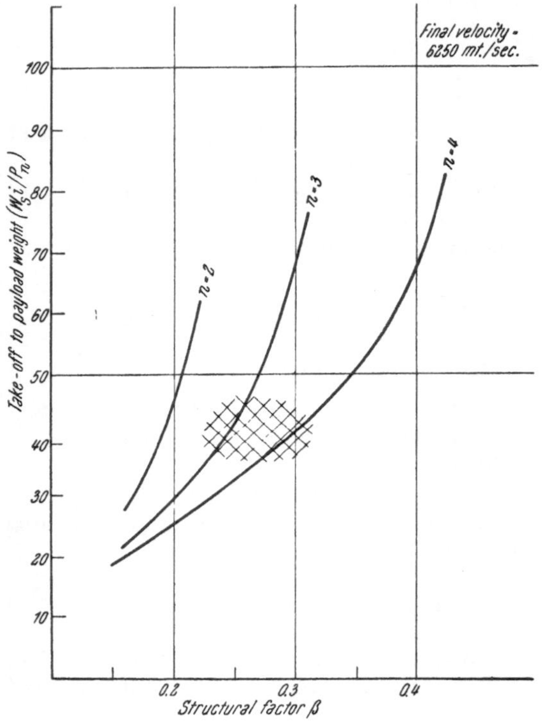

The following expressions are easily obtained:

Initial weight for the step s of n steps:

$$\frac{W_s i}{P_s} = \frac{c}{1 - \beta (\varrho - 1)}$$

referred to the payload of the step into consideration (P_s).

On the contrary, if we refer the initial weight to the payload of the last step (P_n), which is what we are interested in, we get

$$\frac{W_s i}{P_n} = \left[\frac{c}{1 - \beta (\varrho - 1)} \right]^{n - s + 1} \, .$$

For the burn-out weight, we have

$$\frac{W_s b}{P_n} = \left[\frac{1}{1 - \beta (\varrho - 1)} \right]^{n - s + 1} \, .$$

For the propellant we get

$$\frac{w_s}{P_n} = \frac{\varrho^{(n - s)} (\varrho - 1)}{[1 - \beta (\varrho - 1)]^{n - s + 1}} \, .$$

Fig. 2. Take-off weight in dependance of the final payload for vehicles of 2, 3, 4 steps

We can now deduce Fig. 2, giving us the take-off weights referred to the final payload, for vehicles of 2, 3, 4 steps, built with different β, and which can all reach the final velocity of 6250 mt/sec.

It is evident that we must limit ourselves in the field of 3 or 4 steps. With β of 0.25 to 0.3 which may be attained today, and with 3 or 4-step vehicles, the take-off weight is about 40 times the payload.

In other words, for 100 kilos of payload, the take-off weight is about 4000 kilos. A 3-step vehicle having these characteristics would require $\beta = 0.24$ and $\varrho = 2.3$, and should exactly have the following composition:

Steps	I (+ II + III)	II (+ III)	III
Take-off weight, kg	3750	1120	335
Burn-out weight, kg	1630	487	145
Propellant weight, kg	2120	633	190
Vehicle weight, kg	510	152	45
Payload weight, kg	1120=II	335=III	100

We have therefore a total weight of the three steps without propellant and without payload of about 700 kilos, and a total propellant weight of about 3000 kilos, in order to carry 100 kilos of mail at a distance of 6500 kilometers in approximately 40 minutes.

IV. Flight Frequency

In the same way as we did for the European service, let us consider the following:

—time between post office and take-off site and between landing site and post office   1 hr

—sorting time (in the worst case) $\dfrac{24 \text{ hr}}{n}$

—flying time (40 minutes)   0.67 hr

—total time $1.67 + \dfrac{24}{n}$

where n is the flight frequency per day.

Making a diagram of the whole situation (Fig. 3), we can see how it is desirable to launch at least one vehicle per hour, that is, 24 vehicles per day.

In the worst case the letter will reach the post office in the destination city after 2 hours and 40 minutes from its delivery at the post office of the sender's city.

If we envisage this system in connection to the European system described in another report, we have in Paris a maximum possible traffic of 5 vehicles every 15 minutes; that is, 20 vehicles per hour, with a loading capacity of 10 kilos each,

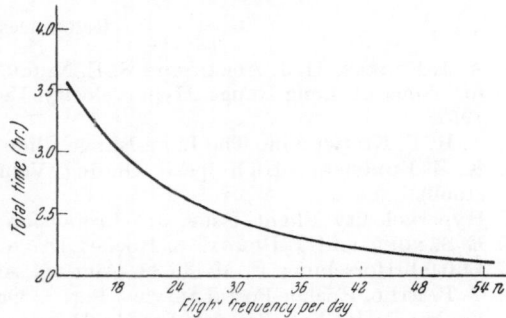

Fig. 3. Flight frequency vs. total time

for a total of 200 kilos of mail. Therefore, we foresee here that only half of the mail arriving in Paris can be forwarded overseas.

To this mail, we have to add the local mail dispatched overseas.

V. Costs

We aim of course to obtain only a general idea, as this report has neither an optimization character nor it is a detailed study.

We can foresee to have liquid propellant stages for the first two steps and to consider the engines as lost. For each flight we are thus going to lose 660 kilos of vehicle, which, at a cost of about 35 dollars per kilo represent 23,100 dollars.

The fuel is about 3000 kilos (the last step weighs little, and it is included at the same price even if this is not correct); at a cost of 1.7 dollars per kilo, we have 5100 dollars.

In this way, we reach a total of 28,200 dollars for expendable flight material to carry 100 kilos of mail, that is 20,000 letters of 5 grams each. The cost per letter is 1.41 dollars, with an utilization factor of 75 per cent.

We must add to these figures the depreciation outlays for the installations and vehicles, the other operating costs and the overhead expenses.

The daily traffic consists of:

$15,000 \times 24 = 360,000$ letters for single trips and an equivalent number for return trips. Therefore, we have about 700,000 letters per day weighing 3500 kilos, with a cost of expendable flight material amounting to 28,200 dollars $\times 48 = = 1,353,600$ dollars per day.

It seems unlikely that the operating, depreciation, and overhead expenses can much increase this amount.

The postage for a letter mailed to the USA from Europe in 2 hours and 40 minutes cannot, already in this case, be more than 2.0 dollars.

Systems are of course conceivable in order to decrease the highest expense, which is for the lost propulsion units. Similarly, the mass production of the propulsion units will further decrease the cost of same.

It is out of discussion that hypersonic postal vehicles of the type and characteristics here outlined will enter service very soon. The task is not difficult; it will become even easier in the immediate future. We have good reasons to believe that the mail transport of the type mentioned will be really pleasant to design and build, attractive to finance and operate, and will constitute a big boon for the world commerce at large.

References

1. A. J. Eggers, H. J. Allen and S. E. Neice, A Comparative Analysis of the Performance of Long-Range Hypervelocity Vehicles. N.A.C.A. T.N. 4046, October 1957.
2. T. R. F. Nonweiler, The Long-Range Glide Rocket. Aeronautics 38, No. 3 (1958).
3. K. M. Fuechsel, High-Speed Gliding Vehicles. Astronautics 3, No. 8, 34, 72 (1958).
4. Hypervelocity Flight Issue. Jet Propulsion 27, No. 11, November (1957).
5. E. Sänger and I. Bredt, A Rocket Drive for Long-Range Bombers. Deutsche Luftfahrtforschung U. M. 35.38, Ainring, August 1944.
6. G. Partel, Rocket Postal Service. Part I: Organizational Summary of a European Rocket Postal Net. Presented at the II International A.E.R.A. Congress on Rocket and Satellites, Paris, June 1959; and to the VI Nuclear and Electronics International Congress, Rome, June 1959.

Results of Experiments on the Biological Effects of Cosmic Radiation on Seeds of Hordeum (Gold Barley) Bonus 01518/B 19 (Gustafsson), with Special Consideration of Heavy Primaries Effects

By

J. Eugster[1] and **David G. Simons**[2]

(Received May 15, 1959)

The present experiments on a plant object represent a continuation of research done in the years 1952/53, in the course of which a mutagenic effect of cosmic radiation on *Artenia salina* eggs was demonstrated[3].

During the years 1953—58, an experimental material of *Hordeum* seeds has been exposed to cosmic radiation in the course of several flights at altitudes of 30—40,000 metres for an average period of 32 hours. The filial generations of these seeds have been subject to thorough observation. The first series were exposed at Payerne, Switzerland, and at Holloman, New Mexico, the two latest series (F and G) at Sault Ste Marie, Michigan (geomagnetic latitude 55° north). The experimental technique employed was the so-called "sandwich method", the main feature of which consists in the experimental object proper and the nuclear plates necessary for monitoring the nuclear events being assembled so as to form a unit. The seeds are placed in such a way that an exactly defined area of the nuclear emulsion corresponds to the embryonic area, so that any nuclear event penetrating the latter would be monitored in the emulsion. Big star figures, meson showers and, above all, heavy primaries penetrating the embryonic area have been considered, in some sporadic cases also their "thin down" having left its track in the biological specimen.

Results: The original material comprised 280 seeds from which 8539 single plants in 4 filial generations were bred. These were examined with respect to the characteristics introduced by GUSTAFSSON. For purposes of evaluation we have divided the material in four categories:

1. Ground controls (seeds not exposed and their offspring)
2. Test field controls (seeds exposed but not shown to have been locally hit)
3. Seeds with central hits (25, or 4,2 % of the total number exposed)
4. Seeds with partial hits (peripheral hits or hits from smaller disintegration figures).

The offspring of the 25 seeds having suffered central hits have shown without exception a reduction by 30 to 40 % of their "Bestockungszahl" (number of "Triebe" and number of "ährentragende Schosse") as compared to the controls.

---

[1] University of Zurich, Switzerland.

[2] Lt. Col., USAF, Holloman Air Force Base, New Mexico, U.S.A.

[3] Mutagenic Effects of Cosmic Radiation. Bull. Schweiz. Ges. Anthropologie 1952/53, p. 49.

Furthermore, the number of grains yielded per plant was diminished to a statistically significant degree, whereas the "1000-grain-weight" showed no difference. We may thus conclude that the inferiority of the yield is connected with the number of grains only, the weight and size of each grain remaining constant.

Of special interest was the offspring of three seeds having suffered central hits by heavy primaries (F III/1, F IV/6 and G III/2, exposed in the summer of 1955 at Sault Ste Marie, geomagnetic latitude 55° north). Here a colour mutation has occurred. The grains derived from F IV/6 and G III/2 show a dark brown—dark grey, those derived from F III/1 a striking light yellow colour.

Zusammenfassung

Untersuchungsergebnisse der biologischen Wirkung der Kosmischen Strahlung auf Samen von Hordeum (Goldgerste) Bonus 01518/B 19 (Gustafsson) mit besonderer Berücksichtigung der Schweren Primären. Nachdem in den Jahren 1952/53 die mutagene Wirkung der Kosmischen Strahlung an Samen von *Artenia salina* nachgewiesen werden konnte[1], wurden hier solche Versuche an einem pflanzlichen Objekt wiederholt.

Während der Jahre 1953—1958 wurde ein Samenmaterial von Goldgerste (*Hordeum*) in mehreren Flügen in Höhen von 30000 bis 40000 m in Holloman während durchschnittlich 32 Stunden der Kosmischen Strahlung ausgesetzt und ihre Nachzuchten eingehend beobachtet. Die ersten Serien kamen in Payerne (Schweiz) und in Holloman (New Mexico) zur Exposition, die letzten 2 Serien F und G in Sault Ste Marie, Michigan, bei einer geomagnetischen Breite von 55° n. Als *Untersuchungsobjekt* wurde die sogenannte Rastermethode angewendet, die darin besteht, daß das zu untersuchende Objekt zusammen mit den zum Nachweis der physikalischen Ereignisse notwendigen Nuclearplatten eine Einheit bildet. Die Lagerung der Samen geschieht derart, daß gegenüber der Keimanlage ein genau markierter Bereich in der Emulsion entspricht, so daß diejenigen physikalischen Ereignisse in der Emulsion registriert werden konnten, welche diese Keimanlage durchsetzt haben mußten. Berücksichtigt wurden große Sternfiguren, Mesonenschauer, und vor allem Schwere Primäre, die durch den Keim hindurchgegangen sind oder wo in ganz vereinzelten Fällen ihr dünnes Ende ("thin down") im biologischen Objekt stecken blieb.

Ergebnisse: Als Ausgangsmaterial dienten 280 Samen, von denen 8539 Einzelpflanzen in 2 bis 4 Filial-Generationen nachgezüchtet, und auf die von Gustafsson angegebenen Merkmale untersucht wurden. Die Auswertung umfaßte folgende vier Kategorien:

1. Erdkontrollen (nicht exponierte Samen und ihre Nachkommen)
2. Sogenannte Rasterkontrollen (exponierte, aber nicht nachweisbar lokal getroffene Samen)
3. Samen mit Volltreffer (25 = 4,2 % der total exponierten Samen)
4. Samen mit Partialtreffern (Randtreffer oder kleinere Zertrümmerungs-Sterne).

Die *Nachkommen der 25 vollgetroffenen Samen* zeigten durchgehend eine 30 bis 40 %ige Abweichung gegenüber den Kontrollen im Sinne einer geringeren Bestockungszahl (Zahl der Triebe und Zahl der ährentragenden Schoße), sowie einer statistisch gesicherten Verminderung der Kornzahl, während die Werte für 1000 Korngewicht keine Unterschiede aufwiesen. Daraus kann geschlossen werden, daß die Unterwertigkeit des Ertrages sich lediglich auf die Anzahl der Körner bezieht, das Gewicht und die Größe des Einzelkorns jedoch unverändert bleiben.

Besonderes Interesse boten die Nachzuchten von drei vollgetroffenen Samen (F III/I, F IV/6 und G III/2), die von Schweren Primären getroffen waren (Exposition Sommer 1955 Sault Ste Marie, geomagnetische Breite 55° n). Es handelt sich hier um eine Farbenmutation. Die Körner von F III/1 sind auffallend hellgelb, diejenigen von F IV/6 und G III/2 dunkelbraun bis dunkelgrau.

---

[1] Die mutagene Wirkung der Kosmischen Strahlung. Bull. Schweiz. Ges. Anthropologie 1952/53, S. 49.

Aus dem im Jahre 1958 ausgewerteten Material geht hervor, daß auch chemische Mutationen nachgewiesen werden konnten, die sich auf die Enzym-Aktivität, den Stickstoffgehalt und die Geschwindigkeit der Keimung in 24 Stunden beziehen.

Eine besondere Aufmerksamkeit wird bei den kommenden Analysen auf die Kopplung morphologischer Merkmale mit chemischen Eigenschaften zu richten sein.

Résumé

Résultats de recherches sur les effets biologiques de rayons cosmiques sur des germes de Hordeum Bonus Gustafsson (orge doré). Un échantillonnage important de ces semences ont été soumises au rayonnement cosmique dans différents vols en ballon à des altitudes comprises entre 30.000 et 40.000 mètres de 1953 à 1958 (Holloman U.S.A.). La descendance des graines atteintes par des rayons primaires lourds montre une mutation caractéristique vers des formes naines. De plus se présentent aussi des changements de couleur, en particulier une variété jaune clair éclatante. Ces changements de caractéristiques ont pu être suivis au cours de quatre générations successives. Dans les analyses futures une attention spéciale sera accordée aux relations entre caractéristiques morphologiques et propriétés chimiques. En particulier l'activité des enzymes, l'importance des constituants azotés et la vitesse de germination par 24 heures doit être investiguée (Institut de botanique E.T.H. Zürich).

Table I. *Total Survey*

Compilation of the average % deviations of the filial generations of centrally hit seeds in all series.

Number of observations (plants) Series A: 3212, Series C: 3155,
Series F/G: 2172
Total: 8539

Characteristics	Series	Deviation of the filial generations of hit seeds	Total	Remarks
Length of Blades (cm.)	A. (4 generations) C. (3 generations) FG. (3 generations)	— 9% — 0.5% — 7.4%	— 5.3%	±8.0
Number of sprouts	A. C. FG.	—60 % —44 % —50 % —23.5%	—51.3% —35.3%	±2.7
Length of ears	A. C. FG.	— 1.8% — 7.0% — 8.7%	— 5.8%	±4.8
Number of grains	A. C. FG.	—44.6% —62.4% —69.7%	—57.2%	±10.2
"Bestockung" (Number of ear bearing sprouts)	A. C. FG.	—16.8% —43.3% —27.0%	—29.0%	±1.0
Number of plants	A. C. FG.	—37.3% —47.2% —29.2%	—37.9%	±5.1
Weight of plants	A. C.	—31.0% —51.3% —23.5%	—35.3%	±0.8

Table II. *Total Survey (B)*

Characteristics	Series	Deviation of the filial generations of hit seeds	Total	Remarks
Number of ears	A. C.	—32.3% —49.9% —22.0%	—34.7%	±2.3
Weight of ears	A. C. FG.	—28.9% —50.6% —26.9%	—35.5%	±1.7
Weight of grains	A. C. FG.	—31.0% —49.1% —24.4%	—34.8%	±1.5
1000 grain weight	A. C. FG.	— 1.7% — 3.7% — 7.4%	— 4.3%	±0.1
Grain % (grain weight per gram of plant weight)	A. C. FG.	— 4.8% — 6.3% — 8.2%	— 6.4%	±2.9

Sur le danger météoritique en astronautique
(2me partie)

(Reçu le 29 juin 1959)

Résumé — Zusammenfassung — Abstract

Sur le danger météoritique en astronautique. Dans sa communication faite à Amsterdam (le 28 août 1958 — IXe Congrès International d'Astronautique) l'auteur a mis en rapport le problème ancien de l'origine des cratères lunaires et le problème nouveau de la sécurité des voyages interplanétaires et il est arrivé à une conclusion optimiste, confirmée par les observations postérieures.

Dans cette seconde partie de sa communication l'auteur fait quelques compléments nouveaux à sa démonstration. Il montre, en particulier, que la forme exactement circulaire de Mercure qui tourne toujours le même côté au Soleil est un indice de la "pureté" relative de l'espace et de la sécurité très élevée des voyages interplanétaires.

Über die Meteoritengefahr in der Astronautik. In seiner am 28. August 1958 beim IX. Internationalen Astronautischen Kongreß in Amsterdam gemachten Mitteilung hat der Verfasser das alte Problem des Ursprungs der Mondkrater in Beziehung mit dem neuen Problem der Sicherheit interplanetarischer Reisen gebracht. Er gelangte dabei zu einer optimistischen Schlußfolgerung, die durch spätere Beobachtungen bestätigt wurde.

Im vorliegenden zweiten Teil seiner Mitteilung fügt der Verfasser seiner Beweisführung mehrere neue Ergänzungen hinzu. Im besonderen zeigt er, daß die genau kreisförmige Gestalt der Planetenscheibe des Merkur, der der Sonne stets dieselbe Seite zukehrt, ein Hinweis für die relative „Reinheit" des Raumes und die große Sicherheit interplanetarischer Reisen ist.

On the Meteoritic Danger in Astronautics. In a communication presented at the IXth International Astronautical Congress (Amsterdam, August 28, 1958), the author looked into the old problem of the origin of lunar craters and the new problem of safety in interplanetary voyages, and an optimistic conclusion was reached, confirmed by later observations.

In this second part the author makes some new observations to his demonstration. He shows, in particular, that the exactly circular form of Mercury, which always turns the same side to the Sun, is an indication of the relative "purity" of space and of the very high security of interplanetary travel.

1. Dans notre communication sous le même titre faite au IXe Congrès International d'Astronautique à Amsterdam (présidence de M. STEMMER) [1] nous avons exposé quelques considérations qui tendent à montrer que le danger météoritique menaçant les voyages interplanétaires n'est pas si grand comme on le pensait en général. Nous avons mis, proprement dit, en rapport le problème

---

[1] Président de la Section d'Astronautique de l'Aéroclub Central, Institut d'Astronomie, Université de Sofia, Bulgarie.

actuel de la sécurité des voyages interplanétaires avec le problème ancien de l'origine des cratères lunaires. Les deux hypothèses principales relatives à cette origine sont, comme on sait, l'hypothèse volcanique et l'hypothèse météoritique. La première hypothèse est plus ancienne et elle semble être plus "naturelle"; la notion de cratère, en effet, est liée dans l'esprit à celle de volcan et, d'autre part, ces formations lunaires ressemblent beaucoup aux cratères volcaniques terrestres. Cependant l'hypothèse météoritique a gagné beaucoup de terrain au cours des dernières années (R. B. Baldwin, F. Hoyle, ...) et V. V. Charonov a constaté dans son livre de 1958 [2] que la lutte entre les deux hypothèses principales relatives à l'origine des cratères lunaires continue.

Si l'hypothèse météoritique correspondait à la réalité, c'est à dire, si les cratères lunaires seraient dûs à la chute de météorites sur la Lune, cela pourrait obscurcir l'horizon et les perspectives de l'Astronautique. Nous devons, par conséquent, lutter contre l'hypothèse météoritique mais, naturellement, par des arguments fondés et solides.

Nous avons fait dans ce but une comparaison statistique entre la distribution des cratères sur l'hémisphère Est et l'hémisphère Ouest de la Lune. L'hémisphère Est est plus exposé au bombardement météoritique que l'hémisphère Ouest et nous devrions discerner cette différence dans la distribution des cratères sur les deux hémisphères de la Lune, si, en réalité, ces formations seraient dues à la chute de météorites. Or, le résultat de notre comparaison statistique, basée sur quelques théorèmes de la théorie des probabilités, n'est pas favorable à l'hypothèse météoritique et est, par conséquent, en faveur de l'ancienne hypothèse volcanique. Ainsi, dans notre communication susdite de l'année passée (le 28 Août 1958) nous avons pu aboutir à une conclusion optimiste concernant la sécurité des voyages interplanétaires.

2. Un peu plus de deux mois plus tard, le 4 Novembre 1958, l'astronome soviétique N. A. Kozirev a donné une confirmation remarquable à notre conclusion susdite en observant nettement l'éruption d'un volcan sur la Lune. Au mois d'Avril 1959, d'autre part, on a confirmé en U.R.S.S. d'une autre manière notre conclusion optimiste susdite, basée sur l'observation du satellite naturel de la Terre. Les observations automatiques à l'aide du 3ième satellite artificiel soviétique ont montré, en effet, que le nombre des corps météoritiques dans l'espace n'est pas si considérable comme on l'a cru jusqu'à maintenant et que, par conséquent, le danger météoritique en Astronautique n'est pas, de même, considérable.

3. On pourrait objecter, quant à notre démonstration, que la vitesse de la Lune autour de la Terre est petite par rapport à celle de la Terre dans son mouvement annuel autour du Soleil et qu'on ne devrait pas espérer de trouver, par conséquent, la dite différence dans la distribution des cratères sur les deux hémisphères lunaires. Cependant dans le passé la Lune a été tout près de la Terre[1]. Sa distance, à cause des marées, s'est accrue et a atteint sa valeur actuelle. Il est permis de croire qu'il a eu donc dans le passé une époque T lorsque cette distance a été sensiblement plus courte, la Lune tournant, tout de même, la même face à la Terre. Il est évident qu'à cette époque T la vitesse de la Lune dans son mouvement orbital autour de la Terre n'a pas été petite, comme actuellement, par rapport à la vitesse de la Terre.

Il est clair cependant que, si à l'époque T du passé l'espace interplanétaire a été caractérisé par une "pureté" relative élevée, il n'y a pas de raison de croire qu'à l'époque actuelle cette "pureté" relative soit moins parfaite. C'est le contraire qui paraît naturel.

---

[1] En faisant même la Terre sensiblement plus aplatie [3].

4. La planète Mercure tourne toujours la même face au Soleil; c'est un état analogue à l'état de la Lune par rapport à la Terre. Si la vitesse de la Lune dans son mouvement autour de la Terre est petite, la vitesse de Mercure autour du Soleil est considérable; elle est même supérieure à celle de la Terre autour du Soleil. Or, sur la Terre on observe "l'effet SCHIAPARELLI" — le nombre des météores après minuit est plus grand que le nombre des météores avant minuit. "L'effet SCHIAPARELLI" devrait donc se manifester sur Mercure plus fort que sur la Terre et, au cours de centaines de millions ou de milliards d'années, l'hémisphère par lequel cette planète avance sur son orbite devrait accumuler sur lui plus de matière météorique que l'autre hémisphère. On devrait donc s'attendre à constater aux époques des "passages" de cette planète sur le disque du Soleil une déviation de sa forme de la forme circulaire. Il ne s'agit nullement de constater une forme élliptique mais une déformation plus forte sur le côté droit que sur le côté gauche du disque planétaire. Il est vrai que cette déformation devrait être sensible pour être observable de la Terre. Nous devrions, d'autre part, avoir la certitude que l'effet dû à l'action de marée du Soleil sur cette planète ne s'y mêle pas, c'est à dire, qu'il est inobservable de la Terre.

Une question analogue peut se poser relativement à la planète Vénus au sujet de laquelle on a émis dernièrement le soupçon qu'elle tourne, de même, toujours le même côté au Soleil. D'après J. GAUZIT [4] "DANJON a observé certaines taches, qui paraissent persistantes, et il en a déduit que Vénus présente toujours la même face au Soleil, comme Mercure".

5. Etudions l'effet de l'action de marée du Soleil sur Mercure.

C'est la plus petite des "grosses planètes". Les petits corps célestes n'ont pas en général une forte condensation au centre. On peut estimer, d'autre part, que les planètes ont été dans le passé plus homogènes. C'est pour cette raison que nous allons appliquer ici la théorie d'une masse fluide homogène en rotation (Mercure), attirée par un corps extérieur éloigné C (Soleil), situé dans le plan de l'équateur [5]. Le centre de gravité de cette masse décrit un cercle ayant son centre en C; on suppose que la durée de révolution est égale à la durée de la rotation de la masse fluide. Sa forme d'équilibre est un ellipsoïde à trois axes a, b, c.

Nous avons l'équation

$$\frac{s\,(1-s)}{s+3+\mu} \int_0^\infty \frac{u\,du}{(1+u)\,(1+su)\,\varDelta} = \frac{t\,(1-t)}{t+\mu} \int_0^\infty \frac{u\,du}{(1+u)\,(1+tu)\,\varDelta} \tag{1}$$

$$s = \frac{a^2}{b^2}, \; t = \frac{a^2}{c^2}; \quad 0<s<1, 0<t<1; \quad \varDelta = \sqrt{(1+u)\,(1+su)\,(1+tu)}$$

a est l'axe de rotation — le petit axe; b — le grand axe — est dirigé vers le Soleil; c — l'axe moyen.

$\mu = \dfrac{M}{M'}$, M — masse de Mercure; M' —masse du Soleil.

Pour les formes stables $t>s$.

Au voisinage du point A (Fig. 15 des Leçons... de POINCARÉ) l'ellipsoïde diffère peu de la sphère, s et t sont presque égaux et, par conséquent, les deux intégrales dans l'éq. (1) sont presque égales. En divisant par elles nous obtiendrons l'équation

$$\frac{s\,(1-s)}{s+3+\mu} = \frac{t\,(1-t)}{t+\mu}. \tag{2}$$

Pour Mercure $\mu = \dfrac{1}{6\,000\,000}$; en négligeant μ nous obtiendrons

$$t = \frac{3 + s^2}{3 + s} \, . \tag{3}$$

Quoique cette équation soit approchée, on peut vérifier facilement que $t > s$ (par exemple si $s = 0,85 \; t = 0,97$).

L'éq. (3) nous donne

$$s_{1,2} = \frac{t \pm \sqrt{t^2 - 12\,(1 - t)}}{2} \, .$$

Les deux valeurs de s, t étant donné, correspondent aux deux ellipsoïdes d'un même t, voisin de 1, qui existent comme formes d'équilibre (la même fig. 15 — Poincaré). Si $t < 0,925$ ces deux formes cessent d'exister.

La distance minimale moyenne de Mercure à la Terre (passage) est de 90.000.000 km environ. Pour qu'une longueur, à une telle distance, puisse être bien visible, il faut qu'elle ait 218 km au minimum (nous estimons qu'un angle de $0'',5$ peut être mesuré avec *certitude*).

Nous avons $c - a = 218$ km, ou

$$1 - \frac{a}{c} = \frac{218}{c} \, , \; c = 2350 \; \text{km (Mercure)}$$

$$1 - \frac{a}{c} = 0,0928, \; \frac{a}{c} = 0,9072, \; t = \frac{a^2}{c^2} = 0,8230 \, .$$

Cependant, comme nous l'avons vu, un tel ellipsoïde $(t = 0,823)$ ne peut pas exister. Il s'ensuit que la marée solaire ne peut pas engendrer sur Mercure une déformation distinctement observable de la Terre.

Un calcul analogue relatif à Vénus (en admettant qu'elle est homogène et tourne la même face au Soleil) nous a donné un résultat incertain.

Conclusion

La sphéricité observée, presque absolue, de Mercure est un indice de la "pureté" relative de l'espace interplanétaire. Le danger météoritique est insignifiant... mais nous devons, tout de même, chercher des moyens pour lutter contre les corps météoritiques tout petits.

Bibliographie

1. N. Boneff, Sur le danger météoritique en astronautique. Comptes Rendus du IXe Congrès International d'Astronautique, Amsterdam 1958, p. 557. Wien: Springer, 1959. — La distribution des cratères lunaires en rapport à leur origine. Un argument nouveau contre l'hypothèse météorique. Ann. Univ. Sofia **50**, 1955/1956, livre 1 (partie I). — La distribution des cratères lunaires en rapport à leur origine. Un nouvel argument contre l'hypothèse météorique (suite). Ann. Univ. Sofia **50**, 1955/1956, livre 1 (partie II). — Sur l'hypothèse météorique de l'origine des cratères lunaires. Rev. Météorit. Acad. Sci. U.R.S.S., Moscou. B. **16**, 115 (1958) (en russe).
2. V. V. Charonov, La nature des planètes, p. 195. Moscou, 1958 (en russe).
3. N. Boneff, Sur le contour des continents terrestres. Influence probable de la Lune. Astr. Nachr. **284**, H. 4, 155 (1958).
4. J. Gauzit, Les grands problèmes de l'Astronomie, p. 50. Paris, 1957.
5. H. Poincaré, Leçons sur les hypothèses cosmogoniques, p. 53. Paris, 1913.

Some New Methods of Satellite Orbit Calculations and Stability Problems

By

Herbert Knothe[1], ARS

(Received July 16, 1959)

Abstract — Zusammenfassung — Résumé

Some New Methods of Satellite Orbit Calculations and Stability Problems. In chapter 1 the equations of motion for a satellite in a rotationally symmetric gravity field are reduced to a system of two ordinary first order differential equations for p and $1/r$ as functions of the angular polar distance θ by first introducing the arc length s of the spherical image of the orbit on the unit sphere as independent parameter. The connection between s and θ is given by a differential relation between inclination i, s and θ.

In chapter 2 a system of three ordinary first order differential equations for p, ε, φ_0 as functions of φ (φ defined by $\cos \theta = \sin i \sin \varphi$; φ_0 corresponds to the apogee) is derived. A rapidly convergent process of iteration for solving these equations is explained and explicit examples are given.

In chapter 3 a differential geometrical approach is applied which leads to simple formulae for the calculation of the regression. A vector integral equation is derived for the deviation of the satellite from the plane of the initial osculating ellipse. This equation together with a simple scalar equation of chapter 1 is equivalent to the equations of motion.

Einige neue Methoden zur Berechnung von Satellitenbahnen und Stabilitätsproblemen. Im ersten Kapitel werden die Bewegungsgleichungen eines Satelliten in einem rotations-symmetrischen Schwerefeld reduziert auf ein System von zwei gewöhnlichen Differentialgleichungen erster Ordnung für p und $1/r$ als Funktionen des Polabstandes θ. Zuerst wird die Bogenlänge s des sphärischen Bildes des Orbits auf der Einheitskugel als unabhängiger Parameter eingeführt. s und θ sind durch eine Differentialbeziehung zwischen θ, i, s miteinander verbunden.

Im zweiten Kapitel werden drei gewöhnliche Differentialgleichungen erster Ordnung für p, ε, φ_0 als Funktionen von φ (φ wird definiert durch $\cos \theta = \sin i \sin \varphi$; φ_0 ist der φ-Wert des Apogäums) abgeleitet. Diese Gleichungen lassen sich mit Hilfe eines rasch konvergierenden Iterationsprozesses lösen. Explizite Beispiele werden angegeben.

Im dritten Kapitel wird eine differentialgeometrische Methode angewandt, die zu einer einfachen Formel zur Berechnung der Regression führt. Eine Vektorintegralgleichung wird abgeleitet für die Abweichung des Satelliten von der Ebene der oskulierenden Anfangsellipse. Diese Gleichung zusammen mit einer einfachen skalaren Gleichung des ersten Kapitels ist dem System der Bewegungsgleichungen äquivalent.

Nouvelles méthodes de calcul d'orbites et de problèmes de stabilité. Au premier chapitre les équations du mouvement d'un satellite dans un champ de gravité à symétrie axiale sont réduites à un système de deux équations différentielles ordinaires

---

[1] U.S. Air Force Missile Development Center, Holloman Air Force Base, New Mexico, U.S.A.

du premier ordre. Elles utilisent commé paramètre indépendant l'arc s d'une orbite image sur la sphère unité. La relation différentielle entre s et la colatitude θ implique l'inclinaison i.

Dans le second chapitre on établit un système de trois équations différentielles du premier ordre p, ε et φ_0 comme fonctions d'un angle φ défini par $\cos \theta = \sin i \sin \varphi$ (φ_0 correspond à l'apogée). On expose un procédé d'itération rapidement convergent pour leur solution avec exemples.

Au chapitre 3 une approche de géométrie différentielle conduit à des formules simples pour le calcul de la régression. Une équation intégrale vectorielle gouverne la déviation au plan de l'ellipse osculatrice initiale. Cette équation associée à une équation scalaire du premier chapitre constitue une formulation équivalente du mouvement.

Symbols

r Distance between center of gravity (c.g.) of the earth and satellite

ϱ $1/r$

$\bar{x}$ Local vector c.g. → satellite

$\bar{a}$ Unit vector $1/r \cdot \bar{x}$ (spherical image of the satellite)

Φ Gravitational potential of the earth

ε Eccentricity of the osculating ellipse

p Parameter of the osculating ellipse

i Angle between instantaneous orbit plane and equator

θ Angle between $\bar{a}$ and earth axis $\bar{z}$ ($\bar{z}$ = unit vector)

φ Angle defined by $\cos \theta = \sin i \sin \varphi$, φ is counted from the ascending node of the osculating ellipse on

φ_0 Angle between apogee of the o.e. and ascending node

$\bar{\xi}$ Unit vector of the direction c.g. → apogee of the o.e.

$\bar{\eta}$ Unit vector perpendicular to $\bar{\xi}$ in the o.e.

$\bar{n}$ Unit vector $\bar{\xi} \times \bar{\eta}$ (vector product of $\bar{\xi}$ and $\bar{\eta}$)

$\bar{c}$ Vector product $\bar{x} \times \dot{\bar{x}}$

1. Reduction of the equations of motion of a satellite to a system of two ordinary first order differential equations, considering rotationally symmetric gravitational anomalies.

We start from the equation of motion

$$\ddot{\bar{x}} = -\frac{\mu_0}{r^3}\bar{x} + \bar{y} \tag{1}$$

where $\bar{y}$ is the gradient of a disturbing potential Φ_1 and the term

$$-\frac{\mu_0}{r^3}\bar{x}$$

is the gradient of the spherically symmetric part μ_0/r of the earth's potential Φ. For many purposes it is convenient to replace the time t by the arc lengths of the spherical image of the orbit as independent parameter. In other words the parameter s means the arc length of the curve described by the unit vector $\bar{a}$ ($\bar{x} = r\,\bar{a}$) on the unit sphere. The familiar Keplerian formula, expressing Kepler's first law,

$$r^2 \frac{d\varphi}{dt} = C = \text{const} = \sqrt{p\,\mu_0}$$

can in the case of a nonspherically symmetric body be replaced by

$$r^2 \frac{ds}{dt} = \sqrt{p\,\mu_0} \tag{2}$$

where p is now no longer necessarily constant. We have operationally

$$\frac{d}{dt} = \frac{\sqrt{p\,\mu_0}}{r^2}\frac{d}{ds}. \tag{3}$$

Characterizing now the derivatives with respect to s by a prime we can transform formula (1) into

$$(r'' \bar{a} + 2 r' \bar{a}' + r \bar{a}'') \frac{p \mu_0}{r^4} + \mu_0 (r' \bar{a} + r \bar{a}') \left(\frac{p'}{2 r^4} - \frac{2 p r'}{r^5} \right) = - \frac{\mu_0}{r^2} \bar{a} + y. \qquad (4)$$

We multiply both sides of eq. (4) by $\bar{a}, \bar{a}', \bar{a} \times \bar{a}'$, and introduce $\varrho = 1/r$ instead fo r. Formula (4) is then equivalent to the following three equations

$$(\varrho + \varrho'') p + {}^1/_2 p' \varrho' = 1 - \frac{1}{\mu_0 \varrho^2} (y \bar{a}) \qquad (5a)$$

$$p' = \frac{2}{\varrho^3 \mu_0} (\bar{y} \bar{a}') \qquad (5b)$$

$$p (\bar{a} \bar{a}' \bar{a}'') = \frac{1}{\varrho^3 \mu_0} (\bar{y} \bar{a} \bar{a}') \qquad (5c)$$

where $(\bar{a} \bar{a}' \bar{a}'')$ means the determinant of the vectors $\bar{a} \bar{a}', \bar{a},''$.

Eq. (5a) can be replaced by the energy equation

$$\frac{p}{2} (\varrho^2 + \varrho'^2) - \varrho - \frac{1}{\mu_0} \Phi_1 = K = \text{const.}$$

which can be derived easily from eqs. (5a), (5b) where

$$\bar{y} = \Delta \Phi_1 . \qquad (6)$$

Thus the eqs. (5a), (5b), (5c) can be simplified to

$$\frac{p}{2} (\varrho^2 + \varrho'^2) - \varrho - \frac{1}{\mu_0} \Phi_1 = K = \text{const.} \qquad (6a)$$

$$p' = \frac{2}{\varrho^3 \mu_0} (\bar{y} \bar{a}') \qquad (6b)$$

$$p (\bar{a}, \bar{a}', \bar{a}'') = \frac{1}{\varrho^3 \mu_0} (\bar{y} \bar{a} \bar{a}'). \qquad (6c)$$

We now represent $\bar{y}$ in formula (6) as a linear combination of the vectors $\bar{a}$ and $\bar{z}$ (unit vector of the earth axis).

$$\bar{y} = \left(\frac{\partial \Phi_1}{\partial r} + \frac{\partial \Phi_1}{r \partial \theta} \cot \theta \right) \bar{a} - \frac{\partial \Phi_1}{\partial \theta} \cdot \frac{1}{r \sin \theta} \bar{z} . \qquad (7)$$

Introducing this expression for $\bar{y}$ into eqs. (6b), (6c) and taking into account that $\bar{z} \bar{a} = \cos \theta$ the eqs. (6b), (6c) assume the form

$$\frac{dp}{ds} = \frac{2}{\varrho^2 \mu_0} \frac{\partial \Phi_1}{\partial \theta} \frac{d \theta}{ds} \qquad (8)$$

$$p (\bar{a} \bar{a}' \bar{a}'') = - \frac{1}{\varrho^2 \mu_0} \frac{\cos i}{\sin \theta} \frac{\partial \Phi_1}{\partial \theta} . \qquad (9)$$

For later problems the following remark is useful: $(\bar{a} \bar{a}' \bar{a}'')$ is the geodesic curvature K_g of the spherical image. Hence it follows from formula (9)

$$K_g = - \frac{1}{p \varrho^2 \mu_0} \frac{\cos i}{\sin \theta} \frac{\partial \Phi_1}{\partial \theta} . \qquad (10)$$

The system of differential eqs. (6a), (6b), (6c) is valid for any arbitrary gravitational anomaly. For the case of a rotationally symmetric body it is advantageous to choose formulae (6a), (8), (9) as equations of motion. In order to reduce this system to a system of two ordinary first order differential equations we introduce θ instead of s as independent parameter in the following way. The well known fact that the component of the angular momentum vector in the direction of the symmetry axis is constant can be expressed as follows

$$p \cos^2 i = \lambda = \text{const.} \tag{11}$$

We now calculate the determinant $(\bar{a}, d\bar{a}, \bar{z})$. It is

$$(\bar{a}, d\bar{a}, \bar{z})^2 = \cos^2 i \, ds^2 = \begin{vmatrix} 1 & 0 & \cos\theta \\ 0 & ds^2 & d(\cos\theta) \\ \cos\theta & d(\cos\theta) & 1 \end{vmatrix}$$

or

$$ds^2 = \frac{\sin^2\theta \, d\theta^2}{\sin^2\theta - \cos^2 i} \, . \tag{12}$$

Elimination of i from eq. (12) by means of eq. (11) yields

$$ds^2 = \frac{p \sin^2\theta}{p \sin^2\theta - \lambda} \, d\theta^2 \, . \tag{13}$$

The relation (13) enables us to write the eqs. (6a), (8) as two differential equations of the first order for p and ϱ as functions of θ:

$$\frac{p}{2} \left\{ \varrho^2 + \left(\frac{d\varrho}{d\theta} \right)^2 \left(1 - \frac{\lambda}{p \sin^2\theta} \right) \right\} - \varrho - \frac{1}{\mu_0} \Phi_1 (\varrho, \cos\theta) = K \tag{14a}$$

$$\frac{dp}{d\theta} = \frac{2}{\mu_0 \varrho^2} \frac{\partial \Phi_1}{\partial \varrho} \, . \tag{14b}$$

For later purposes it is useful to express the time element dt by $d\theta$ and $d\varphi$. Since φ extends from 0 to 2π if the satellite wanders from one ascending node to the succeeding one φ is a particularly valuable parameter.

At first we write formula (14b) in the form

$$dp = \frac{2}{\mu_0 \varrho^2} \frac{\partial \Phi_1}{\partial (\cos\theta)} d(\cos\theta) \, . \tag{15}$$

Taking advantage of the formulae

$$\cos\theta = \sin i \sin\varphi$$

$$p \cos^2 i = \lambda = \text{const.} \tag{16}$$

we derive from eq. (15) after elementary calculations

$$dp = \frac{2 \dfrac{\partial \Phi_1}{\partial (\cos\theta)} (p - \lambda) \cos\varphi}{\mu_0 p^2 \varrho^2 \sqrt{1 - \dfrac{\lambda}{p} - \dfrac{\partial \Phi_1}{\partial (\cos\theta)} \sin\varphi}} \, d\varphi \tag{17}$$

from eqs. (2) and (12) we conclude immediately

$$dt^2 = \frac{r^4}{p \mu_0} \frac{\sin^2\theta \, d\theta^2}{\sin^2 i \cos^2\varphi}$$

or

$$dt = -\frac{r^2}{\sqrt{p \mu_0}} \frac{\sin\theta \, d\theta}{\sin i \cos\varphi} = \frac{r^2}{\sqrt{p \mu_0}} \left(\frac{\tan\varphi \cot^2 i}{2 p} \, dp + d\varphi \right) . \tag{18}$$

Taking dp from formula (17) dt is expressed in the desired form.

2. Reduction of the equations of motion to a system of three ordinary first order differential equations, representing φ_0, p, ε as functions of φ.

Since the detailed calculations of this chapter have already been made accessible to the general public [1, 2], I can confine myself to the explanation of the fundamental ideas. Let us assume that we do not know anything about a spherical unsymmetry of the earth but that we know the coefficient μ_0. The knowledge of $\bar{x}$ and the velocity vector $\dot{\bar{x}}$ would enable us to calculate the elements of a KEPLER ellipse corresponding to a spherically symmetric earth with the characteristic coefficient μ_0. We call this KEPLER-ellipse the osculating ellipse.

In vector representation these elements can be taken from the elementary equations

$$\bar{x} = \frac{p}{1 - \varepsilon \cos \psi} \left(\cos \psi \, \bar{\xi} + \sin \psi \, \bar{\eta} \right) \tag{19}$$

$$\dot{\bar{x}} = \sqrt{\frac{\mu_0}{p}} \left(- \sin \psi \, \bar{\xi} + (\cos \psi - \varepsilon) \, \bar{\eta} \right) \tag{20}$$

where $\psi = \varphi + \varphi_0$.

If we would perform the calculations of these elements for the same satellite at different times we would get different vectors $\bar{\xi}$, $\bar{\eta}$ and different values of p and ε, in other words these elements have to be considered as functions of t. However, the formulae (19) and (20) hold for any time t. This is possible only if the time derivative of the right hand side of eq. (19) in which p, ε, ψ, $\bar{\xi}$, $\bar{\eta}$ are now functions of t equals the right hand side of eq. (20). Thus we obtain three vector equations which can be transformed into three scalar equations by scalar multiplication with $\bar{\xi}$, $\bar{\eta}$, n. The three missing equations are obtained by differentiating eq. (20) with respect to t and substituting these expressions in eq. (1).

After some transformations we obtain the following equations

$$\bar{\xi} \, d\bar{\eta} = \frac{1}{2} \frac{\sin \psi}{\varepsilon \, p} dp - \frac{(\bar{y} \, \bar{\xi})}{\varepsilon} \sqrt{\frac{p}{\mu_0}} \, dt \tag{21}$$

$$\bar{n} \, d\bar{\xi} = - \sqrt{\frac{p}{\mu_0}} \cdot \frac{1}{N} \sin \psi \, (\bar{y} \, \bar{n}) \, dt \tag{22}$$

$$d\varepsilon = - \sqrt{\frac{p}{\mu_0}} \, (\bar{y} \, \bar{\eta}) \, dt + \frac{(\varepsilon - \cos \psi)}{2 \, p} \, dp \, . \tag{23}$$

The different steps to be taken in order to express the right hand sides of eqs. (21), (22), (23) in terms of p, ε, φ, φ_0, $d\varphi$ are:

Step 1: Taking $\bar{y}$ as given by formula (7) and using the relations

$$\bar{a} \, \bar{\xi} = \cos (\varphi + \varphi_0) \tag{24a}$$

$$\bar{a} \, \bar{\eta} = \sin (\varphi + \varphi_0) \tag{24b}$$

$$\bar{a} \, \bar{n} = 0 \tag{24c}$$

$$\bar{z} \, \bar{\xi} = - \sin i \sin \varphi_0 \tag{24d}$$

$$\bar{z} \, \bar{\eta} = \sin i \cos \varphi_0 \tag{24e}$$

$$\bar{z} \, \bar{n} = \cos i \tag{24f}$$

the scalar products $(\bar{y} \, \bar{\xi})$, $(\bar{y} \, \bar{\eta})$, $(\bar{y} \, \bar{n})$ are determined as functions of p, ε, φ, φ_0.

Step 2. Eq. (17) already gives us dp in the desired form:

$$\frac{dp}{d\varphi} = f_1\,(p,\,\varepsilon,\,\varphi,\varphi_0)\,.\tag{25}$$

Step 3. Substituting eq. (25) in eq. (18) gives dt in the form

$$dt = h\,(p,\,\varepsilon,\,\varphi_0,\,\varphi)\,d\varphi\,.$$

We now have

$$\frac{d\varepsilon}{d\varphi} = f_2\,(p,\,\varepsilon,\,\varphi,\varphi_0)\,.\tag{26}$$

In order to find $d\varphi_0$ we begin with the relation (24 d) from which it follows

$$d\bar{\xi}\,\bar{z} = -\cos i \sin\varphi_0\,di - \sin i \cos\varphi_0\,d\varphi_0\,.\tag{27}$$

Representing $d\bar{\xi}$ as a linear combination of $\bar{\eta}$ and $\bar{n}$ ($\bar{\xi}\,d\bar{\xi}=0$) and using formula (24e), (24f) we find

$$\bar{z}\,d\bar{\xi} = -(\bar{\xi}\,d\bar{\eta})\sin i\,\cos\varphi_0 + (d\bar{\xi}\,\bar{n})\cos i\,.\tag{28}$$

We now recall that the scalar products $\bar{\xi}\,d\bar{\eta}$ and $d\bar{\xi}\,\bar{n}$ have been expressed as functions of p, ε, φ, φ_0, $d\varphi$ and that

$$di = \cot i\,\frac{dp}{2\,p}\,.$$

(Consequence of $p\cos^2 i=\lambda=\text{const.}$). Therefore, combining eqs. (27) and (28) and using eq. (17) we obtain finally

$$\frac{d\varphi_0}{d\varphi} = f_3\,(p,\,\varepsilon,\,\varphi,\varphi_0)\,.\tag{29}$$

Eqs. (25), (26), (29) form a system of three ordinary first order differential equations for p, ε, φ_0 as functions of φ. This system can be solved by the following rapidly converging process of iteration. We first substitute for p, ε, φ_0 their initial values and integrate to obtain the first approximation functions $p^{(1)}$, $\varepsilon^{(1)}$, $\varphi_0^{(1)}$. After introducing these functions into the right hand sides of eqs. (25), (26), (29) we obtain the second approximation functions $p^{(2)}$, $\varepsilon^{(2)}$, $\varphi_0^{(2)}$, and so on. It can be shown that for all practical purposes the second approximation yields extremely accurate results. We now consider two explicit examples. Assuming Φ_1 in the form

$$\Phi_1 = \frac{\mu_2}{r^3}\,(3\cos^2\theta - 1)$$

we obtain

$$\varphi_0^{(1)}\,(2\,\pi) - \varphi_0^{(1)}\,(0) = \Delta\varphi_0 = \frac{3\,\pi\,\mu_2}{\mu_0\,p^2}\,(4 - 5\sin^2 i)\tag{30}$$

which describes the motion of the apogee. As second example we consider the variation of p between two succeeding ascending nodes. We obtain

$$\{p^{(2)}\,(2\,\pi)\}^2 - \{p^{(2)}\,(0)\}^2 = 12\,p^2\left(\frac{\mu_2}{\mu_0\,p^2}\right)^2\varepsilon\pi\sin^2 i\,\sin\varphi_0\left[(4-5\sin^2 i)+\right.$$

$$\left. + \varepsilon\cos\varphi_0\left(\frac{15}{4}\sin^2 i - \frac{7}{2}\right)\right].\tag{31}$$

We shall now draw some conclusions from formula (31). Since the difference $p^{(2)} (2\pi) - p^{(2)} (0) = \Delta p$ is very small compared with p formula (31) can be written with sufficient accuracy

$$\frac{\Delta p}{p} = 6 \left(\frac{u_2}{\mu_\bullet p^2}\right)^2 \varepsilon \pi \sin^2 i \sin \varphi_0 \left[(4 - 5 \sin^2 i) + \varepsilon \cos \varphi_0 \left(\frac{15}{4} \sin^2 i - \frac{7}{2}\right)\right]. \quad (32)$$

For an orbit with $p = 7 \times 10^6$ meters the dimensionless quantity

$$\frac{\mu_2}{\mu_0 p^2}$$

equals 0.45×10^{-3}. It is evident that the maximum value of the percentual change of p is reached for approximately $i = \frac{\pi}{2}$, $\varphi_0 = \frac{\pi}{2}$. Assuming $\varepsilon = 0.05$, $i = \frac{\pi}{2}$, $\varphi_0 = \frac{\pi}{2}$, formula (32) shows that the maximum change of p in this case would be about 1.3 meters.

The question arises how p would change during a large number or rotations of the satellite. In answering this question we shall first confine ourselves to inclination angles which differ from $i_0 = 63.5°$ by more than about 3°. Let us divide both sides of eq. (32) by

$$\Delta \varphi_0 = \frac{3 \pi \mu_2}{\mu_0 p^2} (4 - 5 \sin^2 i). \quad (33)$$

We obtain

$$\frac{\Delta p}{\Delta \varphi_0} = 2 \frac{\mu_2}{\mu_0 p} \varepsilon \sin^2 i \sin \varphi_0 \left[1 + \varepsilon \cos \varphi_0 \frac{\frac{15}{4} \sin^2 i - \frac{7}{2}}{4 - 5 \sin^2 i}\right]. \quad (34)$$

In the case where i deviates considerably (about 3°) from i_0 it is justified to consider ε, i, p on the righthand side of eq. (34) as constants, since their changes are extremely small. For the same reason we are entitled to consider eq. (34) as a differential equation for p as a function of φ_0. We obtain by integration

$$p - p_0 = 2 \frac{\mu_2 \varepsilon}{\mu_0 p} \sin^2 i \left[\cos \varphi_0^{(0)} - \cos \varphi_0 + \varepsilon (\cos 2\varphi_0^{(0)} - \cos 2\varphi_0) \cdot \frac{\frac{15}{8} \sin^2 i - \frac{7}{4}}{4 - 5 \sin^2 i}\right]. \quad (35)$$

This formula gives the change of p after a large number of rotations, i.e., the number of rotations necessary for varying φ_0 from $\varphi_0^{(0)}$ to φ_0.

Formula (35) shows that for $i = \frac{\pi}{2}$, $\varepsilon = 0.05$,

$p = 7 \times 10^6$ meters, the maximum change of p is about 13 kilometers. In order to avoid any misunderstanding we mention that these changes of p refer to values of p at the ascending nodes.

For the sake of completeness the variation $\Delta \varepsilon$ of ε between two succeeding ascending nodes shall be derived. It would be cumbersome to determine $\Delta \varepsilon$ by integrating eq. (26) after substituting in the right hand side the first approximation functions $p^{(1)}$, $\varepsilon^{'1)}$, $\varphi_0^{(1)}$. Therefore we take advantage of the equation

$$\varepsilon^2 = 1 + \frac{2 (\mu_0 K + \Phi_1) p}{\mu_0} \quad (36)$$

which can be derived easily from eq. (14a). At a node the equation holds

$$\Phi_1 = -\mu_2 \varrho^3 = -\frac{\mu_2}{p^3}(1 - \varepsilon \cos\varphi_0)^3 .\tag{37}$$

Let ε_1, p_1 be the values of ε, p at the first ascending node and let ε_2, p_2 be the values of ε, p at the succeeding one. We then have

$$\varepsilon_1^2 = 1 + 2 K p_1 - \frac{2\mu_2}{\mu_0 p^2}(1 - \varepsilon_1 \cos\varphi_0)^3 \tag{38}$$

$$\varepsilon_2^2 = 1 + 2 K p_2 - \frac{2\mu_2}{\mu_0 p^2}(1 - \varepsilon_2 \cos(\varphi_0 + \varDelta\varphi_0)^3) .\tag{39}$$

If we neglect third and higher powers of

$$\frac{\mu_2}{\mu_0 p^2}\tag{40}$$

than the third we are entitled to substitute $\varepsilon_1 = \varepsilon_2$ in the last terms of eqs. (38), (39) since the difference $\varepsilon_2 - \varepsilon_1$ contains no linear terms of the dimensionless quantity (40). Therefore we obtain, subtracting eqs. (38), (39)

$$\varepsilon \varDelta\varepsilon = K\varDelta p - \frac{3\mu_2\varepsilon}{\mu_0 p^2}\sin\varphi_0(1 - \varepsilon\cos\varphi_0)^2 \varDelta\varphi_0\tag{41}$$

where K can be replaced by

$$\frac{\varepsilon^2 - 1}{2 p}$$

according to eq. (14a). Since [see formula (30)]

$$\varDelta\varphi_0 = \frac{3\pi\mu_2}{\mu_0 p^2}(4 - 5\sin^2 i)$$

we finally obtain

$$\varDelta\varepsilon = \frac{\varepsilon^2 - 1}{2\varepsilon p}\varDelta p - 9\pi\left(\frac{\mu_2}{\mu_0 p^2}\right)^2 \sin\varphi_0 (1 - \varepsilon\cos\varphi_0)^2 (4 - 5\sin^2 i)\tag{42}$$

where $\varDelta p$ can be derived from eq. (32).

It is important to remark that all derivations have been made in such a way as to trivialize the generalization to any linear combination of rotationally symmetric potentials.

3. Differential Geometrical Approach. Spherical image of the orbit.

The unit vector $\bar{a}$ describes, as mentioned above, a curve on the unit sphere the geodesic curvature of which K_g is connected with the primary orbit elements through the equation

$$K_g = -\frac{1}{p\varrho^2\mu_0}\frac{\cos i}{\sin\theta}\frac{\partial\Phi_1}{\partial\theta}\tag{43}$$

proved in chapter 1. If the orbit would be a KEPLER ellipse this curve which we call the spherical image of the orbit is a great circle which is characterized by the relation

$$K_g = 0 .\tag{44}$$

It is now of interest to investigate how the satellite will gradually deviate from the great circle of the initial osculating ellipse. This effect describes essentially the so-called regression. One might say that his effect has been eliminated in the preceding chapters.

We introduce again the arc length s of the spherical image as independent parameter and map the initial o.e. and the actual orbit on each other by equal arc lengths s of their spherical images. Let $\bar{a}_0(s)$ be the spherical image of the initial o.e. and let $\bar{a}(s)$ be the spherical image of the actual orbit. Denoting derivatives with respect to s by a prime the relations hold

$$\bar{a}_0{}'' = -\bar{a}_0 \tag{45}$$

$$\bar{a}'' = -\bar{a} + K_g\,(\bar{a} \times \bar{a}') \,. \tag{46}$$

Denoting the vector difference $\bar{a} - \bar{a}_0$ by $\bar{d}$ we obtain by subtracting eq. (45) from eq. (46)

$$\bar{d}'' = -\bar{d} + K_g\,(\bar{a} \times \bar{a}') \,. \tag{47}$$

The concept of osculating ellipse demands

$$\bar{a}(o) = \bar{a}_0(o) \tag{48}$$

$$\bar{a}'(o) = \bar{a}_0(o) \tag{49}$$

or

$$\bar{d}(o) = 0, \quad \bar{d}'(o) = 0 \,. \tag{50}$$

Therefore, we conclude from eq. (47)

$$\bar{d}(s) = \int_0^s \sin(s - \sigma)\, K_g\,(\sigma)\,(\bar{a}(\sigma) \times \bar{a}'(\sigma))\, d\sigma \,. \tag{51}$$

We recognize in $\bar{a} \times \bar{a}'$ the unit vector $\bar{n}$ (normal to the plane of the o.e.) of chapter 2. Therefore, eq. (51) can be written

$$\bar{d}(s) = \int_0^s \sin(s - \sigma)\, K_g\,(\sigma)\,\bar{n}(\sigma)\, d\sigma = -\frac{1}{\mu_0} \int_0^s \frac{\cos i}{p\,\varrho^2 \sin\theta}\, \frac{\partial \Phi_1}{\partial\theta}\, \sin(s - \sigma)\, \bar{n}(\sigma)\, d\sigma \,. \tag{52}$$

The integral vector eq. (52) contains the theory of the regression. It is extremely easy to calculate from eq. (52) the angular distance δ of two succeeding ascending nodes for any rotationally symmetric potential if we neglect second and higher powers of $\mu_2/\mu_0\, p^2$. In this case it is justified to assume that in the right hand side of eq. (52)

$$\bar{n}(\sigma) = \bar{n}(o) = \text{constant vector}$$

$$i, p, \varepsilon, \varphi_0 = \text{const.,} \quad \sigma = \varphi \,.$$

We obtain

$$\delta = -\frac{\cot i}{\mu_0\, p} \int_0^{2\pi} \frac{1}{\varrho^2}\, \frac{\partial \Phi_1}{\partial(\cos\theta)}\, \sin\varphi\, d\varphi \,. \tag{53}$$

A negative (positive) value of δ indicates that the direction of motion of the ascending node forms an angle $> \dfrac{\pi}{2}\left(< \dfrac{\pi}{2}\right)$ with $\bar{n}(o)$.

Taking advantage of the developments in the preceding chapters it is principally not difficult to give more precise values for δ containing for instance second powers of $\mu_2/\mu_0\, p^2$. In any case the eq. (52) and eq. (14b) are equivalent to the equations of motion.

References

1. H. Knothe and R. H. Anderson, On Satellite Orbits I. Air Force Missile Development Center, Holloman Air Force Base, New Mexico, A.F.M.D.C.-T.R.-59-11, April 1959.
2. H. Knothe and R. H. Anderson, On Satellite Orbits II. Air Force Missile Development Center, Holloman Air Force Base, New Mexico, A.F.M.D.C.-T.R.-59-19, May 1959.

Problems of Magnetic Propulsion of Plasma[1]

Ralph W. Waniek[2]

(Received July 13, 1959)

Abstract — Zusammenfassung — Résumé

Problems of Magnetic Propulsion of Plasma. In the case where a plasma structure can be made to have ideal or nearly ideal diamagnetic properties a magnetic field will act on the gaseous ionized boundary like a piston and will impart net momentum to the configuration.

This paper will deal with theoretical problems and experimental results obtained during the course of a study aimed at accelerating ionized gases by means of strong transient magnetic fields.

Recently developed strong magnetic field techniques are discussed in view of their possible application to high field plasma thrustors. Special air-core magnet configurations are shown and their characteristics as intermittent plasma propulsors are outlined.

Probleme des magnetischen Plasma-Antriebs. In einer Anordnung, in welcher eine Plasma-Masse von wenigstens nahezu idealen diamagnetischen Eigenschaften vorliegt, wirkt ein magnetisches Feld wie ein Kolben auf die Grenzfläche des ionisierten Gases und erteilt dem Aggregat einen Impuls.

Die Abhandlung befaßt sich mit den dabei auftretenden theoretischen Problemen und den Versuchsergebnissen, die bei einer Untersuchung der Beschleunigung ionisierter Gase durch starke vorübergehende Magnetfelder erhalten wurden.

Neue Methoden zur Erzeugung starker Magnetfelder werden mit Rücksicht auf ihre mögliche Anwendung auf Plasma-Antriebsmotoren besprochen. Magnetanordnungen mit Luftkern und ihre Eigenschaften im intermittierenden Plasma-Betrieb werden beschrieben.

Problèmes posés par la propulsion magnétique des plasmas. Dans un plasma de structure idéale, aux propriétés presque diamagnétiques, l'action d'un champ magnétique sur les surfaces limites ionisées est analogue à celle d'un piston et communique à l'ensemble une quantité de mouvement.

L'article envisage les problèmes théoriques et donne les résultats expérimentaux d'une étude entreprise en vue d'accélérer des gaz ionisés à l'aide de champs magnétiques intermittents.

Les techniques récemment développées pour la production de champs magnétiques intenses et leurs applications possibles aux propulseurs sont discutées. Des configurations spéciales d'aimants à noyau à air sont envisagées et leurs caractéristiques comme propulseurs intermittents sont esquissées.

---

[1] This work was supported by the U.S. Air Research and Development Command through its Office of Scientific Research.

[2] Director of Research, Plasma Physics Laboratory, Giannini Plasmadyne Corporation, Santa Ana, California, U.S.A.

I. Introduction

Plasma has been known to us for a good many decades. W. CROOKES, some 80 years ago, had the foresight to call it the fourth state of matter [1]. It is therefore somewhat astonishing that mankind's everlasting endeavour to move and to propel objects was applied to this form of matter only during recent times (pure ion and electron beams should not be considered in this context and might hence be categorized as another state of matter). In effect, much of the detailed knowledge of plasma physics required for such an application has been provided only recently by the large-scale effort of the international thermonuclear program [2]. These extensive investigations have led to several pertinent experiments on plasma propulsion. The method which all these efforts have in common is that the propulsive forces are provided by a magnetic field which acts on the gas transiently. In such a case, as is well known, the forces are always orthogonal to the direction of the magnetic field and are directed against the plasma boundary to be moved.

The present-date experiments on magnetic propulsion of plasma could be classified into two broad categories. The first of these comprises geometries in which discharges are established between two electrodes or between two linear or coaxial rails. Such discharges heat the plasma by Joule dissipation and impart to it momentum by the field generated by the discharge current itself or by integral parts of the circuit. To this class belong the very early work by FOWLER [3], the basic experiments by BOSTICK [4], and the research performed by ARTSIMOVICH and associates [5], by KOLB [6], by JANES and PATRICK [7], by V. JOSEPHSON [8] and by KORNEFF [9].

The second class of devices makes no use of electrodes but relies on the principle that induced currents can be set up within the plasma structure. These FOUCAULT currents prevent the penetration of the primary applied field for a certain length of time and allow the formation of a defined plasma boundary for magneto-kinetic energy transfer. To this second category belong the early attempts by THONEMANN and co-workers [10], the interesting device by J. MARSHALL [11], and some of the work presented in this paper.

Although it is much too early to pass judgment on the relative merits of these two basic approaches, one statement could perhaps be made already at this stage. Fundamental studies on the behavior of plasma can be made in many different geometries, a magnetic thrustor, however, has to stand the wear of a long period of operation due to the nature of its possible application. This might possibly disqualify systems with exposed electrodes. Furthermore, since the efficiency of magnetic propulsion depends largely on the electrical conductivity of the plasma any contact with material walls should be avoided. These considerations on electrode erosion and on the thermal insulation of the ionized gas would seem to favour devices of the second type over the geometries of the first category.

II. Theoretical Considerations

In the following we wish to consider the action of a transient magnetic field against a plasma boundary. In order to account for propulsive forces it is useful to remember that a magnetic field will behave like a two-dimensional gas exerting pressures corresponding to its energy density $B^2/8\pi$ and numerically identical with it (ergs/cm^3 = dynes/cm^2).

For the case of infinite electrical conductivity the magnetic field would be completely separated from the plasma boundary and the full pressure of its energy density would bear against the impenetrable gaseous barrier. But a well defined plasma-field boundary is only a theoretical expedience. In reality, interpenetration will occur instantly with a penetration depth dependent on the electrical conductivity of the plasma. In such a case, the pressure exerted will have to be evaluated by the gradient of the energy density over the thickness of the boundary. In analogy to metallic conductors, we can define for a plasma acted upon by a transient magnetic field a skin depth δ equal to the $1/e$ distance of field attenuation [11, 13]

$$\delta = \left(\frac{\eta}{2\pi\omega}\right)^{1/2}\text{cm} \tag{1}$$

where η is the resistivity of the gas in e.m.u., ω is the angular frequency of the applied field. The electrical resistivity of an ionized gas is given by:

$$\eta = 3.80 \cdot 10^{12}\left(\frac{Z\ln\Lambda}{\gamma\ T^{3/2}}\right)\text{e.m.u.} \tag{2}$$

where Z is the ionic charge, Λ the ratio of DEBYE shielding distance to the value of the impact parameter for orthogonal deflections, γ is the correction term to account for electron-electron encounters, T is the plasma temperature in °K. The combination of formulae (1) and (2) yields a third one giving the temperature dependence of the skin depth:

$$\delta = \frac{3.12 \cdot 10^5}{T^{3/4}}\left(\frac{Z\ln\Lambda}{\gamma f}\right)^{1/2} \tag{3}$$

where $f = \omega/2\pi$ is the frequency of the field applied.

A simple plot of this functional relationship for three typical frequencies of the magnetic field used and for the case of a hydrogenous plasma (Fig. 1)

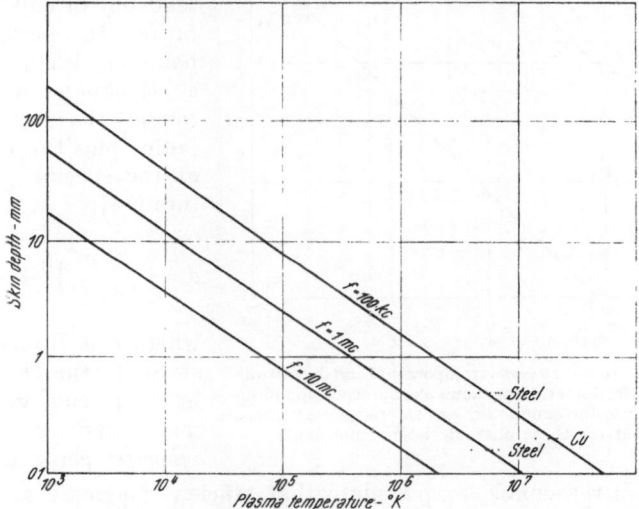

Fig. 1. Skin depth versus temperature for a hydrogenous plasma (for three frequencies of the applied field; corresponding skin depths in metallic conductors are entered)

indicates that skin depths are rather sizeable unless the plasma is preheated to temperatures of several million degrees or unless the frequency of the applied field amounts to several megacycles. The subsequent diffusion of the magnetic field into the plasma is described by the usual differential equation:

$$\frac{\partial H}{\partial t} = \frac{1}{4\pi\sigma}\frac{\partial^2 H}{\partial x^2} \tag{4}$$

where σ is the electrical conductivity of the gas.

It is customary [12] to define a time constant for the process of interpenetration or field decay time:

$$\tau \approx \frac{4 \pi L^2}{\eta} \; \text{sec}. \tag{5}$$

where L is a characteristic dimension of the system, in this specific case, the radius of the plasma thrustor if a cylindrical geometry is adopted. Fig. 2 illustrates the magnitude of the field decay times for three geometrical dimensions and as a function of gas conductivity or temperature (hydrogenous plasma). The field decay time sets a maximum value for the time interval available for the acceleration of the gas at a given temperature. If this time is exceeded the field will have interpenetrated the plasma and little or no propulsive effect will have been achieved.

We should examine now briefly the structure of the plasma boundary. This boundary has obviously finite thickness due to the action of the magnetic field on the charged particles and of the DEBYE forces resisting any tendency of bulk charge separation. In effect, the plasma boundary is supposed to have a thickness amounting to roughly the electron cyclotron radius plus the DEBYE length at the specific gas temperature [14]:

$$d = \left(\frac{k \, T}{4 \pi n_e e^2} \right)^{1/2} + \frac{m_e v v c}{Z \, e \, B} \, \text{cm} \tag{6}$$

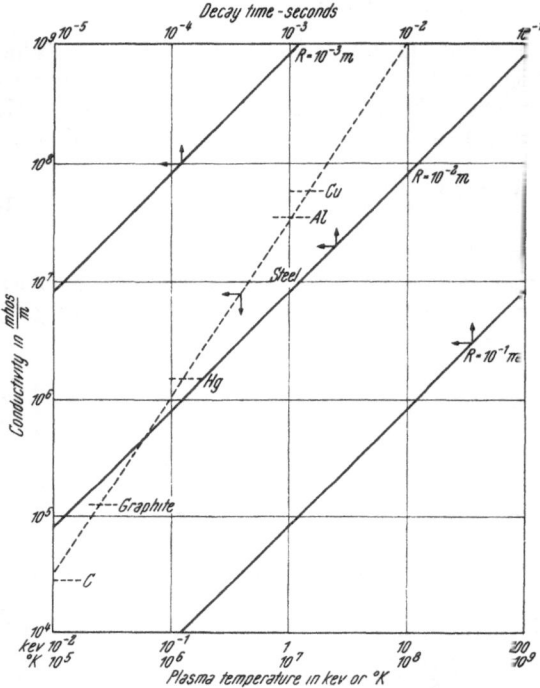

Fig. 2. Plasma conductivity versus temperature and decay time versus conductivity for a hydrogenous plasma (corresponding conductivity points for conductors are entered; decay times are presented for three characteristic dimensions)

where k is BOLTZMANN's constant, n_e the electron density in the plasma, m_e the electron mass and v the electron velocity perpendicular to B.

It becomes reasonable to postulate that efficient magnetic propulsion will require skin depths which are small with respect to the thickness of the plasma boundary itself. This is not an easily realizable situation since, as calculations can easily show, even at 10^6 °K the plasma skin depth will be some 500 microns for a 1 megacycle field, whereas the electron radius will be some 35 microns in a 10 kilooersteds field and the DEBYE shielding distance some fraction of a micron for an electron density of $10^{16}/\text{cm}^3$. Hence, the boundary thickness is only 10 per cent of the skin depth. At 10^7 °K the boundary thickness roughly equals the skin penetration if the other auxiliary parameters are maintained the same. Better performance at lower temperatures can be achieved by applying magnetic fields of faster risetimes. Because of its tendency to expand and to perform work the magnetic field has been likened to a mechanical piston. We can extend the analogy in this connection by adopting the term "porous" for the piston whenever the magnetic penetration is sizeable with respect to the boundary thickness (low gas conductivity) or "solid" when the skin depth is shallow (high conductivity). The difference in the efficiency for these two cases becomes obvious by these pictorial terms and is mathematically accountable by the slip effect of the piston in any given geometry.

All these considerations clearly stress that gas heating above a million °K is a sine qua non condition for efficient conversion of magnetic energy into directed kinetic energy. This prerequisite is by far not as demanding as the temperatures to be produced for thermonuclear reactions which range by one to two order of magnitude higher.

III. Experimental Work

A magnetic accelerator system should combine two distinct functions. First, it should incorporate some mechanism for heating and containing the gas. Second, it should be able to provide the previously outlined piston action to impart net momentum to the plasma. The gas heating function might be accomplished in any one of the many ways presently under study by various groups [15, 16]. At this laboratory we have under theoretical [17] and experimental study a cylindrical geometry in which the magnetic field causes a radial inward collapse of a plasma column (Fig. 3). A somewhat similar geometry was the object of preliminary studies by S. COLGATE at an early stage during the thermonuclear program [18]. Special cycles of shock and relaxation heating with programmed symmetric end-confinement are under investigation. The mirror confinement [19]

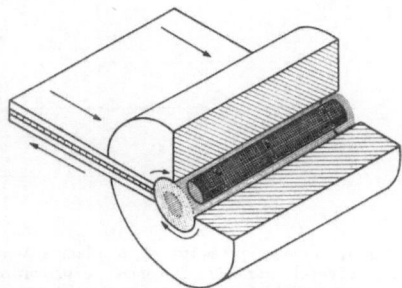

Fig. 3. Basic collapse configuration (primary and induced currents are indicated by arrows in the conductors and in the plasma column. Magnetic field between plasma and conductor is dotted)

can be rendered asymmetric by appropriate high power circuitry so as to transfer the plasma to a different magnetic region. It is mainly on the properties of this second region that we wish to report some preliminary experimental results.

The heated plasma will exhibit a considerable kinetic pressure $(n_e + n_i) kT$ which, even at only 10^6 °K and at a particle density of some $10^{17}/cm^3$, will amount to some 100 atmospheres pressure (see Fig. 4). To contend with such pressures very strong magnetic fields are necessary (for the above case 50,000 oersteds are needed for containment). In order to impart net momentum even higher fields should be used or alternatively a field gradient could be introduced in the geometry. This would cause the almost diamagnetic plasma configuration to move in the direction of the weaker field [13]. In practice, both requirements can be combined, for instance, in the geometry outlined in Fig. 5.

During the past few years we have performed some research on the production of very strong pulsed magnetic fields [20, 21]. This work has indicated that strong magnetic fields can be created in geometries which rely on the mechanical strength of the conductor only up to a point which we have termed the "magnetic yield point." At this point, characteristic for each conductor material, the expansive forces of the "magnetic gas" overcome the tensile strength of the surrounding conductor and cause the matter to "cold-flow." This natural limitation may be circumvented in the future by the use of force-free fields [21], now under investigation. Even without these unconventional configurations, much remains to be done in the submegaoersted region by intelligent use of materials and magnetic configurations. In the geometry outlined in Fig. 5 typical fields of 500,000 oersteds are reached near the apex with some, 150,000 oersteds near the large aperture. The corresponding pressure of these fields range between the 10^3 and 10^4 atmospheres.

Plasma structures ejected from such geometries have been studied by photomulti-

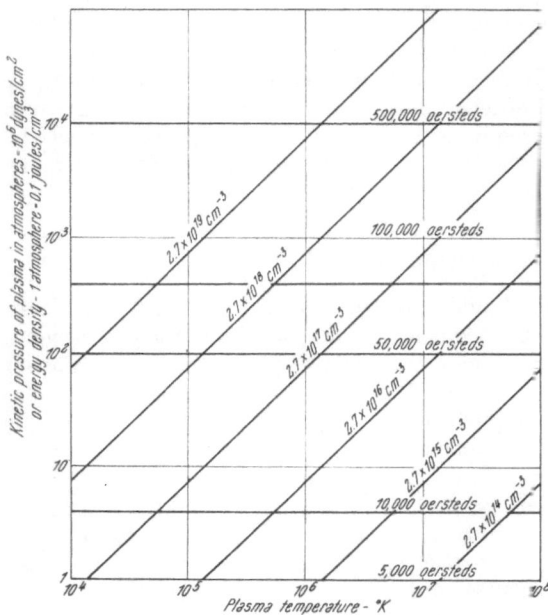

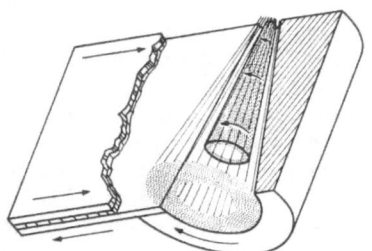

Fig. 4. Kinetic pressure of a plasma versus temperature for several particle densities. Horizontal lines indicate magnetic field necessary for transient confinement of a given pressure

Fig. 5. Asymmetrical collapse configuration (arrows indicate currents in the conductors and in the induced plasma configuration)

pliers, KERR-cells, smear camera, time-resolved spectroscopy, and microwaves. The directed velocities achieved up-to-date with a still insufficient heating process amount to more than 20 cm/microsecond or 200 km/sec at pressures of a few hundred microns Hg. Results from ballistic measurements show that more than 1 microgram of mass is being set in motion during the acceleration process. Such a momentum would indicate efficiencies of electro-kinetic energy conversion of 2 per cent maximum. This low conversion efficiency is attributable to the poor heating produced during these initial studies and higher figures should be attainable in the future.

IV. Conclusion

Magnetic propulsion of plasma might become a method competing with pure ion propulsion in almost the same region of specific impulse and thrust. It seems reasonable to anticipate that specific impulses between 10^2 and 10^5 seconds should be provided by magnetic acceleration at thrust levels of a few tenth of a kg per pulse.

Presently, about 20 per cent of the initially stored electrostatic energy is converted into a useful magnetic field with some allowance for variations of geometry. Of the field used, about 10 per cent have been converted into kinetic energy of the moving plasma. Energy can be stored electrostatically (capacitors) at a volume density of about 1 joule/cubic inch and at a specific weight of roughly 25 joules/lb. These figures should become much better as the techniques progress further in future years. The weight of the air-core geometries with feeder lines is negligibly small in comparison to the weight of the storage devices for any sizeable geometry.

Magnetic plasma accelerators offer the great initial advantage of charge neutrality combined with a wide range of specific impulse at moderately high thrust levels. With the presently rapidly evolving understanding and techniques of the physics of ionized gases magnetic thrustors might be very likely to find a useful place in space vehicles of the future.

Acknowledgements

The author gratefully acknowledges stimulating discussions on the subject with Dr. G. M. GIANNINI and Dr. H. G. LOOS. It is also a pleasure to thank Mr. K. J. PARK for his competent assistance during the above mentioned experiments.

References

1. W. CROOKES, Phil. Trans. 1, 135 (1879).
2. See Proceedings of the 1958 Geneva Meeting, papers on Thermonuclear Problems.
3. R. G. FOWLER, J. S. GOLSTEIN and B. E. CLOTFELDER, Physic. Rev. 82, 879 (1951).
4. W. H. BOSTICK, Physic. Rev. 104, 292 (1956) and 106, 104 (1957).
5. L. ARTSIMOVICH, S. CHUVATIN, S. LAKIJAVOV and I. PODGORNY, Soviet JETP 33, 3 (1957).
6. A. C. KOLB, Physic. Rev. 107, 345 (1957).
7. G. S. JANES and R. M. PATRICK, Conference on Extremely High Temperatures, edited by H. FISHER and L. MANZUR. New York: Wiley, 1958.
8. V. JOSEPHSON, J. Appl. Physics 29, 30 (1958).
9. T. KORNEFF, F. NADIG and J. BOHN, Conference on Extremely High Temperatures, edited by H. FISHER and L. MANZUR. New York: Wiley, 1958.
10. P. C. THONEMANN, W. T. COWLING and P. A. DAVENPORT, Nature 169, 34 (1952).
11. J. MARSHALL, Second UN International Conference on Peaceful Uses of Atomic Energy, P. 355, 1958.
12. L. SPITZER, Physics of Fully Ionized Gases. New York: Interscience Publ., 1956.
13. H. ALFVÉN, Cosmical Electrodynamics. Oxford: University Press, 1950.
14. C. L. LONGMIRE and M. ROSENBLUTH, Ann. Physics 1, 120 (1957).
15. J. BERGER, W. NEWCOMB, J. DAWSON, E. FRIEMAN, R. KULSRUD and A. LENARD, Physics of Fluids 1, 301 (1958).
16. A. SCHLUETER, Z. Naturforsch. 12a, 822 (1957).
17. H. G. LOOS, Physic. Rev. Letters 2, 508 (1959).
18. A. BISHOP, Project Sherwood. The US Program in Controlled Fusion. Reading, Mass.: Addison-Wesley Publishing Co., Inc., 1958.
19. W. C. ELMORE, E. M. LITTLE and W. E. QUINN, Physic. Rev. Letters 1, 32 (1958).
20. H. P. FURTH and R. W. WANIEK, Rev. Sci. Instruments 27, 195 (1956).
21. H. P. FURTH, M. A. LEVINE and R. W. WANIEK, Rev. Sci. Instruments 28, 949 (1957).

Prediction of Man's Performance in Space Using Flight Simulators and Balloon-Borne Systems

By

J. Gordon Vaeth[1]

(Received July 3, 1959)

Abstract — Zusammenfassung — Résumé

Prediction of Man's Performance in Space Using Flight Simulators and Balloon-Borne Systems. The ability of space crews to carry out useful missions and tasks has until now been a matter of speculation, based mainly on extrapolation of aviation experience, philosophical reasoning, and fragmentary research. This paper proposes the development of advanced forms of ground-based flight simulators and the use of long-endurance high altitude manned balloon flights to measure and determine the capability of men to perform in space. By comparing these measurements with corresponding data on the ability of automatic equipment to do the same, it will be possible to specify and predict those tasks which can be achieved better by manned than by unmanned space systems. The characteristics of these simulators and balloon operations will be described in some detail, together with the role each will play in determining optimum instrumentation, data displays, and cockpit layout for manned space vehicles. Their application to the effective training of space crews will also be briefly discussed.

Vorhersage der menschlichen Leistung im Raum mit Hilfe von Flugsimulatoren und ballongetragenen Systemen. Die Fähigkeit von Raumfahrzeugmannschaften zur Ausführung nutzbringender Missionen und Unternehmungen war bis jetzt Gegenstand der Spekulation, und zwar hauptsächlich auf der Grundlage einer Extrapolation von Flugerfahrungen, philoscphischen Überlegungen und fragmentarischer Forschung. In der folgenden Arbeit wird die Entwicklung fortgeschrittener Formen von auf dem Boden befindlichen Flugsimulatoren und die Benützung langdauernder Flüge bemannter Ballons in großen Höhen vorgeschlagen, um die menschliche Leistungsfähigkeit im Raum zu messen und zu bestimmen. Wenn man diese Messungen mit den entsprechenden Daten über die Eignung automatischer Ausstattung, die dasselbe tun soll, vergleicht, wird es möglich sein, jene Aufgaben auszusondern und vorherzusagen, die besser mit bemannten als mit unbemannten Raumfahrtsystemen ausgeführt werden können. Die Charakteristika dieser Simulatoren und Ballonoperationen werden mit einiger Ausführlichkeit beschrieben, zusammen mit der Rolle, die jede Art bei der Auswahl der besten Instrumentation, der Aufzeigung von Daten und der Führersitzauslegung für bemannte Raumfahrzeuge spielen wird. Auch ihre Anwendung für das tatsächliche Training von Raumfahrzeugbemannungen wird kurz erörtert.

La prédiction des performances humaines en vol spatial à l'aide de simulateurs et d'aérostats. Les aptitudes d'un équipage en vol spatial ont été jusqu'à présent l'objet de spéculations basées principalement sur une extrapolation de l'expérience aéro-

---

[1] Advanced Research Projects Agency Technical Staff Member for Man-in-Space; 7709 Massena Road, Bethesda, Maryland, U.S.A.

nautique, sur le raisonnement philosophique et fragmentairement sur la recherche. L'article propose le développement de formes avancées de simulateurs de vol et l'utilisation d'aérostats de haute altitude et grande endurance pour mesurer les aptitudes humaines dans les missions de vol spatial. Une comparaison avec les possibilités des équipements automatiques pour les mêmes tâches devrait permettre de délimiter celles qui sont mieux remplies par l'opérateur humain. Les caractéristiques des simulateurs et les missions des aérostats sont décrites en détail, ainsi que le rôle joué par chacun dans la recherche de la meilleure instrumentation, de la meilleure présentation des données et de la meilleure disposition du poste de pilotage. L'application à l'entrainement des équipages est aussi brièvement discutée.

In the twelve months which have elapsed since last years' Congress of the International Astronautical Federation, substantial progress has been made in almost all areas of space technology. A variety of satellites, lunar probes. and rocket test vehicles have been launched, each serving to clarify and resolve the unknowns and problems of flight beyond the earth.

One of the most significant developments of the past year has certainly been the institution of Project *Mercury*, the United States National Aeronautics and Space Administration's program to place a man into a low orbit around the earth. The details of this project have been quite fully revealed, including the characteristics of its triangular-shaped capsule, its operating sequence following separation from the *Atlas* booster, and its pilot escape and recovery systems. The names of the seven test pilots who are being trained to ride the capsule in orbit have even been made public.

Project *Mercury* is but a few months old. Yet so fast is modern technology moving that even now proposals exist for follow-on programs, for manned vehicles to take up where *Mercury* leaves off. Some of these call for maneuverable satellites, ones capable of changing orbit and possessing enough control to be able to operate routinely from established bases. Others propose the assembly and operation of manned stations or platforms in orbit, relying upon ferry rockets for logistics support and personnel transfer. Still others suggest manned flights to and around the moon and planets, followed by the establishment of scientific bases on those alien worlds.

Such proposals are based on the premise that man can survive and perform in space. Recent medical research supports this survival premise. The more that is learned about bio-astronautics, the more certain it seems that the physiological problems of space flight—some of which have been doubtlessly exaggerated and even imagined—can and will be eliminated, solved, or circumvented.

But although doctors, scientists, and engineers are reaching agreement that man can or can be made able to tolerate and survive the conditions and environments of flight through space, they are far from agreed that he can perform usefully and advantageously there.

These divergent and sometimes warmly disputed viewpoints are characterized by two opinion extremes:

One group, comprised largely of pure scientists and engineers, believes that man is an expensive and quite unnecessary luxury in complex space systems; they claim that anything man can do, automatic equipment can do better; they consider the human body and mind obsolete for the demands of extraterrestrial flight.

The other group, made up of operationally-minded persons, takes just the opposite stand; recalling their problems with the use of automatic equipment in the air and at sea, they express distrust of too great reliance upon it; they are concerned over the frequency of malfunctions and failures; they want no part of any total reliance upon "black boxes" or upon the duplicating back-up equipment which they claim would be necessary for any degree of acceptably reliable performance.

These diametrically opposed views sharply divide informed astronautical opinion today. The fact that they do is a sign that for all the current progress in space technology, the question of human performance has been too little investigated or considered. The problem of survival has dominated the human factors thinking. Yet performance is equally important. For one is dependent upon the other. Unless man can both survive and perform, there is no reason to try to put him into space at all.

Four questions await an answer:

1. Just how necessary is it for man to operate in space himself?
2. Can he usefully, effectively, and advantageously perform there?
3. What are the space-borne tasks which he is best suited to do?
4. Where does the desirability of using automatic equipment end and the requirement for human control begin?

Many attempts have been made to provide the answer. In so doing, the capabilities and advantages of man have often been enumerated as have his limitations and drawbacks:

The Advantages of Using Man in Space

1. Versatility to do many and varied tasks
2. Adaptability to meet the unknown and unexpected
3. Ability to discriminate between the important and the unimportant
4. Ability to filter out redundant or useless information
5. Highly developed powers of perception
6. Ability to provide first-hand on-the-spot observations, thus contributing to their increased accuracy, efficiency, reliability and interpretation
7. Immediate on-the-scenes decision making (time consuming data relay and interpretation not needed)
8. Ability to provide subjective data (impressions, feelings, sensations, opinions) as well as objective information
9. Immense built-in data storage capacity and recall ability (developed by education and training)
10. Ability to provide highly refined and precise piloting or maneuvering skill
11. Ability to troubleshoot, service, maintain, and repair equipment in flight
12. Efficiency as a servo-mechanism
13. Compactness and light weight
14. Availability.

Disadvantages of Using Man in Space

1. Need for life support system (requiring space, weight, power, and involving high costs and complexity if long durations and several crew members are involved)
2. Need for minimum of two or three crew members if flight exceeds half-day
3. Possibility that each major task performed will require separate crew member
4. Cost and complexity of crew selection and training program
5. Susceptibility to confusion if overpowered by unknowns and emergencies
6. Constant possibility of pilot error.

Based on the foregoing pros and cons of using men to perform the exploration, scientific observation, space station assembly, services of supply, maintenance, derelict-sweeping, and retrieval and rescue tasks which must inevitably be performed in space, many opinions have been stated as to why, where, and when man should or should not be so employed.

The trouble with these opinions is that they are just that—opinions. In some cases they are the results of attempts to apply the lessons learned in aviation to the problems of astronautics. In others they are little more than statements of philosophy and the results of wishful thinking.

Until some measurements of crew performances in space can be made and compared with the results of using automatic equipment to accomplish the same tasks, the role of man in space will remain vague and undefined, subject to continued uncertainty, argument, and debate.

It is not necessary, however, to await the completion of the *Mercury* flights to obtain measurements and definitive data on man's ability to perform in orbit or beyond. Actual flights in space are not absolutely essential. Crew performance can be measured—within limits—on the ground by flight simulators and in the air by means of balloon-borne systems.

To the aviator the flight simulator is an old friend. He knows it as an exact replica or mock-up of the cockpit and cockpit equipment of the plane it represents. He moves the controls and its dials and gages respond correctly and realistically. The sounds and vibrations of the actual aircraft are duplicated and reproduced, even to the squeal of the tires as the wheels touch down upon the runway. The control stick has the "feel" of the real thing. Pressure suits can be attached and actuated. Complete flights, with emergencies too dangerous to practice on the air, are simulated. Flight problems and piloting procedures are practiced again and again until coordinated crew proficiency is guaranteed. Although originally intended to be used for training purposes, these devices have proven valuable research tools in testing and evaluating instrumentation and control concepts.

By adapting present-day flight simulator techniques, complete space vehicle cabins and experimental data displays could be mocked up and synthetically activated to determine human ability to use the equipment, controls, instruments, lights, and scopes involved. Entire space missions could be "flown" on a realistic time basis and with the emergencies peculiar to space realistically incorporated. These simulated flights could include usage and tests of closed cycle regenerative life support systems over extended periods. Crew behaviour in confined quarters under synthetic operational conditions could also be examined.

Ground based simulators suffer from one severe limitation: they cannot reproduce weightlessness. Drugs have been proposed as a possible means to give crews an artificial sensation of flight under zero gravity but this approach does not appear very promising, at least not at this time.

The effects of G-forces, on the other hand, can be reproduced on the ground. Either synthetically or actually. The first method entails the dimming of cockpit lights to represent gray- or black-out, also the tightening of restraining straps to simulate the sensations a pilot would feel. The second method is to create the forces themselves by mounting the cabin or cockpit on a centrifuge arm and by swinging it around in what has been called "dynamic flight simulation".

Some very excellent work has been done in this area by the U.S. Navy's Aviation Medical Acceleration Laboratory at Johnsville, Pennsylvania. The instrument panel and controls of the research aircraft X-15 were mocked up and built into the gondola at the end of the Johnsville 50 foot centrifuge arm. The G-forces exerted on the man in the gondola were the results of his control actions. Prospective pilots of the X-15 practiced simulated flights with this device, learning much about the re-entry characteristics of the vehicle and about its flight trajectory. The results left little doubt about the value of using this technique to determine the ability of pilots and crews of space craft to control their vehicles during the acceleration and deceleration phases—launch, speeding up or maneuvering to make contact or rendezvous with another ship, re-entry, and landing—of orbital or deep space operations.

The realism of all space flights simulated on the ground is handicapped by the fact that the subjects know that they are on the ground and that they can usually be released almost at once from the circumstances of the flight. Thus the key factors of stress, tension, and psychological and physiological hazards are to a large degree lacking. Crew reaction to the "break-away phenomenon"—the feeling of complete detachment from the earth—cannot, for example, be determined.

High altitude manned balloons offer an available and inexpensive means to overcome this lack of realism on the ground. They enable simulated space missions to be carried out at heights above the "break away phenomenon" and in a hostile environment which is biologically equivalent to space.

Using existing balloons, it is possible within the current state of the art to make a flight of several days duration (maintaining approximate position by careful use of wind flow and counter-flow), taking a "space crew" of three men to a height of 80,000 feet. During their flight the occupants of the space vehicle-configured and equipped gondola would work at the various tasks which they might be expected to achieve in orbit. Data displays and earth- and sky-observing devices would be tried out and evaluated to determine their usability for the crew.

Balloons have their shortcomings as do all space flight simulation techniques. None, for instance, can reproduce the state of weightlessness, an admittedly serious limitation. Yet, despite this, each can play a significant role in helping to predict and piece together a picture of man's probable performance in space.

The flight simulator and the balloon-borne system are available now. If we start to apply their capabilities to the simulation of missions and operations in space, we shall at long last be on our way to answering statistically and definitely: *Where does man himself fit in the conquest of space?*

Interplanetary Homing

By

E. V. Stearns[1]

(With 7 Figures)

(Received July 7, 1959)

Abstract — Zusammenfassung — Résumé

Interplanetary Homing. Some problems of interplanetary navigation are discussed for the special conditions in which the spacecraft is approaching the destination planet. The work of other investigators is cited in order to establish the accuracy which may be expected from the mid-course phase of navigation. Considering these estimates of accuracy, the performance requirements for terminal guidance are examined.

The forces acting on a space vehicle in the vicinity of the destination planet are discussed briefly with relation to the homing problem. A closed loop system of instrumentation is proposed for the control of the vehicle trajectory to provide thrust control necessary to place it in a suitable orbit for final approach to the destination.

Interplanetarisches Selbstleitsystem. Es werden einige Probleme der interplanetarischen Navigation für die besonderen Bedingungen erörtert, unter denen sich ein Raumfahrzeug an einen Zielplaneten annähert. Die Arbeiten anderer Forscher werden angeführt, um die Genauigkeit zu ermitteln, die während der Unterwegs-Phase der Navigation erwartet werden kann. Unter Berücksichtigung dieser Genauigkeitsschätzungen werden die Erfordernisse der praktischen Ausführung für die Endlenkung geprüft.

Es werden die auf ein Raumfahrzeug in der Nachbarschaft des Zielplaneten wirkenden Kräfte mit Rücksicht auf das Selbstleitproblem kurz besprochen. Ein geschlossenes Instrumentierungssystem (vor allem der Interplanetarische Sextant) wird für die Kontrolle der Fahrzeugbahn vorgeschlagen, um für die Schubkontrolle vorzusorgen, deren es zur Erreichung einer geeigneten Umlaufbahn bei der endgültigen Annäherung an das Ziel bedarf.

Interception planétaire. Quelques problèmes de navigation interplanétaire dans les conditions d'approche d'une planète de destination sont mis en discussion. Les travaux d'autres chercheurs sont utilisés pour établir la précision qui peut être attendue de la phase intermédiare de navigation. Les spécifications du guidage terminal sont examinées sur cette base.

Les forces agissant sur l'astronef au voisinage de la planète de destination sont brièvement discutées en relation avec le problème de l'interception finale. Une boucle d'instrumentation fermée est proposée pour le contrôle de la poussée en approche terminale.

---

[1] Research and Development Scientist, Lockheed Missiles and Space Division, Sunnyvale, California, U.S.A.

List of Symbols

r	position vector relative to planet	p	semi-latus rectum
r_p	position vector relative to planet at periapsis	Φ	flight path angle (see Fig. 1)
		$\theta - \theta'$	the true anomaly
v	velocity of craft relative to planet	r_0	planet radius
v_∞	velocity of craft at infinite distance from planet	d	distance from planet center to asymptote of hyperbolic trajectory
k	gravitational field constant $= GM$		
	$k_{\text{Earth}} = 9.6 \times 10^4 \text{ mi}^3/\text{sec}^2$	$\Delta V_{\parallel}$	velocity correction if component applied parallel to v
	$k_{\text{Venus}} = 7.75 \times 10^4 \text{ mi}^3/\text{sec}^2$	$\Delta V_{\perp}$	velocity correction if component applied perpendicular to v
	$k_{\text{Mars}} = 1.026 \times 10^4 \text{ mi}^3/\text{sec}^2$		
ε	eccentricity	$E(\ )$	error in quantity shown in parenthesis
g_s	acceleration due to solar gravity		
R	position vector relative to Sun		

I. Introduction

While it would, perhaps be considered desirable if an interplanetary vehicle could be given its initial boost at the beginning of its flight and then be left to coast for great distances on trajectories carefully designed to terminate with appropriate landing conditions for the distant planets, this does not now appear to be within the predictable state of the art. Consequently, some attention must be directed towards the problem of steering a space vehicle into a suitable landing trajectory as it approaches its destination. We refer to this as the homing problem.

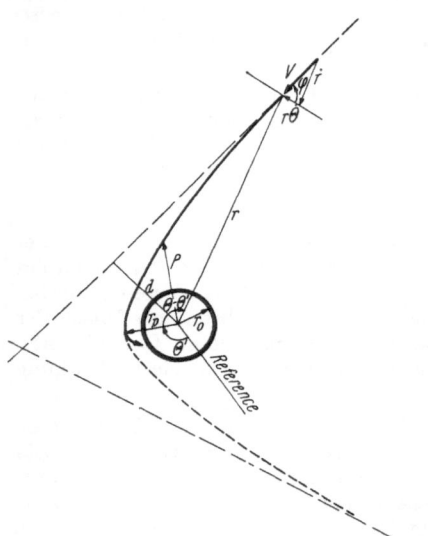

In the example discussed we have chosen to consider only "self-contained" systems; i.e., those requiring no outside aid or cooperation from equipment, personnel, etc. external to the vehicle. This restriction is considered acceptable since, as will be seen, the problem may still be solved in a reasonable manner; furthermore, the restriction seems essential if exploration of unknown planets is to be proposed.

Fig. 1. The approach and landing trajectory. The space vehicle approaches a planet with the velocity, V, along the hyperbolic trajectory as shown. The homing problem is concerned with adjusting the parameters of the motion until the trajectory has periapsis (r_p) such that the drag reduces the energy of motion and the craft lands as shown rather than continue along the dashed trajectory

The discussion here is confined to a planar encounter; however, no loss in generality results since the plane chosen is that which is defined by the vector describing the approach velocity and the center of attractive force of the destination planet. There is no need for maneuvers which will change this plane, and they are not given further consideration here. The entire problem is treated in terms of the relative motion between the spacecraft and the destination planet.

The homing problem may be considered to be concerned with the control of a space vehicle causing it to approach a designated planet in such a way that the Keplerian relationships ensure a periapsis radius, r_p, and velocity, v_p, which is appropriate for landing. (See Fig. 1.) This is to say, there exists for each approach trajectory a range of velocities which will place the vehicle at periapsis within a narrow corridor leading to a suitable landing trajectory. The criteria for these capture trajectories will depend on the specific vehicle and planet considered.

Three major subproblems of the homing problem exist. These are:

1. Application of corrections to the trajectory at a distant point to insure that the midcourse trajectory will close with the destination planet in such a way as to permit a fine correction to be made at a later time.

2. A second correction adjusting the approach trajectory causing it to pass through a narrow landing corridor.

3. Determination of the corrections required and time for application such that a final conversion into orbit or landing trajectory can be made.

The first correction can be performed rather simply and, for best results, should be performed at the greatest possible distance from the destination planet; this distance is limited by the accuracy with which the corrections can be made.

The second correction is made on closer approach when precision measurements of trajectory are feasible. Without the first correction, excessive maneuver energy would be required.

The third subproblem can only be solved when the vehicle is at the suitable point in the desired trajectory. That is to say, the maneuver should be a drag caused velocity change and occurs at a time when the vehicle is at periapsis.

Since the consequences of an inaccurate solution are intolerable, we choose, at the outset, to reject the use of open loop solutions but rather to consider a form of closed loop system in which the spacecraft homes on the destination planet and corrections are continuously monitored. A closed loop system of this kind requires that a suitable prediction system be provided to forecast the motions of the system under control. Corrections to this solution may then be made at the appropriate time through the application of continuous thrust, or through impulses applied at selected intervals.

II. Accuracy Requirements

The approach to the problem can, perhaps, best be discussed if we treat the individual steps in reverse order. The optimal time for conversion from an hyperbolic approach trajectory to a planetary orbit or landing ellipse may be shown to be at the Keplerian periapsis ($r = r_p$) of the approach trajectory [1]. It has also been shown that for all members of the approach trajectory family each with a characteristic parameter ε and having periapses within a narrow range; i.e., within a landing corridor, the velocity at periapsis is $\sqrt{\dfrac{(1 + \varepsilon)\, k}{r_p}}$; if r_p is selected properly for the planet concerned and the vehicle configuration under consideration, then atmospheric drag will cause a safe landing without excessive accelerations [2]. It is necessary that the transition trajectory be defined so as to modify the midcourse trajectory providing an appropriate Keplerian periapsis for the hyperbolic approach arc.

The accuracy of an interplanetary transfer orbit has been studied and error analyses expressing the dispersions in the parameters of an approach trajectory may be found in the literature [3, 4 and 5]. The uncertainty of miss distance in a "field free" encounter has been analyzed by DeBra, who gives the errors in position to be expected from a midcourse transfer where the boosting is confined to the vicinity of the departed planet. This analysis considers the transfer as not being confined to HOHMANN orbits; results are presented in terms of transfer angle. Consideration has been given to the effects of time of transfer in estimating the magnitude of errors. An example of the results using these methods is given in Fig. 2. From this figure, and a knowledge of probable guidance and control accuracies, it is apparent that the approach to a destination planet will be such that the midcourse trajectory will be uncertain on approach to the destination planet with a probable error of about 100,000 miles.

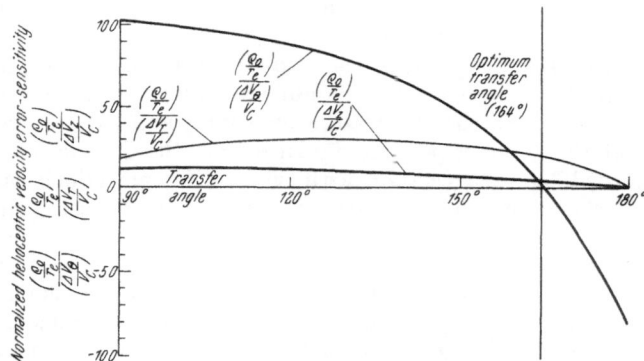

Fig. 2. Mars probe. Guidance accuracy requirements

Geo-centric guidance requirements for 100,000 mile accuracy
(164 deg. transfer 180 deg. transfer)

Altitude	8 miles	0.6 miles
Down range	35 miles	not critical
Cross range	not critical	not critical
Velocity magnitude	45 F.P.S.	3 F.P.S.
Cross range direction	1.2 deg.	not critical
Flight path angle	0.1 deg.	not critical

The accuracy of a midcourse transfer is summarized, errors are nondimensional showing miss distance in units of the Earth's orbit radius (ϱ_0/r_e) and velocity errors are the components' radial, tangential, and out of the plane of ecliptic ΔV_r, $\Delta V\theta$, and ΔV_z respectively; units are the Earth's circular orbit velocity, V_C. The transfer angle is the angle between solar vertical at points of departure and arrival

Techniques for determination of actual craft position and velocity have been described [6]. They are essential to the solution of the homing problem and can be derived on board the spacecraft using an Interplanetary Sextant [6]. The manner of use of this instrument is presented in summary in Fig. 3. Estimates of the accuracy of this device indicate that in the midcourse mode it will provide data on position and velocity of the spacecraft relative to the reference planet within 10,000 miles and 0.5 miles/second. (See Fig. 3 and [6].)

The accuracy of guidance needed for landing control is indicated by the work of CHAPMAN [2], who has shown that narrow landing corridors exist. While these corridors are, to some degree, a function of the vehicle configuration, we shall, for the present and without loss of generality, consider vehicles of the ballistic entry type; i.e., $L/D = 0$. For this class of vehicles, the landing corridors are represented by narrow bands of altitude of about 5 to 10 miles in width[1].

The problem, then is to provide control for a spacecraft as it approaches its destination and to "steer" it to the landing corridor. It may be in a course which differs as much as 100,000 miles from a collision course, and it must now be brought

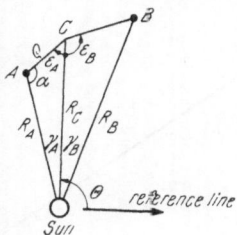

Fig. 3. Interplanetary navigation position triangle

Known: R_A, R_B; $\gamma_A + \gamma_B$; Ω_B, Ω_A, t

Measured: ε_A, ε_B; θ see column I; R_c see column II; $|\varrho|$ see column III

Column I θ Measured	Column II R_c Measured	Column III $|\varrho|$ Measured
$\varrho = \dfrac{R_A \sin (\Omega_A t - \theta)}{\sin \varepsilon}$ $a = \pi - (\Omega_A t - \theta) - \varepsilon$	$\varrho = R_c \left[\cos \varepsilon_A \pm \sqrt{\cos^2 \varepsilon_A - \left(1 - \dfrac{R_A}{R_c}\right)} \right]$ $a = \sin^{-1}\left(\dfrac{R_c \sin \varepsilon_A}{R_A}\right)$	$\varrho = \varrho_0 / \tan \delta$
Errors	**Errors**	**Errors**
$\dfrac{\partial \varrho}{\partial R_A} = \dfrac{\sin \gamma_A}{\sin \varepsilon}$	$\dfrac{\partial \varrho}{\partial R_A} = \varrho \tan \varepsilon$	$\dfrac{\partial \varrho}{\partial \delta} = \varrho_0 \left(\dfrac{\varrho_0{}^2 + \varrho^2}{\varrho_0{}^2}\right)$
$\dfrac{\partial \varrho}{\partial \theta} = \dfrac{R_A \cos (\Omega_A t - \theta)}{\sin \varepsilon}$	$\dfrac{\partial \varrho}{\partial R_c} = \dfrac{R_A}{R_c \cos \varepsilon}$	$\dfrac{\partial a}{\partial \varepsilon} = -2$
$\dfrac{\partial \varrho}{\partial \varepsilon_A} = \dfrac{R_A \sin \gamma_A \cot \varepsilon}{\sin \varepsilon}$	$\dfrac{\partial \varrho}{\partial \varepsilon_A} = \dfrac{1}{\cos \varepsilon}$	
$\dfrac{\partial a}{\partial \theta} = 1$	$\dfrac{\partial a}{\partial R_c} = \dfrac{\sin \varepsilon_A}{R_A \cos a}$	
$\dfrac{\partial a}{\partial \varepsilon_A} = -1$	$\dfrac{\partial a}{\partial \varepsilon_A} = \dfrac{R_c \cos \varepsilon_A}{R_A}$	
	$\dfrac{\partial a}{\partial R_A} = \dfrac{R_c \sin \varepsilon_A}{R_A{}^2 \cos a}$	

The solution of a navigation triangle is summarized. It is desired to locate the position of craft (C) relative to planets (A) and (B). Since solar system configuration is known, the craft position will be defined by the length (ϱ) and orientation (a) of the position vector

into a landing corridor which is perhaps 10 miles wide at periapsis. As if this were not a task of sufficient magnitude, this must be done under the strong influence of the gravity field of the destination planet, thus requiring that account be taken of the effects of this force.

---

[1] In detail, the character of the landing corridor is dependent on the nature of the atmosphere, the vehicle configuration, and the size and density of the planet of interest. The size chosen here is characteristic of the Earth and its neighbor planet Venus. For more detail see [2].

III. Summary of Dynamics

The spacecraft, as it terminates its midcourse flight and enters the homing phase of flight, is of course under the strong influence of both the planet and solar gravity fields. However, as the encounter develops, it will be appropriate to consider only the gravity field due to the planet since both the planet and the spacecraft experience (within a high degree of approximation) the same accelerations

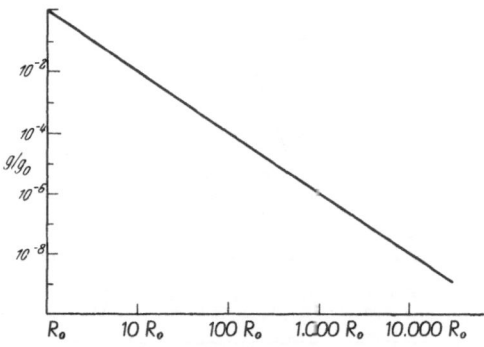

Fig. 4. Planet gravity fields

Solar gravity table

Planet		R(Ast. Units)	Solar grav.	Surf. grav.	Solar grav./surf. grav.
Mercury..............	☿	0.387	$4.07 \times 10^{-3}g.$	$0.28\ g.$	1.4×10^{-2}
Venus	♀	0.723	1.19×10^{-3}	0.85	1.4×10^{-3}
Earth	⊕	1.00	6.1×10^{-4}	1.00	6.1×10^{-4}
Mars.................	♂	1.52	2.6×10^{-4}	0.38	6.8×10^{-4}
Jupiter	♃	5.20	2.26×10^{-5}	2.6	8.7×10^{-6}
Saturn	♄	9.54	6.75×10^{-6}	1.2	5.6×10^{-6}
Uranus	♁	19.18	1.66×10^{-6}	1.1	1.5×10^{-6}
Neptune.............	♆	30.05	6.80×10^{-7}	1.4	4.85×10^{-7}
Pluto	♇	39.52	3.9×10^{-7}	?	

The characteristics of planet and solar gravity fields are shown. The chart is useful in determining the relative magnitude of solar and planet gravity fields at various distances from the planet surfaces

due to the solar gravity (see Fig. 4). For example, a space ship approaching Venus at a distance of 100 planet radii (400,000 miles) will experience a solar gravitational acceleration differing from that at Venus by an amount less than

$$\frac{\Delta g_s}{g} = \frac{2\Delta R}{R} \approx 1\% . \tag{1}$$

The acceleration arising from such a differential in solar gravity is about one tenth of that due to the planet directly. Unlike the planet field, this differential term vanishes as the two bodies approach one another.

The trajectory of the approaching space ship relative to the planet may be considered then to be determined by the planet's gravity field and the initial conditions of motion. This trajectory may be mathematically described relative to the planet in terms of three parameters; we have chosen p, ε, and θ' for reasons of convenience. These parameters may be expressed in terms of the coordinates of the spacecraft as (see Fig. 1):

1. The semi-latus rectum

$$p = r_p(1+\varepsilon) = \frac{r^4 \dot{\theta}^2}{k}. \tag{2}$$

2. The eccentricity

$$\varepsilon = \sqrt{1 + p\left(\frac{\dot{r}^2 + r^2 \dot{\theta}^2}{k} - \frac{2}{r}\right)}. \tag{3}$$

3. The position of the line of apsides

$$\frac{\varepsilon r \sin(\theta - \theta')}{p} = \frac{\dot{r}}{r\dot{\theta}}. \tag{4}$$

With these relations, it is now possible to predict periapsis by measurement of altitude (r), altitude rate $(\dot{r})$, and turning rate of the line of sight $(\dot{\theta})$. By adjustment of these, it will be possible to control the periapsis to coincide with the value appropriate for the landing corridor.

In calculating the corrections necessary for a suitable landing, the equation involving the true anomaly is not required unless a specific landing site is demanded. Since it is felt that such a requirement would unduly complicate the problem and that a post-entry maneuver capability is more practicable, this has not been considered. In the process of preparing this work however, it has become apparent that these same techniques could be applied to the ascent guidance problem for the escape arc of the hyperbola provided that the relations involving the inertial directions of the line of apsides and asymptote were employed permitting the adjustment of the direction of the asymptote of this hyperbola. Discussion of this technique is beyond the scope of the present work.

By differentiation of (2) and (3) above, the error expressions can be written as:

$$\frac{\partial r_p}{\partial r} = \frac{2 r_p(\varepsilon+1)}{r\varepsilon} - \frac{r\dot{\theta}^2}{g_p \varepsilon} - \frac{r_p^2}{\varepsilon r^2} \tag{5}$$

$$\frac{\partial r_p}{\partial \dot{r}} = -\frac{\dot{r}}{\varepsilon g_p} \tag{6}$$

$$\frac{\partial r_p}{\partial \dot{\theta}} = \frac{1}{\dot{\theta}}\left(r_p \frac{\varepsilon+1}{\varepsilon} - \frac{r^2 \dot{\theta}^2}{\varepsilon g_p}\right). \tag{7}$$

These may be combined in the familiar form

$$d r_p = \frac{\partial r_p}{\partial r} dr + \frac{\partial r_p}{\partial \dot{r}} d\dot{r} + \frac{\partial r_p}{\partial \dot{\theta}} d\dot{\theta}. \tag{8}$$

Before discussing the instruments required for control of this maneuver, let us examine first the kind of maneuvers which will be required. It must be recognized of course that we wish to minimize the maneuver in order not to require too great a fraction of the fuel load and thus to jeopardize the success of the mission. Accordingly, it is desirable to make all corrections to velocity direction at the greatest distance possible from the destination planet; i.e., while required direction changes are small, and before the planet's force of attraction has caused too great an increase in the relative velocity thereby making the correction energy requirement excessive.

The limit which we must apply to distance from the destination is that the craft must be close enough to sense the correction needed with sufficient accuracy. However, should we delay until we can make the entire correction in a single maneuver, excessive velocity impulses would be required. The correction must then be made in two steps, as follows.

Step 1: A correction made at great distance ($r/r_0 \approx 100$) which modifies the trajectory to provide a periapsis which is roughly correct for landing (i.e., $r_p \approx r_0$).

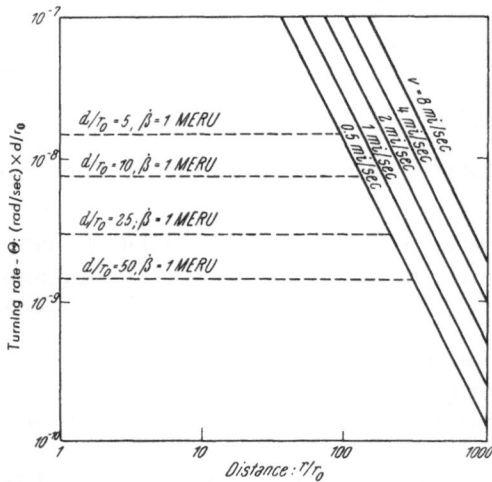

Turning rate - Θ: (rad/sec) × d/r_0

$d/r_0 = 5, \beta = 1\ MERU$

$d/r_0 = 10, \beta = 1\ MERU$

$d/r_0 = 25, \beta = 1\ MERU$

$d/r_0 = 50, \beta = 1\ MERU$

$v = 8\ mi/sec$
$2\ mi/sec$
$4\ mi/sec$
$1\ mi/sec$
$0.5\ mi/sec$

Distance : r/r_0

Fig. 5. Turning rate of line of sight to destination planet

The turning rate has been graphed as $\dot\theta$ times the planet miss distance parameter (d/r_0). The graph illustrates the range of distances, velocities, and miss distance parameters for which a gyro having threshold sensitivity of 1/1000 of Earth rate (1 MERU) would be effective in sensing the planet's approach

Step 2: A correction to make r_p coincide with the appropriate landing corridor. (This represents

$$\frac{\Delta r_p}{r_p} \approx 0.02.)$$

Consider now a vehicle in flight towards its destination planet. As the distance between the two is reduced, the line of sight from craft to destination will begin to turn relative to an inertial reference except in the rare instance that a collision is imminent. At great distances this turning rate is expressed by

$$\dot\theta = \frac{v_\infty d}{r_0{}^2}\left(\frac{r_0}{r}\right)^2. \tag{9}$$

These turning rates have been graphed for several encounters in Fig. 5. From this it may be seen that a gyro capable of sensing 1/1000 of the earth's angular rate (1 MERU) will sense this turning rate and permit the execution of Step 1 at distances of approximately $r = 100\ r_0$.

After executing Step 1, as the vehicle approaches still closer to its destination, it will begin computing the periapsis of its trajectory to determine the final correction maneuver necessary to adjust the periapsis placing it within the indicated landing corridor. This adjustment is computed using the expression of eq. (8). However, since it is not considered feasible to make a maneuver changing r abruptly, Δr is assumed to vanish for purposes of this maneuver, thus

$$\Delta r_p = \frac{\partial r_p}{\partial \dot r}\Delta V_{||} + \frac{1}{r}\frac{\partial r_p}{\partial \dot\theta}\Delta V_\perp \tag{10}$$

which when substitutions are made becomes

$$\Delta r_p = -\frac{\dot r}{g_p\,\varepsilon}\Delta V_{||} + \frac{1}{r\dot\theta}\left[r_p\frac{\varepsilon+1}{\varepsilon} - \frac{(r\dot\theta)^2}{\varepsilon\,g_p}\right]\Delta V_\perp. \tag{11}$$

The above expression may be approximated as

$$\Delta r_p = -\frac{V\sin\Phi}{g_p\,\varepsilon}\Delta V_{||} - \frac{V\cos\Phi}{g_p}\Delta V_\perp. \tag{12}$$

As the encounter develops, the vehicle and the planet grow closer and the flight path angle, Φ, decreases while the velocity, v, increases. Accordingly, these later adjustments are most appropriately made in the form of cross velocity maneuvers.

With reference to eqs. (11) and (12) and Fig. 5, it can be seen that at distances of about 100 planet radii (where the first maneuvers will be made) the approximate values of these coefficients for Earth will be[1]

$$\frac{\partial r_p}{\partial \dot{r}} \approx 700 \text{ miles/mi/sec.}$$

$$\frac{\partial r_p}{r \partial \dot{\theta}} \approx 70{,}000 \text{ miles/mi/sec.}$$

The major difference in these two quantities represents an overwhelming argument in favor of making corrections at great distance by maneuvers which change velocity direction. The magnitude of these maneuvers is estimated for an Earth landing then as

$$\frac{\Delta r_p}{r_p} = 17.5 \, \Delta V_\perp$$

or

$$\Delta V_\perp = .057 \frac{\Delta r_p}{r_p}.$$

Correction maneuvers to bring the periapsis closer by several planet radii can thus be made for a velocity change of less then 1 mi/sec; calculation of the exact maneuver would of course require precision calculation of a transfer and approach trajectory. Furthermore, with reference to eq. (11), it is apparent that changes in periapsis can be controlled by velocity direction changes if we measure only r and θ. This represents a distinct advantage to the instrumentation.

IV. Instrumentation System

In the previous sections we have discussed the general nature of the homing problem and presented a simplified mathematical model for computation and execution of the maneuver required to convert an interplanetary transfer orbit leading to a miss of a few hundred thousand miles into a landing orbit which is suitable for air drag braking without excessive deceleration. It has been shown that a set of instruments capable of measuring r, $\dot{r}$, and $\dot{\theta}$ is needed. In this section of the paper, a configuration of instruments suitable for the task will be described.

Before describing the configuration of a system of instruments, however, let us examine the basic means which are available for measurement of the individual coordinates.

1. *Range to Destination* (r): At first glance, it might appear that this coordinate could be adequately measured using radar methods. However, even the simplest of calculations indicates that the space losses are such that, unless a system of beacons could be distributed on the destination planet, no useful signal strength can be expected for pulse transmissions so high as 0.1 megawatts. At the distances for which we are concerned, optical measurements appear to be quite adequate where range can be sensed as related to the diameter of the disk of the planet. (At $r = 1000\, r_0$, the planet will subtend 2 milliradians; at $r = 100\, r_0$, about 1 degree, etc.)

2. *Range Rate* $(\dot{r})$: We do not consider doppler radio means for this measurement for the reasons set forth above. Provided that adequate accuracy is obtained in measurement of range, it will be satisfactory to note the measured change in

---

[1] Calculations are based on $v = 5$ mi/sec and $\varepsilon = 1.5$.

range during controlled time intervals in order to estimate range rate. In so doing, an adequate measurement of the error in the derivative is given by

$$E(\dot{r}) = \frac{E(r)}{\Delta t} .\tag{13}$$

3. *Turning Rate of Line of Sight* $(\dot{\theta})$: For this measurement it is sufficient if we employ a precision gyroscope for stabilization of the measuring telescope described above and to note the torquing rate signals derived from the tracking servo. This gyroscope must be mounted rigidly to the telescope, and it must also be capable of acting as a command receiver for the tracking servo loop. A full three-axis gyro-stabilized gimbal system will be desired by which the angular motions of the spacecraft may be isolated from the telescope.

A configuration of gyroscopes, servo motors, and a tracking telescope designed to operate in accordance with the principles outlined above is shown in Fig. 6. A simple representation of the computation system is included for clarity. Actual computations would probably (but not necessarily) be performed in a central digital computer.

Since the function of the homing instruments is to compute and apply velocity corrections, and since these corrections must be applied to the vehicle with precise reference to the direction measured by the homing instruments, it is desirable to include accelerometers in the instrument package in order to monitor the actual thrust applied and, if necessary, to command corrections to the thrust direction and magnitude during application.

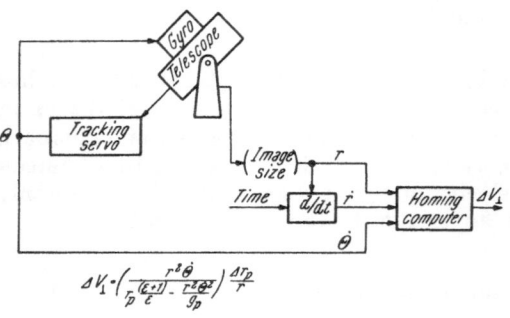

Fig. 6. The interplanetary homing seeker
The application of a tracking telescope to the determination of steering commands is shown. Accelerometers, mounted to a stabilized reference frame (not shown), are employed to monitor and control the application of thrust for corrections

All of the elements described above are contained in the Interplanetary Sextant described in [6]. A photograph of a model of that instrument is shown in Fig. 7.

We may estimate the errors in navigation in terms of the errors in measurement; recalling the expression (8) for the differential in periapsis, we may now state:

$$E(r_p) = \frac{\partial r_p}{\partial r} E(r) + \frac{\partial r_p}{\partial \dot{r}} E(\dot{r}) + \frac{\partial r_p}{\partial \dot{\theta}} E(\dot{\theta}) .$$

The various coefficients may be approximately evaluated

$$\frac{\partial r_p}{\partial r} = 2 \frac{r_p}{r} \frac{(\varepsilon + 1)}{\varepsilon} \frac{r \dot{\theta}^2}{g_p \varepsilon} - \frac{r_p^2}{\varepsilon r^2} \approx 1 \text{ mi/mi}$$

$$\frac{\partial r_p}{\partial \dot{r}} = \frac{\dot{r}}{\varepsilon g_p} \approx \frac{5280 \, v_e}{32 \, \varepsilon} = 1000 \text{ mi/(mi/sec)}$$

$$\frac{\partial r_p}{\partial \dot{\theta}} = \frac{r_p}{\dot{\theta}} \left(\frac{\varepsilon + 1}{\varepsilon}\right) - \frac{r^2 \dot{\theta}}{\varepsilon g_p} \approx 10^{13} \text{ mi/(rad/sec)} .$$

If it is assumed that the errors in measurement are independent of each other and each error source makes approximately equal contribution to the total

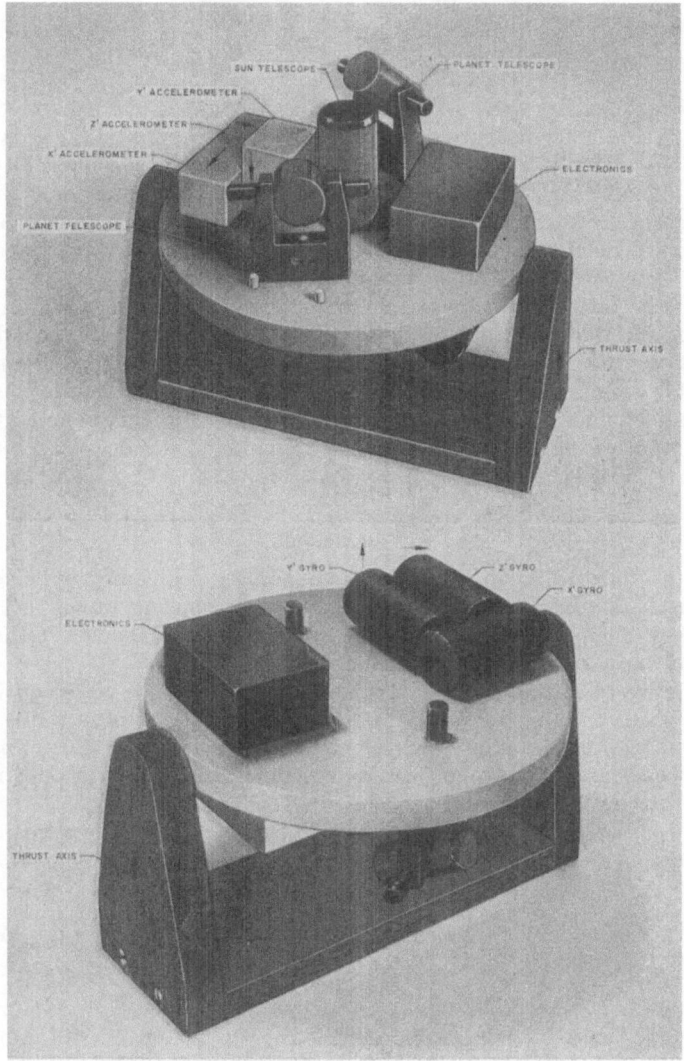

Fig. 7. Interplanetary sextant
The interplanetary sextant (see [6]) may be used for midcourse navigation. The instrument contains all of the basic elements required for the homing application

error, the tolerable errors in measurement then may be stated as

$$E\left(q\right) = \frac{E\left(r_p\right)}{\sqrt{3}\,\dfrac{\partial\, r_p}{\partial\, q}}$$

which have been tabulated below for an assumed 1-mile tolerance in r_p.

Error	Value	% of Quantity Measured
$E\ (r)$	0.6 mi	0.00015 %
$E\ (r_p)$	0.0006 mi/sec	0.00015 %
	$\left\{ \begin{array}{l} \tau = 1000 \text{ sec} \\ \text{See eq. (13)} \end{array} \right\}$	
$E\ (\theta)$	0.6×10^{-13}	0.8 %

While it appears from the tabulated values shown that an excessive precision may be called for in the measurement of r, it must be recognized that r is measured by sensing the planet's disk as an angle [7]. This angle measurement at even relatively great distances, $r/r_0 = 100$, is as much as one degree. An error in disk size of about one second of arc would then correspond to about one mile in measuring distance from a planet having a 7200 mile diameter. The relative importance of such an error would decrease as the planet is approached.

The differentiating time interval indicated in the table is set in order to permit sufficient smoothing of the derivatives. During this 1000 second interval, we should expect the craft to advance perhaps two planet radii; while this might be acceptable at large distances of say $r/r_0 = 100$, at closer distances, say $r/r_0 = 25$, it would be desirable to reduce this smoothing interval to limit the travel to lesser distances. This may of course be done since the $E\ (r)$ will be reduced directly in proportion to the distance to the planet.

V. Conclusions

It has been shown that the interplanetary homing problem may be adequately solved with a system of instruments capable of measuring range to the planet and the rate of change of the line of sight relative to inertial space. Other quantities required may be derived from these.

The precision required of the instruments is of the same order of magnitude as that required in fixing periapsis of the approach trajectory. While measurements made close to the planet will permit more precise control of periapsis, the maneuver requirements may become excessive if corrections are not made and checked at moderate distances; i.e., $r/r_0 \approx 10$.

The Interplanetary Sextant described in [6] contains the basic elements needed for the homing task.

References

1. D. B. DeBra and R. W. Gillespie, Minimum Maneuver Circular Capture Orbits. Astronaut. Sci. Rev. 1959.
2. D. R. Chapman, On the Corridor and Associated Trajectory Accuracy for Entry of Manned Spacecraft into Planetary Atmospheres. Proceedings of the Xth International Astronautical Congress, London, 1959, p. 254. Wien: Springer, 1960.
3. D. B. DeBra, The Effect of Guidance Errors in Astroballistic Trajectories. Proceedings of I.R.E. Third National Convention on Military Electronics. July 1959.
4. K. A. Ehricke, Error Analysis of Keplerian Flights Involving a Single Central Force Field and Transfer between Two Central Force Fields. Navigation 6, No. 1 (Spring 1958).
5. T. A. Magness, J. B. McGuire and O. K. Smith, Accuracy Requirements for Interplanetary Ballistic Trajectories. Proceedings of the IXth International Astronautical Congress, Amsterdam, 1958, p. 286. Wien: Springer, 1959.
6. E. V. Stearns, An Interplanetary Navigation System. Proceedings of the IXth International Astronautical Congress, Amsterdam, 1958, p. 265. Wien: Springer, 1959.
7. L. Larmore, Celestial Observations for Space Navigation. Institute of Aeronautical Sciences, Los Angeles, Calif. July 8, 11, 1958.

The Application of Solid Propellants to Space Flight Vehicles

By

H. L. Thackwell, Jr.[1]

(With 12 Figures)

(Received June 12, 1959)

Abstract — Zusammenfassung — Résumé

The Application of Solid Propellants to Space Flight Vehicles. The author first outlines the reasons why solid propellants were chosen for the terminal stages of the successful U.S. *Farside, Explorer, Vanguard,* and *Pioneer* satellite and space probe rocket vehicles. He then describes some of his own experiences in the design and development of the solid propellant third stage *Vanguard* motors, two of which are now orbiting the Earth.

Presently planned solid propellant systems in the United States Space efforts are then discussed including the NASA four stage, all solid *Scout* satellite vehicle, and the escape and retro-rocket systems intended for the NASA manned satellite *Mercury*. The expected applications of solid propellants to more extensive missions of manned satellite supply, lunar circumnavigation, lunar landings, deep space journeys, and the like are then set forth; and the advantages which can be achieved by the solid propellant rocket's inherent high reliability, instant readiness, and low overall cost are shown.

The performance advantages to be gained by using the solid propellant rocket's intrinsic ability to achieve high motor/mass ratios and high thrust-to-weight ratios, its ease of staging and clustering, its precise control of total impulse, its stability under long-term storage, and other factors are then detailed.

The author concludes his paper by presenting a design of an ultra-high performance, all solid propellant, three stage escape vehicle utilizing a novel staging concept. Moreover, it is shown that this space vehicle which can be either ground or air launched, would be sufficiently inexpensive to be within the budget of even the smallest nation wishing to have its own space flight program.

Die Anwendung fester Treibstoffe in Raumfahrzeugen. Der Autor bespricht zuerst die Gründe für die Wahl fester Treibstoffe in den Erststufen der erfolgreichen amerikanischen Raumsonden und Satelliten *Farside, Explorer, Vanguard* und *Pioneer*. Dann beschreibt er einige seiner eigenen Erfahrungen aus Entwurf und Entwicklung der *Vanguard*-Endstufenmotoren, von denen zwei heute die Erde umkreisen.

Die gegenwärtig geplanten Feststoffsysteme in den Raumfahrtbemühungen der USA werden erörtert: nämlich das völlig mit Feststofftriebwerken ausgerüstete Satellitenfahrzeug der NASA, *Scout*, und das Antriebs- und Bremsraketensystem des bemannten NASA-Satelliten *Mercury*. Die voraussichtlichen Anwendungen fester Treibstoffe für ausgedehntere Aufgaben bemannter Satelliten, Mondumfahrungen, Mondlandungen, tiefere Raumreisen und ähnliches werden besprochen. Die Vorteile, die man durch die überaus große Zuverlässigkeit, Einsatzbereitschaft und die niedrigen Gesamtkosten von Feststoffraketen erreichen kann, werden aufgezeigt.

---

[1] Senior Vice President, Grand Central Rocket Co., Redlands, California, U.S.A.

Dann werden die leistungsmäßigen Vorteile im einzelnen besprochen, die sich aus
den für Feststofftriebwerke typischen Eigenschaften ergeben, wie hohes Triebwerk-
massenverhältnis, hohes Schub-Gewicht-Verhältnis, einfache stufenweise Anordnung
und Bündelung, genaue Bestimmung des Gesamtimpulses, lange Lagerfähigkeit usw.

Der Verfasser schließt das Referat mit dem Entwurf eines völlig feststoffgetrie-
benen, dreistufigen Satellitenfahrzeuges besonders hoher Leistung mit einer neuartigen
Stufenanordnung. Darüber hinaus wird gezeigt, daß dieses Raumfahrzeug, das entweder vom Boden oder aus der Luft gestartet werden kann, billig genug wäre, um
sogar innerhalb des Budgets der kleinsten Nation zu liegen, die ihr eigenes Raum-
flugprogramm aufstellen möchte.

L'utilisation de propergols solides dans les véhicules spatiaux. L'auteur résume
d'abord rapidement les raisons pour lesquelles des propergols solides ont été retenus
pour les derniers étages des satellites et fusées spatiales expérimentales U.S. *Farside,
Explorer, Vanguard* et *Pioneer*. Il relate ensuite quelques-unes de ses expériences
personnelles concernant l'étude et la réalisation du moteur à propergol solide du
3ème étage du *Vanguard*, dont deux exemplaires décrivent actuellement une orbite
autour de la terre.

Il passe ensuite en revue les systèmes à propergol solide actuellement projetés
dans le cadre du programme américain d'engins spatiaux, y compris le véhicule satellite
Scout à 4 étages du NASA, à propulsion entièrement solide et les dispositifs d'évacuation et d'inversion de poussée prévus pour le *Mercury*, Satellite habité du NASA. Puis,
il énumère les applications des propergols solides à des missions plus larges comportant le ravitaillement des satellites habités, la navigation autour de la lune, les atterris-
sages sur la lune, les voyages intersidéraux etc., et montre les résultats qui peuvent
être obtenus grâce aux avantages inhérents aux propergols solides tels que grande
sûreté de fonctionnement, fonctionnement immédiat et coût total faible.

L'auteur montre ensuite l'amélioration des performances qu'on peut attendre de
l'utilisation des propergols solides qui permettent notamment d'obtenir pour les
moteurs des rapports de masse élévés et des rapports poussée/poids avantageux, qui
sont faciles à étager et à grouper, qui autorisent un contrôle précis de l'impulsion totale,
qui restent stables pendant des stockages de longue durée etc.

L'auteur conclut son article en présentant le dessin d'un véhicule à trois étages,
possédant des performances extrêmement élevées, permettant l'évacuation et utilisant
un étagement de conception originale. Il montre en outre, que ce véhicule spatial qui
pourrait être lancé aussi bien à partir d'un aérodyne que du sol serait suffisamment
bon marché pour rester dans les possibilités budgétaires de n'importe quelle nation,
si petite soit-elle, qui désirerait avoir son propre programme de vols spatiaux.

Perhaps the first attempt by man to launch himself into space by rockets
is described in the amusing story related by Willy Ley in his book "Rockets,
Missiles and Space Travel". In about 1500, according to this legend, an adventure —
some Chinese gentleman named Won Hoo strapped himself into his favorite
chair, to which he attached two enormous kites and a large number of the largest
rockets he could find. He then ordered his favorite coolie servants to ignite
the rockets, and when the smoke cleared away, nothing could be found of the
two kites, the chair, the coolies, or the honorable Mr. Won Hoo. Now, because
it is fairly certain that these rockets used gunpowder, we solid propellant enthusi-
asts proudly believe that Mr. Won Hoo, his chair, his two kites, and his coolies
were all launched into a 300-mile orbit and are still up there happily and peace-
fully spinning around our earth!

Then, over four centuries later, a most significant advancement in solid
propellant rockets occurred in 1935 when Dr. H. J. Poole, the eminent English
scientist, originated the internal-burning, star-shaped, solid propellant grain
which enabled the combustion chamber wall to be completely shielded from
the hot gases, thereby permitting a lightweight chamber. This extremely im-

portant contribution, together with the pioneering work in the United States
in 1945-47 done by the Jet Propulsion Laboratory of the California Institute
of Technology (JPL) in developing castable solid propellants capable of being
bonded to and supported directly by the combustion chamber wall, made it
feasible for solid propellant rockets to be built in very large sizes and also to
attain considerably higher burnout velocities than had hitherto been thought
possible (see Fig. 1).

It was the author's good fortune to join JPL at about this time as a project
engineer to aid in the design of a 6-inch rocket to flight-test this new type pro-

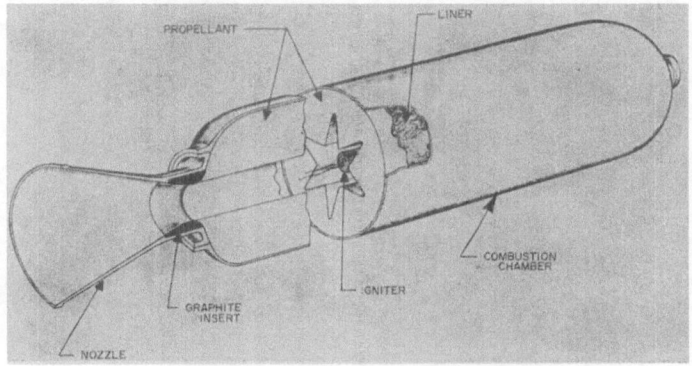

Fig. 1. Internal-burning case-bonded star-shaped grain

pellant and grain design. These flight tests were made at the Naval Ordnance
Test Station in late 1947 and, to our gratification, were so successful that we
quickly published reports predicting that case-bonded, internal-burning, solid
propellant rockets would ultimately attain V-2 size and performance. Needless

to say, these reports created
quite a stir and even some
laughter in liquid propellant
circles at that time.

Shortly thereafter, JPL and
the General Electric Company
joined forces to develop and
flight-test the first large two-
stage test missile to be launched
in the United States, the U.S.
Army *Bumper-Wac*. This missile
utilized a captured German V-2 as
the first stage and the JPL
liquid-propelled *Wac-Corporal* as
the second stage. During this

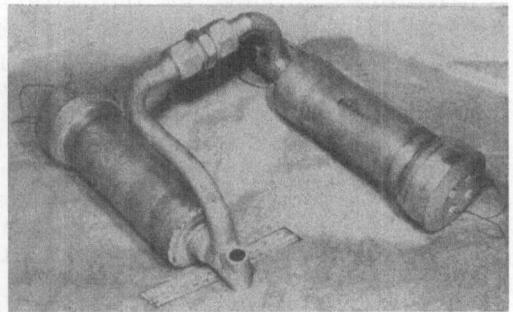

Fig. 2. Solid propellant spin rocket for *Bumper-Wac*
missile

program, the first successful demonstration in the United States of rocket
ignition at extreme altitudes was made at the U.S. Army White Sands Proving
Ground on May 19, 1948 with JPL internal-burning, case-bonded, star grain,
solid propellant rockets which were used to spin the second stage of the
Bumper-Wac vehicle (see Fig. 2). Also for this first flight test, a solid propellant
test rocket was used to form a dummy second stage to check stage separation.
This second stage solid propellant test rocket achieved an altitude of approximately 85 miles, thus being the first U.S. built rocket to reach such an altitude

As a result of the successful work on case-bonded, star-shaped grain solid propellant rockets at JPL, the General Electric Company and the Thiokol Chemical Corporation, under the sponsorship of the U.S. Army, started development in 1950 at Redstone Arsenal of what was then believed to be the largest solid

Fig. 3. First static firing test of large U.S. Army solid propellant motor

propellant motor in the world. The first static firing test of this motor occurred in December 1951 (see Fig. 3). The author had joined the Thiokol Corporation as the Project Engineer for this large motor and was quite gratified with the test results because then we were finally beginning to prove that large high performance solid propellant rocket motors were feasible. This particular motor was successfully flight-tested in 1953; and somewhat later in an improved version, named the *Sergeant*, formed the first stage of the highly successful U.S. Air Force X-17 hypersonic test vehicle, an all-solid propellant, three-

a   b

Fig. 4. The second and third-stage cluster of solid propellant motors for the *Jupiter C*

stage missile. The X-17 was developed by the Lockheed Aircraft Corporation and is believed to have attained altitudes of over 500 miles during test flights in 1956.

Before that time, the author had joined the newly formed Grand Central Rocket Co. (GCR) where one of our first jobs was to manufacture a high perform-ance 3-inch *Loki* rocket which had been recently developed by JPL. Later, as a result of this *Loki* work the author attended a meeting in Dr. WERNHER VON BRAUN's office in 1954 at the U.S. Army's Redstone Arsenal in which Dr. VON BRAUN was describing to a number of high-ranking Army and Navy officers how he proposed to launch a three-stage cluster of *Loki* rockets from the top of the Army's *Redstone* rocket and thus create the first U.S. Earth Satellite. This historic occasion is most vivid in the author's memory because the vision and salesmanship of Dr. VON BRAUN made one imagine that he was transported back to Queen Isabella's court and was hearing Columbus plead for the money and permission to be the first to sail across the Atlantic Ocean and into the unknown world beyond.

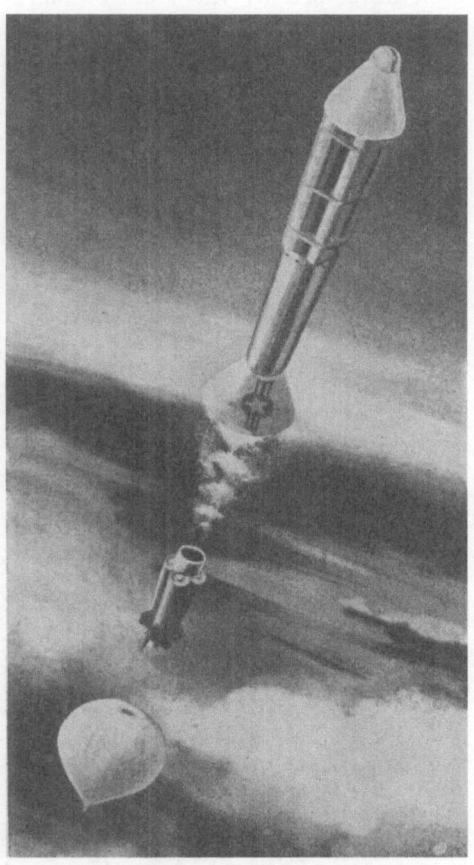

Fig. 5. The U.S. Air Force all solid propellant *Farside* vehicle

This first U.S. satellite program, which was to have been called Project *Orbiter*, was never actually started because of the choice of the U.S. Navy's *Vanguard* satellite program. However, the JPL-Army team proceeded to develop a very similar type vehicle, called the *Jupiter C*, which at that time was intended only for hypersonic test purposes. On this *Jupiter C* vehicle, the 3-inch *Loki* rockets intended for the three upper stages of the *Orbiter* were replaced with clusters of larger, scaled-down *Sergeant* solid propellant rockets six inches in diameter (see Fig. 4). The *Jupiter C* set a new U.S. altitude record of approxi-mately 650 miles in late 1956 during a preliminary test which utilized only three of its four stages. This height remained the U.S. altitude record until 1957, when the U.S. Air Force *Farside* vehicle, a four-stage all solid propellant rocket developed by Aeronutronics Systems, Inc., was launched at 75,000 feet from a balloon (see Fig. 5). On one of these tests, it is believed that the top stage, a *Loki*-type *Arrow II* rocket manufactured by Grand Central Rocket Co., reached an altitude of approximately 4,000 miles. Shortly thereafter, on January 31, 1958, the U.S. Army-JPL team launched the *Explorer I*, the first successful U.S. satellite. The *Explorer I* vehicle was a modified *Jupiter C* which incorporated a single 6-inch *Sergeant* acting as the fourth stage.

A similar type, three-stage solid propellant cluster but with an improved solid propellant final stage and a larger liquid propellant first stage (the *Jupiter*)

was used to launch a moon probe into translunar space by the JPL-U.S. Army team during the successful flight test of the *Pioneer IV* vehicle on March 3, 1959. The final stage solid propellant motor of the *Pioneer IV* therefore became the first U.S. rocket to be placed into an orbit about the sun.

In the meantime, the U.S. Navy, three-stage, *Vanguard* vehicle which had been developed by the Martin Company had also successfully launched two

Fig. 6. U.S. Navy *Vanguard* satellite vehicle

satellites, one in 1958, and one in 1959 (see Fig. 6). The final stages of these *Vanguard* vehicles were GCR solid propellant rockets which are expected to continue orbiting around the earth for the next 2000 years (see Fig. 7).

Thus, from the above short history of United States space flight, it can be seen that, to date, solid propellant powered upper-stage rockets hold most of the U.S. altitude records, injected the first U.S. satellites into orbit, and pioneered the first modest penetration by the U.S. into space. Why were solid propellant rockets used in these applications despite the very great efforts on liquid rocket development in the United States? The answers to this question are:

1. Because of their basic simplicity, solid propellant rockets were intrinsically reliable.

2. Solid propellant rockets could be developed in a short time (for example, the third-stage motor of the Martin *Vanguard* vehicle was developed by the Grand Central Rocket Co. in the brief space of nine months).

3. Solid propellant rockets were inherently adaptable for spinning (spin stability eliminated the necessity for guidance in the terminal stages of the U.S. satellite vehicles).

4. Because of the relatively short time required for development and the simplicity of the inert parts, the cost of the solid propellant rockets was low compared to liquid propellant rockets of the same size and performance.

5. The required high performance could be obtained from solid propellant rockets in the relatively small sizes used (the CGR third-stage

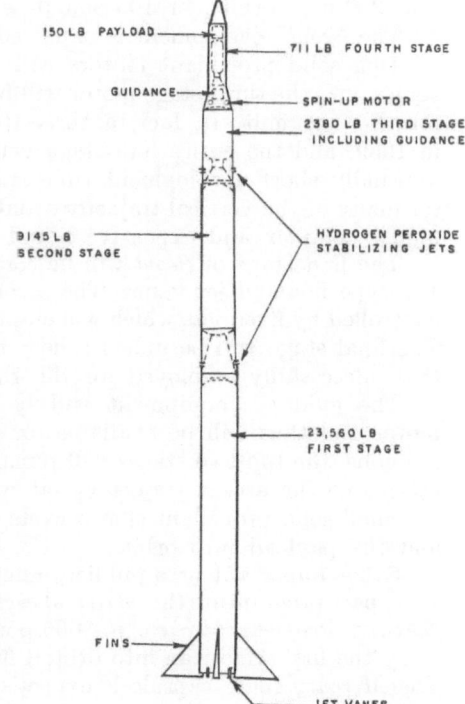

150 LB PAYLOAD

711 LB FOURTH STAGE

GUIDANCE

SPIN-UP MOTOR

2380 LB THIRD STAGE INCLUDING GUIDANCE

9145 LB SECOND STAGE

HYDROGEN PEROXIDE STABILIZING JETS

23,560 LB FIRST STAGE

FINS

JET VANES

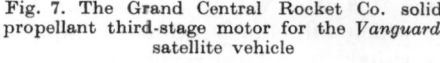

Fig. 7. The Grand Central Rocket Co. solid propellant third-stage motor for the *Vanguard* satellite vehicle

Fig. 8. The NASA solid propellant *Scout* satellite vehicle

rockets contributed more than one-half of the required 25,000 fps incremental orbital velocity for the *Vanguard* satellites).

Regarding the ability of the above mentioned GCR third-stage *Vanguard* motor to attain high performance, to the author's knowledge, this particular motor was the first rocket ever to be signed specifically for the essentially vacuum conditions which exist at extreme altitudes. The design of this motor was considered a challenge by us because the burnout velocity requirement established by the Martin Company was considerably greater than was at that time thought possible by some of our competitors. However, by taking advantage of the fact that the specific impulse of a rocket motor operating in a vacuum is essentially independent of the combustion pressure, we were able to use a very low pressure which in turn made possible an unusually lightweight combustion chamber. This lightweight chamber, together with a star-shaped, case-bonded propellant grain of high volumetric efficiency, resulted in a conservatively designed motor which we were able to develop in the short time allowed, yet which met all specifications.

The success and demonstrated reliability of this and other similar type U.S. solid propellant motors led the National Aeronautical and Space Administration (NASA) in mid-1959 to start development of the *Scout*, a four-stage all solid propellant vehicle intended for low cost space probe missions (see Fig. 8). The *Scout* is expected to have a total gross weight of approximately 36,000 pounds and to cost only $ 500,000 each. It will be used as a sounding rocket to carry a 50-pound payload to a 7000-mile altitude; and also to put 150-pound satellites into 300-mile orbits, or 100-pound satellites into 500-mile orbits.

The *Scout* development is expected to be rapid because improved models of existing solid propellant motors will be used for the first, second, and fourth stages, and the third-stage motor will be a simple scaled-up version of the existing fourth-stage unit. In fact, a three-stage version is expected to be test-flown in 1959, and the entire four-stage vehicle to be flight-tested during 1960. This unusually short development time is expected to enable the *Scout* to be used for many of the vertical trajectory and orbiting missions which presently require highly complex and expensive liquid propelled vehicles.

The first stage of *Scout* will be controlled and stabilized with the traditional V-2 type fins and jet vanes. The second and third stages will be stabilized and controlled by fixed jets which will employ an auxiliary hydrogen peroxide system. The final stage will be uncontrolled, but spin stabilized in a manner similar to that successfully employed for the third-stage *Vanguard* motor.

The guidance equipment will be located just forward of the third-stage motor and thus will be available for use on all three lower stages. For orbital missions, the top two stages will remain together during the coast period to the apogee of the ascent trajectory, at which time the final stage will be rotated by small solid propellant spin rockets and then fired horizontally to inject itself and the payload into orbit.

Other important uses publicly announced for solid propellant rockets in the U.S. space program are the retro and escape rocket systems now being developed for NASA's Project *Mercury*, a 3000-pound space capsule which is expected to carry the first American into orbital flight around the Earth, sometime in 1961. The *Mercury* space capsule is expected to be launched by the liquid propellant *Atlas* rocket into a near circular orbit 100 to 150 miles high to permit at least a 24-hour satellite lifetime. The descent from this orbit will be initiated by the application of solid propellant retro-rockets which will supply sufficient impulse to permit atmospheric entry in less than one-half an orbital revolution. Then, after the capsule is slowed down by aerodynamic drag to approximately sonic velocity, parachutes will be used to stabilize and safely land the vehicle.

Perhaps one of the most interesting space applications of solid propellants is a rocket now under development at Grand Central Rocket Co. for the emergency escape system of the *Mercury* space capsule. Should the liquid propelled *Atlas* start to malfunction during the boost phase, this solid propellant rocket, which will be mounted in a tower at the extreme nose of the capsule, will be fired. The resulting impulse will accelerate the capsule, in one second, to a point 250 feet ahead and to one side of the parent booster (see Fig. 9). After coasting to a safe distance, the retro-rockets and the parachute system of the capsule will then be used in the normal manner to decelerate the capsule and its human inhabitant to a safe landing. A solid propellant device was chosen by NASA for this extremely important and sobering escape task primarily because of the mandatory requirement of maximum reliability. In fact, NASA requires that the developed GCR unit shall have a reliability factor of 0.999 (i.e., no more than one malfunction in one thousand flight tests).

The fact that this type of high reliability has been already demonstrated by many solid propellant rocket developments in the United States is the chief reason why the author believes that solid propellants will be used on an ever-increasing scale for the primary propulsion systems of space flight vehicles.

Fig. 9. The solid propellant escape rocket system for the NASA Project *Mercury* Manned Space Capsule

Regarding this pressing need for reliability, the U.S. "Space Handbook" (prepared by the Rand Corporation and dated 19 December 1958) states that, "it is utterly meaningless to talk about flights to Mars if the equipment being sent there has no reasonable probability of working during the duration of the flight or for a useful period after arrival".

Let us take launching vehicles (boosters) as a prime example. We believe at GCR that solid rocket boosters will eventually be chosen for civilian programs involving manned space flight because of their superior reliability. We visualize no limitations to the size of these solid propellant boosters because they, like liquid propellant rockets, can be loaded with propellant at the launching site. Also, the instant readiness provided by such solid propellant boosters will be invaluable for supplying manned satellites. Consider a satellite or space vehicle which is circling the Earth every one and a half hours in a 200-mile orbit. In each pass over a supply base launching site, it will travel from horizon to horizon in approximately eight minutes. Thus, in order to efficiently intercept the satellite, the supply vehicle must be launched with split-second timing accuracy.

Moreover, for the relatively short space exploration journeys to be expected within the next decade such as for landing on the moon and return to Earth, we believe that solid propellant rockets will be used both for the primary propulsion motors, and also to provide the retro-thrust impulses needed for landing maneuvers.

This latter requirement could be satisfied by a circular cluster of small, highly efficient solid propellant rockets which could be fired and jettisoned in a planned sequence to gradually and safely lower the vehicle to a designated landing spot. The outstanding storage characteristics of the solid rocket (no boil-off) would also make it a preferred choice for such applications.

For the much longer trips to the planets Mars, Venus, and eventually into interstellar space, which may prove feasible within the next 25 years, clusters of extremely large solid propellant motors should see much service as Earth take-off boosters. Because such expeditions will extend over many months, years, and even human lifetimes, it is highly probable that stable solid propellant rockets will be carried along for steering, landing and return take-off purposes. However, the large solid rocket boosters will probably be used only for the purpose of escaping from the launching planet's atmosphere, because much more efficient means (possibly nuclear propulsion) must be employed to accelerate an interstellar vehicle to the ultra-high velocities necessary for reasonable flight times (for example, a chemical rocket escaping the Earth at the relatively high velocity of 53,000 fps would take 70,000 years to reach Alpha Centauri, our nearest star).

Another basic reason why solid propellant rockets will be extensively used for space exploration is their comparatively low cost. Those who must in the end pay for such space activities (you and I and all the other taxpayers in the world) are becoming increasingly aware of the fact that this astronautics business is an inherently high-cost activity. In fact, we begin to see it could grow, if left unchecked, into truly astronomical proportions (no pun intended here!) with corresponding catastrophic economic consequences which could engulf us all.

Why have solid propellant rockets demonstrated themselves to be the least expensive form of chemical jet propulsion? The reasons are:

1. The superior reliability of the solid propellants means more successful missions completed per dollar spent.

2. Because of their basic simplicity, solid rockets can be developed in a comparatively short time with corresponding minimum labor and facilities costs.

3. The inert parts of the solid rocket are not complex, and therefore, are very inexpensive as compared to liquid rocket parts (no pumps, valves, and similar costly hardware items are required for solid rockets).

4. The field handling and pre-launch checkout operations for solid rockets are an order of magnitude cheaper than those for present liquid rockets[1] because

---

[1] I refer here chiefly to those liquid propellant motors which require liquid oxygen, liquid hydrogen and other types of non-storable (cryogenic) fuels and/or oxidizers.

the need for elaborate ground handling equipment and expensive "standby" personnel is eliminated.

5. The cost of the propellant itself (loaded into the rocket) is comparable to that for liquid propellants when one realistically considers the cost of special equipment and highly trained personnel required for fueling (and refueling) some of our present liquid rockets[1].

We have now discussed topics covering the solid propellant space rocket's probable reliability, development time, and cost. In fact, one might conclude from the foregoing discussion that if he were given the task of developing an *entirely new* multi-stage space rocket (using no existing components) to deliver a particular payload into an Earth orbit, to the moon, or to an escape velocity for deep space missions; with the further stipulation that this job be done in the most reliable manner, with the least amount of money, and in the shortest possible time, he would unhesitantly have to choose solid propellants. But, you now may well ask, how do solid propellant rockets compare on the basis of overall performance? Does not the recognized high specific impulse attainable by some liquid propellants, and the proven high ratios of propellant weight to gross mass in existing large liquid motors tend to tip the scales in favor of liquids?

In answering this question, let us first consider that for realistic space work, we must be talking about *very large* rockets in the order of at least three times the thrust of the *Atlas*. Here, as Mr. K. J. Bossart of Convair has stated[2], the problems of controlling liquid propellant sloshing and minimizing ullage will be very severe indeed, and will most certainly cause a decrease in attainable motor mass ratio. Also, when such liquid rockets as the *Atlas* are scaled up to larger sizes, the turbo-pump weight required to maintain sufficient thrust for take-off increases faster than the propellant weight, which also will cause an appreciable reduction in the motor mass ratio.

Conversely, solid propellants have no sloshing or ullage difficulties; and being self-pumping, are not bothered by turbo-pump weight problems. Also, the great density of the solid propellants (from 50 percent to 70 percent higher than most liquid propellants) permits considerably smaller tanks for a given propellant weight. This phenomenon, plus the fact that solid propellants can be effectively utilized at very low combustion pressures (as in the GCR third-stage *Vanguard* motor). should allow large solid propellant rocket motors to achieve mass ratios as great as or even superior to those attainable with comparable liquid systems.

Regarding propellant specific impulse, it is true that the common liquid propellants of today have specific impulse values of from 8 percent to 20 percent better than most solid propellants now in use. However, the solid propellant rocket makes up for this disadvantage because it can be readily designed, with no losses in motor mass ratio, to operate at the relatively high thrust-to-weight ratios necessary to minimize the velocity increment lost to gravity (i.e., $g \times t$ losses) during boost stage vertical flight[3]. On the other hand, most large liquid rockets are at a serious disadvantage in this respect because they must live

---

[1] See footnote 1, p. 164.

[2] K. J. Bossart, Design Problems of Large Rockets. High Altitude and Satellite Rockets Symposium, 1957.

[3] Note that the solid propellant rocket's thrust is proportional to the product of the *burning rate* of the propellant and the *burning area* of the grain, both of which can be readily varied over a wide range.

with the excessively long burning times and attendant exorbitant gravity losses caused by their inherently low thrust-to-weight ratios. To visualize how large these velocity losses due to gravity can be, suppose that we were forced, because of turbo-pump weight problems, to have a liquid booster burning time of 120 seconds instead of a hypothetical value of approximately 45 seconds. For a three-stage satellite vehicle, the burnout velocity of the first stage might be 8000 fps, thus we would effectively gain, assuming vertical flight during boost, (120-45) $32 = 2400$ fps or a 30 percent potential *increase* in velocity by using the faster burning, higher thrust solid propellant motor. This reduction of the gravity deceleration effect would then more than cancel the advantage of the liquid unit's higher specific impulse.

Conversely, by using slow-burning solid propellants, long burning times can be achieved for intermediate stages where the flight path is more horizontal than vertical. Here, where it may be desired to maintain thrust for an appreciable time in order to minimize the remaining time required to reach a desired orbiting altitude, a large saving in gravity losses can again be achieved with such solid rockets by making a tangible reduction in the relatively long time normally spent in upward coasting flight.

Finally, solid propellant rockets, because of their smaller size (due to their greater propellant density) and their basic simplicity, are inherently easier and lighter to stage and cluster than comparable liquid rockets. Here, the superior reliability of the solid rocket will really "pay off" because the overall propulsion system reliability is the product of the individual motor reliabilities. For a hypothetical example, a *four*-stage solid rocket with an individual motor reliability of 0.99 would have an overall reliability of at least 0.96, as compared to only 0.72 for a *three*-stage liquid rocket with an individual motor reliability of 0.90. Moreover, if the two lower stages of each type vehicle were composed of clusters of three separate motors, the reliability of the solid vehicle would still be about 0.93, whereas the liquid vehicle would then be only 47 percent reliable (i.e., at least one out of every two liquid vehicles would fail).

To illustrate the potential of solid propellants for reliable, low cost space flight, Grand Central Rocket Co. has made preliminary design calculations for a simple three-stage, all-solid propellant vehicle called the *Envoy* which, when ground-launched, could send a 50-pound payload to the moon or place a 230-pound payload into a 300-mile high orbit. This vehicle would weigh only 17,000 pounds and be only 37.8 feet high (see Fig. 10). Moreover, after development, it is anticipated that the three-stage *Envoy* would cost considerably less than the $ 500,000 figure for the *Scout* or $ 1,000,000 for a single *Thor* liquid booster[1].

It's intended that the same type simple control and guidance system as the *Scout* would be used for the *Envoy*: fins and jet vances for the first stage, fixed hydrogen peroxide type jets for the second stage, and the third stage spin stabilized. The guidance equipment and the solid propellant spin rockets for the final stage would be carried by the second stage in a forward compartment in a manner similar to that employed for the *Scout* rocket.

The motors of all three stages would employ conventional type case-bonded, internal burning star-type grains of high volumetric efficiency similar to the grain design used for the GCR third-stage *Vanguard* motors. The 8026-pound motor of the second stage would be a direct scale-up of the 401-pound third-stage motor, with both motors operating at the relatively low combustion pressure

---

[1] U.S. Space Handbook, p. 142.

of 350 psi and having nozzle expansion ratios of about 25. The 8138-pound first-stage motor would be identical to that of the second stage except that it would operate at 700 psi and, therefore, would employ a nozzle with a smaller throat and have a greater combustion chamber wall thickness. Also, because the first stage would be operating for some time in the dense part of the atmosphere, its nozzle expansion ratio would be reduced to a value of approximately ten to allow optimum thrust efficiency. With careful attention being paid to the design and testing of the inert parts, and with high strength-to-weight ratio

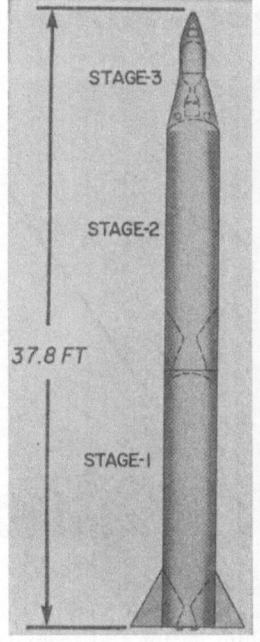

Orbit	Stage 3	Escape
230	Payload	52
50	Structure	50
401	Motor	401
681	Total	503

Orbit	Stage 2	Escape
125 {	Control .... } Spin-up	125
100	Structure	100
8026	Motor	8026
8251	Total	8251

Orbit	Stage 1	Escape
40	Control	40
200	Structure	200
8138	Motor	8138
8378	Total	8378
17310	Gross	17132

Fig. 10. Proposed Grand Central Rocket Co. *Envoy* three-stage all solid propellant escape vehicle

materials being used, it is believed that a motor mass ratio of about 0.927 could be achieved for the first stage, and correspondingly higher ratios of approximately 0.94 attained for the lower pressure second and third-stage motors.

Propellant specific impulse values of about 284 lbf-sec/lbm have been assumed for the upper two high-altitude stages, with a corresponding reduction to an average value of 264 lbf-sec/lbm for the first stage which must operate within the atmosphere. These particular specific impulse values correspond to 250 lbf-sec/lbm at 1000 psi pressure at optimum expansion ratio at sea level, a commonly published value for high energy ammonium perchlorate-based solid propellants (see U.S. Space Handbook, p. 44).

Despite its being of smaller size and only half the weight of the *Scout* vehicle, the *Envoy* would have equal or greater performance because it could have newer type motors especially designed and optimized for its specific task, rather than being comprised of existing motors which were designed originally for other purposes. Also, being a completely new design, the *Envoy* would benefit from the use of optimum staging principles. For example, for a minimum weight missile with a particular required performance and payload, the upper stages should be made much larger than indicated by classical staging theory[1] because of their inherently higher motor mass ratios and specific impulse values. In fact, for the *Envoy* gross weight and required performance, the arrangement chosen (which enable the first-stage motor to utilize the same case dimensions and propellant grain as the thinner wall, lower pressure, second-stage motor) is a near optimum configuration.

This is indeed a happy coincidence because the time and cost required for the development of the first-stage motor would then be greatly reduced as a

Fig. 11. Air launching the *Envoy* escape vehicle

result of being able to use the same fabrication tooling as for the second-stage motor case and grain, and because the development test information on the second-stage motor would also be generally applicable.

Would the *Envoy* be useful only as a ground-launched vehicle? This question brings to mind a pleasant occasion a year ago in San Diego when the author first met that distinguished scientist and world authority on jet propulsion, Dr. THEODORE VON KÁRMÁN, at a technical meeting on space flight problems. In answer to my question as to what he believed would be the chief means of space propulsion during the next decade, Dr. VON KÁRMÁN replied that solid propellant space vehicles would be most commonly used; and furthermore, that he believed such solid propellant space flight vehicles would be launched at high altitudes from existing jet aircraft which could be used for many separate launchings, rather than employ very large expensive boosters which would not be easily recoverable.

In view of Dr. VON KÁRMÁN's remarks on this important subject, let us find out what performance the 17,000-pound three-stage *Envoy* would have when launched at 50,000 feet altitude from a jet aircraft flying at 600 miles per hour (see Fig. 11).

---

[1] See discussion of classical staging theory in H. S. SEIFERT, M. M. MILLS and M. SUMMERFIELD, Physics of Rockets. Amer. J. Physics **15**, 1 (1947).

Since the *Envoy* would be operating entirely at high altitude, it would be best to employ another second-stage motor for the booster in place of the less efficient first-stage motor of the ground-launched version. The use of this higher performance booster, together with the initial 600 mph boost velocity and the somewhat lower drag and gravity losses permitted by the high-altitude launch, would enable the air-launched *Envoy* to escape with at least a 185-pound payload, and also to place 430 pounds of useful payload into a 300-mile orbit. Assuming a total cost of $ 300,000 per flight, this dollar figure amounts to less than $ 700 for each pound of payload put into orbit, a very low cost compared to that presently being spent in the United States for the payloads of liquid propellant boosted orbiting vehicles.

However, there are many interesting escape and satellite missions for which much smaller (and therefore less expensive) payloads would be entirely adequate.

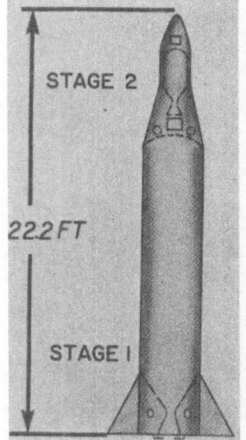

Orbit		Escape
	Stage 2	
60	Payload	5
50	Structure	50
401	Motor	401
511	Total	456
	Stage 1	
100	{ Guidance	100
	{ Spin-up	
70	Control	70
200	Structure	200
8138	Motor	8138
8508	Total	8508
9019	Gross	8964

Fig. 12. The two-stage *Envoy* escape vehicle

With this end in view, let us now investigate what we at GCR term the "stripped-down" economy model, or the two-stage *Envoy* (see Fig. 12). This simplified two-stage ground-launched version of the *Envoy* would be identical to three-stage, ground-launched vehicles, except that the second-stage motor would be omitted. This model, which would be only 22.2 feet high, is expected to escape with a small 5-pound payload and permit a 60-pound payload to be put into a 300-mile orbit. Similarly, the two-stage, air-launched *Envoy* (with one of the two second-stage motors omitted) would be able to orbit a 140-pound payload, or to escape with a 25-pound payload. Here then, in the field of low-cost space probes, is an area where solid propellant rockets are in the ascendancy. In a very short time, these solid space rockets will prove to be sufficiently inexpensive to be within the budget of any nation wishing to pursue its own space program.

Manned space flight is more expensive and here also, the author believes, we will be relying on solid propellants, particularly because of their inherent reliability. We will never subject a human being to an unnecessary risk in our space programs; therefore, the next few years should see the development of very large solid propellant rockets. These will be used for launching tremendous payloads into orbit, and from there, man will eventually penetrate into the far reaches of interplanetary space. We may use low thrust nuclear-powered rockets for interplanetary transfer, but we will be sure to bring reliable high thrust solid rockets with us for the descent maneuvers to the new worlds, and for safe return to our Earth. These events will trace man's first struggling steps into deep space and forever welcome him to his ultimate environment, the Universe.

The "Green" Areas of Mars and Color Vision

By

Ingeborg Schmidt[1]

(Received June 18, 1959)

Abstract — Zusammenfassung — Résumé

The "Green" Areas of Mars and Color Vision. The appearance of the "green" areas of Mars has been discussed from the standpoint of visual physiology. The conditions for perception of colors on Mars are favorable insofar as the observation is foveal and the luminance range is within that of daytime. Because of the smallness of some details normal color vision may reach an area-intensity limit beyond which it becomes very unreliable. In order to elucidate whether the variable blueness, bluegreenness or greenness of the dark areas are real or not the following factors have been discussed: (1) the transmittance of the atmospheres of Mars and Earth in its effect on the appearance of coloration on Mars; (2) the possibility of interpreting the coloration of the dark areas as a contrast phenomenon. Here we have to consider (a) the sun as a chromatic illuminant; (b) the yellow-red bright areas as potential contrast inducing areas.— The production of contrasts was studied experimentally by using colored papers simulating the bright and dark areas. The experiments permit us to deduce that a contrast induction is possible on the surface of Mars, the contrast depending on hue, brightness and saturation of contrast inducing and contrasting area. Blurredness of contours enhances contrasts. An attempt has been made to explain the difference in the induced hue on small (10′ to 20′) and large (1° and up) contrasting neutral gray areas. A method is shown for reducing contrast development which may be helpful in deciding about the real nature of the "green" areas of Mars.

Die „grünen" Gebiete des Mars und die Farbwahrnehmung. Bei der zukünftigen interplanetaren Raumfahrt wird der Planet Mars eine der ersten Landungsstationen sein. Es ist offensichtlich, daß eine vorherige Kenntnis des Geländes für den Raumfahrer von größter Bedeutung ist. Solange wir keine andere Möglichkeit haben, müssen wir uns hauptsächlich auf die Eindrücke verlassen, die unser Auge von der Oberflächenbeschaffenheit des Planeten erhält. Von besonderem Interesse sind die dunklen „grünen" Gebiete des Mars, welche schon früher die widersprechendsten Meinungen hervorgerufen haben und es auch noch heute tun. Die Bedingungen für die Wahrnehmung von Farben auf dem Mars sind günstig insofern, als sein Leuchtdichte-Bereich innerhalb des Tagessehens liegt. Da manche Einzelheiten von sehr geringer Ausdehnung sind, kann es vorkommen, daß eine Ausdehnungs-Intensitätsgrenze erreicht wird, jenseits der das normale Farbsehen sehr unzuverlässig wird (Schwellentritanomalie). Um zu klären, ob die veränderliche Bläue, Blaugrüne oder Grüne der dunklen Gebiete reell ist, werden folgende Faktoren erörtert: 1. die Durchlassigkeit der Atmosphären von Mars und Erde in ihrer Wirkung auf die Erscheinungsweise der Färbung auf Mars; 2. die Möglichkeit einer Deutung der Färbung der dunklen Gebiete als Kontrasterscheinung. Hier muß folgendes berücksichtigt werden: a) die Sonne als farbige Lichtquelle; b) die gelbroten hellen Gebiete als kontrastinduzierende Gebiete. — Die Kontrastinduktion wurde experimentell untersucht unter Verwendung farbiger Papiere in Nachahmung der dunklen und hellen Gebiete des Mars. Die Ergeb-

---

[1] M. D., Division of Optometry, Indiana University, Bloomington, Indiana, U.S.A.

nisse gestatteten die Schlußfolgerung, daß eine Kontrastinduktion auf der Marsober-
fläche möglich ist. Der Kontrast ist abhängig von Farbton, Sättigung und Helligkeit
des kontrastinduzierenden und des kontrastleidenden Feldes. Ein interessanter Befund
war der Unterschied im induzierten Farbton eines kleinen kontrastleidenden Feldes
(10 bis 20 Winkelmin. im Durchmesser) im Vergleich zu einem größeren (etwa 1°);
d. h. in einer bestimmten gelbroten Umgebung zeigte der Kontrast auf einem kleinen
neutralgrauen Feld eine leichte Verschiebung nach längeren Wellenlängen, indem
er in einem grüneren Blaugrün erschien als auf dem größeren grauen Feld. Es wird
versucht, diesen Unterschied zu erklären. Unschärfe der Konturen steigert den Kontrast. Eine Methode wird geschildert, die zur Verminderung der Kontrastentwicklung
dient und die von Nutzen bei der Entscheidung über die wahre Natur der „grünen"
Gebiete auf dem Mars sein könnte.

Les zones "vertes" de Mars et la perception des couleurs. L'apparence des zones
vertes de Mars a été discutée du point de vue de la physiologie de la vue. Les conditions de perception des couleurs de Mars sont favorables pour autant que l'observation soit localisée sur la fovéa et que l'intervalle de luminosité soit comparable à
celui du jour. L'exiguité de certains détails peut descendre sous le seuil à partir
duquel la perception normale des couleurs devient trompeuse. Pour savoir si les
variations de teintes dans les bleus, bleu-verts et verts sont réelles pour les zones
sombres les facteurs suivants ont été mis en discussion: (1) la transmittance des
atmosphères de Mars et de la Terre et son effet sur la coloration apparente; (2) la
possibilité d'interpréter la coloration de zones sombres comme un phénomène de
contraste. Ici il faut considérer (a) le soleil comme source lumineuse chromatique;
(b) les surfaces de couleur vive jaune-rouge comme productrices potentielles de contrastes. — La production de contrastes a été obtenue expérimentalement à l'aide de papiers
colorés simulant les zones vives et sombres. On peut conclure des expériences que
l'induction de contrastes est possible sur la surface de Mars. Elle dépend des nuances,
de la luminosité et de la saturation des surfaces en contraste. Plus les contours sont
indéterminés plus le contraste est fort. Une tentative a été faite pour expliquer les
écarts dans la teinte induite sur des surfaces d'un gris neutre de petites dimensions
(10′ à 20′) et grandes dimensions (1° et plus). Une méthode pour la réduction des
contrastes est présentée. Elle devrait faciliter une décision sur la nature réelle ou
non des zones "vertes" de Mars.

Introduction

We are approaching the age when interplanetary space travel will no longer
be an impossibility, and we shall soon have to choose appropriate target planets.
Therefore our knowledge of the surface qualities of the celestial bodies receives
new emphasis. One of the first target planets will probably be the planet Mars.
Up to now we have had to rely upon visual impressions of its surface features
to a great extent. In obtaining and interpreting these impressions the achievements of visual physiology may be helpful.

Since the end of the last century, the green areas of Mars have been a subject
of controversial discussion. In order to indicate this the word "green" has been
put in quotation marks in the title. The question is this: are these areas actually
or only apparently green? If real, are they the result of vegetation, or are they
the result of some chemical reaction of the soil to increased humidity. If not
real, then the explanation may be a matter of discussion for visual physiology.

Visual Appearance of Mars

Of primary importance for a useful application of the laws of visual physiology
is the knowledge of the size and luminance of the observed areas. Mars appears
as a small disk, subtending at the eye not more than half a degree under usual

conditions of observation. At the highest magnifications used, it may subtend two to three degrees in diameter. According to DE VAUCOULEURS the mean surface luminance of the disc of Mars, at mean opposition and corrected for atmospheric absorption, equals 2350 cd/m² (nit). About three-quarters of the surface of Mars is occupied by fairly uniform bright reddish or yellowish areas whose color and brightness in most places is very stable. Although forming a pattern of more or less constant configuration, the dark areas do vary slightly. At some places they encroach on neighbouring bright areas. They vary in colors in accordance with the seasonal variations of the polar caps. ANTONIADI, who gave us the most elaborate description of Martian colors and their changes in general, listed the following variations: (1) gray or green to brown, (2) green, gray or blue to purple-brown, (3) spots turning maroon, (4) gray to carmine, (5) unchanging green, and (6) unchanging blue. In addition, MARIN during the favorable opposition of Mars in the summer and fall of 1956 saw the bright areas yellow, orange, or infrequently, pink; the dark areas grayish-green. In general the luminance of the dark areas is about one-half the luminance of the bright areas, with a minimum of about two-fifths in the darkest parts (DE VAUCOULEURS).

Some Basic Facts about Color Vision

When our eyes are perfectly adapted to a luminance equal to that of Mars, the conditions would be daytime vision or photopic vision which is present from an adaptation level of 30 cd/m² up. Because of the smallness of the disc and because the eyes would also scan the dark sky, the actual adaptation level of the eye may be lower than expected from the luminance of Mars.

We observe the planet foveally, that is, with the central part of our retina used for fixation. For practical visual purposes the maximum extent of the foveal area can be assumed equal to about 2 degrees in diameter. Both photopic vision and foveal observation are most favorable for color perception.

It will be assumed for our purposes that persons observing and reporting the colors of Mars possess normal color vision. It should be taken into consideration also that the crystalline lens of persons over 50 years of age starts to become yellowish, which makes the distinguishing of green and blue-green shades more difficult.

Our normal color vision has a lower limitation which is an area-intensity problem. From data compiled by FARNSWORTH it appears that for a certain total luminuous energy, color vision becomes similar to that of a "blue-yellow weak", or tritanomalous, observer (threshold tritanomaly). At a somewhat lower value we become "blue-yellow blind" or tritanopic, and finally, at very low levels, totally color blind. In general the tritanomalous effect manifests itself as follows: colors of longer wavelengths than yellow approach orange-red in appearance; colors of wavelengths shorter than yellow approach bluish-green; purple and yellow-green become desaturated, appearing almost gray and almost indistinguishable from each other. Only reddish-orange and blue-green are relatively stable. One can derive from a curve by FARNSWORTH, that a tritanomaly would occur for details on the dark areas of Mars when they subtend 0.7 log square minutes of arc at the eye (assuming 940 cd/m²) as the average apparent luminance of the dark areas when observing at mean opposition, zenith and sea level). Thus at magnifications below 50 × the colors of areas of the extent of Syrtis Major cannot be recognized reliably any more. At a lower luminance the critical angle (hence, critical magnification) will be larger.

The colors perceived on Mars are so-called surface colors which depend on the spectral energy distribution of the incident radiation, upon the spectral reflectance of the illuminated material, and upon the average adaptation level of the observer's eye.

Physical Factors which May Affect the Appearance of Colors on Mars

The incident radiation is that of the sun, modified by the attenuation of the Martian atmosphere. The attenuation would vary, depending on meteorological conditions and on the angle of incidence of the sunlight. The light reflected from the planetary surface traverses both the atmospheres of Mars (twice) and of Earth before reaching our eye.

Because longer wavelengths are transmitted well by the atmospheres of Mars and of Earth, the apparent coloration of the orange and reddish areas may correspond closely to their inherent color. However, they will appear the more desaturated the longer the pathway through the atmospheres traversed by light reflected from these areas. Since both atmospheres are not very transparent for short wavelengths, areas selectively reflecting short wavelengths may not appear in their correct hue. Of special interest for the appearance of the coloration on Mars are the calculations of MIDDLETON about the changes of grass green and of orange in clear air at different horizontal distances on Earth. Green appears bluish from a few tens of kilometers whereas orange does not much change in hue. Entirely dark objects in the daytime appear bluish from a distance because they aquire the bluish coloration from scattered sunlight of the atmospheric layer between the object and the observer. Similar color changes may take place also on Mars. One should expect, for instance, that the green areas of vegetation would appear bluish or bluish green.

It may be mentioned that, for the planet Jupiter, PEEK discusses the possibility of periodic color effects caused by atmospheric dispersion which may introduce spurious seasonal changes.

The "Green" Areas of Mars a Contrast Phenomenon?

A much discussed problem is whether the green and blue coloration of the dark areas of Mars is a simultaneous contrast phenomenon.

Two possibilities of inducing simultaneous color contrasts should be considered. Color contrasts can be induced on entirely achromatic areas when they are illuminated by a chromatic light source. This was especially studied by HELSON. When observing achromatic objects of different brightness on an achromatic background, the eyes are adapted to the average brightness of background plus objects. Areas having this average brightness appear achromatic. Areas of higher reflectance than the average assume the hue of the illuminant. Areas of lower reflectance assume a color complementary to the color of the illuminant. (Complementary colors are color pairs which in proper relative amounts produce an achromatic mixture, white or gray.) Areas of selective reflectance tend to keep their specific hue as long as their dominant wavelength is at least a minor component in the illuminant. Everybody has had the opportunity to observe this type of contrast. Sunlit snowfields appear pale yellow, acquiring the hue of the illuminant, the sun; whereas shadowed areas appear pale blue or violet, complementary to the yellow of the brighter areas. Similar effects may occur also on Mars. The adaptation level of the observer will be determined by the apparent luminance of Mars and to some extent also by that of the surrounding sky.

A second possibility is the induction of a simultaneous contrast by an area with a highly selective reflectance on an adjacent achromatic field or a field of a weak hue, the illumination being non-selective. On Mars the arrangement for contrast induction in the dark areas is favorable insofar as these smaller areas are, at least partly, surrounded by the larger bright areas. Less favorable is the brightness difference since simultaneous contrasts develop best when both contrast-inducing and contrasting areas are of equal brightness. A simultaneous contrast can be perceived in the first second of observation. For instance, within a yellow surround an achromatic area may appear blue. This contrast may be further enhanced by prolonged observation. When the eyes observe the contrast-inducing area, a chromatic adaptation takes place during which the perceived hue changes in a predictable manner. A yellow becomes more greenish-yellow. When the achromatic test field us successively observed on such a "tuned" retina the contrast color appears intensified but its hue is slightly different from the primary contrast hue, since it is now complementary to the inducing hue altered by chromatic adaptation. In the above sample the contrast becomes a more purplish blue. In case the contrasting area also has an inherent hue, mutual effects of the two colored areas ensue.

· A simultaneous contrast on Mars would belong to the type of effects known as gauze contrasts. A gauze contrast can be demonstrated on a colored field containing some neutral gray pattern by covering it with a tissue paper or by viewing it through a strong blurring plus lens held directly in front of the viewing eye. The neutral area shows a very noticeable contrast. The tissue paper or the plus lens diminishes the distinctness of contours and texture, and that favors the appearance of contrasts. On Mars the atmosphere which is usually cloudy plays the role of the tissue paper or plus lens. Also the turbulence of our own atmosphere contributes to the blurredness of the surface features.

Experiments on Contrast Induction

In order to study the possibility of contrast development on Mars several series of experiments were performed, using the 16 papers of the HERING color circle, the 17 gray papers of the MUNSELL neutral scale and also choosing a specially selected set of yellowish and pinkish colors of different hue, saturation and brightness simulating the bright areas of Mars in combination with neutral grays and unsaturated dark greens. A color photo of Mars published by SLIPHER served as a guide for this selection. The papers were specified by MUNSELL notations. In the MUNSELL terminology the hue (H) is designated by integer symbols cited first. The value (V), the aspect responsible for the lightness of the color (i.e. luminous reflectance), and the chroma (C), the aspect responsible for the saturation of the color (i.e. purity), are expressed as a pseudofraction V/C. For instance, the paper 5 YR 5/6 has the hue five yellow-red of value 5 and chroma 6. The higher the value, the higher the reflectance. The higher the chroma, the higher the purity. The MUNSELL notations are valid when the papers are illuminated by a light source of color temperature 6500° K. The following papers were used in the critical experiments:

Table I

Papers simulating the bright areas		Papers simulating the dark areas	
Munsell notation	Color impression	Munsell notation	Color impression
5 YR 4/4	reddish-brown	5 G 3/2	dark green
5 YR 4/6	medium-brown		
5 YR 4/8	dark-orange	10 GY 3/4	dark yellowish-green
5 YR 5/6	reddish-ochre		
5 YR 5/8	medium orange	N 3/ (reflectance 0.07)	dark neutral-gray
5 YR 6/6	ochre		
5 YR 6/8	orange yellow		
5 RP 6/6	purplish-pink	N 4/ (reflectance 0.12)	neutral-gray

One series of tests was performed with a 750 watt projection lamp of color temperature 3000°K, affording 31,200 lm/m² (lux) and producing on the paper 5 YR 5/6 a luminance of 1880 cd/m² which would correspond to the average apparent luminance of Mars at mean opposition, zenithal observation and sea level. Another series was performed with an illuminant of color temperature 7000°K, affording 7530 lm/m² and produced by two Macbeth daylight lamps. It was not possible to obtain the desired high illuminance with the available daylight lamps. The color temperature of the sunlight on Mars is probably of an intermediate value. The illumination was at an angle of 45 degrees, the observation normal to the plane of the papers. The test areas were circular patches 2 mm. and 10 mm. in diameter at the center of colored papers 20 and 30 mm. in diameter. On using the 10 mm. gray or green on the 20 mm. colored disc, the relationship of the areas was about the same as that on Mars. This is important, since the contrast depends on the expanse of the inducing area. A contrast could be perceived even when the neutral area was only half surrounded by the colored area. For simplification in all tests the neutral areas were entirely surrounded by the contrast-inducing color. The observation distance was 42 cm. For patches subtending less than 10 minutes of arc the observation of contrast colors became very difficult. The observations were made monocularly and with no artificial pupil. Before each test the observer adapted for 30 seconds to a neutral gray area of N3/ illuminated by the light source used in the test series. The contrasts were observed for 2 minutes.

Table II
Illuminant: 31,200 lm/m². 3000° K

Contrast inducing color (size 2°42)	Contrasting test patch	Contrast color				
		on contrasting area of size		on large and small contrasting area when observed through		
		1°21'	18'	1 mm. pinhole	—.5 dptr. lens	—.75 dptr. lens
5YR 4/8	N 4/	bluish-green	yellowish green	bluish-green	rather yellowish	bluish-green
5YR 5/6	N 4/	grayish-blue	greenish-blue	bluish	greenish	bluish
5RP 6/6	N 4/	grayish-green	blue-green	no striking effect	no change	no change

The main findings were:

(1) The *contrast colors* induced by a yellow-red surround were grayish-blue, blue, greenish-blue, and blue-green on the larger contrasting neutral grays (1° and up); bluegreen, green and yellowish-green on the smaller neutral grays (10' to 20'). With a red-purple surround the contrast colors were green or grayish-green on large, blue-green or bluish on small. The difference in hue on large and small test areas became more noticeable with increasing value and especially with increasing chroma of the inducing color.

Dark greenish papers of low saturation, e.g. 5G 3/2 and 10 GY 3/4 gained in saturation by the contrast in yellow-red and purplish surroundings, the small green areas showing the same trend observed on small neutral grays.

Of interest may be the following observations: a dark brown surrounded by 5 YR 6/8 became slightly bluish after one minute observation. A purplish pink 5 RP 6/6 in a 5 YR 6/8 surround became bluish when covered by a tissue paper, a result which recalls an observation by TOMBAUGH. He reports that the long dark sash, Sabaeus Sinus is habitually bluish-green. When a blue clearing occurred it suddenly turned to bright lavender or perhaps magenta.

(2) With an *illuminant* of 7000°K the contrasts were similar to those with an illuminant of 3000°K. Thus in respect to color temperature the application of the findings to the conditions on Mars would be justified.

(3) The contrast was more pronounced on neutral N4/ than on N3/, the former being nearer in *value* to the contrast inducing paper.

(4) The saturation of the contrast increased with the *chroma* of the inducing area, which is in accordance with the laws of contrast induction.

(5) The saturation of the contrast was higher on *small* than on large contrasting areas on any expanse of the surround.

(6) The saturation of the contrast increased with the *duration* of the observation and its hue slightly changed, as expected from chromatic adaptation.

(7) When producing small vibratory movements of an amplitude of about 1 minute and a frequency of about 10 per sec., simulating *atmospheric oscillations*, with the total arrangement, the contrasts were intensified. On a yellow-red surround the contrast color increased in greenness as if it were reobserved through a yellow-red filter.

A finding which needs explanation is the *difference* in *hue* of the *contrast* on *small* and *larger* neutral gray areas. Several facts let us assume that the difference may be mainly caused by a phenomenon known as irradiation, in the given case a spreading of the contrast-inducing area over the border into the neutral-gray. This effect, at least in part, may be ascribed to imperfections of the refracting system of our eye, namely spherical and chromatic aberration and diffraction phenomena produced by the edge of our pupil. Such an enlargement at the border would be more noticeable on a small area than on large. That this colored border seems to act like a filter producing a subtractive mixture with the contrast can be shown by observing the large contrasting area through a weak filter of the hue of the contrast-inducing color. The contrast color on the large area then assumes the hue of the small. When a 1 mm. or 2 mm. pinhole was suddenly inserted the contrast on both large and small appeared equal in most instances, namely on a YR surround it was less greenish mainly on small. A possible explanation would be that the pinhole abolishes the ocular aberrations and thus also the irradiation, and now the actual contrast color shows up also on the small areas. A 1 or 2 mm. pupil is regarded optimal to diminish the aberrations of the eye without yet introducing excessively noticeable diffraction phenomena. A 3 mm. pinhole was not effective. Another explanation may be the Stiles-Crawford effect, type II. It is known that some colors change their hues, depending on the direction of the rays entering the pupil. Green rays entering from the edge of the pupil appear more bluegreen than when entering at the center, yellow appears more orange. The colors entering from the edge appear more saturated. Consequently, when using the 1 mm. pupil the contrast-inducing yellow-red would appear yellower, which would mean a more bluish contrast color. That is what was actually observed, although less noticeable on the large area than on the small. The pinhole was not so effective on contrasts induced by red-purple. Its change due to the direction of incidence is not unequivocal.

Besides, the directional effect should also be considered when comparing observations about colors obtained through different eyepieces. A 1 mm. aperture has the great disadvantage that it reduces the brightness of the retinal image by an appreciable amount.

When weak minus or plus lenses were inserted the contrasts showed characteristic changes. On YR both large and small neutral areas appeared greener with a +.5 dptr. lens and both appeared bluer with a —.75 dptr. lens. These effects may be explained by the chromatic aberration of the crystalline lens in the eye, which focusses the short wavelengths in front of the long wavelengths. When red is focussed on the retina, blue has its focus in front of it. The difference is about 1.5 dptr. When the yellow-red surround is in focus on the retina, then so is the yellowish fringe produced by irradiation at the border of the neutral gray. When a minus lens is inserted, the eye, at least for a moment, focusses for shorter wavelengths, the yellow-red comes somewhat out of focus and so does the yellowish fringe which becomes less noticeable. The contrast on small areas then appears similar to that on large. A plus lens has the opposite effect making the contrasting area greener. A purplish color requires two foci, one for red and one for blue, therefore the results are not so consistent. When blue is in focus on the retina, the contrast on small areas would appear bluish, due to fringes from the surround and a minus lens would not change the appearance.

The limiting angle at which the contrast on the small area and on the large area become identical could not be established with certainty. Sometimes a veil of the contrast-inducing hue spreads out over the whole contrasting area, a phenomenon observed by Tschermak with inducing colors of high saturation and intensity only and explained by a scatter in our eye media. Such a yellow-red veil may make the contrasting area appear greener.

The experiments may permit us to deduce that a contrast induction is possible on the surface of Mars, depending on hue, brightness and saturation of the contrast inducing and the contrasting area, on the dimension of the latter and on the blurredness of contours.

That an astronomer never saw green hues except through very small telescopes (cited after de Vaucouleurs) may be explained by the fact that contrasts are more noticeable on small areas. Moreover the contrast sensitivity is individually different. A red-green color blind person would perceive blue instead of blue-green as a contrast to orange. The findings of Dollfus favor the contrast theory, namely that the reflectance of red wavelengths from the dark areas is twice the reflectance of blue light from these same areas which would not be possible if these areas are actually green.

In conclusion it may be safe to assume that at least some apparently green, bluegreen and blue areas on Mars may be due to contrasts, especially when of small subtense, others may actually be green, in some instances enhanced by simultaneous contrasts. A fact which still requires an explanation is the seasonal change of coloration. As indicated, meteorological factors may play a role, causing a variability of brightness on different areas and changing the distinctness of contours.

What Can we Do to Avoid a Contrast Development?

The best method to decide whether the green areas on Mars are due to a contrast phenomenon or not would be an isolated observation of the dark areas, apart from the bright areas, if that were feasible.

We have some means of reducing the contrast although not abolishing it completely since it is already present in the first 0.25 seconds. This method would be an adaptation to a neutral white surface and a succeeding "momentary" observation of the area of interest, since the contrast requires some time to fully develop. In the ideal case this surface would be arranged in the telescope as a white ring with the image of the planet at its center. It is important that the luminance of the neutral surface does not exceed the luminance of the contrasting area. The ring may be helpful also in another respect. During an observation the fixation may wander around the surface of Mars. Prolonged fixation of one point would cause an after-image which may affect the appearance of the next point fixated. The luminous ring may help stabilize fixation (communication by H. W. Hofstetter).

A neutral "tuning" of the eye was suggested by Tschermak by adapting to total darkness and successively observing the colored area shortly. Dark areas are sufficiently available to the astronomer. However with the eye adapted to dark any hue may become desaturated and that would not be desirable in observing the planet.

When by using neutral adaptation the greenness or blueness of the areas in question diminishes significantly, their contrast nature is evident. Now, from the standpoint of biology, Strughold has examined generally the possibility that living matter, as we know it, might exist on Mars. Of course, the existence of vegetation on Mars cannot be denied from visual observation alone. The Martian vegetation may not be green at all. Or the areas may be too small to show their true coloration. Other methods like spectrophotometry, especially as studied by Kuiper and by Dollfus, and a comparison with the spectral analysis of plants as largely applied by Tichoff would be very helpful. It may be noted also that colors of very different spectral composition can nevertheless appear of the same hue to our eye, which should be kept in mind in visual comparison. Physical phenomena depending on the qualities of the magnifiers may play a role in producing spurious "green" areas. But a discussion of all these factors is beyond the scope of this paper.

References

1. E. M. Antoniadi, La Planète Mars. Paris: Libr. Scientif. Hermann et Cie., 1930 (cited after G. P. Kuiper [7]).
2. G. de Vaucouleurs, Physics of the Planet Mars. London: Faber & Faber Ltd., 1954.
3. A. Dollfus, The Nature of the Surface of Mars. Publ. Astronom. Soc. Pacif. 70, 56 (1958).
4. D. Farnsworth, Tritanomalous Vision as a Threshold Function. Die Farbe 4, 185 (1955).
5. H. Helson, Adaptation Level and Frames of Reference. Psychol. Rev. 55, 297 (1940).

6. H. HELSON, Some Factors and Implication of Color Constancy. J. Opt. Soc. Amer. **33**, 555 (1943).
7. G. P. KUIPER, The Atmospheres of the Earth and Planets, Rev. Ed. Chicago: University of Chicago Press, 1952.
8. MARIN, after A. DOLLFUS and J. CAMUS, Commission des surfaces planétaires. Observation de la planète Mars en 1956. L'Astronomie **72**, 16 (1958).
9. W. E. MIDDLETON, Vision through the Atmosphere. Toronto: University Press. Reprinted 1958.
10. B. M. PEEK, The Planet Jupiter. London: Faber & Faber Ltd., 1958.
11. E. C. SLIPHER, New Light on the Changing Face of Mars. Geogr. Mag. **108**, 427 (1955).
12. W. S. STILES, The Luminous Efficiency of Monochromatic Rays Entering the Eye Pupil at Different Points and a New Colour Effect. Proc. Roy. Soc. London, B **127**, 90 (1937).
13. H. STRUGHOLD, The Green and Red Planet. Albuquerque, N. M.: University of New Mexico Press, 1953.
14. G. A. TICHOFF, Astrobiologia. Moscow, 1953.
15. A. VON TSCHERMAK-SEYSENEGG, Einführung in die physiologische Optik, p. 71. Wien: Springer, 1947.
16. C. W. TOMBAUGH, Mars—a World for Exploration. Astronautics **4**, 30—32, 86—93 (1959).

Minimum Energy Requirements for Space Travel[1]

By

Harry O. Ruppe[2], ARS, BIS, DGRR

(With 12 Figures)

(Received June 26, 1959)

Abstract — Zusammenfassung — Résumé

Minimum Energy Requirements for Space Travel. The minimum energy requirement for many space missions is calculated and expressed as velocity requirement of
a rocket vehicle, which is supposed to fulfill them. This gives a fast possibility for a
preliminary outline of an optimum vehicle, or for an approximation of the payload
capability of a given vehicle.

Of course, mission flight times can be reduced by utilizing more than minimum
energy. This is particularly pronounced in lunar and interplanetary transfers.

Minimalenergie-Erfordernisse beim Raumflug. Die Minimalenergie-Erfordernisse
für viele Raumflugprojekte werden berechnet und als Geschwindigkeitserfordernisse
eines Raketenfahrzeuges ausgedrückt, das diese Bedingungen erfüllen soll. Dies bietet
die rasche Möglichkeit für einen vorläufigen Entwurf eines optimalen Fahrzeuges
oder für die angenäherte Berechnung der Nutzlast-Aufnahmefähigkeit eines
gegebenen Fahrzeuges.

Selbstverständlich können die einem Projekt zugeordneten Flugzeiten verringert
werden, wenn mehr als die Minimalenergie verwendet wird. Dies wird besonders
deutlich bei Übergangsbahnen zum Mond oder zwischen Planeten.

Spécifications minimum d'énergie pour voyages interplanétaires. Les spécifications
minimum d'énergie sont converties en spécifications de vitesse de la fusée destinée
à remplir une mission. Ceci permet une estimation rapide de la charge payante et
de la configuration optimale.

Il est évident que les temps de vol peuvent être réduits par un excédent d'énergie,
particulièrement pour les vols lunaires et les transferts interplanétaires.

---

[1] Statements and opinions are to be understood as individual expressions of the
author and do not necessarily reflect the views and opinions of ABMA.

This study is concerned with some of the physical fundamentals of space flight.
It is in no way related to any project now being worked on by the Army, nor should
it be construed as a description of any future project which may be assigned to the
Army.

There is no reference made to any specific hardware now in the making.

[2] Chief, Interplanetary and Lunar Flight Unit; Deputy Chief, Astronautical Engi-
neering Section. Future Projects Design Branch, Structures and Mechanics Laboratory,
Development Operations Division, Army Ballistic Missile Agency, Redstone Arsenal,
Alabama, U.S.A.

I. Introduction

For an idealized step rocket moving along a straight line free of exterior forces holds the well-known equation

$$V_{id} = \sum_{i=1}^{n} c_i \ln r .$$

This velocity shall be called "ideal velocity capability of the vehicle."

A real rocket moving near the Earth's surface does not reach V_{id} as there are velocity losses due to drag and the gravitational field. It is easy to show that despite the increase in altitude, which gives a potential energy gain, there is still an energy loss due to gravity. The magnitude of losses depends upon the geometry of the ascent path and on the flight program. In space-flight missions involving high acceleration propulsion systems (order of one g)—and only such shall be of interest here—the usual flight method is to impart to the vehicle a certain mission dependent energy E_1 (say, escape energy), make a correction on the way to the target, and apply the terminal maneuver at the target (perhaps brake the energy E_2 of impact on the moon). The energy E_1 corresponds now to a velocity V_1 and if we add some empirical or semi-empirical corrective figures for drag and gravity loss, usually taking account of the variation of specific impulse with ambient pressure by taking some convenient mean value for jet velocity in the first stage, we can quote an ideal velocity figure which is necessary for the vehicle to have in order to perform the escape: "ideal velocity requirement of the mission." Adding to this value for escape some empirical number for correction, and a figure representing E_2, we get the ideal velocity requirement for a soft lunar landing.

If a vehicle shall do a certain mission, then, of course, the ideal velocity capability of the vehicle must be larger, or equal to, the ideal velocity requirement of this mission.

In those requirements a correction is necessary since we can get from Earth rotation up to 460 m/sec free; this means that the "ideal requirement" can be 460 m/sec less than one would think from the pure energetic point of view. But often because of flight geometry and other considerations only 300 m/sec can be utilized.

In order to have a yard stick for the vehicle capabilities, where are we now? As known, lunar space probes have been successfully fired, meaning that $V_{id} = 12.45$ km/sec is available. As there is much talk about a Venusian space probe, $V_{id} = 13.0$ km/sec can be expected. What is the ultimate limit of chemical rocket vehicles? Assuming a launch weight of $10 \cdot 10^6$ lb, and a gross payload (including guidance, etc.) of 1000 lb, we have a growth factor of 10^4. For a four-stage vehicle this means a payload ratio per step of 0.1. Assuming a structure ratio of 0.04, we have for one stage a mass ratio of about 7.14, leading with a jet velocity of 4.5 km/sec to $V_{id} = 35.4$ km/sec. Applying such tricks of the trade as orbital technique, the figure may rise to 45 km/sec, with some small increases possible due to the use of orbital technique at the target. As perturbation maneuvers can reduce the requirements slightly, we can understand them to be a small increase in velocity capability. The "feasibility limit", then, can somewhat arbitrarily be set perhaps at $V_{id} = 50$ km/sec.

As the problems of such a vehicle are stupendous (as all of you realize), I feel, therefore, that a practical limit will be lower. Fortunately, we do not need such high ideal velocities within the solar system, as the "principle of mission staging" is a way to avoid too high requirements. For example, you go to the lunar surface using many landing vehicles, and assemble there an Earth-return vehicle out of the useful payload of your landers. So you can return to Earth, using several vehicles of one-way capability instead of one vehicle of two-way capability. Therefore, as an estimate, the practical limit for a chemically-propelled rocket vehicle may be near $V_{id}=25$ km/sec.

Using a "conventional" nuclear propulsion system as it is seen today (nuclear reactor heating a working fluid, which expands through a nozzle), we could perhaps get 80 km/sec instead of the 50 for the chemical system. The more practical limit could conceivably be near 30 km/sec.

The low-acceleration systems could have still higher velocity capabilities, but part of this is used up against the higher gravity losses. Perhaps as ball-park figures, 100 km/sec as ultimate limit, and 50 km/sec as practical limit, can be envisioned.

To meet still higher requirements, very exotic propulsion systems must be used; for example, Prof. SÄNGER's photon propulsion.

II. Velocity Requirements for Earth-Bound Missions

For comparision only, some approximate figures have been computed, which give range versus ideal velocity capability for ballistic-type missiles (Table I).

III. Velocity Requirements for Satellite Missions (Circular Orbits)

A. Nonrotating Earth (Polar Orbits)

Assuming a HOHMANN transfer (Fig. 1), a first kick is necessary to throw the payload up to orbital altitude:

$$\delta_1' = \left(\frac{\gamma M}{R}\right)^{\frac{1}{2}} \left(\frac{2r}{r+R}\right)^{\frac{1}{2}}.$$

Introducing $\dfrac{\delta'}{\left(\frac{\gamma M}{R}\right)^{\frac{1}{2}}} = \delta$, and $\dfrac{r-R}{R} = k$ comes

$$\delta_1 = \left(\frac{2+2k}{2+k}\right)^{\frac{1}{2}}. \tag{1}$$

The arrival velocity at apogee is

$$V_a = \left\{\frac{2}{(2+k)(1+k)}\right\}^{\frac{1}{2}}. \tag{2}$$

So a second kick to circularize must be given

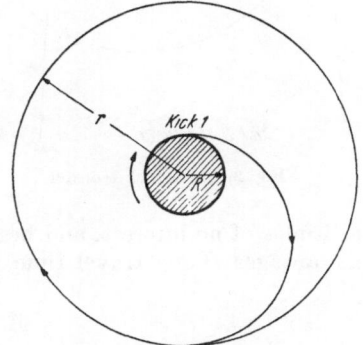

Fig. 1. HOHMANN transfer. Circular velocity for $r = R$ equals $[(\gamma\,M)/R]^{\frac{1}{2}} =$ 7910 m/sec for Earth

$$\delta_2 = \frac{1}{(1+k)^{\frac{1}{2}}} \left\{1 - \left(\frac{2}{2+k}\right)^{\frac{1}{2}}\right\}. \tag{3}$$

The total ideal requirement for the mission then is

$$\Delta_1 = \delta_1 + \delta_2 = \frac{k+(1+k/2)^{\frac{1}{2}}}{(1+k)^{\frac{1}{2}}(1+k/2)^{\frac{1}{2}}} \approx 1+k/2, \text{ for small } k. \tag{4}$$

Introducing for the moment $(1 + k/2)^{\frac{1}{2}} = x$, this can be written

$$\Delta_1 = \frac{2x^2 - 2 + x}{(2x^2 - 1)^{\frac{1}{2}} x} = \frac{2x - 2/x + 1}{(2x^2 - 1)^{\frac{1}{2}}}; \quad \text{from} \quad \frac{\partial \Delta_1}{\partial x} = 0$$

$$\rightarrow x^3 = 3x^2 - 1 .$$

The solution is $x \approx 2.88$, leading to $k = 14.58\_76$ for the most difficult circular orbit with $\Delta_1 \approx 1.5362$.

Using a three-kick transfer (Fig. 2), the first kick is used to throw the payload to infinity.

$$\delta_1 = \sqrt{2} . \tag{5}$$

The second kick at infinity is a zero-adjustment kick; the third kick brakes the arrival speed (escape) to circular speed:

$$\delta_3 = (\sqrt{2} - 1) \frac{1}{(1 + k)^{\frac{1}{2}}} . \tag{6}$$

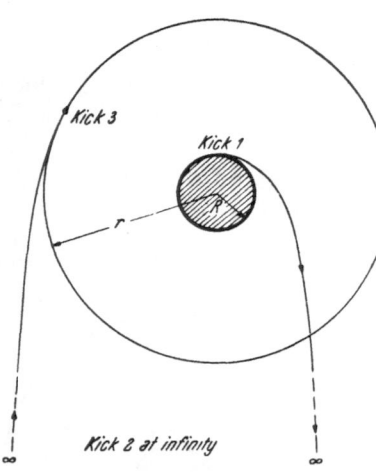

Fig. 2. Three-kick transfer

In total, we have used

$$\Delta_2 = \sqrt{2} + \frac{\sqrt{2} - 1}{(1 + k)^{\frac{1}{2}}} . \tag{7}$$

Comparing Δ_2 and Δ_1, there is $\Delta_2 < \Delta_1$ for $k > 10.94$, which means, that the three-kick transfer is superior for very high orbits.

So, energy-wise, the circular orbit at $k = 10.95$ is the most difficult one, with $\Delta = 1.5340$, if the three-kick transfer is used for larger altitudes. (These considerations are of theoretical value only, since $k = 10.94$ corresponds to satellite altitudes of no interest, and besides the three-kick transfer has other serious disadvantages: Long travel time, and extreme sensitive to the second kick.)

B. Rotating Earth

If the firing occurs from a latitude λ under an azimuth a (East of North) then there is an assistance given from Earth rotation of approximately

$$V_E = 450 \cdot \cos \lambda \sin a \text{ [m/sec]} . \tag{8}$$

C. Change of Orbital Plane

Only a simple case shall be considered here; viz, to go into an equatorial orbit by launching from the latitude φ under an azimuth of $a = 90°$. Ascent is via HOHMANN transfer to a waiting orbit at 200 km altitude. This is necessary in order to place the apogee of the transfer ellipse from the waiting to the final orbit over the equator.

$$\Delta_1 = 1.0157 \approx 1 + \frac{k}{2} .$$

An escape from there would require $\Delta_{Esc} = (\sqrt{2} - 1)\left(1 - \dfrac{k}{2}\right)$.

So the total escape $\Delta = \sqrt{2} + (2 - \sqrt{2})\dfrac{k}{2}$.

The direct escape is $\sqrt{2}$ — so there is a waiting-orbit loss for escape of

$$\Delta_L = (2 - \sqrt{2})\frac{k}{2} \approx 0.0092.$$

Because of maneuvering, I will take $\Delta_L = 0.01$ as "typical waiting-orbit loss."

The upper kick for circularization is given by

$$\delta = \frac{\sqrt{1 + \dfrac{k}{2}} - 1}{\sqrt{1 + \dfrac{k}{2}}\sqrt{1 + k}}. \qquad (9)$$

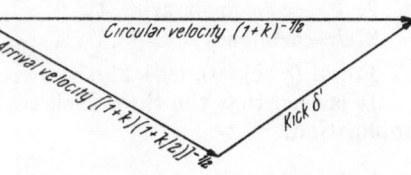

Fig. 3. Change of inclination

If there is a plane change φ involved (see Fig. 3), then the kick to both circularize and accomplish the plane change is

$$\delta^1 = \sqrt{\delta^2 + \frac{4\sin^2 \varphi/2}{(1 + k)\sqrt{1 + k/2}}}. \qquad (10)$$

The total velocity required on a rotating Earth then is

$$\Delta_{1\varphi} = \Delta_1 - \frac{450}{\left(\dfrac{\gamma M}{R}\right)^{1/2}} + (\delta^1 - \delta) + \frac{900}{\left(\dfrac{\gamma M}{R}\right)^{1/2}} \sin^2 \frac{\varphi}{2} + 0.01\, S(\varphi) \qquad (11)$$

where $S(\varphi) = 0$ for $\varphi = 0$ and
$S(\varphi) = 1$ for $\varphi \neq 0$
[δ', δ are given by eqs. (10, 11); Δ, by eq. (4)].

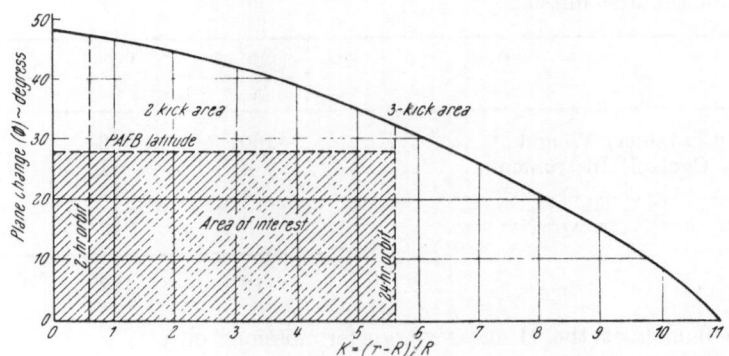

Fig. 4. Three-kick transfer versus two-kick transfer (energy considerations only)

The corresponding expression for the three-kick transfer is

$$\Delta_{2\varphi} = \Delta_2 - \frac{450}{\left(\dfrac{\gamma M}{R}\right)^{1/2}} + \frac{900}{\left(\dfrac{\gamma M}{R}\right)^{1/2}} \sin^2 \frac{\varphi}{2} \qquad (12)$$

[Δ_2 is given by eq. (7)].

Energywise, we consider the function

$$A\,(\varphi,\,k) = \varDelta_{2\varphi} - \varDelta_{1\varphi}\,. \tag{13}$$

If $A > 0$, then the HOHMANN transfer is preferable;
    $A < 0$, then the three-kick transfer is preferable.
It is easy to show:
1. $k < 10.94,\ \varphi = 0 : A > 0$
2. $k = \infty,\ \varphi$ arbitrary : $A = 0$
3. $k = $ arbitrary, $\varphi > 48.3° : A < 0$

For $A\,(\varphi,\,k) = 0$, (see Fig. 4).

It is seen that the three-kick transfer appears not to have a field of practical application.

D. Examples

Circular 568 km-orbit, 96-min, equatorial: .... $(\gamma\,M/R)^{1/2} = $ 7,920 m/sec
Equator-launched, HOHMANN .... $V_{id} = 0.99 = $ 7,841 m/sec
Equator-launched, three-kick ....   1.76 $\ = 13,939$ m/sec
$\varphi = 28°$-launched, HOHMANN ....   1.44 $\ = 11,405$ m/sec
$\varphi = 28°$-launched, three-kick ....   1,766 $= 13,987$ m/sec

Circular 24-hr orbit, equatorial:

Equator-launched, HOHMANN ....   1.45 $\ = 11,484$ m/sec
Equator-launched, three-kick ....   1.52 $\ = 12,038$ m/sec
$\varphi = 28°$-launched, HOHMANN ....   1.51 $\ = 11,959$ m/sec
$\varphi = 28°$-launched, three-kick ....   1.526 $= 12,086$ m/sec
$\varphi = 28°$-launched, 28° inclined orbit ....   55 m/sec to above equa-
                                     tor-launched velocities
Escape: equatorial-launched, Eastward ....   10,736 m/sec
         polar-launched ....   11,186 m/sec

What does this mean in payload? Let us assume the following arbitrary vehicles for the first mission:

	Type		
	A	B	C
Payload (Container) Weight	100	80	70
Guidance, Control, Instruments	4	11	14
Fuselage	4	11	14
Motor, etc.	4	10	14
Total.....................	112	112	112

We will look at the HOHMANN-transfer missions only
28°-launched, 28° inclined orbit: with $I_{sp} = 300$ sec,

$$55 = 300\,g.\ \ \ln \frac{M}{m} \rightarrow \frac{M}{m} = 1.0186\,, \text{ which leads to}$$

$$m = \frac{112}{1.0186} = 109.95, \text{ or fuel used: } 2.05$$

Additional Tankage: 0.2

                                     Payload loss: $\overline{2.25}$ of 100 for Vehicle A.
With an I_{sp} of 400 sec, the payload loss was 1.65 of 100.
In this manner the following table has been computed.

Payloads Including Containers
(HOHMANN transfers only)

	$I_{sp} = 300$ sec			$I_{sp} = 400$ sec		
	A	B	C	A	B	C
Equator-launched Equatorial 96-min Orbit	100.0%	100.0%	100.0%	100.0%	100.0%	100.0%
28°-Launched, 28° Inclined	97.8%	97.0%	96.9%	98.4%	97.9%	97.6%
28°-Launched, Equatorial	13.4%	—	—	26.4%	8.0%	—
Equator-launched, Equatorial 24-hr Orbit	12.4%	—	—	26.0%	7.5%	—
28°-Launched, 28° Inclined 24-hr Orbit	11.9%	—	—	24.9%	6.1%	—
28°-Launched, Equatorial 24-hr Orbit	7.2%	—	—	19.9%	0.0%	—
Escape, Equator-launched, Due East	22.8%	3.5%	—	35.6%	19.5%	8.0%
Escape 28°-launched, Due East	22.0%	2.5%	—	34.9%	18.6%	7.0%
Escape Polar	16.2%	—	—	29.2%	11.5%	0.0%

This table illustrated several well-known facts:

1. If azimuth of firing is 90°, then only small penalties are involved in going up to 30° off the equator.

2. If higher energy missions are flown, then it is not optimum just to exchange propellant for payload in the last stage. But if this is done, then the result is very dependent upon specific impulse.

3. Even for vehicles of similar performance at one mission, the performance at other missions may vary widely.

E. Recovery of Satellites

Here one kick is assumed to brake the orbital velocity so far that a transfer ellipse is entered the pericenter of which is sufficiently deep in the atmosphere, so that further braking is done by atmospheric drag. Final descent could employ lift, or a parachute, or in the case of some types of instruments simply impact.

F. Conversion of "Ideal Required Velocity" to "Required Velocity"

Certain correction figures have to be added, which are of an empirical nature. Those correction figures take care of:

1. Gravity loss and drag loss: 1500 m/sec seem to be a usable figure for escape missions and large vehicles. (Of course, this number depends on vehicle shape, trajectory and acceleration program.)

2. Maneuvering Reserve: This has to be estimated for every mission.

3. Unusable propellant residuals and mixture ratio shifts: With control of mixture ratio and trapped propellants, 3% of the ideal velocity should suffice.

IV. Velocity Requirements for Space Probes

By space probe a vehicle is meant which is used for space research without necessarily approaching a planet or any other celestial body. Therefore, the guidance and space navigation problems are greatly simplified. To go from Earth to the Moon, escape velocity $\left(2\,\dfrac{\gamma M}{R}\right)^{1/2}$ is approximately necessary, the minimum for direct transfers being about 100 m/sec lower. If we look for the circular orbit of the same energy requirement, we have to solve $\dfrac{k+\sqrt{1+k/2}}{\sqrt{(1+k)\,(1+k/2)}}=\sqrt{2}$, from which comes $k\approx 2.303$.

So we can conclude, that, for a simple Earth-escape experiment, about the same payload can be carried as into the circular orbit of $k=2.303$. In practice even more can be carried, as the guidance system should be simpler and, therefore, lighter.

For interplanetary probes, obviously, more energy is required than for probes in near-Earth Space. It is easy to show, that, disregarding Earth, a minimum perihelion velocity of 32.83 km/sec at Earth distance (149.10^6 km) from the Sun is necessary in order to place the aphelion cut to 230.10^6 km (Mars distance). This is $32.83-29.8 = 3.03$ km/sec above local circular velocity (Earth velocity, going around Sun). Therefore, the minimum required launch velocity is

$$\sqrt{11.1862^2 + 3.03^2} = 11.5893 \text{ km/sec}$$

which is only 403.1 m/sec more than the simple Earth escape probe.

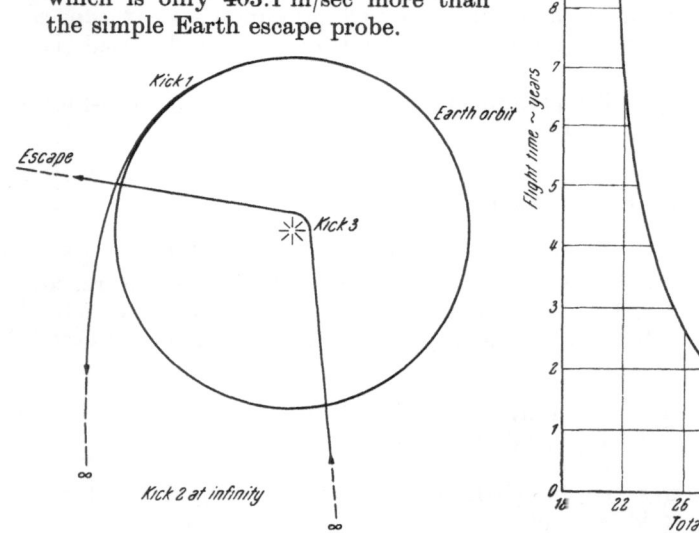

Fig. 5. Solar system escape – Hi-performance vehicle

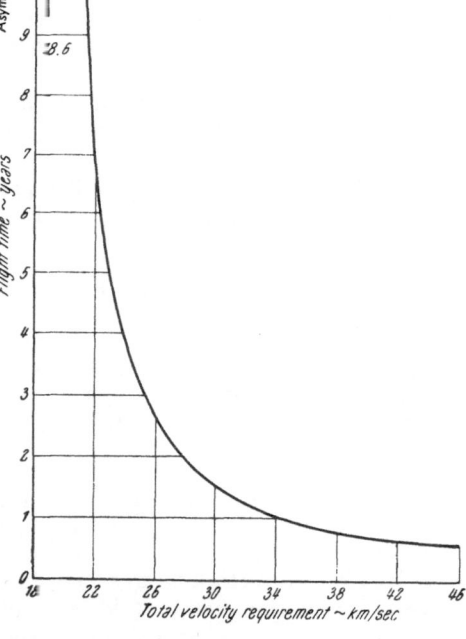

Fig. 6. Solar probe: Flight time versus total velocity

For Venusian probe, put 2.4 km/sec instead of the 3.03 km/sec for Mars, resulting in $V_{min}= 11.4408$ km/sec, only 254.6 km/sec above escape velocity.

The minimum requirement for a Mercury probe is 13.504 km/sec or 2.318 km/sec excess over Earth escape.

The minimum to escape the solar system is to have a residual velocity of $(\sqrt{2}-1) \cdot 29.8 = 12.3436$ km/sec. This leads to a total minimum requirement of 16.6582 km/sec which is 5.472 km/sec over simple Earth escape.

Assuming we had a vehicle of a capability of $V_{id} = 50$ (100) km/sec. To leave the solar system (Fig. 5) it would be best to apply a first impulse in order just to

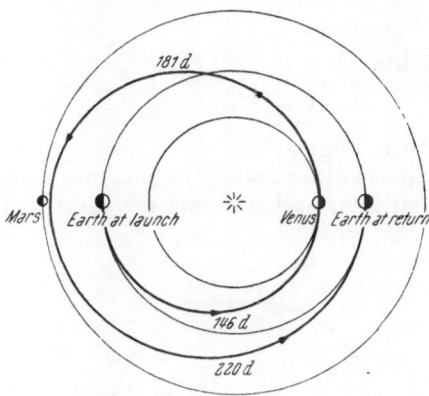

Fig. 7. HOHMANŇ's round-trip trajectory

Maneuvers:
1. Brake original 29.8 km/sec of Earth by .... 2.4 km/sec
2. Increase on passing Venus to 39.4 km/sec by .... 1.8 km/sec
3. Increase on passing Mars to 24.8 km/sec by .... 2.3 km/sec
4. Brake 31.5 km/sec arrival at Earth by .... 1.8 km/sec
   Total velocity requirement .... 17.6 km/sec
   Total flight time .... 547 days

leave the system: $16.66 + \text{losses} \approx 20$ km/sec. At infinity, zero kicks would adjust for a return to just graze the Sun. Arrival velocity would be local escape velocity = $= 617.5$ km/sec at 700,000 km from the Sun's center. To this we add the remaining 30 (70) km/sec, leading to a velocity remaining at infinity of 195 (302) km/sec. Travel time to the nearest fixed star, about 4 light-years distant, would now be about 6150 (3980) years, or total mission time about 6200 (4000) years. Therefore, with the highly advanced vehicles which were assumed, not even the nearest fixed stars are within reach. (In order to get to them in 40 years of transit time, the ideal velocity capability of the vehicle must be of the order of 30,000 km/sec.)

In solar probes we have to differentiate between three classes:

1. Probes in near solar space. A good example of this is the Mercury probe, for which we found $V_{id} \min = 13.504$ km/sec. The flight time is about 111 days.

2. Direct probe to the Sun: The Earth orbital velocity is braked, and the vehicle drops in vertically to the Sun. The required ideal velocity is high— 31.8 km/sec with a flight time of only 65 days.

3. Indirect solar probe. Here a solar system escape must be performed first and then a drop towards the Sun. The ideal energy required is only 16.66 km/sec, but the flight time is infinite. For practical cases, you would, of course, not go quite to infinity, resulting in a higher velocity and lower flight time required (see Fig. 6).

More sophisticated types of probes are those performing round trips, two of which may be of special interest:

HOHMANN Roundtrip (Fig. 7).

CROCCO Roundtrip (Fig. 8).

As these probes can conceivably return to Earth, a manned version of such an expedition may be interesting.

V. Lunar Flights

Six types of flights must be considered:
1. Probes
2. Hard impacts
3. Circumlunar Flights
4. Lunar Satellites
5. Soft Landers
6. Earth-Moon Return Flights.

Probes and hard impacts differ mainly in guidance accuracy required. Therefore, no more additional information seems necessary here.

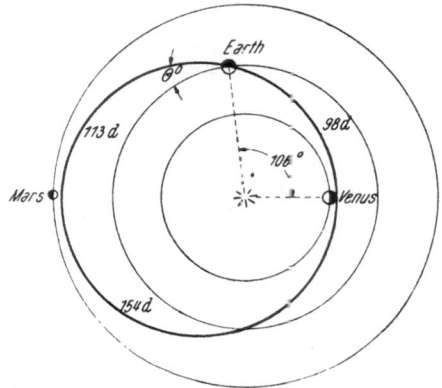

Fig. 8. CROCCO's round-trip trajectory

Total velocity requirement .... 15.968 km/sec
Total flight time .... 12.5 to 13.5 months
Possible launch time .... June 1971
Ø .... 16.04°

For the circumlunar flight the energy requirement, apart from maneuvering, is similar to Earth escape. If a return to Earth is planned, then, because of the critical atmospheric re-entry, ample maneuvering fuel should be provided for.

Let us look for a moment at some characteristic data for transfer trajectories in the Earth-Moon system:

Injection angle: near horizontal

	Cutoff Velocity in Inertial Earth-Centered Space at 200 km Altitude	Flight Time	Unbraked Lunar Impact Velocity
1	10,881 m/sec	∼10 years	2,325 m/sec
2	10,920 m/sec	∼ 5 days	2,500 m/sec
3	10,970 m/sec	2¹/₂ days	2,705 m/sec
4	11,015 m/sec	51 hrs	2,886 m/sec
5	11,100 m/sec	41 hrs	3,179 m/sec

Remarks:

1. Absolute minimum injection velocity to reach the Moon, from JACOBI's Integral.

2. Minimum for direct Earth—Moon transfer.

3. Two-and-a-half day transfer.

4. Injection velocity equal local escape velocity.

5. "Fast" trajectory.

Some Data:

Mean Earth Radius .... 6,371.1 km
Mean Lunar Radius 1,738   km
Escape velocity, Earth, zero altitude .... 11,186 km/sec
Escape velocity, Moon 2,374   km/sec
"Mean" Earth—Moon distance .... 384,412.3 km

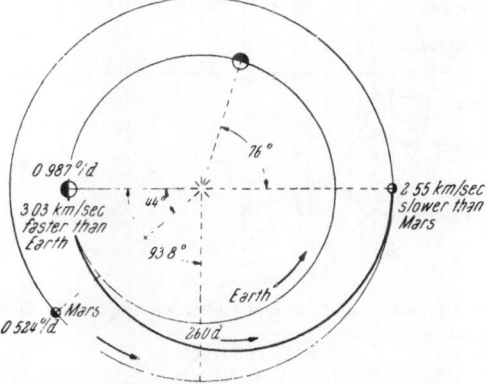

Fig. 9. Earth—Mars mission profile (HOHMANN ellipse)
Opposition occurs (44)/(0.987—0.524) = 95 days = 93.8° after launch. The waiting time is in this case (360—2.76)/(0.987—0.524) = 449 days. So a return mission last (260/449/260) = 970 days

From the guidance point of view, the "slow" trajectories show large deviations for small injection errors. Therefore, trajectory No. 4 seems to be a good compromise, with 3 being a competitor.

Upon arrival for a lunar satellite, we have to brake:

Satellite Altitude	Trajectory	Circular Velocity	To be Braked
$\dfrac{h}{R_{Moon}} = 0$	$2^{1}/_{2}$ days $\Big\}$	1678 m/sec	1027 m/sec
0	51 hrs	1678 m/sec	1208 m/sec
0.25	$2^{1}/_{2}$ days $\Big\}$	1502 m/sec	985 m/sec
0.25	51 hrs	1502 m/sec	1181 m/sec
1	$2^{1}/_{2}$ days $\Big\}$	1188 m/sec	932 m/sec
1	51 hrs	1188 m/sec	1160 m/sec

A soft lunar landing vehicle has to brake the total impact velocity by rocket action.

The return flight is, energy-wise, much simpler; because, at Earth's side probably the atmosphere can be used for re-entry braking. However, corrective fuel should be provided as the ideal re-entry conditions must be met rather closely.

VI. Orbital Technique

It is necessary to say a few words on the use of orbital technique.

a) Upon Departure

For example, in the 96 min-orbit a space vehicle could be assembled. This, then, could take advantage of the energy it already has, and a noticeable structural advantage should result from the possibility of using relatively low accelerations, and from the absence of aerodynamic considerations. Maneuvers into the 96-min

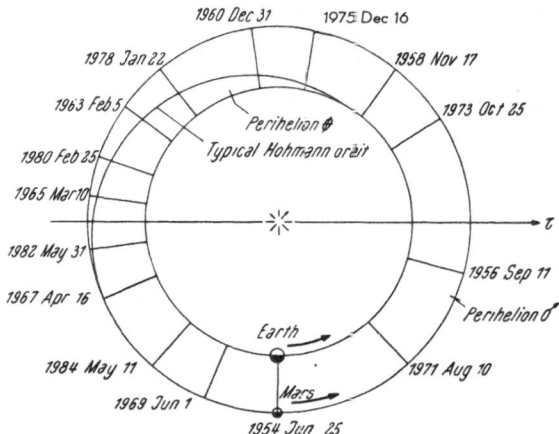

Fig. 10. Opposition of Mars and HOHMANN ellipse injection dates

HOHMANN *ellipse injection dates:*

23 Aug 1958	23 Feb 1969	7 Oct 1977
25 Sep 1960	21 May 1971	29 Oct 1979
16 Oct 1962	7 Aug 1973	4 Dec 1981
9 Nov 1964	12 Sep 1975	28 Jan 1984
25 Dec 1966		

Notes: Maximum error in injection dates: 1958—1971, 3 days; 1971—1984, 5 days. The opposition dates were obtained by extending the ephemeris of the Earth and Mars by a numerical integration method. This ephemeris was extended through 1984 by use of a graph showing synodic period of Mars versus longitude of opposition, prepared from the American Ephemeris and Nautical Almanac. The injection dates were obtained by use of a graph showing opposition time minus injection time versus longitude of opposition. This graph was prepared by choosing a multiple of conjunction times of the HOHMANN ellipse and the orbit of Mars and from this, calculating the period and the theoretical opposition times and longitudes. These were then plotted to obtain the graph

orbit and the resulting high cutoff altitude for the powered phase leaving the orbit will result in some small energy losses.

b) Upon Arrival

At the target site, going first to an orbit and from there down to the surface has potential advantages:

1. Perhaps better control of landing area.

2. Higher safety of mission success, even if the landing fails.

3. Saving of energy, if return fuel is left in orbit and picked up during the return phase.

Disadvantages are:

1. More complicated over-all scheme.

2. All braking could be done aerodynamically if there is an atmosphere; if the braking to orbit is done by rocket, usually much more unfavorable conditions exist.

c) Upon Return

Upon return to Earth, the following possibilities exist:

1. Direct re-entry to the atmosphere.

2. Return to 96-min orbit, thereby using rocket braking. The disadvantage of using fuel is partly compensated for by the advantage of using a special vehicle for transferring and receiving.

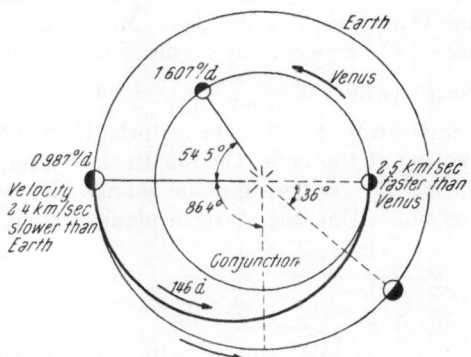

Fig. 11. Earth—Venus mission profile (HOHMANN ellipse)

Inferior conjunction is $(54.5)/(1.607—0.987) = 87.6$ days after launch; the angle is $87.6 \cdot 0.987 = 86.4°$. From this follows the rule: Launch $86.4° = 87.6$ days before inferior conjunction. For the return flight, we need to have Earth leading by $36°$. (In above picture, we simply reverse all directions for the return flight.) So Venus has to catch up $360—2.36 = 288°$, which takes $(288)/(1.607—0.987) \approx 460$ days. So a return mission lasts $146/480/146 = 750$ days

3. Return to another orbit, being picked up there and brought to the 96-min orbit, from where transportation to the Earth's surface is provided. A very interesting return orbit of this type is the following elliptic orbit, which often needs only little rocket braking:

Elliptic Orbit Properties

Perigee: 568 km altitude (circular velocity: 7581 m/sec, escape velocity: 10,721 m/sec).
Velocity: 10,431 m/sec (2850 m/sec above circular velocity).
Apogee: 120,143.4 km from Earth's center.
Eccentricity: 0.89321044.
Major axis: 65,044.7 km.
Minor axis: 29,247.0 km.
Period: 45.848 hr.

So all the braking necessary is $10,721—10,431 = 290$ m/sec. Of course, now the pick-up vehicle has to have an ideal minimum capability of $2 \times 2850 = 5700$ m/sec.

VII. Planetary Flights

Corresponding to paragraph V, six mission types exist. Only planetary satellites, soft landers and return flights are of interest, as probes are already treated elsewhere.

Perhaps advanced propulsion systems—e.g. ionic systems—will be used for large-scale interplanetary operations, but we will limit ourselves here to the conventional chemical impulsive systems.

Only four representative missions via HOHMANN transfers are considered in some detail, bearing in mind that the return flight has somewhat symmetrical demands.

a) Martian Satellite

As shown, the minimum ideal initial (Earth-side) launch velocity equals 11.5893 km/sec. Mars has to be there, when the vehicle reaches the Aphel of its orbit. The transfer time from Earth is 260 days. Mars moves during these 260 days through $0.524 \cdot 260 = 136$ degrees of arc. Therefore, at launch Mars must be 44 degrees ahead of Earth (which is $\dfrac{44}{0.987 - 0.524} = 96$ days before opposition). —The Martian orbit is inclined by 1°51′ to the ecliptic. If the vehicle shall go into the Martian plane of motion at the node, the ideal velocity requirement is about $27 \cdot \sin 1°51′ = 0.872$ km/sec. By giving the plane change kick about halfway and arriving at Mars moving not within the Martian plane, usually (if the node does

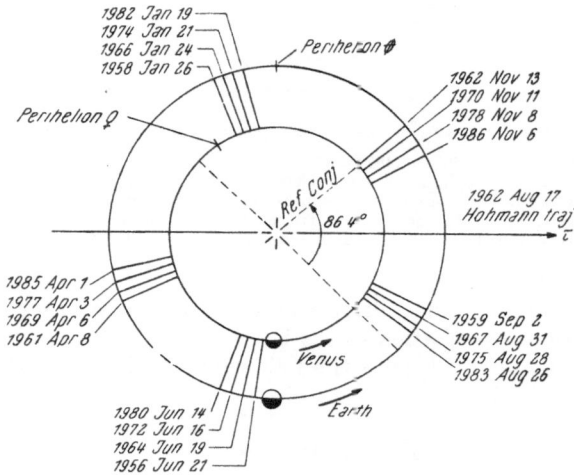

Fig. 12. Inferior conjunction of Venus and HOHMANN ellipse injection date

Hohmann ellipse injection dates:

1956 Mar 26	1967 Jun 4	1978 Aug 13
1957 Oct 31	1969 Jan 8	1980 Mar 18
1959 Jun 6	1970 Aug 15	1981 Oct 23
1961 Jan 10	1972 Mar 21	1983 May 30
1962 Aug 17	1973 Oct 26	1985 Jan 3
1964 Mar 23	1975 Jun 2	1986 Aug 10
1965 Oct 28	1977 Jan 6	

Notes: 1. Estimated error of a conjunction or a HOHMANN injection date is 4 days maximum.
2. Orbits and circles; $r\ (\oplus) = 1.00$ au, $r\ (\female) = 0.723$ au.
3. Orbits and longitudes of inferior conjunctions are: $t = JD\ 243\ 5062.3/583.931\ N$ (days); and $\lambda = 575°$ 5180 N; Reference date = 1954 Nov 15.29 UT.
4. Constant synodic period = 583.921 (days).
5. HOHMANN date and longitude; $t = 87.6$ days, $\lambda = 86°.4$

not happen to be at this place) some savings can be accomplished. If the vehicle travels neither in the ecliptic nor in the Martian plane, and if the central angle between launch and arrival is slightly smaller than 180 degrees, then the loss due to inclination change becomes negligible. (Oral communication from Dr. D. F. LAWDEN.)

Arrival at the Martian orbit occurs with a velocity of 2.55 km/sec less than Mars orbital velocity. At 1000 km above the Martian surface, the vehicle will move 5.15 km/sec; an ideal minimum braking of 2.01 km/sec brings this down to 3.14 km/sec, which is local circular velocity (escape velocity at zero altitude: 5.04 km/sec).

b) Venusian Satellite

At launch, Venus must be 54.5 degrees behind Earth (or 88 days before inferior conjunction). After 146 days, the vehicle approaches Venus, being 2.5 km/sec faster than Venus. As the inclination of the Venusian orbit is about $3°24'$, the maximum requirement of the ideal velocity for the change is $32.5 \sin 3°24' = 1.93$ km/sec.

Escape velocity at Venusian surface is 10.23 km/sec; at 1000 km altitude, this is reduced to 9.49 km/sec (circular velocity: 6.71 km/sec). The vehicle arrives with $\sqrt{(2.5)^2 + (9.49)^2} = 9.814$ km/sec. Therefore, a braking of 3.104 km/sec is necessary.

c) and d) Soft Landings on Mars or Venus

Here I will always assume, that the atmosphere is used for braking and landing. The speed which has to be broken upon arrival is about

> Venus: 10.5 km/sec (from Earth)
> Earth: 11.5 km/sec (from Mars, Venus)
> Mars: 5.64 km/sec (from Earth)

If this is done aerodynamically, then there is only a little fuel used for control. (For the timing and outlay of Martian and Venusian flights, see Figs. 9—12.)

Of some interest might be a manned planetoid mission, because this would give the chance of actually being on another star and doing research there. For a planetoid within Mars' orbit and of negligible gravity, appropriate Mars data can be used.

Appendix

Table I. *Ideal Velocity Requirement for Earth—Bound Missions of Ballistic—Missile Types*

Range (km)	V_{id} (km/sec)
500	3
1,000	3.9
2,000	5
5,000	6.8
10,000	8.5
20,000	8.8

Remarks: According to paragraph III—F, 3% of V_{id} may be added for mixture ratio shifts, trapped residuals and flight performance reserves.

Table II. *Velocity Requirements for Some Missions*
(See Remarks on Table I)

	Values in km/sec	Equatorial Earth Satellites — No Recovery — Equator Launch 96-min	No Recovery — Equator Launch 24-hr	No Recovery — AMR 96-min	No Recovery — Launch 24-hr	+ Recovery — Equator Launch 96-min	+ Recovery — Equator Launch 24-hr	Lunar Missions — No Recovery — Impact Satellite	No Recovery — Soft Landing	No Recovery (Satellite)	+ Return — Circum-Lunar	+ Return — Satellite	+ Return — Soft Landing
Earth launch	Ideal Minimum Launch	8.291	11.934	11.801	12.359	8.291	11.934	11.186	11.186	11.186	11.186	11.186	11.186
	Rotational Gain	0.45	0.45	0.4	0.4	0.45	0.45	0.3	0.3	0.3	0.3	0.3	0.3
	G-Loss	1.4	1.42	1.4	1.42	1.4	1.42	1.42	1.42	1.42	1.42	1.42	1.42
	Drag-Loss	0.15	0.16	0.15	0.16	0.15	0.16	0.16	0.16	0.16	0.16	0.16	0.16
	Maneuvering	0.05	0.05	0.01	0.01	0.05	0.05	0.05	0.05	0.05	0.05	0.05	0.05
	Maneuvering Transfer	—	0.05	—	0.05	—	0.05	0.05	0.05	0.05	0.05	0.05	0.05
At target arrival	Ideal Minimum	—	—	—	—	—	—	—	1.18	2.89	—	1.18	2.89
	Rotational Gain	—	—	—	—	—	—	—	—	—	—	—	—
	G-Loss	—	—	—	—	—	—	—	0.03	0.3	—	0.03	0.3
	Maneuvering	0.05	0.05	0.05	0.05	0.05	0.05	0.1	0.05	0.05	—	0.05	0.05
Target launch	Ideal Minimum Launch	—	—	—	—	0.06	1.49	—	—	—	—	1.18	2.89
	Rotational Gain	—	—	—	—	—	—	—	—	—	—	—	0.2
	G-Loss	—	—	—	—	—	—	—	—	—	—	0.03	0.15
	Drag-Loss	—	—	—	—	—	—	—	—	—	—	—	—
	Maneuvering	—	—	—	—	—	—	—	—	—	—	0.05	0.05
	Maneuvering Transfer	—	—	—	—	—	—	—	—	—	0.05	0.05	0.05
	Earth Landing Maneuver	—	—	—	—	0.05	0.05	—	—	—	0.05	0.05	0.05
	Total	9.50	13.20	13.10	13.75	9.60	14.75	12.55	13.85	15.80	12.75	15.20	19.35

Table II (Continued)

Space Probes

Values in km/sec	No Recovery				Hohmann Probe + Return	Crocco Probe + Return	Simple Solar System Escape	Solar Probe
	Simple Earth Escape	Martian Probe	Venusian Probe	Mercury Probe				
Earth launch								
Ideal Minimum Launch	11.2	11.589	11.441	13.504	11.441	13.948	16.658	17.0—31.8
Rotational Gain	0.4	0.3	0.3	0.3	0.3	0.3	0.4	0.4
G-Loss	1.42	1.42	1.42	1.45	1.42	1.45	1.46	1.46
Drag-Loss	0.16	0.16	0.16	0.17	0.16	0.17	0.18	0.18
Maneuvering	0.05	0.10	0.10	0.10	0.10	0.10	0.10	0.10
At target arrival								
Maneuvering Transfer	—	—	—	—	—	—	—	—
Ideal Minimum	—	—	—	—	—	—	—	—
Rotational Gain	—	—	—	—	—	—	—	—
G-Loss	—	—	—	—	—	—	—	—
Maneuvering	—	0.2	0.2	0.2	0.6+ 1.8+2.3	0.6	—	0.2
Target launch								
Ideal Minimum Launch	—	—	—	—	—	—	—	—
Rotational Gain	—	—	—	—	—	—	—	—
G-Loss	—	—	—	—	—	—	—	—
Drag-Loss	—	—	—	—	—	—	—	—
Maneuvering	—	—	—	—	—	—	—	—
Maneuvering Transfer	—	—	—	—	—	—	—	—
Earth Landing Maneuver	—	—	—	—	0.1	0.1	—	—
Total	12.45	13.15	13.0	15.1	17.60	16.05	18.0	18.55—33.35

Table II (Continued)

Values in km/sec	No Recovery				+ Return				
	Mars Satellite	Venus Satellite	Mars Soft Landing	Venus Soft Landing	Mars Satellite	Venus Satellite	Mars Soft Landing	Venus Soft Landing	Planetoid in Mars Orbit, Soft Landing
Earth launch									
Ideal Minimum Launch	11.589	11.441	11.589	11.441	11.589	11.441	11.589	11.441	11.589
Rotational Gain	0.3	0.3	0.3	0.3	0.3	0.3	0.3	0.3	0.3
G-Loss	1.42	1.42	1.42	1.42	1.42	1.42	1.42	1.42	1.42
Drag-Loss	0.16	0.16	0.16	0.16	0.16	0.16	0.16	0.16	0.16
Maneuvering	0.05	0.05	0.05	0.05	0.05	0.05	0.05	0.05	0.05
Maneuvering Transfer	0.4	0.5	0.4	0.5	0.4	0.5	0.4	0.5	0.4
At target arrival									
Ideal Minimum	2.01	3.104	—	—	2.01	3.104	—	—	2.55
Rotational Gain	—	—	—	—	—	—	—	—	—
G-Loss	0.1	0.15	—	—	0.1	0.15	—	—	—
Maneuvering	0.05	0.05	0.01	0.01	0.05	0.05	0.01	0.01	0.05
Target launch									
Ideal Minimum Launch					2.01	3.104	5.64	10.5	2.55
Rotational Gain					—	0.1	0.2	—	—
G-Loss					0.05	—	0.3	1.5	—
Drag-Loss					—	—	0.15	0.2	—
Maneuvering					0.05	0.05	0.05	0.05	0.05
Maneuvering Transfer					0.04	0.05	0.04	0.05	0.04
Earth Landing Maneuver					0.05	0.05	0.05	0.05	0.05
Total	15.50	16.60	13.40	13.35	18.05	20.40	19.80	26.15	18.95

Table III. *Velocity Requirements for Various Missions (Equator-Launched, Unless Otherwise Specified)*

On this table the total velocity requirements for a number of missions are summarized. "Direct Target" means, that the maneuver at target is employed without going to an orbit around the target first. "Orbit at Target" on the other hand implies the use of orbital technique at target. The "Elliptical Orbit" is the same as is described in paragraph VI of this report. Also, see remarks on Table I.

Mission	AMR Launched 150 km 96 min	AMR 24 hr	AMR 96 min	AMR 24 hr	+ Recovery 150 km 96 min	+ Recovery 96 min	+ Recovery 24 hr	Elliptical 150—35871 km	Elliptical 150—35871 km	Lunar Impact	Lunar Satellite	Lunar Soft Landing	+ Return Circum-Lunar	+ Return Satellite	+ Return Soft Landing
Direct Target															
Earth	9.20	13.20	9.50	13.10	—	—	13.75	11.8	—	12.55	13.85	15.80	12.75	15.20	19.35
Orbit	—	3.85	—	4.40	—	—	—	—	—	3.10	4.40	6.30	3.25	5.75	9.70
Earth—Earth	—	—	—	—	9.25	9.60	14.75	—	11.9	—	—	—	15.90	18.40	22.50
Orbit—Earth	—	—	—	—	—	—	5.4	—	—	—	—	—	6.40	8.85	12.85
Earth—Orbit	—	—	—	—	—	—	17.08	—	—	—	—	—	—	—	—
Orbit—Orbit	—	—	—	—	—	—	7.73	—	—	—	—	—	—	—	—
Earth—Elliptical Orbit	—	—	—	—	—	—	—	—	—	—	—	—	13.05	15.45	19.50
Orbit—Elliptical Orbit	—	—	—	—	—	—	—	—	—	—	—	—	3.55	6.00	10.00
Orbit at Target															
Earth	—	—	—	—	—	—	—	—	—	13.7	—	15.95	—	—	—
Orbit	—	—	—	—	—	—	—	—	—	4.25	—	6.45	—	—	—
Earth—Earth	—	—	—	—	—	—	—	—	—	—	—	—	—	—	19.45
Orbit—Earth	—	—	—	—	—	—	—	—	—	—	—	—	—	—	9.80
Earth—Orbit	—	—	—	—	—	—	—	—	—	—	—	—	—	—	22.60
Orbit—Orbit	—	—	—	—	—	—	—	—	—	—	—	—	—	—	12.95
Earth—Elliptical Orbit	—	—	—	—	—	—	—	—	—	—	—	—	—	—	19.60
Orbit—Elliptical Orbit	—	—	—	—	—	—	—	—	—	—	—	—	—	—	10.10

Column groups: the first seven numeric columns fall under "Equatorial Earth Satellites, Circular" (AMR Launched: 150 km 96 min, 24 hr, 96 min, 24 hr; + Recovery: 150 km 96 min, 96 min, 24 hr); the next two under "Elliptical Equatorial Earth Satellite" (150—35871 km); and the remaining six under "Lunar Missions" (Impact, Satellite, Soft Landing; + Return: Circum-Lunar, Satellite, Soft Landing).

Table III (Continued)

	Space Probes						Return		Planetary Missions				+ Return				
	Earth Escape	Martian	Venusian	Mercury	Solar	Solar System Escape	HOHMANN	CROCCO	Mars Satellite	Venus Satellite	Mars Soft Landing	Venus Soft Landing	Mars Satellite	Venus Satellite	Mars Soft Landing	Venus Soft Landing	Planetoid Soft Landing
Direct Target																	
Earth	12.45	13.15	13.00	15.1	18.55–33.35	18.0	—	—	15.50	16.60	13.40	13.35	—	—	—	—	15.81
Orbit	3.10	3.70	3.55	5.65	9.20–24.0	8.65	—	—	6.05	7.15	3.95	3.90	—	—	—	—	6.45[1]
Earth—Earth	—	—	—	—	—	—	17.60	16.05	—	—	—	—	18.05	20.40	19.80	26.15	18.95
Orbit—Earth	—	—	—	—	—	—	8.15	6.60	—	—	—	—	8.60	10.95	10.35	16.70	9.50
Earth—Orbit	—	—	—	—	—	—	21.00	21.95	—	—	—	—	21.56	23.76	23.43	20.70	22.50
Orbit—Orbit	—	—	—	—	—	—	11.55	12.50	—	—	—	—	12.11	14.31	13.88	20.25	13.05
Earth—Elliptical Orbit	—	—	—	—	—	—	18.15	19.1	—	—	—	—	18.71	20.91	20.48	26.85	19.65
Orbit—Elliptical Orbit	—	—	—	—	—	—	8.70	9.65	—	—	—	—	9.26	11.46	11.03	17.40	10.20
Orbit at Target																	
Earth	—	—	—	—	—	—	—	—	—	—	15.65	16.75	—	—	—	—	16.00[1]
Orbit	—	—	—	—	—	—	—	—	—	—	6.20	7.30	—	—	—	—	6.55[1]
Earth—Earth	—	—	—	—	—	—	—	—	—	—	—	—	—	—	22.00	29.50	19.10
Orbit—Earth	—	—	—	—	—	—	—	—	—	—	—	—	—	—	12.60	20.10	9.60
Earth—Orbit	—	—	—	—	—	—	—	—	—	—	—	—	—	—	25.55	32.90	22.60
Orbit—Orbit	—	—	—	—	—	—	—	—	—	—	—	—	—	—	16.10	23.45	13.15
Earth—Elliptical Orbit	—	—	—	—	—	—	—	—	—	—	—	—	—	—	22.70	30.10	19.80
Orbit—Elliptical Orbit	—	—	—	—	—	—	—	—	—	—	—	—	—	—	13.25	20.60	10.30

[1] No return trip included.

References

1. W. HOHMANN, Die Erreichbarkeit der Himmelskörper. München: R. Oldenbourg, 1925.
2. H. OBERTH, Wege zur Raumschiffahrt. München: R. Oldenbourg, 1929.
3. D. F. LAWDEN, Entry into Circular Orbits I. J. Brit. Interplan. Soc. 10, 5 (1951).
4. E. SÄNGER, Zur Theorie der Photonenraketen. IV. Internationaler Astronautischer Kongreß, Zürich 1953, p. 32. Biel-Bienne: Laubscher & Co., 1955.
5. G. A. CROCCO, Proceedings of the VIIth International Astronautical Congress, Roma 1956, p. 201. Roma: Associazione Italiana Razzi, 1956.
6. G. C. SZEGO, Similitudes and Limitations in Trans-Conventional Propulsion Systems. Proceedings of the IXth International Astronautical Congress, Amsterdam 1958, p. 421. Wien: Springer, 1959.
7. W. E. MOECKEL, Interplanetary Trajectories with Excess Energy. Proceedings of the IXth International Astronautical Congress, Amsterdam 1958, p. 96. Wien: Springer, 1959.
8. H. O. RUPPE, Satellite Technology and Space Navigation. ABMA DSP-TN-9-58, Redstone Arsenal (United States), 1958.
9. R. F. HOELKER and R. SILBER, The Bi-Elliptical Transfer Between Circular Coplanar Orbits. ABMA DA-TM-2-59, Redstone Arsenal (United States).
10. L. LEES, F. W. HARTWIG, and C. B. COHEN, The Use of Aerodynamic Lift During Entry into the Earth's Atmosphere. ASME paper 1959-AV-36 (United States).

On the Technical Realisation of Subgravity and Weightlessness

By

O. Wołczek[1]

(With 9 Figures)

(Received June 5, 1959)

Abstract — Zusammenfassung — Résumé

On the Technical Realisation of Subgravity and Weightlessness. The problem here considered is a technical realisation of subgravity and weightlessness on Earth and under full effects of the force of gravity. Practical methods are given for the realisation of rapid transitions from multi-g field to states of subgravity and weightlessness and vice versa, with the aim of conducting research work in space technics and medicine. A proposal is made for the construction and exploitation of certain apparatus for producing intermittent subgravity and weightlessness lasting for longer periods of time of the order of hours and more.

Über die technische Verwirklichung von Unterschwere und Gewichtslosigkeit. In der vorliegenden Arbeit wird das Problem einer technischen Verwirklichung von Unterschwere und Gewichtslosigkeit auf der Erde und unter voller Wirkung der Schwerkraft geprüft. Es werden praktische Methoden für die Realisierung schneller Übergänge von einem Schwerefeld mit mehreren g Beschleunigung zu Zuständen der Unterschwere und Gewichtslosigkeit sowie umgekehrt angegeben, mit dem Ziel der Ausführung von Forschungen auf dem Gebiet der Raumfahrttechnik und -medizin. Ein Vorschlag für die Konstruktion und Ausnützung bestimmter Apparate wird gemacht, die zur Erzeugung von unterbrochenen Unterschwere- und Gewichtslosigkeits- zuständen während längerer Perioden in der Größenordnung von Stunden und darüber dienen sollen.

Sur la réalisation technique de la subgravité et de l'absence de pesanteur. Des méthodes pratiques sont données pour réaliser des transitions rapides entre des champs de gravité de plusieurs g et des états de subgravité ou d'absence de pesanteur dans le but de recherches en médecine et technique de l'espace. On propose la con- struction et l'exploitation de certains appareils pour la production de subgravité et absence de pesanteur intermittent pendant des périodes de l'ordre de plusieurs heures et davantage.

Introduction

States of subgravity and weightlessness are an important problem not only from the point of view of space medicine, but also in connection with rocket technics and astronautics. In several instances various apparatus and measuring devices react differently in such states than they would in the normal state of

---

[1] Polskie Towarzystwo Astronautyczne (Polish Astronautical Society) and Institute for Nuclear Research of the Polish Academy of Science, Warsaw, Poland.

weightiness which we have on the surface of Earth, when the gravitational field of the planet is not hindered. Due to the weakening of the gravitation force or its complete "inactiveness" brought about by counteraction of other forces, fluid flow, convection etc. are hindered or even completely stopped.

However the problem of a practical realisation of subgravitational and weightlessness states meets with many essential difficulties. It is not possible for us to "eliminate" gravitational forces on Earth and up till now efforts for counteracting these forces by other forces have naturally met with essential difficulties.

Efforts have been made for the realisation of such states on the surface of the Earth by way of a special seesaw [1]. It is also possible to make use of special "lifts" allowing a free fall or comparatively slightly hampered fall. Such methods, it is clear, have essential failings of which the principal, though far from being the only cardinal one, is the considerably short period of time in which it is possible to obtain states of subgravity or even complete weightlessness. Italian experiments [1] for instance, give the period of subgravitation state to be of the order of only 1 second.

A more effective realisation of a subgravitational state is obtainable in the period of time of free jumps from flying objects or else in moments of falling of objects in question preceeding the opening of a parachute especially should be the jumping or falling from a considerable altitude. It has been possible to obtain a subgravitational state lasting a few minutes in falling rocket cones [2, 3, 4]. However it is obvious that scientific observations and experiments with technical equipment in such cases are greatly limited.

A headway was reached when it was possible to obtain states of subgravitation and weightlessness in airplanes flying sections of suitable Keplerian orbits [5, 6, 7, 8. 9], more over, lately it was possible to lengthen the period of time of a state of weightlessness to about 60 seconds.

A state of weightlessness lasting long periods of time of the order of at least weeks and even of many years has become possible with the invention of artificial satellites of the Earth. These satellites even in the early phases of their realisation, if we mention the Soviet "1957 Beta", were used in biological research [3, 4, 10, 11, 12]. However, even than, observation was made especially difficult by there being no possibility of direct observation and control of experimental conditions.

This state of affairs makes it necessary to invent devices which would permit the realisation of states of subgravitation and weightlessness on Earth in as simple and satisfactory a way as multi-g field is realised by means of rocket sleds or centrifuges. As it is impossible to eliminate the force of gravitation even for a very short period of time, in devices for the realisation of subgravitation and weightlessness it is imperative to use counteractive forces for balancement. Here the centrifugal force is among the first to be taken into account.

In this work some possibilities of using the force for the practical realisation of subgravitation and weightlessness have been taken into consideration. This work is only suggesting certain ways in which eventually future experimental devices should develop. The author does not intend to give here detailed computations—that would be the problem of further work to be done in separate instances. That is the reason why only the effects of two cardinal forces are here taken into consideration: gravitation and centrifugal forces, neglecting all other effects like air resistance, mechanical friction, CORIOLIS force [13] etc., also spacial extension of experimental device. However, for constructing such particular devices, it will be necessary to take these effects into consideration.

Vertical Centrifuges

It seems that a comparatively simple method of an effective way of obtaining smaller gravitation and weightlessness on Earth is the use of centrifuges or devices operating on the same basis. All such devices should then be stationed in a vertical position so that their axis of rotation would be paralleled to the surface

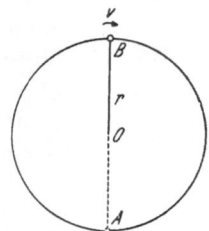

of the Earth. In this way the centrifugal force generated by rotation is added to the gravitation force alternatively weakening or strengthening its effects.

Choosing a suitable speed of the rotation in the highest point of the circular orbit in the circumference of the device a complete compensation of the force of gravitation by centrifugal force may be obtained, in which case a state of weightlessness would be obtained in the capsule moving in the circumference (Fig. 1, pt. B). In the lowest point of the capsule orbit (Fig. 1, pt. A) the force effecting it would, in this instance, double the weight of the capsule.

Fig. 1. Scheme of the rotational motion in a vertical centrifuge

For the clearness of the aspect of the relations dominating in the rotation under consideration let us look at the most simple formula:

$$P = m\, g, \tag{1}$$

with P — weight of capsule (with the point-size)
    m — mass
    g — Earth acceleration

$$F = \frac{m\, v^2}{r}, \tag{2}$$

where:

    F — centrifugal force
    v — linear velocity
    r — radius of the circular orbit.

In the case under consideration (Fig. 1) the condition to be fulfilled is

$$F = P$$

Hence:

$$v = \sqrt{g\, r}. \tag{3}$$

Further—the path s on the orbit of each rotation of the device with the angle 2π amounts to:

$$s = 2\pi r. \tag{4}$$

Hence the time t, necessary for one full rotation is equal:

$$t = 2\pi \sqrt{\frac{r}{g}}. \tag{5}$$

Using the above quite primitive formula it is possible to compute suitable parameters v, s, t, setting them for comparison in a table (Table I).

As is seen from the table, at a very small radius of the device of, for example, 1 m. the changes in the resulting acceleration responsible for the appearance of an adequate resulting force at interval: $2\,g \searrow 0 \nearrow 2\,g$ take place in a period of time of only 2 seconds. The linear velocity in this case is very small, of about 3 m/sec. Should we wish to lengthen the period of acceleration to a given interval,

Table I. *Parameters of the Point Capsule Rotating around a Stable Point During which at the Highest Point of the Orbit the Weight of the Capsule is Balanced by the Centrifugal Force*

No.	r m	v		s m	t sec.
		m/sec.	km/h		
1	1	3,2	11,4	6,3	2
2	2	4,5	16,2	12,6	2,8
3	5	7,1	25,6	31,4	4,4
4	10	10	36	62,8	6,3
5	50	22,4	81	314	14
6	100	31,6	114	628	19,9
7	200	45	162	1260	28

to say, 20 sec. the radius of the orbit would grow to 100 m, and the linear velocity on this orbit would grow to nearly 32 m/sec., thus over 110 km/h. Generally for an n-fold increase of a said period of changes of acceleration it is necessary (on the basis of the accepted simple assumptions) as much as a n^2-fold enlargement of the radius of the circular orbit which entails as much as an n-fold increase of the linear velocity of the movement. These conditions decidedly limit the possibility of a technical realisation. This is to be regreted—at a considerable radius of devices there would occur a certain compensation of the influence of spacial extension of capsule with experimental material.

On the other hand, we should take the opportunity for stressing the meaning of rapid changes of the effective acceleration for both technical and biological research, the more so, because in the cases here discussed it is a question of transition from multi-g field to states of subgravitation and weightlessness and vice-versa. Research in this line is at present highly developed and the more serious results of experiments in this field were referred by von BECKH [9] at the last Congress of the IAF in Amsterdam. Using the centrifugal force in the way mentioned above it would be possible to realise analogical experiments in suitable vertical centrifuges obtaining milder effects at larger radius of orbit of rotating capsules and vice versa.

Milder effects are also to be observed in narrower acceleration intervals obtained by way of a suitable decrease of rotation velocity of the devices. Reducing the velocity of the capsule moving on a 50 m long arm (compare Table I) with 22,4 m/sec. to 15,8 (10) m/sec. we gain a prolongation of the time of one full rotation to 20 (31,4) sec. while the interval of the produced acceleration is narrowed to 1,5 g ↘ 0,5 g ↗ 1,5 g (1,2 g ↘ 0,8 g ↗ 1,2 g). The here quoted periods of rotation of the order of already several seconds can be regarded as an actual basis for a programme of suitable research.

The achievement of a minimum of resulting acceleration, as can be easily proved, is. alas, possible only at the highest point of the orbit. It is therefore imperative to consider how to best use this condition. To simplify the estimation of different situations we can make use of a simple and easy formula for the resulting acceleration which is the sum of Earth and centrifugal acceleration depending from the angle a formed by the capsule leading-arm and the vertical assuming that the centrifugal acceleration thus obtained is equal as to absolute value to the Earth acceleration (Fig. 2). Then the resulting acceleration a_w is equal:

$$a_w = 2\,g\,\sin\frac{a}{2}\,.\qquad\qquad(6)$$

Engaged only in subgravitation and weightlessness it is necessary to consider only the upper part of the orbit contained between the points in which $a_w = g$. The orbit arc corresponds then to the 60° angles on both sides of the vertical (Fig. 3) and is thus indicated by an angle of 120°. For the orbit radius $r = 50$ m. the time of transition of a point capsule from acceleration: $1\ g \searrow 0 \nearrow 1\ g$ is $\frac{14}{3} = 4{,}7$ sec. Narrowing the range of changes of acceleration to: $0{,}5\ g \searrow 0 \nearrow 0{,}5\ g$, it is necessary to limit ourselves to an arc corresponding to $2 \times 29° = 58°$ which is only 16 % of the length of the capsule circular orbit equivalent to one full rotation. With the orbit radius $r = 50$ m this corresponds to a period of 2,2 sec., at a radius of 200 m—4,5 sec. These are rather short periods of time—uncomparable with periods necessary from the point of view of practical realisation.

Should we wish to increase the change of acceleration in the range: $0{,}5\ g \searrow 0 \nearrow 0{,}5\ g$ to 60 sec., then the leading arm of the capsule should be of an absurd length of 35760 m and the velocity of the capsule would grow to an equally absurd

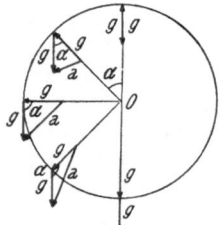

Fig. 2. Accelerations during various phases of the rotational motion in a vertical centrifuge

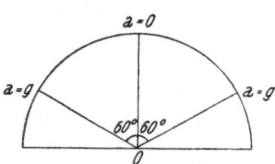

Fig. 3. Acceleration interval corresponding to subgravity in the rotational motion in a vertical centrifuge

value of about 600 m/sec. In such a troublesome situation a rational way out would be to use for experiments not one but two vertical centrifuges suitably placed and coupled. The same capsule would run alternatively the same position of the orbit in one or the other of the centrifuges with a special automatic device for the tossing over of the capsule.

However there will be considerable changes of the resulting acceleration in the short periods of time during the tossing over of the capsule. Perhaps a very rapid tossing over would be without a greater importance especially in the case of non-biological objects thanks to a given inertness of matter. On the other hand one may take some profit of such rapid changes of the resulting acceleration conducting appropriate experiments.

The tossing over would be especially easy in vertical centrifuges with small radius where the linear velocity of the circumference is comparatively small. It is of course not possible to avoid the fact that the tossing over from one centrifuge to the other would be accompanied by additional accelerations in the direction perpendicular to the plane of the motion of the devices. These accelerations would not be great at small circumference velocities. It is also important that due to small velocities it would be easy enough to stop the device if some defect of operation would show. It would be safer however to equip the centrifuges with an automatic stopping device. This matter is especially important for the phase of the tossing over of the capsule from one centrifuge into the other.

As regards the practical realisation of the mentioned devices there are two possibilities of construction. One of them would consist of two parallel counter-current compound centrifuges (Fig. 4), equipped with a channel or several horizontal channels appropriate for receiving and keeping in their interior and giving back of a suitable capsule. The channel would be placed on the circumference. In case of several channels the space between them would depend primarily from the acceleration intervals in which the research work would have to be done and this in turn would be connected in a suitable way with the circumference velocity and radius of the device.

The capsule in its first stage inside in the centrifuge channel *A* (Fig. 4) in position 1 having crossed the arc into position 1′ would be automatically tossed over into the channel of centrifuge *B* into position 2. Now it would progress on the arc path in the direction opposite to the direction it had in centrifuge *A*—to position 2′. Here again it would be tossed over into centrifuge *A* again into the channel in position 1. It is true that in this position the same channel could serve as in the first phase. In this case however, the centrifuge would need to function with a changeable velocity.

The channel in centrifuge *A* from position 1 to 1′ (and in centrifuge *B* from position 2 to 2′) would have a movement of a certain stable velocity and for trav-

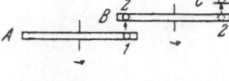

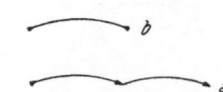

Fig. 4. Scheme of the counter-current compound centrifuges

Fig. 5. Scheme of the in-line compound centrifuges

Fig. 6. Directions of motion in the in-line (*a*) and counter-current (*b*) compound centrifuges

ersing this upper path the time *t* would be used. On the lower path from position 1′ to 1 (and from position 2′ to 2) the motion of the channel would have to be much quicker so that the time used for travelling this arc so much longer than arc 1—1′ (and arc 2—2′) would take the same period of time—*t*. To conform with these conditions it is necessary for the compound centrifuge to be equipped with an additional device which would in fact greatly complicate its construction.

The situation would be much simplified if the compound centrifuges were placed not next to each other but in a suitable line one after the other (Fig. 5). It would form an in-line compound centrifuge. Such an assembly has its additional merit that the motion of the capsule "tossed" from one centrifuge into the other (and eventually into next ones) would be repeated continually in one direction (Fig. 6 *a*), and, not as firstly said, in a to and fro direction (Fig. 6 *b*).

Further improvements in the efficiency of the compound centrifuges could be made by a suitable construction of the channels and their outlets permitting the capsules to be transferred from one rotating disc on to the other as gently as possible, thus reducing additional and interferring side accelerations to a minimum. The channels and their outlets should be in a way that the transferring process could function without shocks in the direction possibly close to the new motion by way of a slip at a possibly small angle with the plane of the rotating disc.

In connection with the types of devices here discussed, we may consider the problem of a vertical, single centrifuge with an oscillating rotor of some fixed velocity between two suitable positions 1—1′ (Fig. 7). In such a device rapid changes of acceleration should be realised—changes which undoubtedly would bring about different other physiological effects than the comparatively slow changes obtained in devices mentioned before.

Returning to compound centrifuges one more interesting item of construction can be mentioned—the in-line building of the devices on the circle circumference (Fig. 8). The centrifuges would be of a cyclical compound type.

In this case the capsule could undergo rhytimical changes of acceleration for any fixed periods of time. This would permit to carry on suitable experiments and series of experiments in a proper way without being limited to short investigations in compound centrifuges consisting of only a few rotors. On the other hand, should long lasting experiments be conducted in ordinary in-line coupled centrifuges, the number of rotors would then have to grow to unattainable values, thus practically having no sense.

And, lastly, one more proposition radically simplifying the problem of construction: the replacement of rotating discs by adequate arms (Fig. 9) the movements of which would be coupled. Then the arms would not have to make full rotations—but only a short motion within an angle of the circumference returning to the initial position and then repeating the movement.

In practice the said arm would carry the capsule with the experimental material across a fixed arc, then this capsule would be taken over by a catching device of the next arm and carried on. There would be no necessity of applying

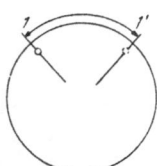  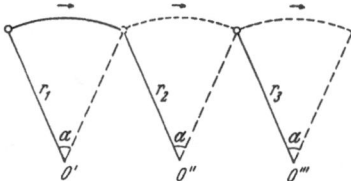

Fig. 7. Scheme of the verti-   Fig. 8. Scheme of the cycli-   Fig. 9. Scheme of the vertical compound
cal centrifuge with an oscil-   cal compound centrifuges   centrifuges with oscillating arms.
lating rotor r_1, r_2, r_3 — oscillating arms
                                                       α — angle of rotation

any special automatic devices transferring the capsule from one arm to the next one—analogical to the formerly proposed and described coupled centrifuges—with discs equipped with proper channels. In this case a vital problem could be avoided, namely additional and noxious side accelerations typical for coupled centrifuges equipped with capsule channels.

The intervals of the realised accelerations and the tempo of their changes would depend on the choice of the length of the carrying arms, their speed, and length of separate arcs of orbits. The shorter the arc the more gentle would be the transfer but as much narrower would be the limits of acceleration changes. The greater the radius of arcs, the longer would be the single periods of motion along the arcs as has been mentioned before. In practice there should be means of regulating the given factors by way of changing the length of the carrying arms and the speed of their movements. This could be obtained with comparatively negligible technical means. Of course the building of the proposed devices in-line on a circle circumference, as mentioned before, in the case of cyclical compound centrifuges, would allow just as long lasting experiments.

In the described cyclical devices with carrying arms the principle of an automatic orientation of capsules should be applied in a suitable required position with regard to the direction of the acceleration effects and in accordance with von Beckh [14].

It is worth mentioning that, especially in cyclical devices, it would be possible and advisable to put into simultaneous motion, instead of one, a few if not several capsules with experimental material due to which fact the efficiency of the apparatus would greatly grow. The possibility of wider scope for regulating experimental conditions would be an additional and great advantage.

For obtaining efficiency and flexibility in the realisation of the most universal experimental cycles it would be good to couple the lastly proposed cyclical devices with an adequate electronic computer. The computer would be supplied with only a programme of investigation either in shape of punched cards or perforated tape or the like. Further operations of the experiments would be automatically directed by the computer and in accordance with supplied instructions. In this way the highest stage of vertical centrifuge development would present programmed cyclical compound devices with arms.

It is clear that the proposed devices would first of all serve to carry on investigations of smaller objectives, like samples of construction materials, miniaturised measuring devices and apparatus and smaller experimental animals like mice for instance. In all these instances it would be possible to carry on investigations on a large number of objects simultaneously—a valuable aspect from the point of view of statistics and reliability of the results of measurements.

The matter of utilising the last cyclical devices for experiments with humans is a difficult problem. Quite a superfluous evaluation points to the fact that in view of the considerable spacial extension of experimental capsules in this case the length of the carrying arms would have to be extensive nearing probably 10 m or even longer. Let us note that using only 10 m long carrying arms and being limited to a rather wide acceleration interval: $0,5\,g \searrow 0 \nearrow 0,5\,g$ it would be necessary to utilise arcs corresponding to about 60° consequently of a length of about 10 m, besides the time used for the crossing of each separate arc would be only 1 sec. It is not difficult to imagine that the construction of a cyclical device with carrying arms of such length holding, carrying, transferring and receiving capsules with humans—of a mass of the order of 200 kg with the speed of many miles per hour—would be indeed a very complicated technical problem. The evaluation of the feasibility and economy of an eventual construction of such a cyclical device is a problem in itself and needs theoretical studies as well as adequate experiments on models.

References

1. T. LOMONACO, M. STROLLO, and L. FABRIS, Comportamento della coordinazione motoria in soggetti sottoposti a valori di accelerazione varianti da 3 a 0 G (Behaviour of Motor Coordination in Subjects Exposed to 3 to 0 Acceleration Values). Rendiconti del VII. Congresso Internazionale Astronautico, Roma 1956, p. 825. Roma: Associazione Italiana Razzi, 1956.
2. H. STRUGHOLD, Medical Problems Involved in Orbital Space Flight. Jet Propulsion 26, 745—748, 756, 788 (1956).
3. I. HERSEY, Soviet Biological Experiments. Astronautics 4, No. 2, 31, 80—81 (1959).
4. Predvaritielniye itoghi nautchnikh issledovaniy s pomoshtchyu piervikh soviec-kikh iskusstviennikh sputnikov Zemli i rakiet. XI razdiel programmy MGG (rakiety i sputniki) No. 1. Moskva: Izdatielstvo Akademyi Nauk SSSR, 1958.
5. H. HABER and F. HABER, Possible Methods of Producing the Gravity Free State for Medical Research. Aviat. Med. 21, 395 (1951).
6. H. VON DIRINGSHOFEN, in: ,,Wie wird sich der menschliche Organismus voraussichtlich im schwerefreien Raum verhalten?" Weltraumfahrt 2, 83 (1951).
7. S. J. GERATHEWOHL, Weightlessness. Astronautics 2, No. 4, 32—34, 74—75 (1957).
8. A. E. SLATER, The Problem of Weightlessness. Spaceflight 1, 109 (1957).
9. H. J. VON BECKH, Weightlessness and Space Flight. Astronautics 4, No. 2, 26—27, 84, 86 (1959).
10. G. A. SKURIDIN and L. V. KURNOSOVA, Nautchnyie issledovanya pri pomoshtchi iskusstviennikh sputnikov Zemli. Priroda 46, No. 12, 7 (1957).

11. Sovieckiye iskusstvienniye sputniki — Niekotcriye itoghi nautchnikh issledovaniy na dvukh piervikh sovieckikh iskusstviennikh sputnikakh Zemli. Pravda, No. 117 (14511), 4 (27. 4. 1958).
12. Iskusstvienniye sputniki Zemli, No. 1, 2. Moskva: Izdatielstvo Akademyi Nauk SSSR, 1958.
13. D. W. Lawden, The Simulation of Gravity. J. Brit. Interplan. Soc. 16, 134 (1957).
14. H. J. von Beckh, Multi-Directional G-Protection in Space Vehicles. J. Brit. Interplan. Soc. 16, 525 (1958); Proceedings of the VIIIth International Astronautical Congress, Barcelona 1957, p. 37. Wien: Springer, 1958.

The Motion of an Orbiting Vehicle Subjected to Continuous Radial Thrust, Including a Study of Planetary Encounters[1]

Bernard Paiewonsky[2], ARS

(With 6 Figures)

(Received June 18, 1959)

Abstract — Zusammenfassung — Résumé

The Motion of an Orbiting Vehicle Subjected to Continuous Radial Thrust, Including a Study of Planetary Encounters. The use of continuous radial thrust has been mentioned as a suitable scheme for braking spaceships in planetary encounters since the guidance requirements appear to be simple. The problem of a vehicle approaching a planet is investigated in order to compare different schemes for braking, adjustment of perigee altitude, and assurance of capture. In this regard, continuous radial thrust, retro-rockets, and atmospheric lift and drag are discussed.

The method of equivalent one-dimensional potentials has been applied to the motion of an orbiting vehicle subjected to a continuous radial thrust. The application of this method depends on the fact that the angular momentum remains constant if the thrust is always radial. It is shown that the shape of the potential curve is changed by the addition of radial thrust. Several possibilities exist for the motion· depending upon the initial conditions and the applied radial acceleration, (F/m). For outwardly directed thrust it is shown that unbounded motion will occur for certain values of (F/m) and certain initial conditions. For initially circular orbits this is shown to correspond to limiting conditions obtained in previous papers [7, 9]. It is also shown that escape from a closed orbit is possible if the direction of the radial thrust is changed from inwards to outwards, or vice versa, at the proper places, even if (F/m) is less than the value required for escape determined by considerations of outward thrust alone.

The braking and control of perigee altitude for a vehicle approaching a planet at high speeds is examined by the potential method. This method is used to study the effective potential barrier produced by the application of continuous radial thrust.

The braking of a vehicle by means of impulsive thrust is examined by the energy and angular momentum equations. Charts are presented that allow a rapid estimation of the effects of impulsive control application or errors on the perigee altitude and velocity.

The use of atmospheric lift in connection with the capture of high speed vehicles is discussed with respect to its application in combination with continuous and impulsive thrusting.

Die Bewegung eines Raumfahrzeuges in einer Umlaufbahn unter Einwirkung eines kontinuierlichen Radialschubes (einschließlich einer Studie über Begegnungen mit Planeten). Die Anwendung eines kontinuierlichen Radialschubes ist als geeignetes

---

[1] The discussion presented in this paper resulted from a more general study of the guidance and control of space vehicles being carried on by Aeronautical Research Associates of Princeton for the General Precision Equipment Corporation.

[2] Associate Research Engineer; Aeronautical Research Associates of Princeton, Inc.; Visiting Lecturer, Department of Aeronautical Engineering, Princeton University, Princeton, New Jersey, U.S.A.

Schema für die Bremsung von Raumfahrzeugen bei Begegnungen mit Planeten genannt worden, da die Steuerungserfordernisse einfach zu sein scheinen. In der vorliegenden Arbeit wird das Problem der Annäherung eines Fahrzeuges an einen Planeten untersucht, um verschiedene Systeme der Bremsung, der Einstellung der Höhe im Perigäum und der Gewißheit eines Einfanges zu vergleichen. In dieser Beziehung werden kontinuierlicher Radialschub, Rückraketen und atmosphärischer Auftrieb und Widerstand erörtert.

Die Methode äquivalenter eindimensionaler Potentiale wurde auf die Bewegung eines Fahrzeuges in einer Umlaufbahn angewendet, das einem kontinuierlichen Radialschub unterworfen ist. Die Benützung dieser Methode hängt von der Tatsache ab, daß das Winkelmoment konstant bleibt, wenn der Schub stets radial ist. Es wird gezeigt, daß sich die Gestalt der Potentialkurve durch das Hinzukommen eines Radialschubes ändert. Es gibt mehrere Möglichkeiten für die Bewegung, die von den Anfangsbedingungen und der angewendeten Radialbeschleunigung (F/m) abhängt. Für nach außen gerichteten Schub wird gezeigt, daß eine unbestimmte Bewegung für gewisse Werte von (F/m) und bestimmte Anfangsbedingungen auftritt. Es konnte ermittelt werden, daß dies für anfänglich kreisförmige Bahnen den Grenzbedingungen entspricht, wie sie in früheren Veröffentlichungen [7, 9] erhalten wurden. Es stellte sich ferner heraus, daß das Entkommen aus einer geschlossenen Bahn möglich ist, wenn die Richtung des Radialschubes von innen nach außen oder umgekehrt an den geeigneten Stellen geändert wird, auch wenn (F/m) kleiner als der Wert ist, der für das Entweichen erforderlich ist und durch Berücksichtigung allein des nach auswärts gerichteten Schubes bestimmt wird.

Die Bremsung und Kontrolle der Perigäumshöhe für ein Raumfahrzeug, das sich einem Planeten mit hoher Geschwindigkeit nähert, wird nach der Potentialmethode geprüft. Diese Methode wird benützt, um die wirkende Potentialschranke zu studieren, die durch die Anwendung kontinuierlichen Radialschubes erzeugt wird.

Die Bremsung eines Raumfahrzeuges mittels Impulsschubes wird an den Energie- und Winkelmoment-Gleichungen geprüft. Es werden Diagramme gezeigt, welche die rasche Bestimmung der Wirkungen der Anwendung impulsiver Kontrolle oder von Fehlern in der Perigäumshöhe und -geschwindigkeit gestatten.

Die Benützung des atmosphärischen Auftriebs zum Einfangen von Fahrzeugen von hoher Geschwindigkeit wird hinsichtlich ihrer Anwendung in Kombination mit kontinuierlichem und impulsivem Schub diskutiert.

Le mouvement orbital d'un engin soumis à poussée radiale continue y compris l'étude de la rencontre. La simplicité du guidage a fait mentionner l'utilisation de la poussée radiale continue pour le freinage des astronefs avant la rencontre. Le problème de l'approche est analysé en vue de comparer différentes méthodes de freinage, d'ajustement du périgée pour l'obtention d'une rencontre certaine. La poussée radiale continue, les fusées de freinage, la portance et traînée atmosphériques sont envisagées dans cette optique.

La méthode utilisée est celle des potentiels unidimensionnels équivalents; elle s'appuie sur la constance du moment angulaire quand la poussée est radiale. La forme de l'équipotentielle est modifiée par la présence de la poussée radiale. Les trajectoires dépendent des conditions initiales et de l'accélération radiale appliquée. Si la poussée agit vers l'extérieur, la trajectoire diverge sous certaines conditions initiales et certaines valeurs de l'accélération radiale. Dans le cas d'orbites initialement circulaires ceci correspond à des conditions limites obtenues antérieurement [7, 9]. La libération d'une orbite fermée est possible par inversion de la poussée radiale à des endroits appropriés, même si l'accélération radiale est inférieure à la valeur requise pour la libération déterminée uniquement par des considérations de poussée radiale extérieure.

Le freinage et le contrôle du périgée pour une approche à grande vitesse sont aussi examinés par la méthode du potentiel. Celle-ci est utilisée pour la détermination de la barrière de potentiel effective produite par poussée radiale continue.

Le freinage par impulsion est analysé par les équations d'énergie et de quantité de mouvement angulaires. Des diagrammes permettent une estimation rapide des effets d'impulsions et des erreurs sur l'altitude du périgée ou sur la vitesse.

Nomenclature

r radial distance from center of force to rocket
m mass of rocket
M mass of celestial body
F radial thrust
h angular momentum/unit mass
E total energy
μ gravitational constant $= \gamma M$ ($\mu = 1.4078 \times 10^{16}$ ft^3/sec^2 for Earth)
Φ angular variable
χ angle between velocity and radius vectors

Subscripts

o initial or en route condition
p perigee

Introduction

One of the most critical problems facing space travelers is the approach to a celestial body for the purpose of establishing a close parking orbit or descending to a landing on the body. In many cases the approach path will be locally hyperbolic, so that braking will be necessary if a closed parking orbit is to be established. Many schemes have been proposed [1, 2, 3, 5] to allow a safe transfer from the approach orbit to a parking orbit or to an atmospheric entry path. In all of these methods the perigee altitude, or distance of closest approach, is a critical factor.

It has been suggested that vehicles capable of producing sustained thrust and using continuous steering will be able to carry out simple guidance programs and will be less sensitive to errors in control application than vehicles using impulsive thrust for control. It is the purpose of this paper to study the motion of a spaceship subjected to continuous radial thrust and to determine whether continuously thrusting vehicles of this type have any advantage over vehicles using retrorockets or aerodynamic braking with regard to the accuracy required of the navigation and guidance systems.

The method of equivalent one-dimensional potentials [11] has been applied to the case of a spaceship subjected to continuous radial thrust. The principal advantage of the potential method lies in its ability to provide simple and rapid analyses of the boundaries associated with the motion.

In the analysis to follow it is important to distinguish between the effects of errors in determining the approach orbit parameters and the effect of errors in carrying out the commands, such as errors due to thrust misalignment. This distinction is necessary if comparisons between continuous and impulsive thrusting systems are to be made.

NONWEILER [1, 2], BAKER [3], EHRICKE [4], XENAKIS [5], and others have discussed the problems associated with braking ellipses and other forms of atmospheric braking. The use of down-ward directed lift to keep the vehicle in the vicinity of the body once contact with the atmosphere has been established, as suggested by NONWEILER, seems particularly appealing compared with a succession of braking ellipses. This is particularly attractive for reconnaissance missions where a close approach is desired and the capability for subsequent escape from the body is required. R. M. L. BAKER, JR. [3] has discussed the nature of the cross sections for capture and atmospheric entry. The latter problem has been compared with that of hitting the cellophane wrapping on a golf ball with a dart without touching the ball.

If the energy and angular momentum equations for the ballistic path are differentiated, the changes or errors to the first order, in perigee distance and velocity due to controlled changes or errors in the en route position and velocity vectors are obtained. These relations have been considered, in different forms, by several authors [4, 5]. It will be shown that the angular momentum determination en route plays a key role in predicting the distance of closest approach for both impulsive and continuous thrusting systems. In examining the results of these studies the tremendous sensitivity of the perigee conditions to changes in the approach conditions becomes apparent.

It seems reasonable to expect the accuracy of the measurement of the position and velocity of the vehicle to change as the vehicle approaches the body. These accuracies will probably improve as the distance from the body decreases. This will naturally depend upon the instrumentation characteristics.

However, for a given ΔV obtainable from an impulsive thrust type control rocket, the ability to produce changes in the angular momentum and, hence, changes in the distance of closest approach, decreases as the distance from the body decreases. Thus, depending on the measuring devices used, there will exist an optimum region for the application of impulsive control thrust and also a region within which the required correction cannot be made.

If the purpose of a planetary approach is to make a single reconnaissance run or establish a parking orbit, then the distance of closest approach must be such that the vehicle does not collide with the surface of the body or plunge too deeply into an atmosphere. Hard landings or collisions with airless bodies require the distance of closest approach to lie below the body's surface. If aerodynamic braking, alone, is to be employed, the distance of closest approach must lie within a narrow band defined by the characteristics of both the atmosphere and the vehicle in question.

In each case, the distance of closest approach is the parameter that the space navigator must know with great certainty.

Discussion

I. Continuous Radial Thrust

1. The Method of Equivalent One-Dimensional Potentials

In recent papers, DOBROWOLSKI [9] and COPELAND [10] have examined various aspects of the behavior of a spaceship experiencing a continuous radial thrust and obtained results in terms of elliptic integrals. It is possible to develop this problem in a slightly different manner, yielding qualitative information and allowing an interesting physical interpretation of the behavior of the vehicle. The possibility of developing a simple method for analysing this type of problem depends on the fact that the angular momentum remains constant for vehicles with radial thrust only.

Let us consider the problems posed by DOBROWOLSKI and COPELAND with the removal of the restriction that the initial orbit be circular. A vehicle is in orbit and at time $t = 0$ the radial thrust commences. For simplicity in developing the method let (F/m) be constant. If desired, this restriction can be removed and (F/m) is required then to be a function only of r. The equations of motion are the same as those shown in [9] where $+F$ is directed outwards.

$$m \ddot{r} = \pm F - \frac{m \mu}{r^2} + \frac{m h^2}{r^3}. \tag{1}$$

Eq. (1) may be put in the following form

$$\ddot{r} = -\frac{\hat{c}}{\hat{c}\,r}\left(-\left(\frac{F}{m}\right)r - \frac{\mu}{r} + \frac{h^2}{2\,r^2}\right).\tag{1a}$$

The conservation of angular momentum is expressed by eq. (2).

$$r^2\,\dot{\Phi} = h\,.\tag{2}$$

The problem of describing the motion of the vehicle will be treated by the one-dimensional equivalent potential method. The theory underlying this method is presented in detail in GOLDSTEIN's "Classical Mechanics" [11].

According to this method, we may define a fictitious potential V, such that:

$$V = -\frac{\mu}{r} \mp \left(\frac{F}{m}\right)r + \frac{h^2}{2\,r^2}\,.\tag{3}$$

The potential in the absence of radial thrust, V_0, is the familiar $V_0 = -\frac{\mu}{r} + \frac{h^2}{2\,r^2}$ which leads to the classical KEPLER problem. We have thus added a term describing the constant radial acceleration ($\pm F/m$) to the original potential V_0.

If V_0 vs. r is plotted for a given value of h then the fictitious potential V may be obtained graphically by adding ($\pm F/m$) r to V_0.

Fig. 1 illustrates graphically the addition of the ($-F/m$) r term to the original potential.

This family of curves corresponds to a given value of angular momentum h. For each value of (F/m), a different potential curve is obtained.

An examination of these curves will show that as (F/m) directed outward is increased in magnitude from zero, a second extremum, a local maximum, appears whose location will be denoted by r_b. The bottom of the new potential well is at r_a, while the extremum of the original well was at $r_{a\,0}$. As (F/m) increases, the maximum point (b) moves in toward the minimum point (a) until they meet at (c). This defines $(F/m)_{max}$. At (c), there is an inflection point with zero slope and the well has disappeared. Further increases in (F/m) increase negatively the slope of the potential curve but do not change its features (i.e., no more extrema appear).

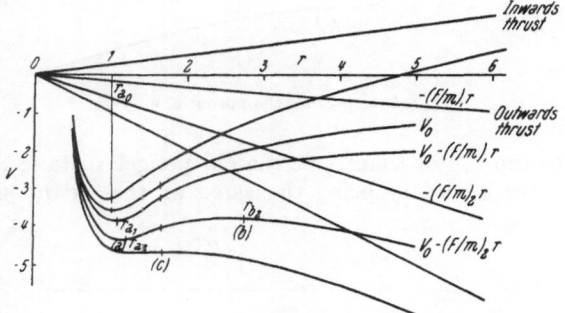

Fig. 1. One-dimensional potential curves

2. Interpretation of the Potentials

There is an interesting method of interpreting these potential curves as pointed out in GOLDSTEIN. The original orbit is characterized by values of its total energy and angular momentum. Since the thrust is always radial, the angular momentum remains constant. Thus, for a given initial orbit, only one family of curves need be examined. In fact, all of the possible orbits may be examined by specifying only the initial angular momentum.

In the figures, a horizontal line corresponds to a line of constant total energy. The bounds on an orbit are given by the intersection of the constant total energy line E, and a potential curve V. The difference between the horizontal E line and V at any r, is a measure of the radial velocity $\dot{r}$.

The curve O represents the power off potential, V_0, for a given value of angular momentum h. Lines 1 and 2 represent lines of constant total energy. The motion corresponding to (h, E_1) is seen to be a bounded motion (i.e. it is an elliptical orbit with perigee a_1 and apogee b_1 as indicated in figure 2).

The initial or power off motion is circular with radius $r = r_3$, when the initial energy equals E_2. Note that the Energy E_2 line is shown tangent to the V_0 line at r_3. For non-negative initial energies (i.e., $E \geqslant 0$) the motion will be unbounded.

The curve 4 represents the equivalent potential for a value of $(F/m) > (F/m)_{max}$.

As an example, suppose the vehicle were in an elliptic orbit corresponding to E_1 in the figure. The rocket can be turned on at any position. Suppose this is done at perigee $r = r_1$. The initial conditions are $r = r_1$, $\dot{r} = 0$. This is shown by point a_1. The turning on of the rocket corresponds to the addition of the new potential term and the motion of the vehicle is determined by curve 3 and the initial point corresponds to point a_2.

The motion is still bounded; however, it is not a simple ellipse. The bounds are given by the radii corresponding to a_2 and b_2. It is clear that the rocket for this value of (F/m) can never approach closer to the center than r_1 or the old perigee distance. Firing at apogee results in the outer boundary being the same as the old apogee distance (for this particular value of (F/m) and initial energy). If, for example, the initial orbit were circular then the initial point in the new well would be at a_3.

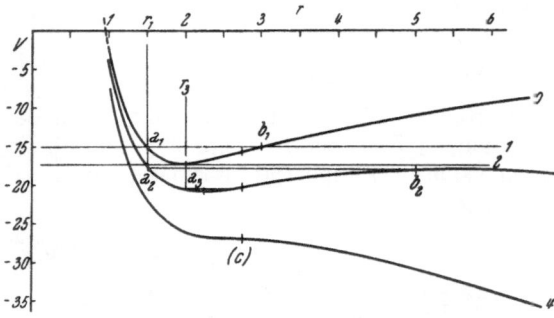

Fig. 2. Interpreting the potential curves

From the previous discussion, it can be seen that several distinct possibilities exist for the resultant motion, depending on the shape of the new potential.

For outwardly directed thrust, if the motion was originally bounded, then

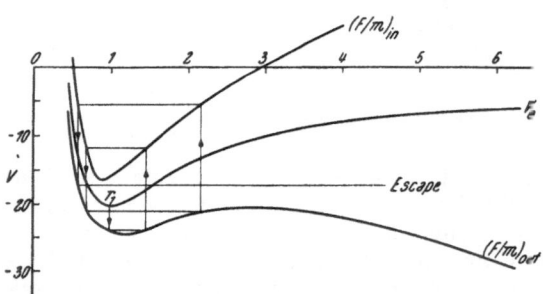

Fig. 3. Escape using cyclic thrusting

(a) There exists a value of (F/m), i.e., $(F/m)_{critical}$ depending on the initial energy and position, such that for values of $(F/m) > (F/m)_{critical}$ the motion may be unbounded.

(b) There exists an $(F/m)_{max}$ such that unbounded motion will occur for $(F/m) > (F/m)_{max}$, for all values of initial energy and initial position.

$(F/m)_{max}$ is the maximum value of (F/m) for which bounded motion is possible. For values of $(F/m) < (F/m)_{max}$, boundedness of the motion depends on the initial conditions and the value of (F/m).

It can be shown that for outwardly directed thrust the new minimum point always lies to the right of the original one ($r_a > r_{a_0}$). For orbits that are originally circular it is easy to show that the critical condition $\lambda = 1/8$ obtained by previous authors [7, 8, 9] corresponds to $(F/m) = (F/m)_{critical}$. This means, physically, that the top edge of the new potential well (determined by $(F/m)_{critical}$) is at the same height as the initial point (point a_3 on Fig. 6) on the new well. Thus, for $\lambda = 1/8$ the rocket can just reach the edge of the new well and then escape from it.

If the thrust is turned off, the vehicles motion is determined by the conditions $(r, \dot r)$ at cut-off and the original potential curve V_0.

For inwardly directed thrust a similar analysis can be made and the corresponding fictitious potential is shown in Fig. 3.

It is interesting to note that while escape is not possible with outwardly directed thrust for $(F/m) < (F/m)_{critical}$ (as for example in the special case of initially circular orbit with $\lambda < \dfrac{1}{8}$) escape becomes possible if the direction of the radial thrust is alternately changed from inwards to outwards at the proper locations in the orbit.

In particular, if the thrust direction is reversed, at each apogee or perigee, a condition analogous to a resonance occurs and the outer bound on the motion increases, slowly perhaps, with each cycle. Several cycles are shown in Fig. 3. The inner boundaries must be checked to see that a collision will not occur with the planet.

3. Braking with Continuous Radial Thrust

The method of analysis by equivalent potentials can be applied to the study of the motion of vehicles approaching a planet or other celestial body. For a given energy and angular momentum the V_0 potential curve yields the information obtained in the section on ballistic approaches. Changes in r_p resulting from changes in h_0 can be seen to result from an upwards or downwards stretching and shifting of the V_0 curve.

For fixed h, changes in E_0 are represented by vertical translation of the E_0 line.

Suppose a vehicle is approaching a body with zero or positive energy and braking is desired. The motion is described by the one-dimensional potential and the behavior of the vehicle near perigee can be determined.

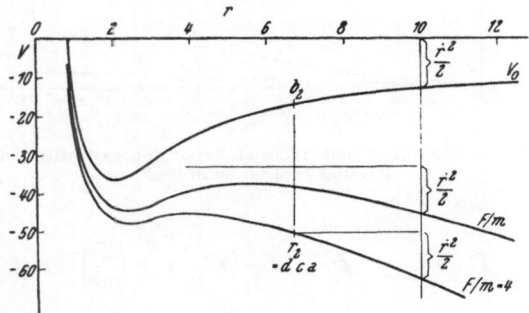

Fig. 4a. Approach to planet using continuous radial braking

Let the initial energy by E_1. If the thrust is turned on at $r = r_1$, then the motion proceeds along the V curve until it strikes the potential barrier at $r = r_2$ and is deflected back outwards again. A distance of closest approach can be found which, in many cases, is not critically dependent upon h.

For example, a vehicle approaching earth with energy $E_0 = 0$ and angular momentum h_0 is shown in Fig. 4. If $(F/m) = 4$, and the continuous radial thrust is turned on near $r = 20,000$ st. mi. (10^8 feet), then the vehicle will be repelled from the potential barrier at $r = 5.75 \times 10^7$ feet or 10,900 st. mi., giving a distance of closest approach on the order of 6,900 st. mi. If no braking were applied the vehicle would collide with the earth.

If the thrust were shut down at $r = r_2$, then the motion would proceed from the point b_2 along the V_0 curve, and a bounded orbit is assured. It is possible to study the effect of cycling inward and outward thrust and periods of coasting on the approach paths by this method.

There are values of r_0 such that, for certain values of E_0 and (F/m), the rocket passes over the peak of the potential barrier and goes down into the well. This is a case of particular interest since this motion will involve a passage close to the planet.

It is seen from Fig. 4 that the r_p for outward thrust is less than r_p for coasting for the same (E_0, h_0). Thus, there is some benefit from the use of continuous thrust. If a close approach is desired then the effect of an error in the initial angular momentum must be considered.

The distance of closest approach has been plotted against the radius at which the thrust is initially turned on (i.e., r_1) for two values of (F/m). The scales are loga-

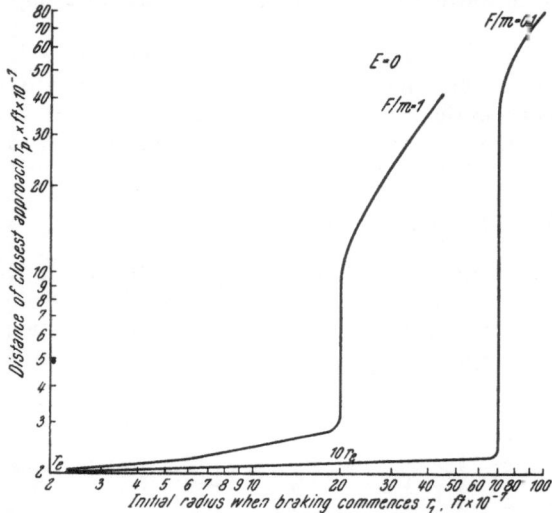

Fig. 4b. Variation in distance of closest approach with point of initial application of braking

rithmic. The sharp drop in distance of closest approach corresponding to the condition of the rocket passing over the outer peak of the well is clearly seen (Fig. 4b).

The relation between Δh and Δr_p for continuous radial thrust is derived below.

$$V = -\frac{\mu}{r} - \left(\frac{F}{m}\right) r + \frac{h^2}{2 r^2}$$

$$E = \frac{1}{2} (\dot{r})^2 + V$$

$$E_0 = -\frac{\mu}{r_p} - \left(\frac{F}{m}\right) r_p + \frac{h^2}{2 r_p}.$$

Suppose that there is an error Δh in h, and we also require $\Delta E = 0$. This corresponds to an error in measuring the angle between the velocity and radius vectors.

Then

$$0 = \Delta E = \mu \frac{\Delta r_p}{r_p^2} + \left(\frac{F}{m}\right) \Delta r_p + \frac{2 h \Delta h}{2 r_p^2} - \frac{2 h^2}{2 r_p^3} \Delta r_p$$

or

$$\Delta r_p = h_0 r_{p0} \left[\frac{1}{-\mu r_p - (F/m) r_p^3 + h_0^2} \right] \Delta h. \tag{4}$$

For example, if $h_0 = 7.2 \times 10^{11}$ ft²/sec, and $r_{p0} = 2.10 \times 10^7$ ft, $E_0 = 0$, $(F/m = 1)$, then

$$\frac{\Delta r_p}{\Delta h} = 7.05 \times 10^{-5} \text{ ft/ft}^2/\text{sec}$$

$$(r_0 = 40,000 \text{ n. mi.}, \quad \Delta \chi = .057 \text{ deg.}).$$

Thus, an error in measuring h_0 on the order of 2.1×10^9 ft²/sec gives an error in r_p on the order of 147,000 ft.

Let us suppose, now, that the initial conditions are known with great precision. How accurately must the thrust vector be aligned with the radius vector to keep the error within specified bounds? Alternatively, we may ask what effect an alignment error ε will have on the distance of closest approach.

The change in angular momentum due to the error angle is given by the expression:

$$\Delta h = \int_{t_0}^{t} (F/m) \sin \varepsilon \cdot r \, dt .$$

If F/m and ε are assumed constant and ε is small, we may write

$$\Delta h = (F/m) \varepsilon \int_{t_0}^{t} r \, dt$$

or

$$\Delta h = (F/m) \varepsilon \int_{r_0}^{r} \frac{r \, dr}{\sqrt{2 \left[E_0 - \mu/r - (F/m) \, r - \dfrac{h_0{}^2}{2 \, r^2} \right]}} . \tag{5}$$

For a typical approach similar to the one examined for the previous case, $h_0 = 7.2 \times 10^{11}$ ft²/sec, $E_0 = 0 \cdot r_0 = 40{,}000$ n.mi.

For an angular error $\varepsilon = .01$ deg. the change in h is on the order of 10^9 ft²/sec, and the change in the distance of closest approach to earth is on the order of 14.0 st. mi.

II. Impulsive Thrust Corrections

The subject of impulsive thrust corrections and the associated error analysis has been much discussed in the past [3, 4, 5]. The objective here is to briefly examine the effects of errors in measurements and control application on the conditions at perigee, in particular, the changes in the distance of closest approach, so that a comparison with a continuous thrusting system can be made.

The energy and angular momentum equations for coasting or ballistic flight are given below.

$$E/m = \frac{1}{2} V_0{}^2 - \frac{\gamma M}{r_0} = \frac{1}{2} V_{P0}{}^2 - \frac{\gamma M}{r_{P0}} \tag{6}$$

$$h_0 = r_0 V_0 \sin \chi_0 = V_{P0} \, r_{P0} . \tag{7}$$

These equations can be differentiated and rearranged. Thus, the sensitivities of the perigee conditions are related, to the first order, to changes in the angular momentum and energy. These changes in h and E are then related to errors or controlled changes in the radius vector and velocity vector. Some of the results are shown below.

Assume

$$\frac{V_P{}^2}{2} >> E/m$$

$$\frac{\Delta r_P}{\Delta \chi} = 2 \, r_{P0} \cot \chi_0; \qquad \frac{\Delta r_P}{r_P} = \frac{2 \, \Delta \chi}{\tan \chi_0} \tag{8}$$

$$\frac{\Delta V_P}{\Delta \chi} = \frac{2 \, \gamma M}{h_0} \cot \chi_0 \tag{9}$$

$$\frac{\Delta r_P}{\Delta V_0} = 2 \left[1 - \left(\frac{V_0}{V_{P0}} \right)^2 \right] \frac{r_{P0}}{V_0} \tag{10}$$

$$\frac{\Delta r_P}{r_{P0}} = \frac{\Delta h}{h_0} - \frac{\Delta V_P}{V_{P0}} . \tag{11}$$

These sensitivities are plotted for some typical cases in the next set of figures.

Notice that errors in measuring the angle γ between the velocity vector and radius vector, on the order of .057 deg or .001 rad. measured at $r = 40,000$ nautical miles for a vehicle with $E_0 = 0$ produce changes in h on the order of 2.1×10^9 ft²/sec, causing an error in the predicted distance of closest approach of 138,000 ft. Thus, relatively small errors in the quantities determined en route produce intolerable uncertainties in the critical factors for a close approach. This means that the position and velocity measurements taken en route require a high degree of accuracy if the predicted value of the distance of closest approach

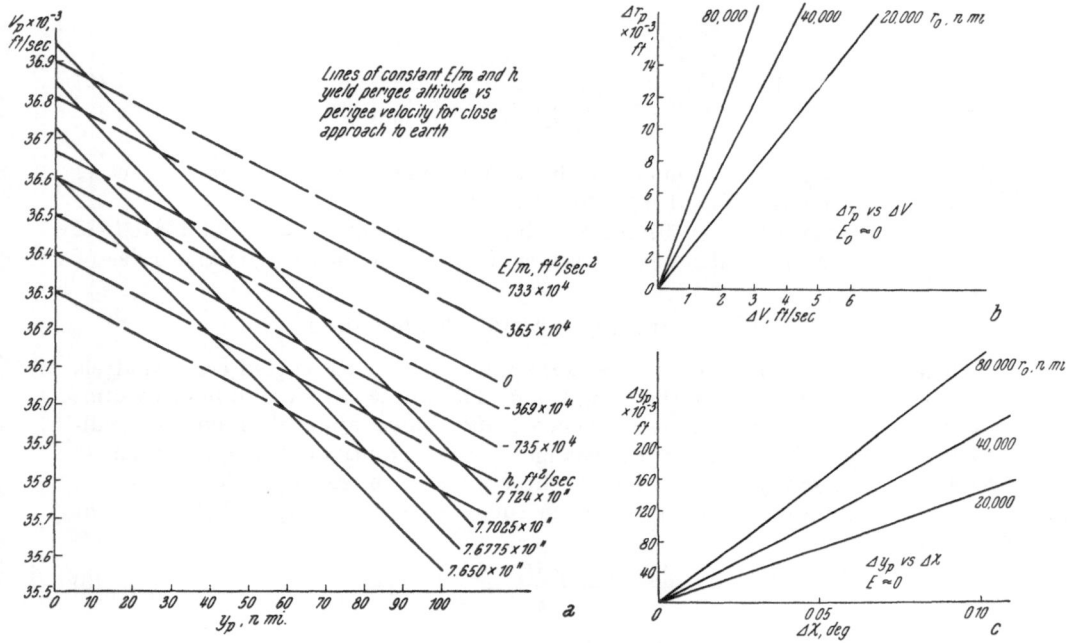

Fig. 5. Parameter sensitivities for ballistic paths

is to be trusted and used during the approach maneuver. It seems clear that approach paths that are aimed close to the body should be avoided. BAKER's suggestion [3] that the vehicle aim for a high orbit and let down later has considerable merit.

The fact that the perigee conditions are very sensitive to the angular momentum does not always lead to gloomy conclusions. This behavior can be used to advantage in applying a control as it makes the control very powerful. Suppose one knew that a change in angular momentum was necessary to avoid a collision with a planet, and suppose also that the amount of the required change was known accurately. It is a simple computation to determine the ΔV necessary, particularly if the correction is to be applied by firing perpendicular to the radius vector. The question that comes to mind now is how accurately does the thrust vector have to be aligned with the commanded direction in order to keep the error in the change in perigee altitude below some specified value. It is again, a simple task to show that the angular momentum, to the first order, is insensitive to small errors in the thrust direction for firing perpendicular to the radius vector.

In general, for the case of the ballistic approach with impulsive thrust corrections we have, in a sense, a conflict between our ability to predict the distance of
closest approach and our ability to control and change the distance of closest
approach if our path is not satisfactory. The problem of finding the optimum
locations for application of control under these conditions, as a function of the
instrumentation characteristics, deserves and is receiving much study.

Results and Conclusions

A study of the motion of an orbiting vehicle subjected to continuous radial
thrust has been performed using one-dimensional potentials. The application of
this method depends on the fact that the angular momentum remains constant
if the thrust is always radial. It is shown that the shape of the potential curve is
changed by the addition of radial thrust. Several possibilities exist for the motion
depending upon the initial conditions and the applied radial acceleration, (F/m).
For outwardly directed thrust it is shown that unbounded motion will occur for
certain values of (F/m) and certain initial conditions. For initially circular orbits
this is shown to correspond to limiting conditions obtained in previous papers
[7, 9]. It is also shown that escape from a closed orbit is possible, if the direction
of the radial thrust is changed from inwards to outwards, or vice versa, at the
proper places, even if (F/m) is less than the value required for escape determined by
considerations of outward thrust alone. It is seen that for close approaches the
errors in the predicted distance of closest approach due to errors in the en route
determination of the angular momentum are of the same magnitude for both the
impulsive thrusting and continuous thrusting vehicle.

The use of radial braking produces an increase in the perigee altitude compared
with the perigee of the coasting path. This generally is a desirable result.

Due to the extreme sensitivity of the predicted perigee distance, for close
approaches, the en route measurements of the navigation and guidance equipment
for both the continuous radial braking system and the impulsive thrust correction system, will require the same high degree of accuracy. In view of these results
there does not seem to be any overwhelming advantage of one system over the
other for close approaches with the exception of the increase in r_p due to the radial
braking.

For moderately distant approaches the radial braking can provide a means of
insuring an inner bound for the distance of closest approach by adjusting the
potential shape if certain initial conditions are satisfied.

Appendix I

Calculation of $(F/m)_{critical}$ for the case of initially circular orbits

The rocket thrust is turned on at $r = r_1$. The vehicle was originally in a circular
orbit. The new total energy corresponds to the line E_2.

$$V = -\left(\frac{F}{m}\right)r - \frac{\mu}{r} + \frac{h^2}{2r^2}.$$

For initially circular orbits $h^2 = V^2 \cdot r_1 = \mu r_1$. The values of r that we seek are
given by the equation.

$$-\left(\frac{F}{m}\right)r - \frac{\mu}{r} + \frac{\mu r_1}{2r^2} = -\left(\frac{F}{m}\right)r_1 - \frac{\mu}{r_1} + \frac{\mu}{2r_1^2}.$$

This may be put in the form:

$$\frac{\mu\, r_1}{2\, r^2} - \frac{\mu}{r} + \frac{\mu}{2\, r_1} = (F/m)\, (r - r_1).$$

Clearly $r = r_1$ satisfies this equation.

The equation may be written again, yielding:

$$\frac{\mu}{2\, r^2 r_1}\, (r_1{}^2 - 2\, r_1\, r + r^2) = (F/m)\, (r - r_1)$$

or

$$\frac{\mu}{2\, r_1\, r^2}\, (r - r_1) = (F/m).$$

Thus

$$(F/m)\,^2 - \frac{\mu}{2\, r_1}\, (r - r_1) = 0.$$

Solving for r gives

$$r = \frac{\dfrac{\mu}{2\, r_1} \pm \sqrt{\dfrac{\mu^2}{4\, r_1{}^2} - 2\mu\, (F/m)}}{2\, (F/m)}.$$

$(F/m)_{critical}$ occurs when the quantity $\left(\dfrac{\mu^2}{4\, r_1{}^2} - 2\mu\, (F/m)\right)$ vanishes, corresponding to a double root for r.

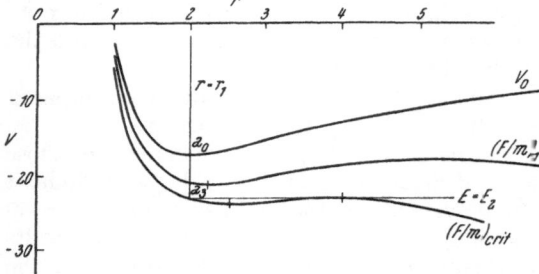

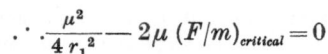

Fig. 6. Definition of $(F/m)_{critical}$

$$\therefore \; \frac{\mu^2}{4\, r_1{}^2} - 2\mu\, (F/m)_{critical} = 0$$

$$(F/m)_{critical} = \frac{\mu}{8\, r_1{}^2}$$

but

$$\lambda = (F/m)\, \frac{1}{g}\left(\frac{r}{R}\right)^2 = (F/m)\, \frac{r^2}{\mu}$$

$$\therefore \; \lambda_{max} = (F/m)_{critical} \cdot \frac{r_1{}^2}{\mu} = \frac{1}{8}.$$

If the orbit is not initially circular and we fire the rocket at the perigee, a solution for $(F/m)_{critical}$ can be obtained.

$$- (F/m)\, r - \frac{\mu}{r} + \frac{h^2}{2\, r^2} = - (F/m)\, r_1 - \frac{\mu}{r_1} + \frac{h^2}{2\, r_1{}^2}.$$

Again, $r = r_1$ must be a solution.

$$\frac{h^2}{2\, r^2} - \frac{\mu}{r} + \frac{\mu}{r_1} - \frac{h^2}{2\, r_1{}^2} = (F/m)\, (r - r_1)$$

or

$$- \frac{h^2}{2\, r^2 r_1{}^2}\left[r + r_1 - \frac{2\, \mu}{h^2}\, r_1\, r\right] = (F/m).$$

This equation can be solved for r. As in the previous case, the condition for a double root gives the value of $(F/m)_{critical}$. From this equation, one can show that

$$(F/m)_{critical} = \frac{2\, r_1}{h^2}\left(\frac{h^2}{2\, r_1{}^2} - \frac{\mu}{r_1}\right)^2.$$

Appendix II

The value of $(F/m)_{max}$ is found by obtaining that value of (F/m) causing the potential curve to have a point of inflection with zero slope. This corresponds to the complete disappearance of the potential wall.

$$V = -\frac{F}{m}r - \frac{\mu}{r} + \frac{h^2}{2\,r^2}$$

$$V' = -\frac{F}{m} + \frac{\mu}{r^2} - \frac{h^2}{r^3}$$

$$V'' = -\frac{2\,\mu}{r^3} + \frac{3\,h^2}{r^4}\,.$$

The condition that $V'' = V' = 0$ yields $r_c = \frac{3}{2}\frac{h^2}{\mu}\,.$

Note that the location of the inflection point is independent of F/m.

$$(F/m)_{max} = \frac{\mu}{r_c^2} - \frac{h^2}{r_c^3}$$

$$(F/m)_{max} = \frac{4}{27}\frac{\mu^3}{h^4}\,.$$

Bibliography

1. T. NONWEILER, Descent from Satellite Orbits Using Aerodynamic Braking. J. Brit. Interplan. Soc. **10**, 258 (1951).
2. T. NONWEILER, Problems of Missiles Entering the Atmosphere. J. Brit. Interplan. Soc. **10**, 26 (1951).
3. R. M. L. BAKER, Navigational Reqs. for the Return Trip from a Space Voyage. Navigation **6**, No. 3, Autumn (1958).
4. K. A. EHRICKE, Error Analysis of Keplerian Flights. Navigation **6**, No. 1, 5, Spring (1958).
5. G. XENAKIS, Some Flight Control Problems of a Circumnavigating Lunar Vehicle. WADC TN 58-82.
6. D. F. LAWDEN, Dynamic Problems of Interplanetary Flight. Aeronaut. Quart. **6**, Part 3, 165, August (1955).
7. H. S. TSIEN, Take-off from a Satellite Orbit. Jet Propulsion **23**, 233 (1953).
8. D. J. BENNEY, Escape from a Circular Orbit Using Tangential Thrust. Jet Propulsion **28**, 167 (1958).
9. A. DOBROWOLSKI, Satellite Orbit Perturbations Under a Continuous Radia Thrust of Small Magnitude. Jet Propulsion **28**, 687 (1958).
10. J. COPELAND, Interplanetary Trajectories Under Low Thrust Radial Acceleration. J. Amer. Rocket Soc. **29**, 267 (1959).
11. H. GOLDSTEIN, Classical Mechanics, pp. 58—80. Reading, Mass.: Addison-Wesley, 1951.

Additional References

1. L. WALTERS, Lunar Trajectory Mechanics. Navigation **6**, 51 (1958).
2. D. F. LAWDEN, Minimal Trajectories. J. Brit. Interplan. Soc. **9**, 179 (1950).
3. S. HERRICK, Space Rocket Trajectories. J. Brit. Interplan. Soc. **9**, 235 (1950).
4. C. C. KIESS and K. LASSOVZSKY, The Known Physical Characteristics of the Moon and the Planets. Georgetown College Observatory. WADC TR 58-41.
5. C. R. FAULDERS, Low Thrust Steering Programs for Minimum Time Transfer Between Planetary Orbits. SAE Preprint 88A.

6. E. Rodriguez, A Method of Determining Steering Programs for Low Thrust Interplanetary Vehicles. ARS Preprint 645-58.

7. E. Stuhlinger, Flight Path of an Ion Propelled Spaceship. Jet Propulsion 27, 410 (1957).

8. G. M. Low, Nearly Circular Transfer Trajectories for Descending Satellites. NASA Rep. 3, 1959.

9. D. Chapman, An Approximate Analytical Method for Studying Re-entry into Planetary Atmospheres. NASA TN 4276, 1958.

10. R. M. L. Baker, Jr., Practical Limitations on Orbit Determination. IAS Preprint 842, July 1958.

Secular Variation in the Inclination of the Orbit of Earth Satellite (1957 β) and Air Drag

By

L. N. Rowell[1] and M. C. Smith[1]

(Received June 12, 1959)

Abstract — Zusammenfassung — Résumé

Secular Variation in the Inclination of the Orbit of Earth Satellite (1957 β) and Air Drag. Recent orbital studies made by the Smithsonian Astrophysical Observatory of the artificial earth satellite 1957 β indicate a decrease in the inclination of the orbit of the order of 10^{-3} degrees per day [5].

One plausible explanation of this phenomenon is the component of drag acceleration normal to the orbital plane arising from the rotation of the earth's atmosphere in the same sense as the satellite. This component causes the inclination of the orbital plane to decrease at a rate which is found from an approximate relation to be proportional to $\cos^2 \omega \sin i$, where ω is the argument of perigee and i the angle of inclination of the orbital plane. The magnitude of the decrease in inclination as given by the approximate relation for satellite 1957 β is 4×10^{-4} degrees per day.

The third and fourth harmonics in the gravitational potential of the earth also produce a long period change in the inclination, but in each case it is considerably smaller than that of the rotating atmosphere.

Säkulare Variation der Bahnneigung des Erdsatelliten 1957 β infolge des Luftwiderstandes. Jüngst ausgeführte Bahnstudien an dem künstlichen Erdsatelliten 1957 β, die vom Smithsonian Astrophysical Observatory vorgenommen wurden, zeigen eine Abnahme der Bahnneigung in der Größenordnung von 10^{-3} Grad je Tag [5].

Eine einleuchtende Erklärung dieses Phänomens ist durch die zur Bahnebene normale Komponente der Widerstandsbeschleunigung gegeben, die von der im selben Sinn wie die Satellitenbewegung erfolgenden Rotation der Erdatmosphäre herrührt. Diese Komponente bewirkt, daß die Neigung der Bahnebene mit einer Geschwindigkeit abnimmt, die, wie sich aus einer angenäherten Rechnung ergibt, proportional zu $\cos^2 \omega \sin i$ ist. Dabei ist ω das Argument des Perigäums und i der Neigungswinkel der Bahnebene. Die Größe der Abnahme der Neigung ergibt sich durch eine angenäherte Rechnung für den Satelliten 1957 β zu 4×10^{-4} je Tag.

Die dritte und vierte harmonische Konstante im Gravitationspotential der Erde bewirken auch eine langperiodische Änderung der Neigung, die aber in jedem Fall beträchtlich kleiner als die der rotierenden Atmosphäre ist.

Influence de la traînée sur les variations séculaires de l'inclinaison de l'orbite du satellite 1957 β. Des études récentes faites par le Smithsonian Astrophysical Observatory sur le satellite artificiel 1957 β révèlent une diminution dans l'inclinaison de l'orbite de l'ordre de 10^{-3} degrés par jour [5].

Une explication plausible réside dans une composante de l'accélération, normale au plan de l'orbite, due à la traînée aérodynamique; l'atmosphère étant entraînée dans le même sens de rotation que le satellite. Un calcul approximatif montre que cette

---

[1] The RAND Corporation, 1700 Main Street, Santa Monica, California, U.S.A.

composante provoque une modification d'inclinaison proportionnelle à $\cos^2 \omega \sin i$, où ω est l'argument du périgée et i l'angle d'inclinaison du plan orbital. Pour 1957 β ce calcul fournit la valeur 4×10^{-4} degrés par jour.

Les troisième et quatrième harmoniques du géopotentiel sont aussi responsables d'une modification d'inclinaison à longue période mais d'amplitude considérablement inférieure.

In addition to the well known secular variations in the orbital node and perigee, arising from the first order effect of oblateness of the earth, there are additional secular variations in the orbital elements due to air drag. The component of drag acceleration normal to the orbit plane produces a secular variation in the inclination of the orbital plane. This normal component to the orbital plane is due to the rotation of the earth's atmosphere with the earth. If the earth is assumed to be spherical, the rate of change in the inclination may be expressed as [1]:

$$\frac{di}{dt} = - \frac{C_D A \varrho (r) r \cos (\nu + \omega) V_{rel} V_{rel_n}}{m n a^2 \sqrt{1 - e^2}} \tag{1}$$

where the components of relative velocity of the satellite and earth atmospheres are:

$$V_{rel} = \{\dot{r}^2 + [r\dot{v} - \Omega r \cos i]^2 + (\Omega r)^2 \sin^2 i \cos^2 u\}^{1/2}$$
$$V_{rel_n} = \Omega r \sin i \cos u . \tag{2}$$

In Eqs. (1) and (2), $\varrho(r)$ is the atmospheric density at geocentric distance; r, m, and A the mass and effective cross-sectional area of the satellite; C_D, the drag coefficient; i, e, and a, respectively, the inclination, eccentricity, and major semi-axis of the orbit; n, ω, and v the orbital period, argument of perigee, and true anomaly, respectively; and Ω the angular velocity of the earth.

Assume in Eq. (1) that the important contribution of drag to the integral comes only from the portion of the orbit near perigee. Then, for the density at perigee $\varrho(r_p)$, the density at altitudes above perigee is given by

$$\varrho = \varrho (r_p) \varepsilon^{-kz} \tag{3}$$

where z is the altitude above perigee and k is the logarithmic gradient of density near perigee,

$$k = - \frac{d}{dz} \ln \varrho . \tag{4}$$

With these assumptions concerning density, it is possible to simplify Eq. (1) and arrive at an expression for the change in inclination of each revolution as a function of eccentric anomaly E,

$$\Delta i = - \frac{\Omega C_D A \varrho (r_p) \varepsilon^{-c} a \sin i}{n m \sqrt{1 - e^2}} \int\limits_0^{2\pi} \varepsilon^{c \cos E} (1 - e^2 \cos^2 E)^{\frac{1}{2}} \times$$
$$\times [(\cos E - e)^2 \cos^2 \omega + (1 - e^2) \sin^2 E \sin^2 \omega -$$
$$- 2 \sqrt{1 - e^2} (\cos E - e) \sin \omega \cos \omega \sin E] dE \tag{5}$$

where $c = k a e$.

Integrating Eq. (5) gives,

$$\Delta i \simeq -\frac{2\pi\,\Omega\,C_D A\,\varrho\,(r_p)\,\varepsilon^{-c}\,a\sin i}{n\,m\,\sqrt{1-e^2}}\,\{I_0\,(c) + \cos 2\omega\,I_2\,(c) - \\ - 4e\cos^2\omega\,I_1\,(c) + 0\,(e^2) + \ldots\},\tag{6}$$

where I_0, I_1, I_2 are BESSEL functions of purely imaginary argument ([2], p. 584).

The uncertainties is perigee density $\varrho(r_p)$, in drag area to mass ratio $C_D A/m$ in Eq. (6) are avoided by substituting Eq. (3), in [3], for these quantities. It follows that the change in inclination is related to the change in orbital period Δp by the equation,

$$\Delta i \simeq -\frac{\Omega}{6\pi}\,\Delta p\,\frac{\sin i}{\sqrt{1-e^2}}\,\left\{\frac{I_0 + I_2\cos 2\omega - 4e\,I_1\cos^2\omega + (0)\,e^2 + \ldots}{I_0 + 2e\,I_1 + (0)\,e^2 + \ldots}\right\}.\tag{7}$$

Using the asymptote expansion for the BESSEL integrals

$$I_n\,(c) = \frac{\varepsilon^c}{\sqrt{2\pi c}}\,\left\{1 + \frac{1-4n^2}{1!\,8c} + \frac{(1-4n^2)\,(9-4n^2)}{2!\,(8c)^2} + \ldots\right\},$$

the approximate Eq. (7) was evaluated for orbital elements of 1 December 1957 of the Satellite 1957β [4]. This calculation gave for the change in inclination per day 3.7×10^{-4} degrees. As a check, a numerical solution was made on the RAND Johnniac digital computer, using perturbation techniques. Including the first order oblateness, as well as air drag, this latter solution gave for the change in inclination 4.2×10^{-4} degrees per day.

Recently, a digital computer study performed at the Smithsonian Institution Astrophysical Observatory indicated such an effect [5]. Their analysis showed a secular change in inclination in 1957β orbit of the order 1×10^{-3} degrees per day.

Conclusions

The approximate Eq. (7) appears to be sufficiently accurate to predict the magnitude of the secular motion of the orbital inclination. If indeed the motion of the atmosphere of the earth is responsible, the motion of the inclination should vary as the square of the cosine of the argument of perigee, since Eq. (7) reduces to,

$$\Delta i \simeq -\frac{\Omega}{3\pi}\,\Delta p\,\sin i\cos^2\omega\,\{1-4e + \ldots\}$$

where, ignoring errors of the order e, the approximation $I_0 \cong I_1 \cong I_2$ is possible. Further experimental results are needed to reveal this functional relation.

References

1. D. Ye. Okhotsimski, T. M. Eneyev and G. P. Taratynova, Uspekhi Fiz. Nauk 63, 1a, 33 (1957).
2. H. B. Jeffreys and B. S. Jeffreys, Methods of Mathematical Physics. Cambridge, Massachusetts, 1956.
3. T. E. Sterne, Special Report No. 11, IGY Project No. 30.10, Smithsonian Astrophysical Observatory, Cambridge, Massachusetts, March 31, 1958.
4. D. G. King-Hele and R. H. Merson, J. Brit. Interplan. Soc. 16, 446 (1958).
5. Conference on Contemporary Geodesy, Harvard College Observatory, Smithsonian Astrophysical Observatory, Cambridge, Massachusetts, December 1, 2, 1958.

Sur deux types de propulseurs électriques

Par

Juliusz Ulam[1], SFA

(Avec 9 Figures)

(Reçu le 16 juillet 1959)

Résumé — Zusammenfassung — Abstract

Sur deux types de propulseurs électriques. Cet exposé décrit les principes de fonctionnement et les propriétés de deux types nouveaux de propulseurs électriques.

1. Le premier type de moteur utilise le phénomène du "jet électrodique", qui apparait, dans certaines conditions, dans les arcs électriques et qui consiste en l'évaporation du matériau de l'électrode et l'éjection de la vapeur à une vitesse élevée. Le corps qu'on soumet à l'évaporation peut être soit le solide constituant l'électrode, soit — de préférence — un liquide injecté dans la tache électrodique. Le dispositif travaille comme un propulseur électrothermique à plasma, la vitesse d'écoulement — de l'ordre de quelques km/sec — étant assurée par la détente de vapeur surchauffée; cependant, avec une construction appropriée, l'appareil est aussi capable de produire un écoulement électrostatique, à débit de masse très faible et avec une vitesse d'écoulement élevée, de l'ordre de quelques dizaines de km/sec.

2. Le principe employé dans l'autre type de propulseur est celui de "l'explosion électrique", c. à d. de la volatilisation d'un conducteur par le courant de haute intensité. Pour l'utilisation en propulseur, l'explosion d'un fil est remplacée par l'évaporation d'un liquide conducteur, la vapeur surchauffée par la décharge étant expulsée à haute vitesse. L'appareil travaille en principe en régime intermittent comme un pulsoréacteur; par un choix convenable des paramètres électriques et de ceux du système d'alimentation, on parvient à remplacer ce régime par un écoulement continu. Le jet à haute vitesse, produit par la détente, peut être soumis à une accélération supplémentaire par l'action d'un champ magnétique transversal.

Über zwei neue Arten elektrischer Strahlantriebsgeräte. Es werden die Wirkungsweisen und Eigenschaften zweier neuer Arten elektrischer Strahlantriebsgeräte behandelt.

1. In dem ersten der beschriebenen Apparate ist die Erscheinung der sogenannten Elektrodenströmung ausgenutzt, die unter gewissen Umständen in den elektrischen Bögen auftritt und die auf Verdampfung des Elektrodenkörpers und Ausströmung des erzeugten Dampfes mit hoher Geschwindigkeit beruht. In dem entsprechenden Strahlantriebsgerät wird entweder der (feste) Elektrodenkörper oder — vorzugsweise — eine Flüssigkeit als Arbeitsmasse angewandt, deren Verdampfung in dem Elektrodenfleck stattfindet. Seinem Prinzip nach ist das Gerät ein elektrothermisches Triebwerk, in dem die Ausströmungsgeschwindigkeit — von der Größenordnung einiger km/sek — durch die Ausdehnung eines überhitzten Dampfes erreicht wird. Bei geeigneter Anpassung kann jedoch dasselbe Gerät auch eine elektrostatische Ausströmung erzeugen, die durch einen kleinen Massendurchsatz, aber eine hohe Ausströmungsgeschwindigkeit (einige Zehner von km/sec) gekennzeichnet wird.

---

[1] 20 Rue de Varenne, Paris 7e, France.

2. In dem zweiten Apparat fand der Vorgang der sogenannten elektrischen Explosion Anwendung, d. h. der explosiven Verdampfung eines elektrischen Leiters durch einen Hochstromstoß. Zur Strömungserzeugung wird als Arbeitsmasse eine leitende Flüssigkeit verwendet, auf deren Verdampfung Überhitzen und Ausdehnung folgt. Das Gerät liefert im Prinzip eine pulsierende Ausströmung, die jedoch durch geeignete Wahl der Parameter des elektrischen Kreises und derjenigen des Massenförderungssystems in eine kontinuierliche umgewandelt werden kann. Die Gesamtanordnung des Apparates ermöglicht es, magnetische Felder zur zusätzlichen Beschleunigung der Ausströmung anzuwenden.

Two New Types of Electric Propulsion Devices. The object of this paper is to present the principles and properties of two new types of electric propulsion devices.

1. The phenomenon utilized in the first type of motor is that of the "electrode jet", which appears in certain circumstances in the electric arcs and consists in evaporation of the electrode material accompanied by a high velocity output of the vapour produced. The working body in the described motor type may be either the solid constituting the electrode or a liquid supplied to, and evaporated in the electrode spot. The device is, in virtue of its principle, an electrothermic plasma motor, the output velocity—of the order of several km/sec—being due to the expansion of a superheated vapour; however, if suitably adapted, it can produce also an electrostatic output at a very low mass flow rate and a high velocity of the order of tens of km/sec.

2. Another type of an electric jet motor is based on the "electric explosion", i.e. the evaporation of a conductor by a high current impulse. In the motor the explosion of a wire is replaced by evaporation of a conducting liquid, the superheated vapour being ejected at a high velocity. The process in the device is intermittant, like that in a pulse-jet engine; by a suitable choice of the parameters of the electric circuit and those of the supply system a continuous output may be realized. The high velocity output of the superheated vapour may be subjected to an additional acceleration by the action of a transversal magnetic field.

Notations

A. Propulseur à jet électrodique
Lettres

V	tension	m	débit massique
J	intensité du courant	λ	chaleur d'évaporation
W	puissance	c	chaleur spécifique
φ	travail d'extraction de	T	température
	l'électron	δ	densité de vapeur
S	surface de l'anode	v	vitesse d'écoulement

Indices

X_0	valeur initiale de X	X_a	X rapportée à l'anode
X_f	valeur finale de X	X_d	X rapportée à la colonne de l'arc
X_T	valeur de X rapportée à la température T	X_c	X rapportée à la cathode

B. Propulseur à explosion électrique
Lettres

E	intensité de champ électrique	X	distance
γ	densité du liquide	T	température
c	chaleur spécifique du liquide	w	vitesse du liquide
ϱ	résistivité du liquide	t	temps
$\dot{m}$	débit massique	p, P	pression

Indices

T_0	valeur initiale de T	T_e	température d'ébullition du liquide

Introduction

La propulsion électrique par réaction est basée sur la transformation de l'énergie électrique en énergie cinétique d'une masse éjectable (propulsif); il existe, en général, deux méthodes pour effectuer cette transformation.

La première méthode consiste à appliquer un champ électrique ou magnétique pour accélerer le propulsif. Deux genres de dispositifs ont été étudiés, notamment les propulseurs ioniques, qui accélèrent les ions par l'action d'un champ électrique [30—36], et les propulseurs magnétiques, dans lesquels le plasma gazeux acquiert une vitesse élevée soit par l'action d'un champ magnétique, soit sous l'action des forces exercées par des champs électrique et magnétique croisés [25—29]. Les propulseurs de ce genre sont caractérisés par de grandes vitesses d'écoulement (de l'ordre de plusieurs dizaines ou centaines de km/sec), donc par un rapport $\frac{\text{puissance}}{\text{poussée}}$ très élevé; le rapport $\frac{\text{débit massique}}{\text{poussée}}$ reste, par contre, très petit (grande impulsion spécifique); le domaine d'application des moteurs ioniques et magnétiques serait, sans doute, surtout la propulsion orbitale, c'est à dire l'accélération d'un véhicule spatial sur une orbite d'équilibre par une poussée faible mais prolongée, effectuée à une consommation réduite du propulsif.

La deuxième méthode, électrothermique, emploie l'échauffement du propulsif par des moyens électriques, jusqu'à une température élevée; la transformation de l'énergie thermique en énergie cinétique est réalisée ensuite par le procédé habituel de détente. Les vitesses d'écoulement atteintes par cette méthode sont plus basses, de l'ordre de quelques km/sec, avec un chiffre de 15—20 km/sec comme limite prévisible dans l'état actuel des études. L'impulsion spécifique est ici plus petite, mais le rapport $\frac{\text{puissance}}{\text{poussée}}$ plus favorable; les dispositifs de ce genre pourraient être destinés à fournir des poussées importantes pendant des périodes assez courtes.

La méthode que l'on envisage de prime abord pour réaliser l'échauffement électrique du propulsif est offerte par les décharges dans le gaz; les recherches dans ce domaine ont été basées surtout sur l'emploi du principe de l'arc stabilisé [37—48], qui permet d'obtenir aisément des températures élevées. Il n'est pas sans certains inconvénients, les plus importants sont l'usure des électrodes et l'encombrement du système d'alimentation en propulsif.

Le propulseur à arc stabilisé n'est qu'un des systèmes possibles de la propulsion électrothermique. Nous allons considérer ici deux autres dispositifs de ce genre, dont l'étude théorique et expérimentale a été poursuivie depuis longtemps et qui ont montré des propriétés intéressantes; en particulier, ils offrent la possibilité d'appliquer des champs électriques et magnétiques, soit pour remplacer, soit pour modifier l'écoulement thermique. Cette propriété parait importante, puisqu'elle permet d'envisager la construction de moteurs capables — tout en conservant un certain niveau de puissance du générateur — de travailler tantôt comme un propulseur de lancement (à grand débit massique et haute poussée, mais pour un temps assez court), tantôt comme un propulseur orbital (à petit débit et faible poussée, mais pendant un temps assez long).

A. Propulseur à jet électrodique

I. Principe

Le phénomène du jet électrodique apparaît dans certaines conditions dans un arc électrique, il a fait depuis longtemps l'objet de nombreuses études [1—24]. Il consiste en l'évaporation du matériau de l'électrode et l'éjection de la vapeur surchauffée à une vitesse élevée; c'est sur l'anode de l'arc à courant continu qu'on observe le plus souvent la formation du jet, on a cependant étudié aussi des jets cathodiques et les jets sur les électrodes d'arcs à courant alternatif [12—19].

Le jet — qui possède une forme bien définie — est créé dans la tache électrodique et donne naissance à une force de réaction agissant sur l'électrode [1—3]; les caractéristiques en sont déterminées, d'une part, par les paramètres du circuit électrique de l'arc, d'autre part par les propriétés physiques du matériau de l'électrode.

La construction du propulseur basé sur ce phénomène doit, évidemment, réaliser les conditions les plus favorables à la formation du jet et à l'obtention d'un rendement suffisant (c.à.d. à la conversion d'une fraction aussi grande que possible de l'énergie dissipée dans l'arc en énergie cinétique du jet).

Nous allons considérer ici brièvement les conditions de la formation du jet sur l'anode de l'arc à courant continu et les propriétés du propulseur utilisant cet effet.

II. Mécanisme de la formation du jet anodique

Les recherches faites sur les propriétés du jet anodique ont permis d'établir une théorie qualitative de ce phénomène, exposée dans les articles de FINKELN-BURG, PODSZUS, COBINE etc. et complétée récemment par MARQUIS, MEAD, KORMAN et SHEER [23]; elle a d'ailleurs été appliquée aussi à la formation du jet

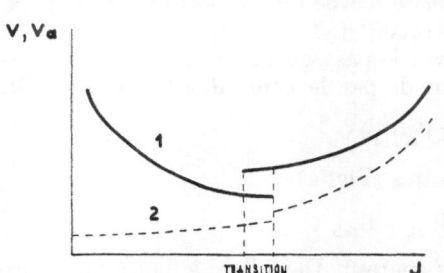

Fig. 1. Caractéristique de l'arc à courant continu. *1* tension totale, V; *2* chute anodique du potentiel, V_a

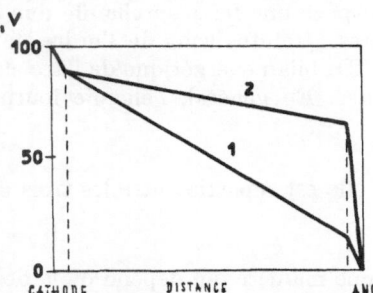

Fig. 2. Répartition de la tension dans l'arc à courant continu. *1* basse puissance; *2* haute puissance

cathodique [13, 14, 19]. Cette théorie considère l'apparition du jet comme un phénomène thermique; il semble pourtant que le rôle du facteur électrostatique [12] ne doive pas être entièrement négligé.

La complexité du processus et le manque de données empiriques rendent toujours difficile une formulation quantitative de la théorie du jet, donc, à plus forte raison, son application à des cas particuliers. Sans entrer dans le détail, nous allons résumer ici les constatations principales concernant le mécanisme du jet anodique.

La formation du jet reste en connexion avec un changement du régime électrique auquel travaille l'arc, et notamment avec la transition au régime de haute puissance.

L'arc travaillant dans le régime normal (basse puissance) montre la caractéristique décroissante ("résistance négative", $\dfrac{dV}{dJ} < 0$, Fig. 1); la majeure partie de la chute de potentiel dans l'arc correspond à la colonne (Fig. 2), les chutes sur les électrodes sont relativement indépendantes de l'intensité totale du courant;

l'énergie dissipée dans l'arc est convertie surtout en rayonnement et en chaleur
évacuée par les électrodes (Fig. 3); l'érosion de l'anode est petite et la température
de l'anode reste en dessous de la température d'ébullition, tandis que la température dans la colonne ne dépasse pas 3000°. L'augmentation du courant entraîne
seulement l'accroissement de l'énergie rayonnée et dissipée par conduction.

Cependant, à partir d'une certaine intensité du courant, le caractère de la
décharge subit des changements, qui établissent un équilibre nouveau, déterminé
par les phénomènes survenant sur l'anode. La chaleur dégagée sur l'anode devient
trop grande pour être effectivement évacuée par conduction; simultanément, le
champ magnétique propre du courant, très augmenté, entraîne une diminution
de la surface d'entrée du courant dans l'anode ce qui donne naissance à la tache
anodique. La température de la tache — grâce à la densité de courant très élevée
— dépasse la température d'ébullition du matériau et l'érosion de l'anode augmente brusquement. La vapeur produite dans la tache, fortement ionisée, forme
une charge d'espace à proximité de l'anode, ce qui augmente la chute anodique;
c'est pourquoi la chute anodique devient, dans le nouvel équilibre, la partie la
plus importante de la chute totale dans l'arc (Fig. 2). L'augmentation du courant
se traduit maintenant par une évaporation plus rapide du matériau de l'anode,
donc par une augmentation de la chute anodique — et totale — de l'arc; la caractéristique est donc maintenant croissante ("résistance positive", $\frac{dV}{dJ} > 0$, Fig. 1).
La vapeur produite dans la tache anodique subit dans la région de la charge
d'espace une forte surchauffe due à la densité élevée du courant et les températures atteintes sont de l'ordre de 6000—15000° [13, 20, 23].

Le bilan énergétique de l'arc change avec le passage au régime de haute puissance. En général, l'énergie fournie à l'anode par le courant est, par seconde,

$$W_a = J (V_a - \varphi)$$

et elle est répartie entre les trois composantes (Fig. 4):

$$W_a = W_{a1} + W_{a2} + W_{a3}$$

d'une manière qui dépend de la densité du courant. Dans l'arc à haute puissance
la composante W_{a1} correspondant au rayonnement et à la conduction joue un rôle
secondaire, qui diminue encore avec l'intensité du courant; par contre, l'énergie
utilisée pour l'évaporation et pour la surchauffe est considérable et la partie
qu'elle représente dans le bilan croît avec le courant. Comme on a déjà indiqué,
la conséquence immédiate de l'évaporation et de la surchauffe est une augmentation de la chute anodique V_a; or, la chute totale dans l'arc est

$$V = V_c + V_{cl} + V_a;$$

pour un courant donné, J, on a

$$W = W_c + W_{cl} + W_a$$
$$= J (V_c - p) + J V_{cl} + J (V_a + \varphi) .$$

L'énergie $W_a = J (V_a + \varphi)$ dégagée sur l'anode croît rapidement avec J puisqu'une
augmentation de J entraîne aussi l'accroissement de V_a; par contre, les chutes
V_c et V_{cl} restent à peu près inchangées et W_c et W_{cl} croissent plus lentement que
W_a. De cette manière W_a devient de plus en plus prépondérant dans le bilan total
de l'énergie de l'arc (Fig. 3).

En somme, on peut affirmer qu'avec l'augmentation du courant la puissance
de l'arc se concentre sur l'anode et qu'une partie toujours plus grande de cette

"puissance anodique" est transformée en évaporation du matériau de l'anode et surchauffe de la vapeur produite.

L'intensité du courant pour laquelle commence la transition au régime de haute puissance dépend, évidemment, des paramètres du circuit, mais surtout des propriétés physiques du matériau de l'anode. En général, la transition sur-

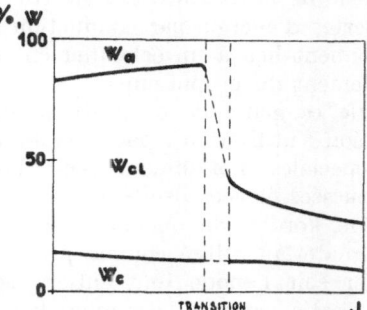

Fig. 3. Bilan énergétique de l'arc à courant continu. W_c puissance dégagée sur la cathode; W_{cl} puissance dégagée sur l'anode; W_a puissance dégagée dans la colonne

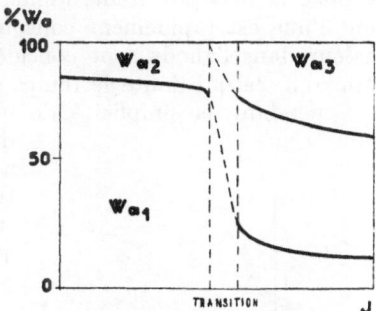

Fig. 4. Bilan énergétique de l'anode de l'arc. W_{a1} puissance dissipée par rayonnement et conduction; W_{a2} puissance utilisée pour l'évaporation du matériau de l'anode; W_{a3} puissance utilisée pour la surchauffe de la vapeur

vient à une intensité de courant d'autant plus basse que la chaleur d'évaporation de ce matériau est plus petite. Cette chaleur détermine aussi le taux d'évaporation et la vitesse d'écoulement; il n'est pas possible, cependant, d'indiquer les relations exactes existant sous ce rapport.

On a, évidemment

$$\dot{m} = \frac{W_{a2}}{\lambda}$$

mais les bases théoriques pour calculer W_{a2} $\left(\text{ou le rapport } \frac{W_{a2}}{W_a}\right)$ ne sont pas suffisantes; on ne peut donc pas déterminer a priori $\dot{m}$ pour un cas particulier donné et c'est au contraire en mesurant $\dot{m}$ qu'on évalue W_{a2} [20].

La formule utilisée [7, 13] pour la vitesse d'écoulement dans le jet

$$v = \frac{W_a/S}{(\lambda + \bar{c}_{\Delta T})\,\delta_T} \qquad\qquad \Delta T = T_f - T_o$$

doit également être considérée comme une première approximation. On admet notamment dans cette formule la conversion totale de W_a en évaporation et surchauffe; la température T_f à laquelle est portée la vapeur est mesurée avec une marge d'erreur considérable et les valeurs de la chaleur spécifique $\bar{c}_{\Delta T}$ à haute température ne sont pas, en général, exactement connues. Néanmoins, cette formule permet d'évaluer la vitesse d'écoulement pour diverses puissances et divers matériaux, et d'apprécier l'influence de ces facteurs sur les performances d'un propulseur [13].

III. Construction et fonctionnement du propulseur

Les méthodes qu'on peut envisager pour l'application du jet anodique à la propulsion diffèrent surtout par l'état physique du propulsif.

1. La méthode la plus simple est d'employer en tant que masse éjectable le matériau de l'électrode, donc un solide; le fonctionnement de l'appareil peut

être interprété dans ce cas entièrement par le mécanisme qui vient d'être exposé ci-dessus. Du point de vue des performances un tel appareil présente pourtant des inconvénients assez sérieux.

Par exemple l'évaporation et l'érosion d'une anode solide sont très difficilement contrôlables et subissent souvent des variations irrégulières; elles entraînent, de plus, la nécessité d'une régulation continuelle de la distance des électrodes, dont l'une est rapidement consommée. Les pertes d'énergie par conduction de chaleur dans l'anode sont considérables et donnent lieu à un échauffement de l'appareil, ce qui limite le temps de fonctionnement du propulseur.

Cependant, la simplicité de construction de ce genre de dispositif permet d'envisager son utilisation pour certaines applications spéciales, notamment pour l'obtention de poussées élevées pendant des temps très courts (de l'ordre de quelques secondes au plus). Le matériau utilisé comme propulsif peut constituer soit l'anode tout entière, soit seulement sa partie centrale, correspondant à la "mèche" des charbons des arcs d'éclairage.

2. Dans l'autre type de propulseur à jet, un liquide est utilisé comme propulsif ([25], Fig. 5). Par un système des canaux le liquide est amené jusqu'à la surface de l'anode, où il subit l'évaporation; lors de son passage dans ou autour de l'anode, le même liquide sert à refroidir celle-ci, ce qui permet non seulement de récupérer une partie de la chaleur et de chauffer préalablement le propulsif, mais favorise aussi la formation de la tache anodique [8, 9]. A l'endroit où le liquide émerge sur la surface de l'anode la tache s'accroche facilement, comme cela se produit toujours sur les inhomogénéités de cette surface; la formation de la tache est notamment facilitée partout où la chaleur de l'évaporation est localement réduite et où le travail d'entrée du courant est localement abaissé [8, 9, 19]. Ces circonstances existent à l'endroit où arrive le liquide, dont les propriétés physiques diffèrent de celles du corps environnant (solide) de l'anode.

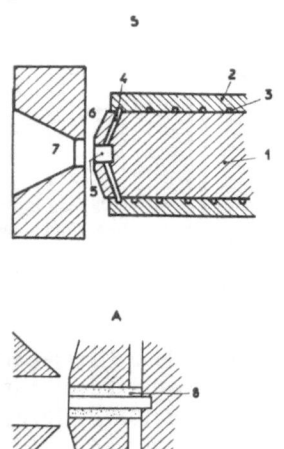

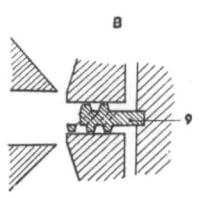

Fig. 5. Propulseur à jet anodique à propulsif liquide. S schéma principal; A et B deux exemples de construction de l'évidement. 1 anode; 2 manche du courant et de la réfrigération; 3 hélice de réfrigération; 4 conduits centripètes d'alimentation; 5 évidement; 6 cathode; 7 tuyère; 8 charbon poreux; 9 tige réfractaire hélicoïdale

Les conditions de formation du jet sont déterminées dans une grande mesure par la structure choisie pour l'évidement par lequel le liquide arrive sur la surface de l'anode; il existe, sous ce rapport, des solutions très diverses.

La Fig. 5 montre le schéma principal de l'appareil (Fig. 5 S) et deux exemples de construction de l'orifice de sortie du liquide; dans la première variante (Fig. 5 A) le liquide traverse, en arrivant à la surface, un cylindre poreux, tandis que dans la deuxième (Fig. 5 B) il forme une flaque sur la surface, où il arrive par un conduit hélicoïdal.

Un des avantages du système décrit est l'absence d'usure de l'anode; c'est seulement — en principe — le liquide qui subit l'évaporation, l'érosion de l'anode n'ayant pas lieu. Une augmentation de la puissance doit être équilibrée par un débit plus grand du liquide arrivant sur la surface (Fig. 6). Si le débit du liquide est trop grand par rapport à la puissance utilisée (zone 1 de la Fig. 6), la formation du jet n'a pas lieu, l'évaporation du liquide est incomplète (c. à d. le liquide s'écoule de l'anode) et l'arc peut s'éteindre ou passer dans le régime de basse puissance. Dans le cas ou le débit est trop faible par rapport à la puissance (zone 3 de la Fig. 6), l'évaporation peut affecter les parties solides de l'anode, détruisant la surface et, en particulier, la structure de l'évidement. Dans des limites assez étroites (zone 2 de la Fig. 6) le débit et la puissance sont en équilibre et le jet est formé ; une certaine latitude dans le rapport débit/puissance est admissible et des changements suffisamment petits de ces quantités entraînent seulement des variations de la température de surchauffe, donc de la vitesse d'écoulement, sans empêcher la formation du jet.

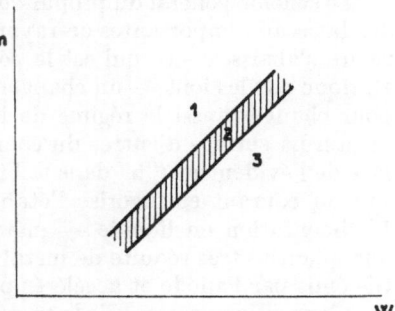

L'érosion de la cathode est beaucoup plus faible que celle de l'anode, dans le régime de haute puissance. Elle peut être diminuée, voire supprimée par l'emploi d'une des méthodes connues dans la technique (p.ex. déplacement de la tache cathodique).

Fig. 6. Relation entre débit et puissance pour le propulseur à jet anodique. *1* zone de suralimentation; *2* zone de fonctionnement; *3* zone de surcharge

Le choix du liquide est un problème important et complexe; les propriétés physiques du liquide sont en rapport étroit avec toutes les caractéristiques du fonctionnement de l'appareil. Elles déterminent, en particulier, la transition au régime de haute puissance, la température de la tache et celle de la région de surchauffe, la vitesse d'écoulement et les conditions d'ionisation dans la colonne de l'arc. Il n'existe, sous ce rapport, aucune solution optimale, et le choix du propulsif doit être basé sur les performances qu'on s'efforce d'obtenir. Au cours des études, le mercure et divers amalgames ont été utilisés, aussi bien que des alliages à bas point de fusion.

Les recherches sur le propulseur à jet anodique ont été faites avec plusieurs appareils de ce genre, qui diffèrent entre eux par des détails de construction: propulsif auquel ils étaient adaptés et puissance moyenne de fonctionnement; celle-ci variait de 2 à 50 kW. L'impulsion spécifique maximum atteinte jusqu'ici était de l'ordre de 600.

Une série d'expériences commencée récemment concerne la formation du jet cathodique, dont les propriétés diffèrent essentiellement [13, 14, 19] de celles du jet anodique.

Le développement ultérieur du dispositif dépend de la possiblité de réaliser un écoulement électrostatique comme une des formes de fonctionnement du propulseur à jet.

IV. Ecoulement électrostatique dans le propulseur à jet

L'appareil qui vient d'être décrit produit, en principe, un écoulement thermique, c.à.d. un écoulement de gaz subissant une détente après échauffement. Le même appareil est cependant en mesure de travailler aussi comme un propulseur ionique, en fournissant une poussée faible à un débit très reduit, mais à une impulsion spécifique élevée. Cette possibilité existe notamment grâce à l'utilisation du phénomène des rayons anodiques. Ce phénomène consiste en l'émission d'ions positifs par l'anode lors de la décharge dans le gaz, surtout notamment d'une décharge à lueurs. Certaines formes de cette décharge favorisent particulièrement l'apparition des rayons anodiques; ils sont très forts dans le régime anormal de la décharge à lueurs, quand la densité du courant est grande et la surface d'entrée dans l'anode petite. Dans ces conditions la chute anodique est élevée, de l'ordre de dizaines ou centaines de volts, l'émission des ions est donc facilitée par le champ électrique qui, de plus, accélère les ions à des vitesses considérables. Pour obtenir une émission importante des ions l'anode doit contenir des corps qui sont capables d'en fournir facilement, p.ex. des métaux légers ou leurs sels.

Le schéma général du propulseur à jet anodique est bien adapté pour envisager des faisceaux importantes de rayons anodiques. En effet, quand la pression extérieure s'abaisse — ce qui est la condition d'application de la propulsion électrostatique par les ions — un changement des paramètres du circuit électrique suffit pour changer aussi le régime de la décharge et passer de l'arc à la décharge à lueurs. La surface d'entrée du courant dans l'anode se limite facilement à la surface de l'évidement ("5" dans la Fig. 5), ce qui entraîne l'augmentation de la densité du courant et favorise l'établissement du régime anormal de la décharge. L'alimentation en liquide — métal ou alliage — doit être fortement abaissée; une quantité très réduite de métal sur la surface suffit pour fournir des ions positifs émis par l'anode et accélérés par la chute anodique.

L'exécution pratique de la transition de l'arc à la décharge à lueurs et de l'écoulement thermique à l'écoulement électrostatique entraîne de nombreuses difficultés et des problèmes. Néanmoins, dans l'état actuel des essais il est déjà possible d'affirmer qu'avec une adaptation convenable de l'appareil un rendement satisfaisant peut être obtenu dans les deux régimes de fonctionnement.

B. Propulseur à explosion électrique

I. Principe

Le fonctionnement du propulseur à explosion électrique est basé sur les phénomènes accompagnant l'évaporation explosive d'un fil conducteur, dans lequel on fait passer un courant de haute intensité [49, 50]. L'évaporation est suivie par une décharge de courte durée dans la vapeur, par un fort échauffement de cette vapeur

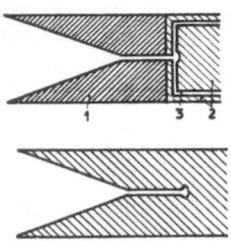

et par la formation d'une onde de choc, qui se propage à une vitesse élevée; des températures de l'ordre de 30000° et plus ont été enregistrées, tandis que les vitesses d'expansion atteignent plusieurs km/sec [49, 50].

Dans le propulseur [51] le fil est remplacé par un liquide conducteur qu'on introduit, sous une pression convenable, entre les électrodes d'un circuit électrique (Fig. 7); établissant le contact, le liquide ferme le circuit et subit l'évaporation par le courant; une décharge électrique s'amorce dans la vapeur, la température et la pression augmentent brusquement et l'arrivée du liquide est arrêtée par la haute pression dans la zône de la décharge. Cette zône étant ouverte (par la tuyère) vers l'extérieur, la vapeur est éjectée à une grande

Fig. 7. Propulseur à explosion électrique. Schéma du type linéaire, sections longitudinales perpendiculaires. *1* électrodes; *2* isolants; *3* canaux d'alimentation

vitesse et au fur et à mesure de son écoulement la pression s'abaisse; à partir d'un certain moment l'arrivée du liquide peut reprendre et le cycle recommence.

Le dispositif travaille donc, en principe, en régime intermittent, comme pour le pulsoréacteur; il est cependant possible d'établir aussi un processus continu, notamment par un choix convenable des facteurs déterminant les propriétés et le fonctionnement du propulseur.

Les facteurs dont il est question peuvent être classés en quatre groupes :

1. paramètres du circuit électrique
2. paramètres du système d'alimentation en liquide
3. propriétés physiques du liquide
4. géométrie de la chambre de décharge et détente.

L'ensemble de ces facteurs détermine les caractéristiques du propulseur, p.ex. la fréquence (c. à d. le nombre de cycles par seconde), la vitesse d'écoulement, le débit massique, la puissance moyenne et maximum etc.

Le propulseur à explosion électrique se passe de tout moyen d'amorçage ; la mise en marche s'effectue toujours par l'introduction du liquide conducteur entre les électrodes et cette propriété constitue un avantage considérable, comparé aux conditions d'amorçage des propulseurs à arc.

Les appareils de ce genre peuvent être conçus pour fonctionner soit sur courant continu, soit sur courant alternatif.

II. Fonctionnement du propulseur

Il est évident que les considérations concernant l'explosion électrique d'un fil ne sont pas applicables dans le cas d'un liquide arrivant continuellement, et évaporé dans une chambre d'explosion. Les phénomènes survenant dans ces circonstances forment un processus complexe, qui comprend l'échauffement et l'évaporation du liquide, la décharge dans la vapeur produite, la formation de l'onde de choc et l'expansion de la vapeur surchauffée.

Une étude détaillée de ce processus — qui fera l'objet d'un article ultérieur — permet de déterminer les rapports entre les constantes du circuit, les constantes du système d'alimentation, les constantes physiques du liquide et les constantes géométriques de la chambre d'explosion et d'établir ainsi une base théorique pour l'étude expérimentale des propulseurs du type décrit ; ici, nous nous bornerons à l'analyse qualitative d'un simple cas particulier, ce qui suffira cependant pour décrire le comportement d'un propulseur à réaction.

Nous supposons que le liquide est alimenté, sous une pression constante p, dans un passage rectiligne de longueur l, formé par deux électrodes parallèles, qui créent dans ce passage un champ homogène d'intensité E (Fig. 8). Le processus qui s'établit est périodique et chaque cycle est composé des quatre phases suivantes :

a) Le liquide avance entre les électrodes à une vitesse constante w (celle-ci étant liée à la pression p par la relation $^1/_2 \gamma w^2 = p$). Au fur et à mesure de son avancement le liquide subit l'échauffement par le courant qui le traverse et un gradient de température s'établit dans la colonne qu'il forme dans le passage. En première approximation on peut négliger les variations que subissent, avec l'échauffement, les constantes physiques γ, c, ϱ du liquide, aussi bien que les pertes de chaleur par conduction ; on arrive ainsi à une expression simplifiée de ce gradient :

$$\frac{dT}{dx} = \frac{E^2}{\gamma c \varrho w} .$$

D'après cette formule, la température d'ébullition du liquide sera atteinte à une distance

$$\Delta_1 x = \gamma c \varrho \, \frac{w \, (T_e - T_o)}{E^2}$$

et après le temps

$$\Delta_1 t = \gamma \, c \, \varrho \; \frac{T_e - T_o}{E^2} \text{ (Fig. 8a)}.$$

b) Tandis que l'arrivée du liquide froid dans le passage se poursuit, l'ébullition commence dans la partie antérieure de la colonne du liquide. En admettant que la chaleur développée par le courant est totalement utilisée, dans cette partie de la colonne, pour évaporer du liquide, on obtient l'expression

$$\Delta_2 t = \frac{\gamma \, \lambda \, \varrho}{E^2}$$

pour le temps pendant lequel la masse du liquide contenue dans une tranche antérieure de la colonne s'évapore totalement; la longueur de cette tranche est égale à

$$\Delta_2 x = \gamma \, \lambda \, \varrho \; \frac{w^2}{E^2} \, .$$

En réalité, les phénomènes survenant dans cette phase sont plus compliqués. L'évaporation de la masse du liquide donne lieu à des interruptions locales du courant, donc à des échauffements et à la création de champs électriques locaux très forts, ce qui conduit, sur les surfaces électrode/vapeur et liquide/vapeur, à l'émission d'électrons et, finalement, à l'amorçage d'une décharge dans la vapeur (Fig. 8 b).

c) La vitesse à laquelle se produisent les phénomènes décrits ci-dessus est telle que, d'ordinaire, une expansion de la vapeur ne parvient pas à se manifester avant l'amorçage de la décharge; c'est pourquoi cette décharge, dans sa phase initiale, a lieu dans de la vapeur fortement comprimée, occupant un volume à peine supérieur à celui qu'elle occupait à l'état liquide. Cette circonstance favorise le développement des températures très élevées dans la décharge (plus élevées, évidemment, que celles qu'il serait possible d'obtenir dans la vapeur à pression ambiante). L'expansion de la vapeur ainsi surchauffée prend la forme d'une onde de choc; la pression arrête l'avance du liquide et repousse même la colonne liquide en arrière. Une certaine partie de cette colonne subit encore l'évaporation et la décharge se propage en arrière, englobant une quantité plus grande du propulsif. Dans l'autre sens la vapeur surchauffée s'écoule au dehors à vitesse élevée, subissant encore détente et accélération dans la tuyère (Fig. 8 c).

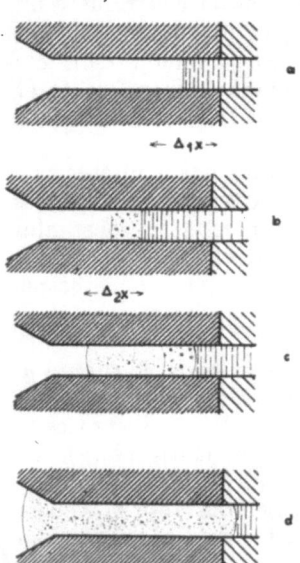

Fig. 8. Les phases d'un cycle du fonctionnement dans le propulseur à l'explosion électrique. Legende v. texte, B II

d) Le liquide ayant été partiellement évaporé et partiellement repoussé, le courant est soutenu seulement par la décharge dans la vapeur; cette décharge, emportée par l'écoulement de vapeur, continue encore le long des électrodes formant le passage et peut se prolonger dans la tuyère, si les parois de celle-ci comportent aussi des électrodes (comme c'est le cas dans la Fig. 7). Cependant, l'expansion de la vapeur entraîne une baisse de la température, favorise la déionisation et supprime les ions de l'espace entre les électrodes. Selon les circonstances, la décharge et le courant peuvent subsister ou s'arrêter pendant les temps où la pression est tombée suffisamment pour permettre un nouvel afflux de liquide et le recommencement de la phase a (Fig. 8d).

C'est la dernière phase du processus qui joue un rôle très important dans le contrôle de l'écoulement; pour le comprendre, nous allons considérer les changements qui surviennent dans le fonctionnement de l'appareil lorsque la pression d'alimentation p augmente, tous les autres paramètres (en particulier E) restant les mêmes.

Supposons d'abord que la pression p soit très basse — et, avec elle, la vitesse w et le debit $\dot{m}$. La colonne de liquide qui subit alors l'échauffement et l'évaporation

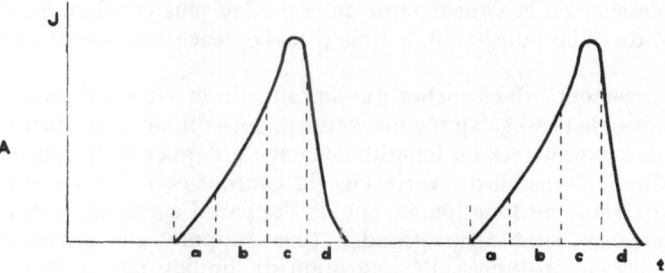

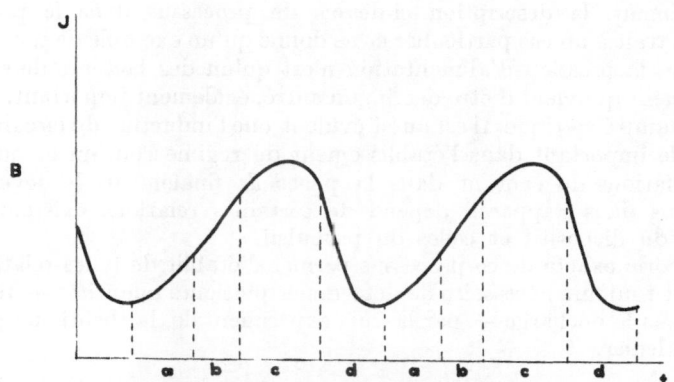

Fig. 9. Courant dans le propulseur à explosion électrique en fonction du temps. A régime intermittent; B transition au régime continu

est très courte, comme l'indiquent les formules pour $\varDelta_1 x, \varDelta_2 x, \varDelta_1 t, \varDelta_2 t$. La pression qui règne dans la chambre d'explosion après la décharge est de loin supérieure à p, la différence $P-p$ est donc très grande au début du processus. Le temps nécessaire pour la réduction de cette différence (par écoulement de la vapeur) est plus long que le temps de disparition de la décharge; un nouvel afflux de liquide ne peut recommencer que lorsque l'évacuation de la vapeur du passage est presque complète, la vitesse d'écoulement ayant été réduite sensiblement. Les explosions consécutives seront donc espacées, le courant intermittent, et la durée d'un cycle grande (Fig. 9A).

Avec l'augmentation de la pression p (donc aussi de la vitesse w et du débit $\dot{m}$) les valeurs de $\varDelta_1 x$ et $\varDelta_2 x$ croissent et il en est de même de $\varDelta_1 t$ et $\varDelta_2 t$; pourtant, la différence $P-p$ est maintenant réduite dès le début et l'afflux de liquide peut commencer plus tôt, avant l'évacuation totale de la vapeur. Les variations de la vitesse seront diminuées et la fréquence du processus plus élevée.

A partir d'une certaine pression p un phénomène nouveau survient; avant l'extinction totale de la décharge (phase d) le liquide apparaît de nouveau entre les électrodes et la phase a est amorcée. Les disparitions totales du courant entre les phases d et a sont supprimées, et remplacées par des baisses du courant, dont l'intensité va passer par des maxima et minima, sans jamais tomber à zéro. Simultanément, la durée d'un cycle va diminuer, les explosions se suivant de près; l'écoulement de la vapeur aura un caractère plus continu, les variations de la vitesse et du débit subsistant toujours, mais présentant une amplitude réduite (Fig. 9 B).

"L'aplatissement" des courbes du courant, de la vitesse d'écoulement et du débit caractérise le passage au régime continu, qui s'établit sans aucun changement discontinu des paramètres de fonctionnement de l'appareil et qui ne correspond qu'à une réduction sensible des variations du courant et de la vitesse d'écoulement.

Un cas extrême de fonctionnement de l'appareil survient, quand la pression p et la vitesse w sont trop grandes (par rapport aux autres paramètres); alors, selon les circonstances, l'évaporation du liquide peut être incomplète ou même n'avoir pas lieu, l'échauffement du liquide étant insuffisant pour le porter à l'ébullition. En effet, si $\Delta_1 x + \Delta_2 x > l$, mais $\Delta_1 x < l$, l'évaporation sera incomplète; si déjà $\Delta_1 x > l$ l'ébullition n'aura pas lieu et le dispositif va débiter seulement du liquide échauffé (Fig. 8).

Evidemment, la description ci-dessus du processus dans le propulseur à explosion a trait à un cas particulier et ne donne qu'un exemple de son fonctionnement. Ainsi, la pression d'alimentation n'est qu'un des facteurs déterminant le développement qui vient d'être décrit; un autre, également important, est l'intensité E du champ électrique. Il est aussi évident que l'induction du circuit électrique joue un rôle important dans l'établissement du régime continu et qu'elle influe sur les variations du courant dans la phase d; finalement, le développement du processus dans l'appareil dépend de certaines relations existant entre les constantes du dispositif et celles du propulsif.

Une théorie exacte de ce processus permet d'établir de telles relations, néanmoins, il est toujours nécessaire de déterminer plusieurs quantités — telle la température dans la décharge — par la voie expérimentale, la théorie ne permettant pas de la calculer.

III. Eléments et propriétés du propulseur

Nous allons considérer ici brièvement les divers problèmes de construction et de contrôle des appareils en question, suivant le point de vue adopté en I.

1. Circuit électrique.

Comme indiqué auparavant, les propulseurs à explosion électrique peuvent être conçus pour travailler soit en courant continu, soit en courant alternatif; certaines constructions — les plus simples — sont même adaptables à l'un comme à l'autre genre d'alimentation en énergie. Dans chaque cas, le circuit doit être capable de supporter des pointes importantes de puissance, du fait que le fonctionnement de l'appareil dans le régime intermittent comporte de telles pointes et que même dans le régime continu, l'amplitude maximum du courant peut être assez importante. Une autre règle générale est que l'induction du circuit doit être élevée afin de soutenir la décharge dans sa dernière phase et faciliter ainsi la transition au régime continu. Le choix des paramètres du circuit dépend, évidemment, des performances exigées dans chaque cas particulier.

2. Système d'alimentation en liquide.

Ce système est, lui aussi, nécessairement adapté aux performances du dispositif pour lequel il est destiné, en particulier à la configuration géométrique de la chambre d'explosion. La solution la plus simple, montrée par la Fig. 7, peut (et doit souvent) être remplacée par d'autres schémas. Ainsi, par exemple, le liquide peut être amené par les électrodes, soit par des canaux aménagés dans celles-ci, soit même à travers des parties poreuses des électrodes; l'explosion survient alors dès l'établissement du contact entre deux portions du liquide. Il est aussi indiqué d'utiliser le liquide pour la réfrigération régénérative des électrodes.

3. Propulsif.

Le choix du liquide pour le propulsif constitue un facteur important dans la construction du propulseur, plus encore que dans le cas du propulseur à arc, du fait que les propriétés physiques du liquide restent ici en rapport étroit avec les paramètres du dispositif; ce choix est limité, bien entendu, aux liquides conducteurs, donc aux métaux et alliages liquides et solutions électrolytiques. Le mercure a été utilisé le plus souvent dans les travaux expérimentaux; cependant les essais faits avec des solutions électrolytiques (aqueuses et non-aqueuses) ont donné des résultats positifs.

4. Géométrie de la chambre d'explosion.

La disposition géométrique la plus simple du propulseur est indiquée sur la Fig. 7, où deux électrodes planes et deux parois isolantes forment le passage pour le liquide et la décharge. D'autres formes géométriques ont été aussi expérimentées, telle la forme "coaxiale" comportant deux électrodes cylindriques coaxiales et assurant une symétrie nécessaire pour l'application du courant continu, ou la forme "polygonale", plusieurs électrodes formant le passage (dont la section transversale est un polygone); pour les débits massiques élevés l'emploi des systèmes à plusieurs électrodes s'impose, la tension totale étant distribuée entre celles-ci. Ces arrangements divers, adaptés aux exigences des cas particuliers, entraînent, bien entendu, des complications considérables du point de vue théorique et pratique.

Les vitesses d'écoulement réalisables dans les propulseurs à explosion électriques sont de l'ordre de quelques km/sec dans le régime continu; les valeurs les plus grandes sont pourtant atteintes dans certaines formes du régime intermittent; dans ce régime, pourtant, il faut distinguer la vitesse maximum et une vitesse effective moyenne de l'écoulement.

Les relations existant entre la vitesse d'écoulement, le débit massique et la puissance sont fort compliquées dans le cas du propulseur à explosion et leur caractère change essentiellement avec la forme du régime auquel l'appareil travaille.

Le rendement du dispositif est satisfaisant, de l'ordre de 60—80 %; les pertes sont dues surtout à la conduction de la chaleur par les électrodes, tandis que l'énergie rayonnée et la conduction par le liquide jouent dans le bilan énergétique des rôles secondaires.

Le phénomène d'érosion des électrodes constitue parfois un problème important; cependant, les électrodes ordinaires, par exemple en charbon dur, sont capables de supporter quelques dizaines de minutes un travail normal (sans surcharges excessives) sans que leur érosion soit excessive. Pour des régimes particulièrement durs de travail (surcharges fréquentes et prolongées) il est nécessaire d'envisager des solutions spéciales, par exemple, l'emploi d'électrodes poreuses à travers lesquelles le liquide passe sous pression, ou encore d'électrodes en matériaux contenant un constituant facilement fusible et volatil dispersé dans un conducteur poreux réfractaire (charbon-cadmium, charbon-argent etc.).

IV. Application du champ magnétique

Le processus dans le propulseur à explosion peut être sensiblement modifié par l'application d'un champ magnétique, en général d'un champ transversal par rapport à la décharge et à l'écoulement. Les modifications réalisables par ce moyen sont diverses; nous en indiquerons les plus importantes.

1. L'accélération de l'écoulement équivaut, évidemment, au soufflage magnétique de la décharge et peut être effectué lors du passage dans la décharge et dans la tuyère, dont la forme géométrique doit être modifiée et adaptée pour convenir à cette opération. En principe, l'accélération de l'écoulement tend à interrompre la décharge tôt dans la phase d, elle favorise donc l'apparition du régime intermittent. Ce développement n'est pourtant pas inévitable, au contraire, on peut employer aussi le champ magnétique pour faciliter le passage au régime continu. Et notamment:

2. Le passage au régime continu exige que l'afflux du liquide commence aussitôt que possible dans le cycle, c. à d. que la phase a remplace la phase d avant qu'un affaiblissement sensible de l'écoulement se soit produit. Un champ magnétique placé là où survient l'explosion peut freiner la propagation de la décharge en arrière, diminuant de cette manière la pression dynamique exercée par l'explosion sur la colonne de liquide et limitant le temps pendant lequel cette pression prend des valeurs élevées; on arrive ainsi à créer des conditions favorables pour le recommencement du cycle.

3. Le champ établi comme en 2. va évidemment agir aussi sur la colonne de liquide, en l'entraînant dans le passage, ce qui favorise la transition au régime continu; il semble même possible de remplacer l'alimentation sous pression par l'application d'un tel champ, donc par un genre de pompe électrodynamique. Cependant, les relations compliquées qui existent entre les conditions d'alimentation et le fonctionnement du propulseur n'ont pas encore permis d'établir une solution satisfaisante basée sur cette méthode.

L'emploi idéal d'un champ magnétique consisterait dans une réalisation simultanée de 1, 2, et 3.

Le champ magnétique employé peut être simplement le champ propre du courant de la décharge; dans ce cas l'action de ce champ s'apparente aux phénomènes connus de l'autosoufflage d'un arc, et une disposition géométrique particulière des électrodes doit être envisagée pour rendre cette méthode efficace (ce qu'elle est toujours dans une certaine mesure). L'emploi d'aimants permanents n'est avantageux que dans certains cas particuliers et, dans la plupart des problèmes l'utilisation d'électroaimants s'impose. Un tel électroaimant étant alimenté par le circuit du propulseur (par exemple, en série avec la décharge) joue, en même temps, le rôle de l'induction nécessaire pour soutenir la décharge et provoquer la transition au régime continu.

Conclusion

L'étude et le développement des appareils décrits ci-dessus étant toujours en cours, il est difficile de formuler une opinion définitive sur leurs propriétés, ainsi que d'indiquer les solutions optimum des problèmes qu'ils présentent. Les performances obtenues jusqu'à présent permettent néanmoins de considérer comme certain l'intérêt des méthodes qui viennent d'être exposées dans la recherche d'un système efficace de propulsion électrique.

Remerciements

L'étude sur le sujet des propulseurs électriques décrits ci-dessus a été faite dans les laboratoires de M. ALBERT DUCROCQ, mises à la disposition de l'auteur par M. DUCROCQ. L'auteur a l'honneur de lui exprimer la reconnaissance la plus profonde pour sa bienveillance et l'intérêt manifesté dans cette étude.

L'éxécution des appareils et prototypes se trouvait entre les mains de M. ALBERT LECOMTE de L'Atelier de Précision LECOMTE et DEGLISE à Paris. Les connaissances et le vif intérêt apportés par M. LECOMTE envers ces problèmes ont rendu possible l'aboutissement des recherches; qu'il veuille accepter les remerciements les plus sincères de l'auteur.

Références

Jet électrodique

1. R. LAMAR, Physic. Rev. **43**, 169 (1933).
2. A. EASTON, Electr. Engng. **53**, 1454 (1934).
3. S. LÜDI, Helvet. Physica Acta 8, 272 (1935).
4. W. FINKELNBURG, Z. Physik **112**, 305 (1939).
5. W. FINKELNBURG, Z. Physik **113**, 562 (1939).
6. W. FINKELNBURG, Z. Physik **114**, 734 (1940).
7. W. FINKELNBURG, Z. Physik **116**, 214 (1940).
8. E. PODSZUS, Z. Physik **116**, 352 (1940).
9. E. PODSZUS, Z. Physik **116**, 651 (1940).
10. A. ROHLOFF, Reichsber. Physik 1, 47 (1944).
11. W. FINKELNBURG, The High Current Carbon Arc. FIAT Report 1052-PB-81644, Washington, 1947.
12. S. HAYNES, Physic. Rev. **73**, 891 (1948).
13. W. FINKELNBURG, Physic. Rev. 74, 1475 (1949).
14. A. ROHLOFF, Z. Physik **126**, 175 (1949).
15. A. ROHLOFF, Z. Physik **126**, 224 (1949).
16. W. FINKELNBURG, Der Hochstromkohlebogen. Berlin-Göttingen-Heidelberg: Springer, 1949.
17. W. FINKELNBURG, J. Appl. Physics 20, 486 (1949).
18. N. GALLAGHER, Physic. Rev. **79**, 231 (1950).
19. J. PARISOT, L'Arc Electrique Intensif. Paris: Gauthier-Villars, 1950.
20. J. COBINE et E. BURGER, J. Appl. Physics **26**, 895 (1955).
21. H. MAECKER, Z. Physik **141**, 198 (1955).
22. W. FINKELNBURG, Proceedings of the Symposium on High Temperatures. Stanford Research Inst. 1956.
23. M. MARQUIS, L. MEAD, S. KORMAN et C. SHEER, Energy Transfer in High Intensity Arcs in Inert Atmospheres and in Vacuum. Electrochem. Soc., New York, 1956.
24. J. ULAM, Demande de Brevet Franç., 1958.

Ecoulement magnétique du plasma

25. M. CHAMPION, Proceedings of the Physic. Soc. **B-66**, 169 (1953).
26. A. KOLB, Lockheed Symposium on Magnetohydrodynamics. Stanford, Calif.: University Press, 1957.
27. T. KORNEFF, F. MADIG et J. BOHN, Conference on Extremely High Temperatures. New York: John Wiley, 1958.
28. W. H. BOSTICK, Conference on Extremely High Temperatures. New York: John Wiley, 1958.
29. W. H. BOSTICK, Comptes rendus, IXᵉ Congrès International d'Astronautique, Amsterdam 1958, p. 794. Wien: Springer, 1959.

Propulsion ionique

30. L. R. Shepherd et A. Cleaver, J. Brit. Interplan. Soc. **1**, 184 (1948).
31. L. R. Shepherd et A. Cleaver, J. Brit. Interplan. Soc. **8**, 59 (1949).
32. L. Spitzer, J. Brit. Interplan. Soc. **10**, 249 (1951).
33. D. C. Romick, Bericht über den V. Internationalen Astronautischen Kongreß, Innsbruck 1954, p. 81. Wien/Innsbruck: Springer, 1955.
34. E. Stuhlinger, Bericht über den V. Internationalen Astronautischen Kongreß, Innsbruck 1954, p. 100. Wien/Innsbruck: Springer, 1955.
35. E. Stuhlinger, J. Astronautics **2**, 149 (1955).
36. L. R. Shepherd, Comptes rendus, IXe Congrès International d'Astronautique, Amsterdam 1958, p. 932. Wien: Springer, 1959.

Arc stabilisé

37. H. Gerdien et A. Lotz, Wiss. Veröff. Siemens-Konz. **2**, 48 (1922).
38. H. Maecker, Z. Physik **129**, 108 (1951).
39. H. Maecker, F. Burhern et T. Peters, Z. Physik **131**, 28 (1951).
40. R. Larenz, Z. Physik **129**, 343 (1951).
41. R. Jürgens, Z. Physik **134**, 21 (1952).
42. H. Maecker, Z. Physik **136**, 119 (1953).
43. R. Weiss, Z. Physik **138**, 170 (1954).
44. T. Peters, Naturwiss. **41**, 571 (1954).
45. T. Peters, Internationale Tagung über Staustrahlen und Raketen in Freudenstadt, 1956. München: Oldenbourg, 1956.
46. T. Peters, Z. Physik **139**, 448 (1954).
47. H. Maecker, Z. Physik **141**, 198 (1955).
48. M. G. Malin, R. John, W. Bade, R. Schweiger et J. Yos, Comptes rendus, IXe Congrès International d'Astronautique, Amsterdam 1958, p. 445. Wien: Springer, 1959.

Explosion électrique

49. B. Eiselt, Z. Physik **132**, 54 (1952).
50. J. Abramson, S. Drabkina, M. Gegetschkori et S. Mandelstamm, Zhurn. Eksper. Teoret. Fiz. **17**, 862 (1947).
51. J. Ulam, Demande de Brevet Franç., 1959.

Measurement of Jupiter Re-entry Radiation

By

David D. Woodbridge[1] and **Warren N. Arnquist**[1, 2]

(With 10 Figures)

(Received June 18, 1959)

Abstract — Zusammenfassung — Résumé

Measurement of Jupiter Re-entry Radiation. The problem of the re-entry of an aerodynamically unstable body into the atmosphere is being approached by studying the radiation emitted by the booster of the Jupiter rocket as it returns to earth. During the flight the booster and an intermediate instrument compartment separate successively from the nose cone so that three objects re-enter separately. With the participation of a number of working groups, preliminary radiometric and photographic measurements have been made from the PbS infrared limit to the near ultraviolet and extensions to cover the 3—5 micron band are in progress. Although the radiation measured generally includes that from all three bodies, the effect of the booster usually dominates the situation because of its size. Re-entry velocities begin at about Mach number 14 corresponding to an adiabatic shock front temperature of nearly 4000° K. However, the actual temperatures behind this front are considerably less than this because of the effects of radiation and heat dissipation. Spectrographic records in the visible with low dispersion gratings and prisms indicate characteristic line spectra superposed on a continuum. Motion pictures help correlate the fluctuations and give the spatial relationships of the bodies. Meteor type cameras provide a continuous streak record. Best results are obtained with instruments located on board ships near the impact area although preliminary work has been done from an airplane and from a distant island. Difficulties with such operations are mentioned and the coordination of instruments at the several locations with appropriate timing circuits is described. Some of the interesting results from the first firings are given and plans for future tests are indicated.

Strahlungsmessungen bei der Rückkehr der Jupiter-Raketenspitze. Beim Wiedereintritt eines Flugkörpers in die dichtere Erdatmosphäre treten physikalische Erscheinungen auf, die sowohl für den Höhenforscher wie für den Ingenieur von großem Interesse sind. Eine Möglichkeit, die Erscheinungen des Wiedereintritts zu untersuchen, besteht in der Messung und im Studium der emittierten Strahlung, die der Hochgeschwindigkeitskörper in der hohen Atmosphäre erzeugt. Möglichkeiten für solche Messungen bieten die Flüge von Mittelstreckenraketen, deren Bahn genügend genau bekannt ist, um Meßstationen zu errichten.

„Operation Gaslight" ist die Bezeichnung eines Strahlen-Meßprogramms der US Army Ballistic Missile Agency, bei dem verschiedene Gruppen der Aerojet-General Corp., des Air Force Cambridge Research Center, des AVCO Research Laboratory und der Barnes Engineering Company, sowie der Army Ballistic Missile Agency und

---

[1] Research Projects Laboratory, Physics and Astrophysics Section, Army Ballistic Missile Agency, Redstone Arsenal, Alabama, U.S.A.

[2] Present Affiliation: Engineering Directorate, System Development Corporation, Santa Monica, California, USA.

der Army Rocket and Guided Missile Agency, Redstone Arsenal, Huntsville, teilge-
nommen haben.

Bei den Messungen selbst wurden sowohl Filmkameras wie Weitwinkel-Meteor-
Kameras mit und ohne Maßstabgitter verwendet. Strahlungsmesser und Photometer
zeichneten die Energien in verschiedenen Spektralbändern auf, von Infrarot bis Ultra-
violett. Diese Instrumente messen normalerweise die Strahlen aller drei Körper
zusammen (Spitze, Steuerungs- und Tankteil der Rakete), wenn man Systeme mit
5° verwendet. Versuche zur Verwendung von Geräten mit kleineren optischen Win-
keln stellten sich als sehr schwierig heraus, da der automatische Verfolger sich nach
dem hellsten Objekt richtet, und das ist der Booster. Photographische Aufzeichnungen
zeigen deutlich die drei Körper, aber hier ist es schwierig, quantitative Bestimmungen
der Energie auszuführen. Um den Nasenkegel zu untersuchen, muß der Tank- und
Instrumententeil stark überentwickelt werden.

Es wurde dabei ein starker Halo-Effekt aufgezeichnet, wahrscheinlich durch die
dünnen Zirruswolken, die bei einem solchen Versuch immer gerade vorhanden sind.
Ebenso störten niedrige Kumuluswolken, die im Schlußteil der Bahn als Vorhang
wirkten. Aus diesem Grund ist es sehr wünschenswert, die Registriergeräte auf zwei
oder mehr Stationen möglichst weit voneinander in der Nähe der Wiedereintritts-
Bahn aufzubauen. Die besten Ergebnisse wurden erzielt, wenn die Schiffe in 30 bis
60 km Entfernung auf einer Seite der Bahn stationiert wurden. Kleinere Entfernungen
ergaben Beobachtungsschwierigkeiten wegen des großen Elevationswinkels, größere
dagegen vermehrten die Störungen durch Wolken.

Die Festlegung der geometrischen Örter erfolgte optisch mit Hilfe eines manuell
eingestellten Teleskops.

Die Daten-Auswertung der „Operation Gaslight" ist noch im Gange. Die vorge-
tragenen Ergebnisse sind daher mehr qualitativ, als wir es wünschten. Bei den ersten
Versuchen, Strahlungs- und photometrische Daten zu erhalten, wurden die Instrumente
durch die Intensität der Strahlung über ihre Empfindlichkeit hinaus beansprucht.
Als erste Näherung kann dies als ein Ergebnis der heißen Oberflächenstrahlung des
Boosters im Gegensatz zur Gasstrahlung betrachtet werden. Schwankungen der Intensität der empfangenen Signale lassen eine Schwingung um einen mittleren Wert vermuten.

Abschätzungen der effektiven Strahlungstemperaturen wurden auf Grund der experimentellen Ergebnisse vorgenommen und die Ergebnisse mit Überschlagsrechnungen
verglichen. Daraus ergaben sich Temperaturen zwischen 2000 und 2500° K. Wenn eine
größere Zahl experimenteller Daten vorliegt, kann dieser Vergleich noch genauer
gemacht werden.

Mesures de rayonnement de la fusée Jupiter à la rentrée. Le problème de la rentrée
d'un engin aérodynamiquement instable est approché par l'étude du rayonnement
émis par la fusée d'appoint de Jupiter. Durant le vol la fusée d'appoint et un comparti-
ment instrumenté se séparent successivement du cône, de sorte que les trois objets
font leur rentrée séparément. Plusieurs groupes de travail ont participé à des mesures
préliminaires radiométriques et photographiques depuis la limite infra-rouge du
PbS jusqu'à l'ultra-violet proche. L'extension de ces mesures à la bande de 3 à 5
microns est en progrès. Quoique les mesures incluent le rayonnement des trois objets,
celui de la fusée d'appoint domine. Les vitesses de rentrée débutent aux environs de
Mach 14 correspondant à une température de choc adiabatique de 4000° K. Cependant
les températures réelles en aval sont considérablement inférieures par suite du rayonnement et de la convection. Les enregistrements spectrographiques dans le visible
avec des réseaux à faible dispersion et prismes montrent des lignes caractéristiques
superposées au mouvement continu. Les photographies sont une aide dans la correlation des fluctuations et donnent la disposition relative des trois objets. Des cameras
du type Meteor donnent un enregistrement continu sur bande. Les meilleurs résultats
sont obtenus par les instruments situés sur les bateaux au voisinage de l'aire d'impact,
quoique le travail préliminaire ait été fait d'avion et d'une île distante. Les difficultés
opératoires sont mentionnées et la coordination des instruments à l'aide de circuits
appropriés décrite. Quelques résultats intéressants des premiers essais sont donnés
ainsi que la planification des essais futurs.

The radiation emitted when a high speed body re-enters the atmosphere can be measured to provide an important source of information concerning the physical processes taking place. Both the upper atmosphere research worker and the missile designer can profit from such measurements. The effects on the bodies themselves are important to the space technologist in designing for the successful recovery of space bodies, or in making sure that such bodies are safely disintegrated before reaching the earth when recovery is not attempted. Thus the phenomena pertaining to the re-entry of both aero-dynamically stable and unstable bodies are of interest. Opportunities for such measurements are provided by the firing of long range ballistic missiles where the trajectory is known sufficiently for the location of instrument stations, and where both stable bodies (the nose cone) and unstable bodies (the booster stages) are involved.

Operation Gaslight is the name given to the re-entry radiation measurement program of the U.S. Army Ballistic Missile Agency. It was begun shortly after the first recovery in the United States of an intermediate range ballistic missile nose cone in August 1957. This firing, a scaled version of the intermediate range ballistic missile Jupiter nose cone, was made at the Atlantic Missile Range, Cape Canaveral, Florida, and the recovery took place at sea in an area north of the Leeward Islands with the help of several vessels of the U.S. Navy. Since the firing took place at night it was possible to judge the intensity of the visible radiation both by eye and by photographic means. As a result, it was decided to proceed with a measurement program using readily available equipment. In order to facilitate the physical interpretation of the phenomena, as much of the radiation as possible in the spectral range from the infrared to the ultraviolet has been recorded. This paper gives a progress report of the early operations, with some of the results of interest including recent tests.

A number of groups have participated in Operation Gaslight. These have included personnel from the Aerojet-General Corporation, Azusa, California, the Air Force Cambridge Research Center, Lexington, Mass., the Avco Research Laboratory, Boston, Mass., and Barnes Engineering Company, Stamford, Conn., as well as from the Army Ballistic Missile Agency and the Army Rocket and

Fig. 1. Jupiter missile AM-5 on launching table

Guided Missile Agency at the Redstone Arsenal. This cooperative program has been possible as a result of the efforts of a number of individuals from these organizations.

The first Gaslight radiation measurements were made as part of the successful recovery operations of the full scale Jupiter fired in May, 1958. The instrumentation was located on board two ships: the Navy destroyer USS Stickell and the Air Force telemetry ship, US MV Swordknot. Since this time, additional Jupiter

Fig. 2. Jupiter nose cone being hoisted aboard the recovery ship

firings at night have been utilized to obtain further data. A beginning has been made with land based instrumentation on Antigua, the nearest accessible island, and with airborne instruments.

The Jupiter missile consists of three parts, a power plant or booster rocket, a considerably smaller instrument compartment, and a still smaller nose cone. Fig. 1 shows a view of the assembled Jupiter on the firing pad. By far the largest part of the unit, the cylindrical section, is the booster. It is separated from the conical section just after burn-out. The instrument compartment is the base of the control section. Its chief function is to properly orient the nose cone for re-entry, after which it separates from the nose cone. Thus, there are three principal bodies that re-enter the atmosphere. Of these the nose cone is expected to be relatively stable, because of its design and orientation, while the other two should begin to tumble as soon as appreciable air friction is encountered. From qualitative energy and drag considerations, the booster is expected to be the largest radiation source and the first to be visible, while the nose cone should start radiating last and be the smallest source. Furthermore, the booster and instrument compartment should finally disintegrate, but the nose cone, equipped with a parachute for the final descent, is recovered. A view of the nose cone is given by Fig. 2 showing it on board the recovery vessel.

In the field operations both motion picture and wide angle meteor or ballistic cameras have been used with and without replica gratings. Radiometers and photometers have recorded the energy in a number of spectral bands, defined by filters

Fig. 3a. Ten seconds after first visual sighting the tank unit (lower) and instrument compartment (upper) begin to glow intensely

Fig. 3b. Three seconds later, the nose cone heats to incandescence and is shown just ahead of the tank unit

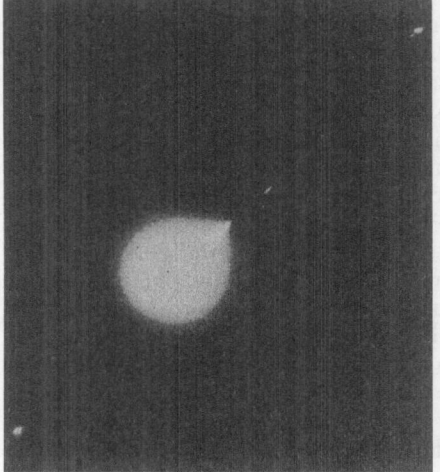

Fig. 3c. At 15 seconds the instrument compartment (upper right) is breaking up and the tank unit (center) is also losing parts. The nose cone is seen in the lower left corner

Fig. 3d. Three seconds later only the tank unit (upper right) and the nose cone (lower left) remain visible

or receiver cut-offs, from the infrared to the ultraviolet. The latter instruments usually record the radiation from all three bodies when optical fields about 5° are used. Attempts to use smaller fields to pick up only the radiation from the nose cone, for example, have been unsuccessful so far as the operator tends to follow the brightest object, which is the booster. The photographic records clearly show the separate bodies but here it is difficult to make quantitative determinations of the energy. In order to record the nose cone it is necessary to greatly over-expose for the booster and instrument compartment. A strong halation effect has been encountered in the photographic records which has been traced to forward scattering in the atmosphere, probably as the radiation passes through the thin cirrus clouds which seem to be always present at the time of a test. Also low cumulus clouds have been encountered which have acted as a curtain for the later parts of the trajectory. Because of this it is highly desirable to locate the recording instruments at two or more widely separated locations in the vicinity of the re-

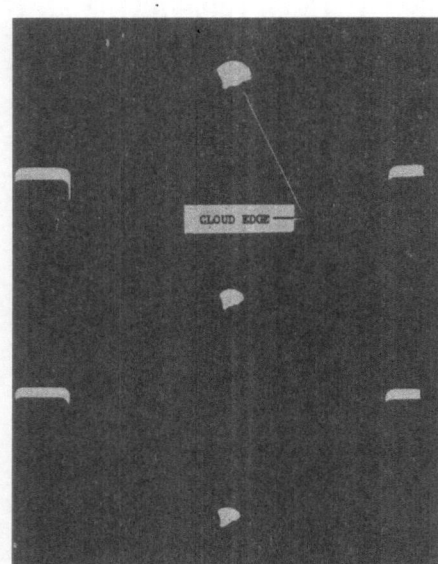

entry trajectory. Best results have been obtained by locating the ships about 20 to 40 miles to one side of the trajectory. Closer distances result in tracking difficulties with the high elevation angles involved, and much greater distances involve more interference from clouds. One cumulus cloud was a blessing in disguise as will be shown. Acquisition and tracking have been manual with visual sighting through a low power boresighted telescope.

The preliminary phase of Operation Gaslight is still in progress. Thus the results to be reported are more qualitative than we would like. Fig. 3 shows four frames from a 35 mm motion picture camera located about 50 miles away and to the side on the first test. The booster and instrument compartment are shown first about 10 seconds after visual sighting when the nose cone is just making its appearance. The succeeding frames are after 3, 5 and 9 seconds respectively. They

Fig. 4. 16 mm boresight camera flames of AM-6A re-entry. Re-entry bodies obscured by cumulous cloud. Approximately 7.7 sec after first sighted

show the gradual development of the nose cone radiation, the relatively enormous and variable intensity from the booster, and the disintegration and burnout of the instrument compartment. Examination of the film shows that fragments detach themselves from the booster and instrument compartment, lag behind because of their relatively greater drag to weight ratios, and suddenly become extinguished presumably as they burn up. By measuring the separations of the objects on this film, it is confirmed that the lower left hand object is the nose cone since it moves away from the other two. The decleration of the two larger bodies relative to the nose cone is progessively greater until final disintegration when it rises sharply. The extreme brilliance of the booster shown in the third frame is probably the result of the explosion of the almost empty fuel tanks.

Fig. 4 shows the booster passing behind a cumulus cloud in the second test. The instrument compartment has disappeared by this time and the nose cone has already been eclipsed. From the sharpness of the cloud edge we have concluded that most of the halation observed on the photographic records is due to atmospheric scattering and not an over-exposure effect on the film. This seems reasonable because of the cirrus clouds. The halation appears to be of the order of a degree. It is present not only with the over-exposed booster image but with the nose cone as well.

Fig. 5. Re-entry photographed from Antigua, BWI

Fig. 6. Cluster of six F-8 aerial reconnaissance cameras, 12 inch focal length f/4.5 lens, modified for use as spectral meteor cameras using 300 lines/mm Bausch & Lomb transmission gratings with 200 cycles/minute time chopper, attached to 5 inch gun barrels

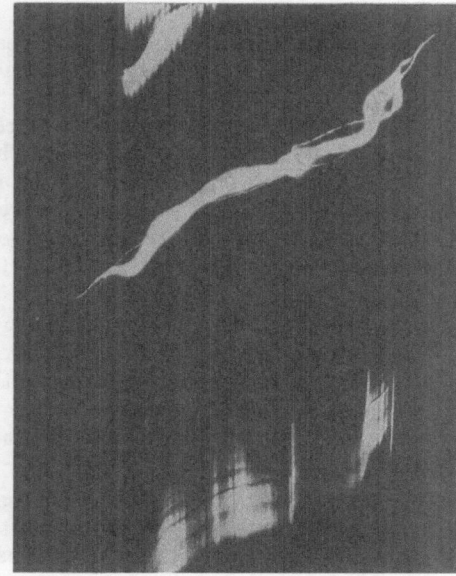

Fig. 7. Spectrogram recorded during re-entry of AM-5

The wide angle meteor type ballistic cameras give a complete record on a single picture. When ballistic cameras are located on Antigua, pictures such as Fig. 5 are obtained. This shows the streaks due to each of the three bodies, with a fourth faintly visible line apparently due to a sizable fragment that became detached from the booster, possibly before re-entry. Again the halation effect is present and evidence of breakup may be seen on close inspection.

Fig. 8. Aerojet General's M2B portable radiometer with collimated 16 mm cine camera on tripod. From left to right are the power supply, amplifier and oscillograph recorder

Similar ballistic cameras have been used on board ship. Fig. 6 shows a cluster of six wide angle aerial reconnaissance cameras arranged to cover an area of sky approximately 100° in azimuth and 70° in elevation. Each of these cameras was fitted with a 300 line/mm. replica transmission grating and a two blade 100 rpm chopper for a local time reference, the group being mounted between two 5 inch guns on the destroyer. Fig. 7 shows a record of the first test and so may be compared with Fig. 3. The zero order traces are shown, the one due to the booster showing that the extremely high intensity persists almost to the very end. As before the trace due to the instrument compartment dies out near the middle of the record and the nose cone starts late. The ship's motion accounts for the wavy lines. Near the beginning the rather sharp motion serves as a time marker to differentiate the three bodies. Thus we see the same spatial relationship as in Fig. 3 namely the nose cone followed by the booster and instrument compartment in that order. The first order spectrum shown is practically all from the booster, of course. It shows both emission and absorption lines and bands with a continuum background developing in the later stages.

Characteristic lines, such as those from aluminum oxide and sodium and the strong atmospheric absorption due to water vapor, are evident. The analysis of this and other spectra is in progress.

The radiometer and photometer records have been due mostly to the booster radiation as indicated. Fig. 8 shows the Aerojet radiometer and four channel recorder together with a boresighted 16 mm cine camera and tracking telescope. Fig. 9 shows some typical results in the 1.75 to 2.6 micron band using a lead sulphide receiver. The four channels have sensitivities spaced by approximate factors of 10. Even this large dynamic range was not large enough as about 1/2 sec. of the record was lost off scale. As a first approximation this may be considered

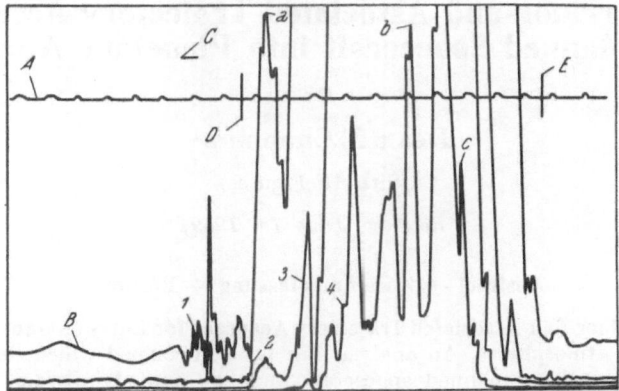

Fig. 9. Trace of oscillograph radiation data of re-entry bodies Jupiter AM-5 (U).

A Time pulse correlated with Patrick Air Force Base standard timing pulses: 1 pip per sec; *B* Background noise level; *C* Time when re-entry bodies were visually sighted; *D* Time when re-entry bodies were first acquired with M2B radiometer; *E* Time at which re-entry bodies IR radiation was attenuated due to clouds.
Radiant intensity levels: *a* 14 kw/ster, *b* 737 kw/ster, *c* 429 kw/ster.
Radiometer recording channels (attenuation ranges): *1* 0—10, *2* 0—100, *3* 0—1000, *4* 0—10.000

due to the hot surface radiation from the booster as contrasted with the gas radiation. The fluctuations suggest a changing aspect of the booster as in oscillating about a mean value. Similar results have been obtained in the other channels as

Fig. 10. AVCO Research Laboratory's MK II radiation recording system

recorded with the Avco apparatus for example as shown in Fig. 10. Using photomultipliers and appropriate interference filters, eight spectral bands have been covered centered at from 3600 Å to 10,000 Å.

Estimates of the effective radiation temperatures have been made from the experimental results and compared with rough calculations. These indicate an effective gas temperature between 2000 and 2500° K. As more experimental data are acquired, this comparison will be made more closely.

On the Corridor and Associated Trajectory Accuracy for Entry of Manned Spacecraft into Planetary Atmospheres

By

Dean R. Chapman[1]

(With 10 Figures)

(Received June 18, 1959)

Abstract — Zusammenfassung — Résumé

On the Corridor and Associated Trajectory Accuracy for Entry of Manned Spacecraft into Planetary Atmospheres. An analysis has been developed which determines the corridor through which manned spacecraft must be guided in order to avoid excessive deceleration for human occupants and yet to encounter sufficient deceleration for completing entry. The analysis introduces a dimensionless parameter coupling the aerodynamic characteristics of the vehicle with certain planetary characteristics evaluated at the perigee altitude corresponding to the approach conic trajectory. This perigee parameter conveniently bridges the two-body orbit equations to the re-entry motion equations, and provides a general basis for specifying the corridor width for entries from either elliptic, parabolic, or hyperbolic approach trajectories. The results apply to vehicles of arbitrary weight, shape, and size entering a planetary atmosphere. Illustrative calculations are presented for Earth, Venus, Mars, Jupiter, and Titan.

It is shown that the altitude of an entry corridor depends strongly on the vehicle weight, size, and drag coefficient, but that the corridor width between its overshoot and undershoot boundaries is independent of these characteristics. For certain planets (Earth, Venus, Jupiter) the corridor width is much greater for vehicles with aerodynamic lift than for nonlifting vehicles, but for other planets (e.g., Mars, Titan) aerodynamic lift cannot effectively broaden the entry corridor. For example, the 10-Earth-G corridor width for single-pass parabolic entry of a non-lifting vehicle can be increased from 0 kilometers for Jupiter, 11 for Earth, and 13 for Venus, to 83, 82, and 83 kilometers, respectively, by employing a vehicle with a lift-drag ratio of 1; the corresponding corridor widths for Mars and Titan cannot be similarly increased very much beyond the values of 650 and 2,200 kilometers, respectively, corresponding to vehicles without lift. For any lift-drag ratio the corridor width decreases rapidly as the entry velocity is increased (e.g., for nonlifting vehicles entering the earth's atmosphere, the corridor decreases from about 290 kilometers wide at circular velocity, to 11 at parabolic velocity, to 0 at hyperbolic velocities greater than $1.8 \times$ circular).

The guidance requirements on accuracy of velocity and flight-path angle as determined by the corridor width are compared with the corresponding guidance requirements for other technological missions such as those for putting a vehicle into orbit, for hitting the moon from the earth, and for achieving intercontinental ballistic missile accuracy.

---

[1] Aeronautical Research Scientist, National Aeronautics and Space Administration, Ames Research Center, Moffett Field, California, U.S.A.

Über die Genauigkeit von „Korridor"- und verwandten Bahnen beim Eintritt bemannter Raumfahrzeuge in Planetenatmosphären. In der vorliegenden Arbeit wird eine Analyse zur Bestimmung des Korridors ausgeführt, durch den ein bemanntes Raumfahrzeug geführt werden muß, um übermäßige Verzögerung für die menschliche Bemannung zu vermeiden und trotzdem ausreichende Verzögerung für das Gelingen des Eintritts in die Atmosphäre zu erreichen. Die Analyse führt einen dimensionslosen Parameter ein, der die aerodynamischen Kennzeichen des Raumfahrzeuges mit bestimmten planetaren Charakteristiken verbindet, die bei der Höhe des Perigäums ausgewertet werden, das der konischen Annäherungsbahn entspricht. Dieser Perigäumparameter verbindet in geeigneter Weise die Zwei-Körper-Bahngleichungen mit den Bewegungsgleichungen für den Wiedereintritt und schafft eine allgemeine Grundlage für die Feststellung der Korridorweite für die Rückkehr entweder aus elliptischer bzw. parabolischer oder hyperbolischer Annäherungsbahn. Die Ergebnisse beziehen sich auf Raumfahrzeuge von willkürlich anzunehmendem Gewicht, Gestalt und Größe, die in eine Planetenatmosphäre eindringen. Illustrative Rechnungen werden für die Erde, Venus, Mars, Jupiter und Titan angeführt.

Es wird gezeigt, daß die Höhe des Eintrittskorridors sehr von Gewicht, Größe und Widerstandskoeffizient des Fahrzeuges abhängen, daß aber die Korridorbreite zwischen seinen obersten und untersten Grenzen unabhängig von diesen Kennzeichen ist. Im Falle bestimmter Planeten (Erde, Venus, Jupiter) ist die Korridorbreite viel größer für Fahrzeuge mit aerodynamischem Auftrieb als für Fahrzeuge ohne Auftrieb, aber bei anderen Planeten (z. B. Mars, Titan) kann der aerodynamische Auftrieb den Eintrittskorridor nicht wirksam verbreitern. Beispielsweise kann die Korridorbreite von 10 irdischen g für einen parabolischen Einpaßeintritt eines auftriebslosen Fahrzeuges von 0 km für Jupiter, 11 für die Erde und 13 für die Venus auf 83, bzw. 82 und 83 km erhöht werden, wenn ein Fahrzeug mit einem Auftrieb-Widerstand-Verhältnis von 1 verwendet wird. Die entsprechenden Korridorbreiten für Mars und Titan können nicht in ähnlicher Weise sehr stark über die Werte von 650 bzw. 2200 km erhöht werden, welche auftriebslosen Fahrzeugen entsprechen. Für irgendein Auftrieb-Widerstand-Verhältnis nimmt die Korridorbreite schnell ab, wenn die Eintrittsgeschwindigkeit vergrößert wird (z. B. nimmt für auftriebslose Fahrzeuge, welche in die Erdatmosphäre eintreten, der Korridor von ungefähr 290 km Breite bei Kreisgeschwindigkeit auf 11 bei parabolischer Geschwindigkeit und auf 0 bei hyperbolischen Geschwindigkeiten ab, die größer als das 1,8fache der Kreisbahngeschwindigkeit sind).

Die Lenkungserfordernisse hinsichtlich der Genauigkeit von Geschwindigkeit und Flugwegwinkel, wie sie durch die Korridorbreite bestimmt werden, werden mit den entsprechenden Lenkungserfordernissen für andere technologische Aufträge verglichen, wie z. B. dem, ein Fahrzeug in die Umlaufbahn zu bringen, von der Erde aus den Mond zu treffen, und für die Verwirklichung genauer interkontinentaler ballistischer Geschosse.

Précision de la trajectoire et corridor de rentrée d'un astronef piloté dans une atmosphère de planète. Analyse du corridor dans lequel un astronef doit être guidé pour obtenir une décélération suffisante et cependant tolérable par les occupants. Un paramètre sans dimension couple les caractéristiques aérodynamiques du véhicule avec certaines caractéristiques planétaires évaluées au périgée de la conique d'approche. Il assure la transition entre les équations du problème des deux corps avec celles de la trajectoire de rentrée et forme la base générale permettant de spécifier la largeur du corridor dans une approche du type elliptique, parabolique ou hyperbolique. Les résultats sont applicables aux véhicules de forme, poids et dimensions arbitraires pénétrant une atmosphère planétaire. Application numérique est faite pour la Terre, Venus, Mars, Jupiter et Titan.

L'altitude d'un corridor de rentrée dépend fortement du poids, des dimensions et du coefficient de trainée du véhicule. Mais la largeur du corridor est indépendante de ces caractéristiques. Pour certaines planètes (Terre, Venus, Jupiter) cette largeur est plus grande en présence de portance; pour d'autres (Mars, Titan) la portance ne peut élargir appréciablement le corridor. Par exemple la largeur d'un corridor de

10 g (terrestres) pour une passe d'approche simple parabolique peut passer de 0 kilomètres pour Jupiter, 11 pour la Terre et 13 pour Venus à respectivement 83, 82 et 83 par l'emploi d'un rapport portance/trainée de 1. Les largeurs correspondantes pour Mars et Titan ne peuvent être accrues sensiblement au delà des valeurs de 650 et 2200 kilomètres obtenues sans portance. Quel que soit le rapport portance//trainée la largeur décroit rapidement avec la vitesse d'approche. Par exemple pour une approche sans portance dans l'atmosphère terrestre, elle décroit de 290 kilomètres, à la vitesse orbitale, à 11, à la vitesse de libération, et 0 pour des vitesses hyperboliques, supérieures à 1,8 fois la vitesse orbitale.

La précision du guidage en vitesse et angles, déterminée par la largeur du corridor, est comparée à celle requise par d'autres performances telles que la mise en orbite, la capture lunaire et la précision balistique intercontinentale.

Introduction

It is generally anticipated that the first entry of a manned space vehicle into the earth's atmosphere will be from a near-circular orbit. Because of this the entry problems of deceleration and heating have been studied for near-circular velocities much more than for supercircular velocities. In the hopefully near future, entry at essentially parabolic velocity will be of practical concern (e.g., upon return from a manned moon mission), as will be entry at hyperbolic velocities in the more distant future.

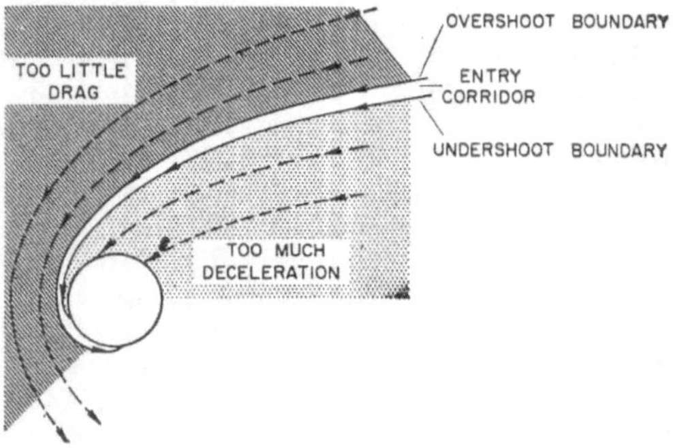

Fig. 1. Entry corridor for manned spacecraft

One problem which is important for supercircular entries, yet is relatively unimportant for near-circular entries, is that associated with the guidance requirements for avoiding excessive deceleration or heating during entry. In terrestrial flight of aircraft, an undershoot approach in landing is easily corrected by a brief application of a small amount of power, and an overshoot approach is easily corrected by making a return pass. In space flight, undershoot or overshoot approaches can also be corrected by analogous procedures, but the consequences are much more severe. If a guidance error causes the spacecraft to undershoot the intended trajectory too much, as illustrated by the inner two dashed trajectories in Fig. 1, the vehicle will enter the atmosphere at an excessively steep angle and experience excessive deceleration for human occupants. It would take a relatively large amount of fuel to correct the trajectory once the vehicle is within or near the planet's atmosphere. If, at the other extreme, a guidance error results in overshooting the intended trajectory too much, as illustrated by the outer dashed trajectories in Fig. 1, then the vehicle will not encounter enough atmosphere for slowing sufficiently to complete entry, and will have to make a return pass. Return passes not only would require additional traverses of any radiation belt around the planet, but also would require in most cases an additional retrorocket thrust to complete entry in other than a large number of such passes. After excluding approach trajectories representing overshoot and excessive undershoot (shaded portions in Fig. 1), all that is left for some planets is a meagerly narrow corridor through which a spacecraft must be guided. The outer and inner boundaries of this single-pass entry corridor are referred to herein as the overshoot and undershoot boundaries, respectively.

This paper summarizes some results of a general study of the entry corridor and its boundaries. Entry guidance and aerodynamic heating problems are considered for spacecraft having various lift-drag ratios entering various planets at velocities between circular and twice circular velocity. A more detailed account of the research summarized herein, including development of the mathematical aspects of the analysis, may be found in [1]. The present paper attempts only to describe certain results; essential equations are stated and explained, but not derived. The results described are based on several hundred solutions to a transformed, nonlinear, differential equation representing entry motion in a planetary atmosphere.

Results and Discussion

Dimensionless Parameters Characterizing Entry

Many of the symbols employed in characterizing supercircular entries and in presenting results which follow are illustrated in Fig. 2. A spacecraft of mass m enters a planet's atmosphere along the trajectory represented by the solid curve. This curve intercepts the "sensible atmosphere" at (i), at which point the flight-path angle is equal to the "initial angle" γ_i. This angle conventionally is employed to characterize an entry motion but is not employed herein. For shallow supercircular entries which just graze an outer edge of atmosphere, appropriate initial conditions are cumbersome to specify because, at all points along that portion of approach trajectory for which aerodynamic forces are negligible (that is, at points for which $y > y_i$), the flight-path angle γ, velocity V, and altitude y change continuously as the initial altitude y_i is approached. Moreover, the appropriate initial γ_i and

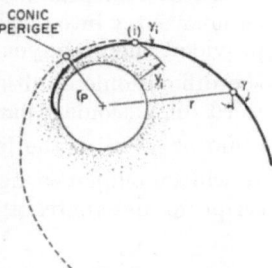

Fig. 2. Sketch illustrating notation

y_i are different for each vehicle and each approach angle. If the atmosphere were not present and the spacecraft encountered no drag, the vehicle would continue along a conic trajectory and pass through a perigee point designated in Fig. 2 as the conic perigee. The actual entry trajectory does not pass through this conic perigee point because it is diverted by aerodynamic forces. The velocity V_p, altitude y_p, and flight-path angle $(\gamma_p = 0)$ at the conic perigee, however, are unique and unchanging for all points along the approach trajectory; thus the conic perigee does not depend on where the sensible atmosphere begins. Primarily for this reason, a parameter based on conditions at the conic perigee, and not on γ_i, is employed for convenience in characterizing supercircular entries.

The mathematical basis for the significance of the particular perigee parameter F_p used throughout this report is not established herein (see [1]); it will suffice to note that this parameter can be applied to any exponential planetary atmosphere and to vehicles of arbitrary weight, size and shape. The perigee parameter couples characteristics of the vehicle with characteristics of the atmosphere at the conic perigee. It is defined as

$$F_p \equiv \frac{\varrho_p}{2\,(m/C_D A)}\,\sqrt{\frac{r_p}{\beta}} \qquad (1)$$

where r_p is the radius to the conic perigee altitude, ϱ_p is the density at this altitude, C_D is the vehicle drag coefficient based upon a fixed reference area A, and β is the exponential decay parameter for the atmosphere defined by

$$-\frac{d}{dy}(\ln \varrho) = \beta . \tag{2}$$

The parameter β is the reciprocal of the scale height of the atmosphere, and is related to the characteristics of the atmosphere through the well-known equation

$$\beta = \frac{Mg}{RT} . \tag{3}$$

Here M is the molecular weight of the atmosphere, T the temperature, R the universal gas constant, and g the gravitational constant. The dimensionless initial velocity used in presenting subsequent results is defined as

$$\overline{V}_i \equiv \frac{V_i}{\sqrt{gr}} . \tag{4}$$

This has the physical significance that $\overline{V}_i = 1$ represents circular entry, $1 < \overline{V}_i < \sqrt{2}$ elliptic entry, $\overline{V}_i = \sqrt{2}$ parabolic entry, and $\overline{V}_i > \sqrt{2}$ hyperbolic entry. For shallow entries $\overline{V}_i$ is equal to $\overline{V}_p$, the dimensionless velocity at conic perigee.

Three dimensionless parameters completely characterize any shallow super-circular entry into a spherically symmetric, nonrotating, exponential atmosphere, provided the aerodynamic coefficients are sensibly constant. One parameter is the dimensionless initial velocity $\overline{V}_i$; another is the perigee parameter F_p. The third dimensionless parameter combines certain characteristics of the planetary atmosphere with the lift-drag ratio of the vehicle, and is equal to $\sqrt{\beta r}\,(L/D)$. It will be employed here in the normalized form $\sqrt{(\beta r)_\oplus}\,(L/D)$, where the subscript $\oplus$ designates a value relative to that for the earth, for example,

$$\sqrt{(\beta r)_\oplus} \equiv \frac{\sqrt{\beta r}}{\sqrt{(\beta r)_{Earth}}} . \tag{5}$$

These three dimensionless quantities determine the overshoot boundary of the entry corridor, and also the undershoot boundary for any given value of the dimensionless maximum deceleration:

$$\overline{G}_{max} \equiv \frac{G_{max}}{g_\oplus \sqrt{(\beta r)_\oplus}} \frac{\sqrt{1 + [\sqrt{(\beta r)_\oplus}\,L/D]^2}}{\sqrt{1 + (L/D)^2}} . \tag{6}$$

The dimensionless deceleration $\overline{G}$ is equal to G, the deceleration in earth sea level G's, only for entry into the earth's atmosphere. Inasmuch as the entry corridor width depends upon the maximum deceleration arbitrarily selected, it follows from eq. (6) that the calculation of corridor widths for various planets requires a knowledge of $g_\oplus$ and $\sqrt{(\beta r)_\oplus}$. Approximate values of these parameters for different celestial objects of the solar system are listed in the following table (also listed for later use are values of the altitude increment $\Delta_{10}y$ over which the density changes by a factor of 10):

	$g_\oplus$	$\sqrt{(\beta r)_\oplus}$	$\Delta_{10}y$ for 10:1 change in ϱ, km
Venus	0.87	1.0	14
Earth	1.00	1.00	16
Mars	.38	.47	42
Jupiter	2.6	2.0	42
Titan	.22	.27	70

In the calculations of aerodynamic heating, additional equations and planetary constants would be needed. These are given in [1]. The main qualitative aspects of the aerodynamic heating problem are discussed with the aid of two simple equations presented later.

Corridor Width

The entry corridor width can be calculated readily from the distance between the overshoot and undershoot boundaries on a plot wherein $\log_{10} F_p$ is used as one of the coordinates. From the defining eq. (1), the corridor with $y_{p_{ov}} - y_{p_{un}}$ between conic perigee altitudes at overshoot and undershoot is simply

$$\frac{y_{p_{ov}} - y_{p_{un}}}{\Delta_{10}\, y} \equiv \frac{y_{p_{ov}} - y_{p_{un}}}{(2.303/\beta)} = \log_{10} \frac{(F_p m/C_D A)_{un}}{(F_p m/C_D A)_{ov}}. \tag{7}$$

It follows that, if $m/C_D A$ is the same at overshoot and undershoot, then the corridor width represented by an increment of one unit in $\log_{10} F_p$ is simply equal to $\Delta_{10}\, y$, the quantity tabulated above.

The importance of guidance problems for supercircular entries, and the relative unimportance for near circular entry, may be seen from the curves in Fig. 3 which show overshoot and undershoot boundaries of the entry corridor as a function of the initial velocity. In the dimensionless coordinates employed, these curves can be applied to any planet. The spacing between values of $\log_{10} F_p$ which represents a 100 km corridor width in the earth's atmosphere (for constant $m/C_D A$) has been indicated for reference in Fig. 3. The curves apply for the case $\sqrt{(\beta r)_\oplus}\, |L/D| = 1$, which represents $|L/D| = 1$ for the earth's atmosphere ($L/D = -1$ along the overshoot boundary, and $L/D = +1$ along undershoot boundary). Curves for other lift-drag ratios show exactly the same trends, namely, a pronounced decrease in the corridor width (proportional to the spacing between the overshoot boundary and the undershoot boundary in Fig. 3) as the entry velocity is increased. In the case of $\overline{G} = 5$, for example, the corridor width for parabolic entry ($\overline{V}_i = \sqrt{2}$) is only about 1/13th as wide as for circular entry, and for hyperbolic entry at $\overline{V}_i = 2.0$, is only a few percent of the corresponding width for circular entry.

Although the curves in Fig. 3 are designated as applying to a fixed value of $\sqrt{(\beta r)_\oplus}\, |L/D|$, the L/D need not be held constant during an entire entry. At overshoot L/D would have to be constant only until the point of local circular velocity is reached, and at undershoot, only until the point of maximum deceleration is reached. Beyond these points L/D could be varied and the curves would still apply.

One qualitative result evident from a study of the three different undershoot boundaries shown in Fig. 3, concerns the influence of the arbitrarily selected maximum deceleration on the corridor width. The corridor width for $\overline{G}_{max} = 10$

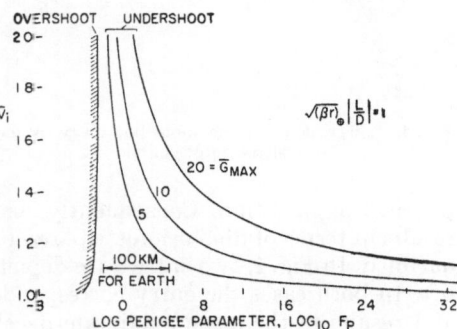

Fig. 3. Effect of initial velocity on overshoot and undershoot boundaries of entry corridor

is approximately twice that for $\overline{G}_{max} = 5$, and likewise the corridor width for $\overline{G}_{max} = 20$ is about twice that for $\overline{G}_{max} = 10$. This rough proportionality of corridor width to $\overline{G}_{max}$ applies throughout the range of entry velocities considered in Fig. 3, but it applies only to vehicles with substantial lift-drag ratios, that is, with lift-drag ratios such that $\sqrt{(\beta r)_{\oplus}}\ L/D > 0.5$ approximately (see [1]). For vehicles without lift the corridor widths are not even approximately proportional to the value of $\overline{G}_{max}$; for example, the values of $(y_{p\,ov} - y_{p\,un})/\Delta_{10}\,y$ for $\overline{G}_{max}$ of 5, 10, and 20 are 0, 0.7, and 2.0, respectively. Nonlifting vehicles do exhibit, however, the same trend of decreasing corridor width with increasing entry velocity as exhibited by lifting vehicles; for example, the values of $(y_{p\,ov} - y_{p\,un})/\Delta_{10}\,y$ with $\overline{G}_{max} = 10$ are 18, 0.7, and 0, for $\overline{V}_i$ of 1.0, 1.4, and 1.8, respectively. The corresponding corridor widths for entry into Earth $(\Delta_{10}\,y = 16$ km), for example, would decrease from 290 km at $\overline{V}_i = 1$, to 11 at $\overline{V}_i = 1.4$, to 0 at $\overline{V}_i \geq 1.8$. These values apply to the case where $m/C_D\,A$ is the same at overshoot as at undershoot.

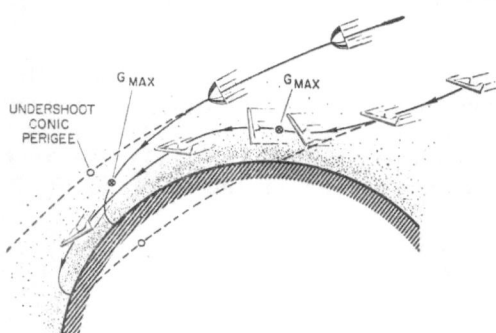

Fig. 4. Lowering the undershoot boundary by means of positive aerodynamic lift

The results just discussed, illustrating how the corridor shrinks as the entry velocity increases and broadens as the permissible deceleration increases, are understandable from simple physical considerations. The results to be presented shortly concerning the influence of positive and negative L/D on the corridor boundaries, however, are less easily understood without some simple physical explanation. Consequently, before presenting additional mathematical results in terms of dimensionless parameters, a simple physical explanation will be outlined. In Fig. 4, two entries are depicted, one without lift and one with positive lift. In both cases the entry corresponds to that along the undershoot boundary and results in the same maximum deceleration. The dotted lines represent what the trajectories would have been had there been no aerodynamic force; they pass through their respective conic perigee points corresponding to the undershoot boundary. The nonlifting vehicle descends in a smooth monotonic fashion, experiencing its maximum deceleration at the point indicated. The lifting vehicle has the same value of $m/C_D\,A$, but, as it enters the atmosphere, the transverse lifting force deflects the trajectory away from the planet's center. Maximum deceleration is experienced at roughly the same altitude as for the nonlifting vehicle, even though the lifting vehicle enters at a much steeper angle and would pass through a conic perigee much closer to the planet's center. Once the vehicle passes the point of maximum deceleration it would skip out of the atmosphere if the L/D were held constant. Consequently, in order to avoid this skip and complete entry on the first pass, L/D must be reduced essentially to 0, and in some cases to small negative values, shortly after experiencing maximum deceleration (a mathematical demonstration of this has been given by Lees, Hartwig, and Cohen in [2]). A reduction in L/D can be accomplished either by reducing the angle of attack α of a lifting surface if the vehicle is operating in the low-drag portion of its drag polar, or by increasing α if operating in the high-drag portion. The latter, or high-drag mode of reducing L/D is preferable because it results in much less aerodynamic heating. Hence the reduction in L/D in Fig. 4 is represented by an increase in angle of attack of the lifting surface just following the point of maximum deceleration. After the vehicle has slowed substantially in this high-drag attitude of small lift-drag ratio, it can then begin to reduce its angle of attack, thereby increasing the lift-drag ratio and entering into an extended terminal glide phase, as represented by the last attitude sketched for the lifting vehicle. By the use of lift in this manner the undershoot boundary for entry into certain planets can be lowered considerably from that for nonlifting vehicles, as will be evident subsequently.

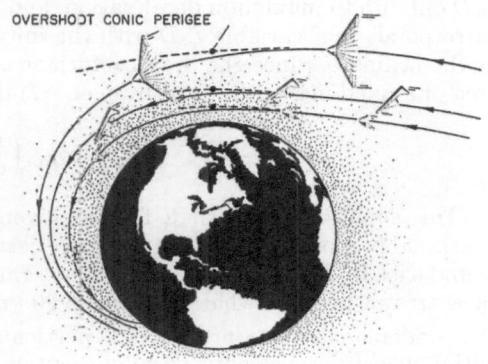

In addition to the capability of lowering the undershoot boundary of the corridor, the judicious use of aerodynamic lift can also raise the overshoot boundary. In this case, however, the direction of application of the lift force with respect to the planet's center must be inverted. This is illustrated by the sketches in Fig. 5. The three trajectories shown are intended to represent those of the overshoot boundary in each case. At overshoot aerodynamic forces are relatively small and the actual

Fig. 5. Raising the overshoot boundary by means of negative aerodynamic lift or drag device

trajectory will pass only a small distance below the conic perigee point, as illustrated by each of the three trajectories in Fig. 5. The nonlifting, projectilelike configuration has the lowest perigee altitude of the three. If this nonlifting

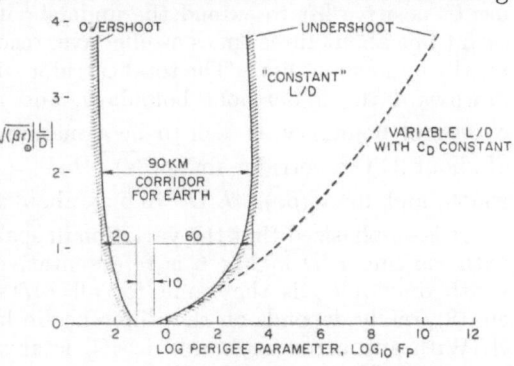

vehicle aimed at a conic perigee altitude which was slightly higher than that of the overshoot boundary, the vehicle would pass through an outer edge of atmosphere, exit from the atmosphere, orbit, and then have to make a return pass in order to complete entry. If the vehicle aimed at any perigee altitude lower than that for overshoot, entry would be completed on the first pass. A vehicle with aerodynamic lift can aim at a higher conic perigee altitude, as illustra-

Fig. 6. Effect of lift-drag ratio on parabolic entry corridor

ted by the middle trajectory in Fig. 5 and still complete entry in a single pass, provided the aerodynamic lift is directed toward the planet's center (negative lift). This upward extension of the overshoot boundary through the application of negative lift is not really large, however, because there is not much atmosphere at the higher altitudes to work with in deflecting the vehicle's path. As we will see later, the overshoot boundary can be extended upward to a considerably greater degree by the use of a light, high-drag device, than by the use of negative aerodynamic lift. This is indicated schematically by the outer trajectory in Fig. 5. The drag device may be thought of as consisting of metallic cloth fabricated from very thin threads of a high-temperature alloy, and as being deployed over as large an area normal to the direction of flight as is practical. In this way the drag chute encounters a sufficient mass of air to complete entry in a single pass even though it aims at a much higher conic perigee altitude than the other vehicles.

The trends anticipated from the foregoing physical considerations are illustrated by the curves representing overshoot and undershoot on a plot of the dimensionless lift-drag parameter $\sqrt{(\beta r)_\oplus} \, |L/D|$ versus the perigee parameter $\log_{10} F_p$, as presented in Fig. 6. The solid undershoot boundary corresponds to constant L/D only up to maximum deceleration, and the dotted boundary (discussed later) corresponds to a variable L/D with the initial L/D at entry equal to that plotted as the ordinate. Since the drag coefficient usually is referred to a fixed reference area of a fixed mass, it follows from eq. (7) that the corridor width in such cases is

$$ y_{p\,ov} - y_{p\,un} = \varDelta_{10}\, y \left[\log_{10}\left(\frac{F_p}{C_D}\right)_{un} - \log\left(\frac{F_p}{C_D}\right)_{ov} \right]. \tag{8} $$

This equation shows that, for any given value of $\sqrt{(\beta r)_\oplus} \, |L/D|$, the corridor width is proportional to the spacing between the overshoot and undershoot boundaries in Fig. 6, provided C_D is the same at these two boundaries. The reference arrows indicate what would be a 90 km corridor for the earth's atmosphere. The undershoot boundary in this particular plot represents $\overline{G}_{max} = 10$ and the initial velocity represents parabolic entry. As is evident from this figure, the corridor width for lifting vehicles is considerably broader than for nonlifting ones having the same C_D. For the earth's atmosphere the corridor width would be 10 km for all nonlifting vehicles, irrespective of the value of $m/C_D A$ (as long as it is the same at undershoot and overshoot). The use of negative lift to extend the overshoot boundary toward smaller values of F_p (corresponding to smaller densities and higher altitudes) broadens the earth's corridor by only about 20 km. The use of positive lift to extend the undershoot boundary to higher values of F_p is seen to be about three times as effective, resulting in an extension of about 60 km for the undershoot limit. The total corridor width, with positive lift used to extend downward the undershoot boundary, and negative lift to extend upward the overshoot boundary, is seen to be a maximum of about 90 km at $\sqrt{(\beta r)_\oplus} \, |L/D|$ of about 2. The corridor for $\sqrt{(\beta r)_\oplus} \, |L/D| = 1$ is almost as wide as the maximum width, and for $\sqrt{(\beta r)_\oplus} \, |L/D| = 0.5$ is about 2/3 as wide as the maximum width.

It is emphasized that the variation in spacing between the corridor boundaries with varying L/D in Fig. 6 is representative of the actual variation in corridor width only if C_D is the same for all L/D's. The corridor width, according to eq. (8), really depends on the difference in $\log F_p/C_D$ for vehicles of fixed m and A. With any aerodynamic device C_D is always coupled to L/D; large L/D's in hypersonic Newtonian flow, for example, are obtained only with slender configurations having small C_D, whereas large C_D's are obtained with blunt configurations having small L/D. If the aerodynamic coupling between L/D and C_D is taken as that corresponding to lifting surfaces wherein both L/D and C_D are changed by varying the angle of attack, then the actual overshoot extension attains a maximum at L/D's of about -0.5 and amounts to about 10 km for Earth. This is one half of the extension possible if C_D could be maintained the same for $L/D = -2$ as for $L/D = 0$.

Extending overshoot by applying negative lift is less effective than increasing drag. By keeping $L/D = 0$, for example, and deploying a light high-drag device, $m/C_D A$ can be changed by a factor of about 1000, which corresponds to an extension upward of the overshoot boundary by $3\varDelta_{10}y$, which is about 50 km for the earth's atmosphere. This is five times the extension possible if negative lift were employed, and is comparable to the 60 km extension in undershoot boundary provided by positive lift with constant L/D and C_D.

When the aerodynamic coupling between L/D and C_D for lifting surfaces is considered, the actual lowering of the undershoot boundary by use of positive L/D is somewhat greater than the apparent trend of the solid boundary in Fig. 6 would suggest. At $L/D = 2$ the 60 km apparent extension would really be about 80 km.

The dotted undershoot boundary in Fig. 6 labeled "variable L/D with constant C_D" represents entry with an initial L/D corresponding to the value plotted, and with lift modulated in a particular fashion which keeps both G and C_D constant during the modulation period, in which the lift force is reduced to alleviate the resultant deceleration. This dotted curve showing the greatest apparent extension in the undershoot boundary with increasing L/D, is based on the original analysis of modulated lift given by LEES, HARTWIG, and COHEN in [2]. It is applicable only to the earth's atmosphere. In order to realize this desirable extension in undershoot, C_D would have to be maintained substantially constant (or decreased) during the lift modulation period when L/D is reduced from its initial value at entry to essentially zero. If L/D is reduced by increasing the angle of attack of a lifting surface operating in the high-drag range of the drag polar, then C_D increases markedly in the process and thereby increases the horizontal component of deceleration more than the decrease in L/D alleviates the transverse component, so that no net gain is achieved. If, on the other hand, L/D is reduced by decreasing α of a lifting surface operating in the low-drag portion of the drag polar, then C_D is reduced moderately, and the apparent extension represented by the dotted undershoot boundary can be fully realized. A complication arises, however. Sizable extensions in corridor width over that for constant L/D are possible only with relatively large L/D which have small C_D, and, for the low-drag portion of a polar, result in a large heating penalty. The aerodynamic heating can be one to two orders of magnitude greater under these conditions than for operation in the high-drag portion with a small, constant L/D.

Aerodynamic Heating

It is unfortunate that vehicles cannot be designed with large lift-drag ratios (the order of 4, say) and simultaneously large drag coefficients. Large L/D is desirable in order to maximize the entry corridor width and large C_D is desirable in order to minimize the aerodynamic heating. As a lifting vehicle changes its attitude within the high-drag regime from essentially normal entry with $L/D = 0$ and maximum C_D (as indicated by the configuration sketched in the lower left portion of Fig. 7), through progressively smaller angles of attack producing higher lift-drag ratios, the effective slenderness of the configuration necessarily changes, and C_D decreases. Since laminar convective heating varies as $C_D^{-0.5}$, and turbulent convective heating as $C_D^{-0.8}$ (see [3] and [1], for example), a reduction in C_D brought about by the use of larger L/D ratios would result in an increase in aerodynamic heating for both types of convection. This is illustrated by the two curves in Fig. 7 which represent heating for lifting surfaces. These curves are normalized with respect to the amount of heating for the maximum drag attitude at $L/D = 0$. Depending on the relative amount of laminar and turbulent flow, a slender vehicle designed to produce $L/D = 3$, for example, would have between about 6 and 16 times as severe a heating problem as a blunt vehicle designed to produce no lift. In any practical case the beneficial effect of employing lifting vehicles to broaden the entry corridor would have to be tempered by its adverse effect on aerodynamic heating. For L/D's of 0.5 to 1, this penalty does not appear to be too severe, and within this range the trade-off between a broadened corridor and a more severe heating problem would favor the lifting vehicle over the non-lifting one. For larger L/D's the trade-off is much more difficult to assess.

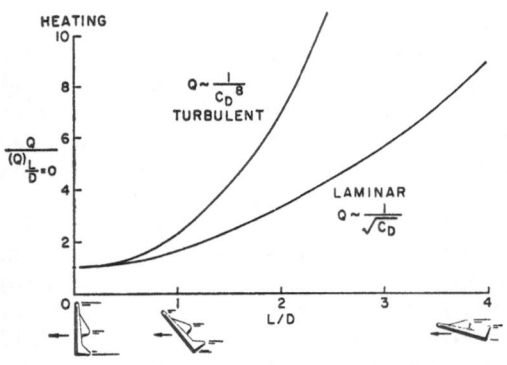

Fig. 7. Increase in convective heating with increasing L/D

In discussing the aerodynamic heating problem associated with the entry corridor, a distinction must be made between the total heat absorbed during entry and the rate at which it is absorbed. For an ablation-type heat shield, the total heat absorbed is of principal importance, but for a radiation-cooled vehicle, the maximum rate of heating is important. The type of heating problem which exists at the overshoot boundary is different from that at the undershoot boundary, as may be deduced from a general relationship between aerodynamic heating and deceleration discussed in [1]. This qualitative relationship for the maximum rate of laminar heating states that

$$\left(\frac{dQ}{dt}\right)_{max} \sim \sqrt{G_{max}}$$

and, for the total laminar heat absorbed, that

$$Q = \frac{1}{\sqrt{G_{mean}}}.$$

In view of this relationship we see that the total heat absorbed will be much less at the undershoot boundary where the deceleration is large than at the overshoot boundary where it is small. A typical variation within the earth-atmosphere corridor is that the decelerations are about nine times as large at the 10-G undershoot boundary as at overshoot, so that the total laminar heat absorbed at this undershoot boundary is about one third of that at overshoot. The maximum heating rates, however, follow an opposite relationship. They are largest at undershoot where decelerations are maximum, and smallest at overshoot where decelerations are minimum. A study of the approximate numerical values involved [1] indicates the situation to be about as follows: At undershoot where the heating rate is relatively large, pure radiation cooling for parabolic entry into the earth's atmosphere is currently impractical, but the total heat absorbed is within practical bounds of present heat absorption techniques. At overshoot, on the other hand, where the heating rate is relatively small, pure radiation cooling is practical, but the total heat absorbed is about three times that at undershoot. For efficient heat protection of spacecraft, therefore, it is desirable to develop versatile shields which can radiate efficiently if the vehicle happens to enter near overshoot, ablate efficiently if it enters near undershoot, and blend these functions efficiently if it enters anywhere in between.

Guidance Requirements

In order to provide a visual picture of the wide variation in entry corridors of different planets in the solar system, and of the corresponding wide variation in guidance requirements, the sketches in Figs. 8 and 9 have been prepared. These sketches are approximately to scale showing each entry corridor in proper proportion to the diameter of its parent planet (or parent celestial satellite in the case of Titan). The corridors for Earth and Venus are sufficiently similar that they have been represented by a single sketch in Fig. 8. The actual width of the corridor for the particular conditions considered ($\overline{V}_i = 1.4$, $G_{max} = 20$, and $|L/D| = 1$) varies from about 170 km at $r/r_0 \cong 1$ to about 400 km at $r/r_0 = 4$. This corridor

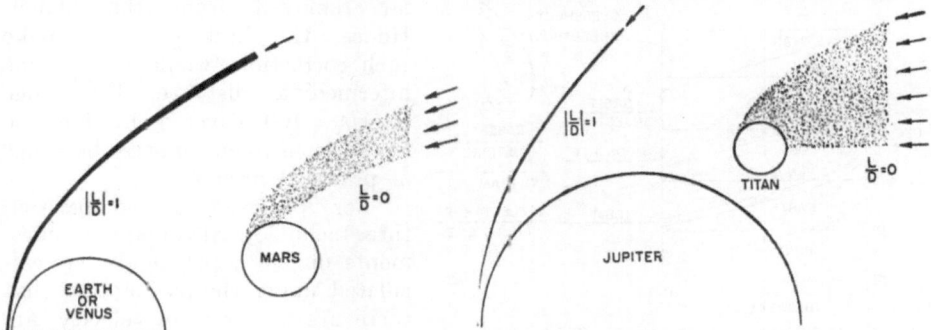

Fig. 8. Parabolic entry corridors for Earth,   Fig. 9. Parabolic entry corridors for Jupiter
Venus, and Mars; $G_{max} = 20$ and Titan; $G_{max} = 20$

appears to be neither impractically narrow nor pleasingly broad. It is much narrower, though, than the corresponding corridor for Mars, also shown in Fig. 8. The Mars corridor would not be expected to impose a really severe guidance problem. For such broad corridors, the entry angle at undershoot is sufficiently steep (47° for Mars) that aerodynamic lift is not effective in broadening the corridor; hence L/D is taken as zero for such corridors. Corridor widths calculated for nonlifting entry into Mars are 340 km for $G_{max} = 5$, 650 km for $G_{max} = 10$, and 2,000 km for $G_{max} = 20$. These values include an allowance of 130 km for the overshoot altitude.

The corresponding parabolic entry corridors for Jupiter and Titan, as sketched in Fig. 9, illustrate the very wide variations encompassed by different objects in the solar system. The corridor for Jupiter is so narrow that it is difficult to illustrate by other than a relatively narrow line. At the opposite extreme, the corridor for Titan is so broad that it includes all possible trajectories which would "hit" this satellite. Even direct vertical entry into Titan's atmosphere at $\overline{V}_i = \sqrt{2}$ would result only in $G_{max} \cong 5$.

In order to determine the trajectory which passes along the center of an entry corridor it would be necessary to make precise three-dimensional orbit calculations giving full consideration to a number of perturbations, such as those due to planetary oblateness, the sun, moon, and perhaps other planets. In calculating the small deviations permissible from this desired centerline trajectory, however, the effects of the perturbations on these deviations will be disregarded, and the entry guidance tolerances calculated as those of a two-body problem. This procedure appears reasonable inasmuch as the terminal correction to an entry approach would presumably be made relatively near the target planet where the trajectory is mainly in one plane and is essentially a conic trajectory.

Guidance tolerances imposed by different parabolic entry corridors can vary widely, as illustrated in Fig. 10. Here the ordinate represents the permissible deviation in flight-path angle $\pm \Delta \gamma$ from that of the centerline trajectory which bisects the overshoot and undershoot boundaries at any given distance r/r_0. These calculations are based on the assumption that there are no errors in velocity or position. The maximum deceleration is taken as $10\,G$ for these curves. As would be expected, the more distant a spacecraft is from the planet, the more severe the requirement is for aligning a trajectory to pass through the entry corridor. At distances close to the planet the required accuracy on $\pm \Delta \gamma$ is least, but the fuel which would have to be expended in making corrections close to the planet

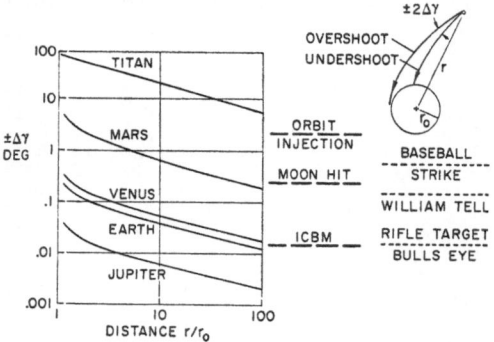

Fig. 10. Guidance requirements on flight path angle for various parabolic entry corridors ($G_{max} = 10$)

is much larger than at distances far removed from the planet. Hence, the best place to make such corrections would be at some intermediate distance. The range $1 < r/r_0 < 100$ covered in Fig. 10 would seem to encompass the range of practical interest.

For purposes of comparison, three technological guidance requirements on $\pm \Delta \gamma$ (which also are calculated under the assumption that there are no errors in velocity) are shown in Fig. 10 just to the right of the $r/r_0 = 100$ line. These correspond to typical requirements for orbiting an earth satellite, for hitting the moon from the earth, and for aligning the azimuthal angle of an ICBM to achieve a miss distance of 1/5000 of the range. As would be expected, the requirements on $\pm \Delta \gamma$ to enter the corridor of Titan are not very severe, and are even much less severe than the comparison guidance requirement on $\pm \Delta \gamma$ of approximately 2° to eject a satellite into orbit around the earth. The corresponding requirements for Mars are seen to be considerably less severe than the comparison requirement of about $\pm 0.25°$ for hitting the moon from the earth (see [4]). The corresponding requirements for Venus and Earth are seen to be somewhat less severe than the comparison ICBM requirement.

It is to be noted that the approximate requirements on $\pm \Delta \gamma$ discussed above are based on the unrealistic assumption that no errors in velocity exist. When simultaneous velocity and position errors are considered, the practical requirements on $\pm \Delta \gamma$ for guiding a vehicle through an entry corridor would be several times as severe as those illustrated in Fig. 10. Since the corridor widths are specified in terms of the difference in conic perigee altitudes, it is an easy matter to compute the approximate permissible errors in velocity and position, as well as in flight path angle, from the well-known Newtonian equations for two-body orbits. It may suffice here, perhaps, to observe that the study of velocity tolerances in reference 1 for parabolic entry into the corridor of Earth or Venus indicates $\pm \Delta V/V$ (for no error in γ) to be about 0.003, which is a less severe requirement than the corresponding velocity tolerance of 0.001 for hitting the moon from the earth [4].

For further comparison purposes, three nontechnological, but equally illuminating guidance requirements on $\pm \Delta \gamma$ are included in the right-hand portion of Fig. 10. These may serve to place the corresponding technological requirements in more balanced perspective. Thus, entry into the corridor of Mars requires about the same guidance accuracy as that with which a skilled man can throw a ball, as exemplified by the accuracy required of a baseball pitcher to pitch a strike; entry into the corridors of Venus and Earth require about the same accuracy as required of William Tell to hit an apple at 20 paces. To enter the corridor of Jupiter requires an accuracy almost an order of magnitude greater than required to hit the bull's eye of a rifle target (± 1 minute of arc), which is achieved essentially 100 percent of the time by skilled individuals although under relatively comfortable and favorable conditions.

The most important difference between these various technological, non-technological, and entry-corridor requirements probably lies not so much in the numerical difference between requirements, but in the difference in reliability with which the individual guidance requirements must be obtained. The consequences of failing to achieve the guidance requirements associated with injecting a satellite into orbit, hitting the moon, scoring a rifle target bull's-eye, or pitching a baseball strike, may be no more severe than that of bad publicity, but the consequences of failing to achieve the guidance requirements associated with the entry corridor for manned spacecraft could result in a catastrophe, as could have resulted if William Tell had failed to achieve his guidance requirements. In such light the reliability with which the guidance requirements on flight-path angle must be obtained may impose perhaps the most challenging technological problem.

References

1. D. R. CHAPMAN, An Analysis of the Corridor and Guidance Requirements for Supercircular Entry into Planetary Atmospheres. NASA Report, 1959.
2. L. LEES, F. W. HARTWIG, and C. B. COHEN, The Use of Aerodynamic Lift During Entry Into the Earth's Atmosphere. Space Technology Labs. Inc., GM-TR-0165-00519, Nov. 1958.
3. H. J. ALLEN and A. J. EGGERS, JR., A Study of the Motion and Aerodynamic Heating of Ballistic Missiles Entering the Earth's Atmosphere at High Supersonic Speeds. NACA Rep. 1381, 1958. (Supersedes NACA TN 4047.)
4. H. A. LIESKE, Accuracy Requirements for Trajectories in the Earth-Moon System. RAND Rep. P-1022, Feb. 1957.

Lunar Exploration by Photography from a Space Vehicle[1]

By

Merton E. Davies[2]

(With 10 Figures)

(Received July 15, 1959)

Abstract — Zusammenfassung — Résumé

Lunar Exploration by Photography from a Space Vehicle. A description of a photographic system that may be used to obtain pictures of either the visible or hidden side of the moon from an early space vehicle. Existing components are suitable for the space vehicle. A panoramic-type camera can be designed in which the spin that stabilizes the vehicle also performs the scanning function for the camera. Tracking can be accomplished by radio. The recovery of the film after its return to earth appears feasible.

Photographische Monderforschung von einem Raumfahrzeug aus. Die vorliegende Arbeit beschreibt ein System der Photographie, mit dessen Hilfe man Bilder sowohl der sichtbaren wie auch der von der Erde abgewandten Seite des Mondes von einem in naher Zukunft konstruierbaren Raumfahrzeug aus erhalten können wird. Für ein solches Fahrzeug existieren bereits geeignete Komponenten. Eine Panorama-Kamera kann gebaut werden, wobei der das Fahrzeug selbst stabilisierende Spin auch die richtende Funktion für die Kamera leistet. Die verfolgende Beobachtung kann über Radio erfolgen und die Bergung des Films nach der Rückkehr des Raumfahrzeuges zur Erde scheint ausführbar.

Exploration lunaire par photographie à partir d'un engin spatial. Description d'un système photographique utilisable aussi bien pour la partie visible que pour la partie invisible de la Lune. Des éléments existants conviennent à l'engin spatial. Une conception de camera panoramique utilise le spin de stabilisation du véhicule pour remplir également la fonction de balayage photographique. Le contrôle de trajectoire peut se faire par radio. La récupération du film après la rentrée apparaît possible.

I. Introduction

The principal tool for the exploration of the moon has long been the earth-based telescope; now, space vehicles provide a powerful new tool. The Pioneer rockets, launched in late 1958 and early 1959, collected data from space and then telemetered these data back to the earth. These rockets, and the Soviet Mechta, have begun a new era of lunar exploration.

The telescope, in answering questions, provided information which led to the asking of more penetrating questions; so it will be with exploration by space

---

[1] RAND P-1671. The views expressed in this paper are not necessarily those of the Corporation.

[2] Engineering Division, The RAND Corporation, 1700 Main Street, Santa Monica, California, U.S.A.

vehicles. There will always, for example, be a demand for better pictures of the moon.

The present paper describes a camera which, as the payload of an early space vehicle, could photograph the moon. The resulting pictures would be superior to those obtained either by telescope, such as the 200-in. Mt. Palomar telescope (as in Fig. 1), or by television camera, such as those carried by Pioneers I and III. The film would be returned to earth in a recoverable capsule.

II. The Vehicle

The U.S. Air Force's Pioneer I [1] used a Douglas Thor rocket as the first stage, and a liquid propellant Aerojet-General as the second stage. This second stage was a further development of the second stage Vanguard rocket and had

Fig. 1. Moon, region of Clavius. 200-inch photograph, Mount Wilson and Palomar Observatories

been mated to the Thor for the Thor-Able project; it contained accelerometer cutoff and a spin rocket table for roll stabilizing the third and successive stages. The third stage was a solid propellant rocket developed by the Alleghany Ballistic Laboratory for the third stage of the Vanguard. The fourth or payload stage contained light, small, solid rockets for vernier velocity control and a Thiokol retro-rocket to slow down the vehicle in the vicinity of the moon so it would become a lunar satellite. This 84 lb payload [2] contained 30 lb of instruments, including a radiation counter, magnetometer, micrometeorite counter, temperature recorder, and a photocell television camera. The retro-rocket weighed about 35 lb and, like the vernier rockets, was fired by command from the ground. The payload was spun at about 110 rpm for stabilization during firing of the solid rockets and for scanning purposes during the operation of the television system.

The U.S. Army's Pioneer III [3] used a modified Jupiter rocket as the first stage, a cluster of eleven scaled Sergeant rockets as a second stage, a cluster of three scaled Sergeants as a third stage, and a single scaled Sergeant rocket as a fourth stage. The 13 lb payload contained a radiation counter, a photoelectric sensor, and a television camera. The sensor is activated 20 hrs after launch when the probe is due to be 140,000 mi from earth; it consists of two photoelectric cells which, when the moon is at a predetermined distance, turn on the camera.

Before the first stage is fired, electric motors spin the other stages to a speed of 250 rpm for stability; the rate is increased to 400 rpm after the first stage is fired. This high spin rate, though desirable during the burning of the solid propellant stages, is much too great for proper operation of the frame type television camera. Consequently, the fourth stage contains a 'de-spin' mechanism which begins to operate about 10 hrs after launch; it consists of two 7-gm weights fastened to two 60-in. wires which unwind and thus increase the moment of inertia and decrease the spin rate to about 6 rpm. When the wires are fully extended they slip off their hook fasteners, and much of the rotational energy is permanently dissipated; the payload continues its flight at the much reduced spin rate.

Both Pioneer I and III used light weight tracking and command systems. The Pioneer I used the command link, whereas the Pioneer III used a self-contained photoelectric sensor to warn of the lunar passage. The Pioneers also used intermediate range ballistic missiles as first stage boosters; they were thus able to place payloads of a few tens of pounds near the moon. A study made at RAND in 1956, however, indicates that if an intercontinental ballistic missile, such as the Atlas or Titan, is used as a first stage booster, payloads of a few hundreds of pounds can be placed near the moon [4]. The Pioneers as well as the one in the RAND study used spin stabilization for maintaining the proper direction of the thrust vector during the burning of the solid stage rockets. This spin rate can be changed by simple mechanical methods as was done with Pioneer III.

III. The Camera

Since the axial orientation of the early lunar vehicles is maintained by spin stabilization, it is desirable to use a camera, such as a panoramic camera, that will not experience image smearing due to this motion. A panoramic camera would use the spin to perform the scanning. At the time of exposure, the film is moved past a slit at a rate which just compensates for the spin, so the image of the ground on the film is not moving relative to the film (i.e., no blurring results from this operation). Modern panoramic cameras are usually designed so that there is a slack loop of film which is accelerated during the picture-taking cycle of operation. The inertial loads involved in accelerating a loop of film are much smaller than those involved in starting and stopping spools of film.

For the camera to operate properly, the missile must sense its spin rate and the approximate direction of the ground. A device similar to that used in the Pioneer III to measure the proximity of the moon could be used to measure the spin rate and lunar direction as well as to initiate camera operation. Another method would be to measure the passage of the sun through a slit with a photo-

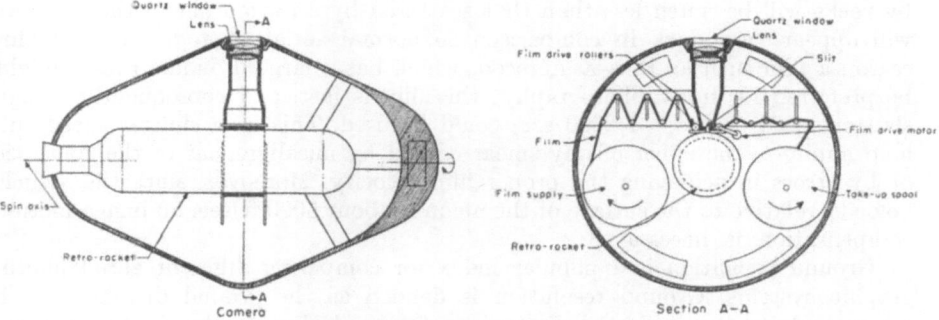

Fig. 2. Schematic of spinning panoramic camera

cell; a simple computer could then adjust the film velocity to compensate for the measured spin rate. To maintain the focus of the lens and to keep the film flexible it is probably desirable to pressurize the camera and to control the temperature and humidity near the lenses and film. Fig. 2 shows the camera operation.

The main factor contributing to the loss of contrast in high-altitude aerial

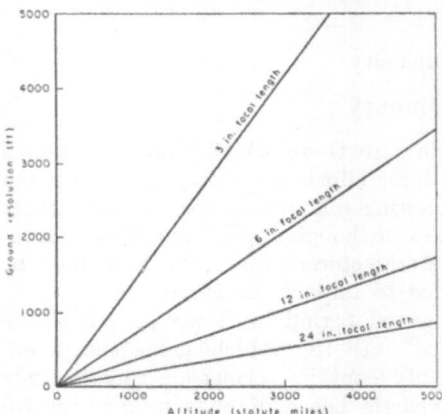

Fig. 3. Ground resolution as a function of altitude—vertical view. $x = 207.7 \dfrac{h}{RF}$

x = ground resolution (ft)
h = altitude (statute miles)
$R = 50$ = film resolution (lines/mm)
F = lens focal point (in.)

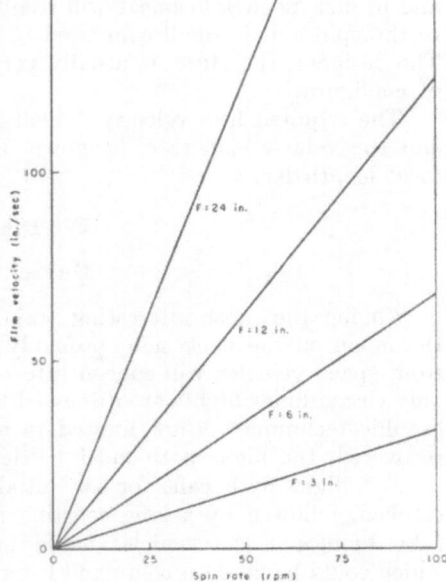

Fig. 4. Spin rate, lens focal length and required film velocity

$V = \cdot 10472\ F$ (rpm)
V = film velocity (in./sec)
F = focal length (in.)
rpm = spin rate (rpm)

photography of the earth is the scattering of non-image forming light into the optical system by the earth's atmosphere. If, for example, the original contrast is 10:1 and scattered light increases the exposure of both black and white areas by 2 units, the resulting contrast is $(10+2):(1+2)$, or $4:1$. This is one reason that contrasty (high gamma) emulsions, such as Microfile, are excellent for aerial use. Since the moon, however, has essentially no atmosphere, the natural ground contrast should be available on the film, and since the light scattered by rocks will be much less than that scattered by an atmosphere, the shadows will appear very dark in comparison to normal aerial photographs. For this reason a film such as Plus-X Aerecon, which has a large dynamic range, might be preferred for lunar photography; this film is fast, and consequently a high shutter speed (perhaps 1/2000 sec) could be used. This high shutter speed will help minimize the effect of any smear caused by misalignment of the spin axis or by errors in achieving the proper film velocity. Moreover, since the vehicle velocity relative to the surface of the moon is about 5000 ft/sec, no image-motion compensation is necessary.

Ground resolution is a popular index for comparing different aerial photographic systems. Ground resolution is defined as the ground distance which corresponds to the projected dimension of a single line at the limiting camera system resolution. The formula for ground resolution is given in Fig. 3, which shows ground resolutions for 3 in., 6 in., 12 in., and 24 in. focal-length lenses with a film resolution of 50 lines/mm for various altitudes.

Theoretically, the spin axis of the camera should pass through the rear nodal point of the lens. In practice this usually results in severe packaging problems, and in fact, negligible smear will result if the distance from the rear nodal point to the spin axis is small compared to the desired or designed *ground* resolution. The designer, therefore, is usually permitted considerable freedom in his choice of configuration.

The required film velocity, which depends upon the focal length of the lens and the vehicle spin rate, is shown in Fig. 4 for 3 in., 6 in., 12 in., and 24 in. focal length lenses.

IV. The Trajectory

Basic Geometry

Among the most interesting trajectories are those which closely approach the moon on the back side; probably all early efforts to photograph the moon from space vehicles will concentrate on viewing this unseen area. Consequently only circumlunar flights are discussed here even though application of the photographic techniques is not limited to these trajectories. Since the film must be recovered, the flight path must be designed to impact the earth.

The flight path calls for an initial powered period until escape velocity is reached, followed by a long coasting period until the vehicle nears the moon. Like Pioneer I the vehicle should probably contain vernier or retro-rockets which could be fired on command from the earth. These might be used to control the velocity at cutoff; they could also modify the free-flight trajectory near the moon to assure a close approach; and on the return path to the earth, they could alter the trajectory to assure impact with the earth, or they might even control, to some extent, the exact impact area.

Lieske [5, 6], has reported on trajectory studies which consider variations of the vehicle's velocity and angle as it leaves the vicinity of the earth. These studies treated only unpowered trajectories and did not consider path shaping by the introduction of thrust near the moon or on the return path to the earth. They do, however, give an indication of the required guidance accuracy if retro-rockets are not used; for instance, a lunar altitude uncertainty of about 250 s mi is indicated for simultaneous initial uncertainties of 1 ft/sec in velocity and 1 mil in path angle for trajectories defined by an initial velocity of 34,860 ft/sec at an altitude of 350 s mi above the earth [6]. For this example, the design lunar altitude is varied by specifying the path angles of the velocity vector at the 350 s mi altitude point.

Interesting circumlunar flights which return to the earth are limited to those which move in retrograde motion relative to the moon. These are the figure eight paths which cross the moon's orbit before the moon arrives. Fig. 5 shows

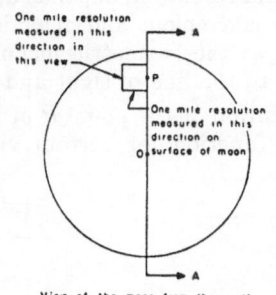

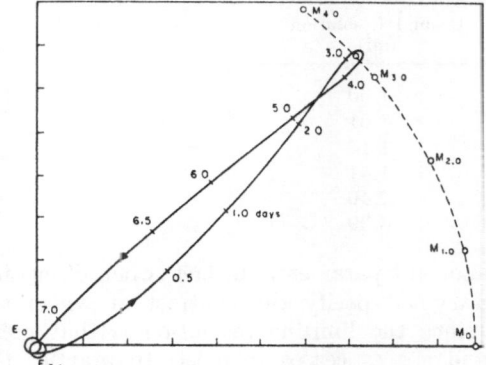

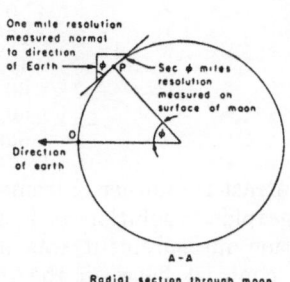

Fig. 5. Typical trajectory of a lunar photographic vehicle

Fig. 6. Ground resolution as measured on the surface of the moon

an example of a trajectory which satisfies the requirement [5]: the path passes within 3000 mi of the moon; the total flight time is 7.1 days. Total flight times of 6 to 12 days are possible for reasonable trajectories, depending upon the particular initial conditions.

Sun-Moon Attitude

In discussing the geometrical relationships between the moon's surface, the vehicle spin axis, and the direction of the sun, it is necessary to examine the photographic objectives. Certainly, one of the major purposes of the first few flights will be to maximize the area of photographic coverage so as to completely map the back side of the moon. Later priorities may shift to obtaining better quality pictures of particular interesting areas; these should include the front, as well as the back side of the moon.

The ground resolution (i.e., resolution measured at the surface of the moon) of pictures taken from the surface of the earth is determined by three factors: (1) the earth's atmosphere, (2) the distance to the moon, and (3) the angle between the tangent plane at the point in question and the normal to the line joining the centers of the earth and the moon. The turbidity of the atmosphere limits the resolution of earth-based telescopic photography to about one second of arc, which at the distance of the moon corresponds to about one mile in the plane normal to the light ray. For this reason, as illustrated in Fig. 6, the ground resolution of excellent moon pictures is about one mile in the direction normal to a plane passing through the center of the moon's face and one mile times the secant of the latitude angle in the plane of the radial cut at the point in question. When the object plane is not perpendicular to the line of sight (as in the case of lunar photography, as well as in oblique aerial photography), ground resolution depends upon the direction in which the measurement is made. It is convenient and meaningful to define average ground resolution for this case as the geometric mean of the two ground resolutions measured perpendicular to the line of sight and in the line of sight. Using this definition, the ground resolution at the point P in Fig. 6 has the value of $\sqrt{\sec \varphi}$ mi. Table I evaluates this function for certain values of φ.

Table I. *Values of* $\sqrt{\sec \varphi}$

φ (deg)	Ground Resolution (mi)
0	1 00
20	1 01
40	1 14
60	1 41
80	2 40
85	3 39

Contrast is another extremely important parameter; in fact, when discussing photographic resolution it is necessary to specify the contrast at which the resolution measurement was made since the limiting resolution is that point where contrast between the lines and spaces ceases to exist. In practice the resolution at high contrast (30:1) might, with a particular lens and film combination, be twice that available at low contrast (2:1). Moreover, as pointed out by MACDONALD [7], when considering information on film it is necessary to specify the type of film used as well as the resolution and contrast of the system, since different films record different information at the same resolution and contrast. This is partly because of the processing of different emulsions to different contrast (gamma) as well as to variations in the grain characteristics of different emulsions. Because of these complexities, the experienced photointerpreter prefers to estimate quality rather than to discuss resolution, contrast, blur, emulsion, etc. [8].

Although high contrast ground resolution, when used as an index of quality, might at times be in error by as much as a factor of two, it is a convenient tool for the comparison of pictures taken under vastly different conditions [9]. Fig. 7 is a comparison of available lunar photography with aerial photography. The picture of Clavius, which was taken with the 200 in. telescope at Mount Palomar, is usually considered to be one of the finest of this region of the moon. The ground resolution in this photograph is probably a little over one mile. The aerial views of Washington, D. C., presented in the same figure have been reproduced at ground resolutions of 200 ft, 1000 ft, and 5000 ft by varying the distance of the copy camera and controlling the focus on the enlarger. The figure illustrates how small the surface detail is on even the best lunar photographs.

The position of the sun during the picture taking operation and the direction of the spin axis of the camera are, to some extent, free parameters that can be selected on the basis of the specific objectives of the mission. If, for example, we wish to obtain pictures of the back side of the moon, then it is desirable to photograph as large an area as possible at a quality level sufficient to record the major information. Since illumination must be furnished for the entire back side, the sun should lie approximately on the line joining the centers of the earth and the moon. The date of launching should therefore be selected so that the pictures are obtained at 'new moon' when viewed from the earth. With

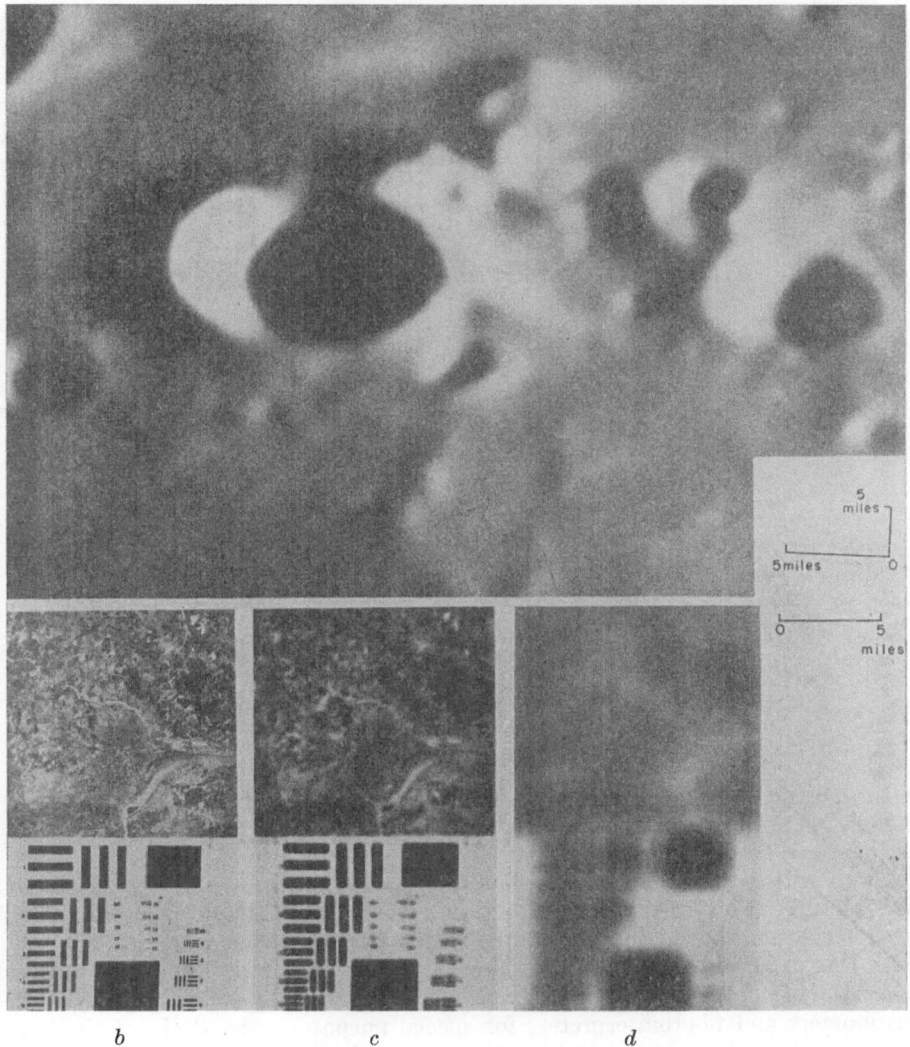

Fig. 7. Lunar photography and aerial photography: resolution comparison. *a* The floor of Clavius. 200-inch photograph, Mount Wilson and Palomar Observatories; *b* 200 ft ground resolution; *c* 1000 ft ground resolution; *d* 5000 ft ground resolution. Washington, D.C.

this lighting, most of the shadowed areas will lie near the horizon, and most of the resulting pictures will be flatter and of lower contrast than when illuminated by oblique lighting. For maximum coverage it is also desirable to have the camera spin axis approximately normal to the line joining the centers of the earth and moon. The vehicle should not come too near the moon, however, since the low horizon would limit the area photographed.

18*

Fig. 8, an enlargement of a portion of the trajectory shown in Fig. 5, indicates the relative motion of the vehicle as it passes around the moon. Its velocity is about 5000 ft/sec relative to the surface of the moon and is about 3000 mi from it at the point of closest approach. The sun and the vehicle spin-axis are shown in a preferred orientation to record maximum-area pictures of the back of the moon.

After obtaining the first pictures and naming the newly discovered lunar formations, there will remain the task of obtaining better quality photographs for scientific study. For this purpose it is very important to make a close approach to the moon with a long focal length camera so that surface detail will be recorded. The ability to make a close approach depends greatly upon the performance

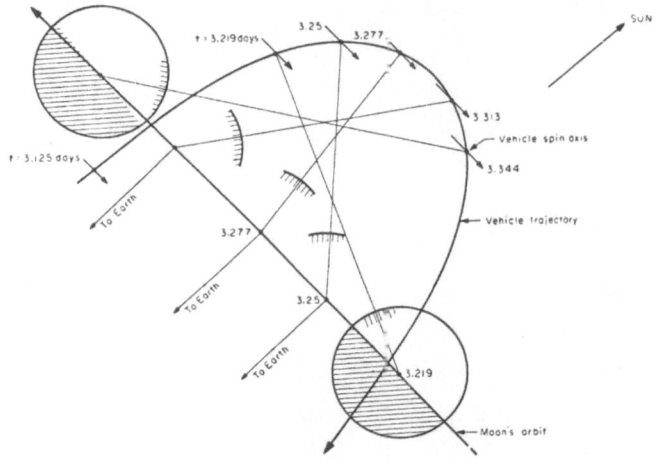

Fig. 8. Trajectory of vehicle in the vicinity of the moon — New Moon
Note: Vehicle spin and sun position for photographing the back side

of the particular vehicle. Design planning figures of about 3000 mi ± 2000 mi altitude at the point of nearest approach seem reasonable for vehicles available in the near future.

Both astronomers and aerial photointerpreters are aware of the fact that 'flat' pictures, where the sun is behind the camera, are less interesting and have lower resolution due to the low contrast than pictures taken with side lighting. With this in mind, Figs. 9 and 10 show typical sun-moon positions and camera spin axis for obtaining excellent lunar pictures. Another effect used by both astronomers and photointerpreters for special purposes is the study of the long shadows when the sun is near the horizon. These high contrast shadows are larger than the objects causing them and so are much easier to resolve. Astronomers find such pictures useful in computing the altitude of lunar formations and in contouring. The photo-interpreter sometimes uses such pictures for computing the height of objects on the ground and for studying the texture of terrain. High resolution lunar pictures of this type will prove to be extremely interesting.

Since the best results will be obtained where the spin axis is parallel to the lunar surface, it is desirable to select the preferred direction of the spin axis. The spin axis shown in Fig. 8 is approximately normal to the line joining the

centers of the earth and moon. In this attitude most of the illuminated surface
will be surveyed. The spin axis is tilted about 45° with the line joining the earth
and moon in Figs. 9 and 10 so that the region photographed from the best camera

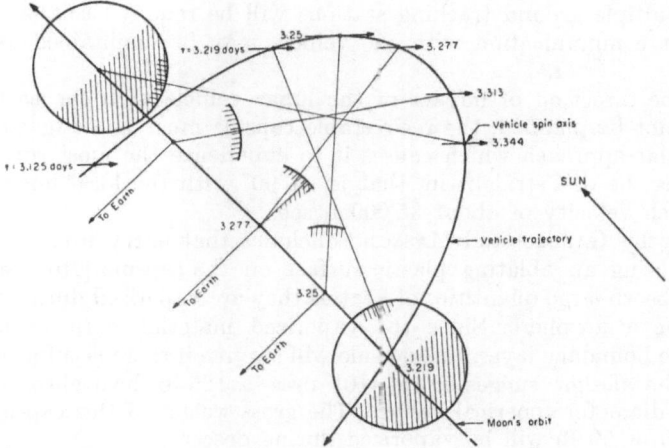

Fig. 9. Trajectory of vehicle in the vicinity of the moon —

orientation corresponds to the region of most interest.
the spin is to control the direction of thrust of a solid

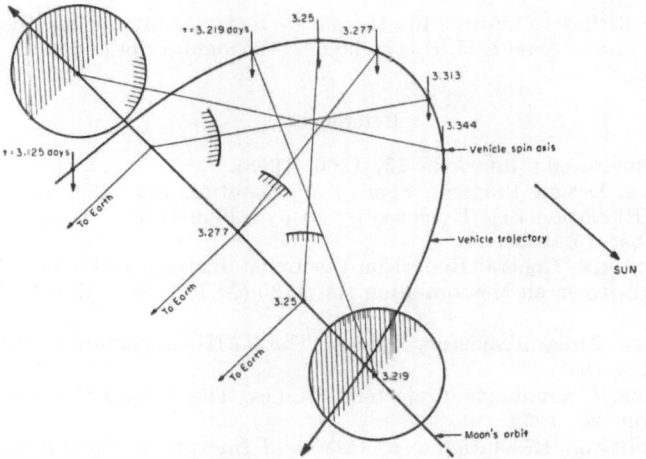

Fig. 10. Trajectory of vehicle in the vicinity of the moon — First Quarter

the case in Pioneer I and Pioneer III), then the tilt angle of the spin axis in
the region of the moon will be between 55° to 65° with the line joining the moon
and earth for reasonable trajectories and in the direction shown in Fig. 9. Obtaining different tilt angles, such as shown in Figs. 8 and 10, will require a pitch
and spin-up program after final stage burnout [4].

V. The Recovery

Tracking the payload on its flight around the moon will be done by radio. A low-power transponder (perhaps one watt) should be adequate if a directional antenna can be used. The spin should provide sufficient stabilization for this purpose. Multiple ground tracking stations will be required so that continuous line-of-sight communication with the vehicle may be maintained as the earth rotates.

Since the direction of impact of the lunar vehicle with the earth's atmosphere can not be planned, the recoverable capsule must be designed to survive the particular approach which causes it to experience the most severe heating. This path is the one straight in, that is, at 90° with the local horizontal, with the approach velocity of about 35,000 ft/sec.

A study by Gazley and Masson concludes that entry from a lunar flight is feasible using an ablating plastic surface on the capsule [10]. These special materials absorb large quantities of heat as they are vaporized during the descent through the atmosphere. Since the vaporized material is removed from the body by the boundary layer, the capsule will survive if there is sufficient material present. The design suggested in [10] uses a 125-lb heat-absorbing surface on the 3-ft-diameter spherical vehicle. The gross weight of this capsule is 400 lb of which 40 or 50 lb will be vaporized during descent.

Since 75 per cent of the earth's surface is water, an ocean recovery is most likely, although the capsule must be built to survive any impact. The tracking stations should be able to predict the point of landing within a few hundred miles so an aircraft search could be made.

Acknowledgments

The author wishes to express his thanks to Hans A. Lieske for the trajectory data presented, and to Amrom H. Katz who read the manuscript and made many helpful suggestions.

References

1. 'Pioneer Illustrated.' Interavia 13, 1260 (1958).
2. I. Stambler, Design Progress. Space / Astronautics, 178—179, January (1959).
3. E. Clark, Radiation Belt Explored by Army's Pioneer III Probe. Aviat. Week, 28, December 15, 1958.
4. R. W. Buchheim, General Report on the Lunar Instrument Carrier. The RAND Corporation, Research Memorandum RM-1720 (ASTIA No. AD-105533), May 28, 1956.
5. H. A. Lieske, Lunar Trajectory Studies. The RAND Corporation, Paper P-1293, February 26, 1958.
6. H. A. Lieske. Circumlunar Trajectory Studies. The RAND Corporation, Paper P-1441, June 25, 1958.
7. D. E. Macdonald, Resolution as a Measure of Interpretability. Photogrammetric Engineering, March, 1958.
8. A. H. Katz, Contributions to the Theory and Mechanics of Photo-Interpretation from Vertical and Oblique Photographs. Photogrammetric Engineering, 339, June, 1950.
9. Optical Image Evaluation, National Bureau of Standards Circular 526, April 29, 1954, pp. 200—202.
10. C. Gazley, Jr., and D. J. Masson, Recovery of a Circum-Lunar Instrument Carrier. Proceedings of the VIIIth International Astronautical Congress, Barcelona 1957, p. 137. Wien: Springer, 1958.

Voyage aux étoiles

Etude de la fusée d'après le principe de relativité

Par

Vicente Roglá Altet[1]

(Avec 6 Figures)

(Reçu le 5 juin 1959)

Résumé — Zusammenfassung — Abstract

Voyage aux étoiles. Etude de la fusée d'après le principe de relativité. Les voyages interstellaires exigeront des vitesses voisines de celle de la lumière et la dynamique de relativité devra-t-être appliquée à leur étude.

Le présent travail met sur pied une théorie générale des fusées adiabatiques en vol rectiligne, en absence de tout champ de gravitation, quel que soit le système de propulsion, dont le caractère est exprimé au moyen d'une constante h de conversion de l'énergie. Le rapport des masses est obtenu de trois façons différentes : par l'application du principe de conservation de la quantité de mouvement par rapport à l'observateur resté sur la Terre, par la considération de la poussée et en établissant la balance de l'énergie.

Les lois de vitesses les plus favorables pour des fusées simples ou à étages sont étudiées et portées sur des graphiques. Les cas particuliers de la fusée à poussée constante et de la fusée photonique sont aussi traités.

Le paradoxe du temps limite, qui ne peut être dépassé pour un parcours fini quelconque, est exposé.

Reise zu den Sternen. Studium der Rakete gemäß dem Relativitätsprinzip. Die interstellaren Reisen werden Geschwindigkeiten nahe der Lichtgeschwindigkeit erfordern und die Dynamik der Relativitätstheorie wird zu ihrer Erforschung angewandt werden müssen.

Die vorliegende Arbeit begründet eine allgemeine Theorie der adiabatischen Raketen in geradlinigem Flug, in Abwesenheit jedes Schwerefeldes, unabhängig vom Antriebssystem, dessen Charakteristika mittels einer Konstante h der Energieumwandlung ausgedrückt werden. Das Massenverhältnis wird auf drei verschiedene Arten erhalten: durch Anwendung des Prinzips der Erhaltung der Bewegungsgröße in Beziehung zum Beobachter auf der Erde, durch die Betrachtung des Schubs und unter Einführung des Energiegleichgewichtes.

Die Gesetze der günstigsten Geschwindigkeiten für einfache oder Stufenraketen werden studiert und in Diagrammen eingezeichnet. Ebenso werden die Sonderfälle der Raketen mit konstantem Schub und der Photonenrakete behandelt.

Des weiteren wird das Paradoxon der Zeitgrenze, die nicht für eine endliche durchlaufene Strecke überschritten werden kann, dargelegt.

Trip to the Stars. Study of a Rocket According to the Principle of Relativity. Interstellar travel requires velocities close to that of light, and relativistic dynamics should be applied to its study.

---

[1] Professeur de Physique Théorique à l'Ecole Technique Supérieure des Ingénieurs des Ponts et Chaussées de Madrid, 27 Rue O'Donnell, Madrid, Espagne.

This treatise considers a general theory of adiabatic rockets travelling linearly in the absence of gravitational fields, whatever be the propulsion system whose character is expressed by an energy conversion constant h. The mass ratio is obtained in three different manners: by the application of the principle of conservation of the momentum in terms of an Earth-bound observation, by the consideration of the thrust, and by establishing the energy balance

The most favorable velocity laws for single or staged rockets are studied and shown in graphic form. The particular cases of the constant thrust rocket and of the photon propelled rocket are also given.

The time limit paradox, which cannot be by-passed in any finite path, is considered.

Prologue

Ce travail n'avait pas été écrit dans le but d'être présenté au Xème Congrès International d'Astronautique; il avait été rédigé par plaisir, comme un simple exercice d'application des Principes de Relativité restreinte.

Après avoir résolu le problème qu'il s'était proposé, l'auteur a fait des recherches bibliographiques en vue de comparer sa théorie avec celles qu'auraient pu éventuellement développer d'autres auteurs.

L'étude comparative réalisée avec les ouvrages qu'il a été à même de consulter, étude qui se trouve à la fin de ces pages (chapitre XVII), a encouragé l'auteur à présenter son travail.

Malgré que certaines formules ici exposées sont déjà connues, l'ensemble constitue un corps logique, une série de conséquences, enchaînées les unes aux autres, d'une théorie très simple et générale qui comprend, comme cas particuliers, les travaux publiés jusqu'à présent, dont il a eu connaissance.

Introduction

L'ennui d'une longue réclusion et la courte durée de la vie humaine exigent des vitesses prochaines de celle de la lumière pour les voyages aux étoiles et demandent par conséquent l'application des corrections de Relativité à l'étude mécanique de la fusée.

Le présent travail est une étude dynamique générale de la fusée sur trajectoire rectiligne, d'après les Principes de Relativité restreinte ou spéciale.

Le fait d'admettre ces Principes suppose un raccourcissement du temps de l'équipage ou temps "propre", par rapport au temps de l'observateur terrestre, ce qui, si cette supposition était exacte, rendrait les futurs voyages interstellaires extraordinairement aisés. Au moment de leur retour sur la terre, les voyageurs se trouveraient, en conséquence, plus jeunes que ceux qu'ils y auraient laissés (il s'agit du célèbre paradoxe de Langevin). Il y a plus, aux chapitres XIV et XV, nous arrivons à des lois régissant le mouvement qui nous donnent un "temps propre limité"; dans ces conditions, n'importe quel parcours imaginable, y compris les voyages aux Galaxies les plus lointaines, serait réalisé en un temps propre inférieur à cette limite, laquelle est uniquement fonction de l'accélération initiale.

A ce sujet, celui des paradoxes du temps, il convient de faire quelques observations.

Le temps qui est appelé temps "propre" en Relativité restreinte n'est pas strictement celui de l'équipage, mais celui que posséderait un système de référence se déplaçant par rapport à l'observateur terrestre à la vitesse de la fusée au moment considéré; mais avec une vitesse constante, sans aucune accélération.

L'accélération de la fusée donne lieu à l'introduction d'un champ qui produit une courbure de "l'espace-temps" et situe le problème en dehors du champ d'application de cette théorie; il faudrait appliquer les Principes de Rélativité généralisée pour calculer correctement le temps de l'équipage.

Si les accélérations ne sont pas trop élevées, et elles ne peuvent pas l'être pour des fusées transportant des voyageurs, l'erreur commise est petite; nous le supposerons ainsi dans cette étude, et nous nous en tiendrons à l'application des principes de relativité restreinte.

D'autre part, la Relativité ne nous dit pas que le temps "en lui-même" s'allonge ou se raccourcit. Elle nous dit seulement que lorsque deux pendules identiques ont un mouvement relatif, le tic-tac de l'une d'elles est plus rapide, pour l'observateur qui se déplace avec cette pendule, que le tic-tac de l'autre pendule, qu'il est à même de mesurer.

Si nous imaginons la fusée immobile et que c'est la Terre, et avec elle tout l'Univers, qui se déplace, le phénomène contraire aurait lieu: ce serait le temps terrestre qui subirait un raccourcissement par rapport à celui de la fusée.

Mais en Relativité il n'y a pas de mouvements absolus et tous les systèmes de référence sont également valables. Par conséquent un doute surgit, celui de savoir si l'équipage, à son retour, se trouverait plus vieux ou plus jeune que les personnes restées sur la terre.

Cependant, quand il faut avoir recours à l'accélération de l'Univers entier pour envisager le phénomène réciproque, dont un remarquable exemple, différent de celui-ci, est donné par l'apparition de forces centrifuges dans un corps immobile, il nous semble que la grandeur disproportionnée des moyens que l'on veut employer prouve qu'il y a tout de même un système privilégié et que la supposition d'un Univers accéléré n'a pas de sens au point de vue dynamique. Quelle que soit la réalité des choses, nous acceptons comme hypothèse pour notre travail, le raccourcissement du temps de la fusée que nous confondrons avec le temps "propre".

Symboles

Grandeurs mesurées par l'observateur solaire

t temps écoulé depuis le moment du départ de la fusée (T temps exprimé en années)

v vitesse de la fusée à l'instant t

m masse de la fusée à l'instant t

M masse initiale de la fusée

x distance parcourue par la fusée pendant l'intervalle t (X distance exprimée en années-lumière)

v_p vitesse par rapport au soleil et à l'instant t des particules éjectées

$-dm_p$ masse infiniment petite qui semble être éjectée pendant l'intervalle dt

Grandeurs mesurées par l'équipage de la fusée

t_0 temps "propre" écoulé depuis le moment du départ (T_0 temps exprimé en années)

u vitesse d'éjection, supposée constante

m_0 masse de la fusée à l'instant t_0

h rapport constant entre la masse transformée en énergie et la perte totale de masse par transformation et par éjection

E énergie totale dépensée pendant l'intervalle t_0

$W = \dfrac{dE}{dt_0}$, puissance à l'instant t_0

F poussée du moteur (grandeur indépendante du système de référence)

$\mu_0 = \dfrac{dm_0}{dt_0}$, perte totale de masse par unité de temps

j accélération sensible, c'est-à-dire gravitation apparente à l'intérieur de la fusée

Grandeurs auxiliaires importantes

c 3×10^{10} cm/sec, vitesse de la lumière
a variable qui remplace v d'après la relation: $v = c \ \mathrm{th} \ a$
β paramètre qui remplace u au moyen de: $u = c \ \mathrm{th} \ \beta$
g 980 cm/sec², accélération de la pesanteur
n rapport entre l'accélération initiale de la fusée et celle de la gravitation terrestre
N même rapport pour le maximum admissible de l'accélération j

Nombres caractéristiques de l'engin

r rapport entre la masse initiale et la masse finale de la fusée. S'il s'agit d'une
        fusée multiple, le rapport est supposée constant pour chacun des étages de la
        fusée.
q nombre d'étages d'une fusée multiple
p rapport, supposée constant, des masses de la fusée multiple immédiatement
        avant et après qu'un corps usagé a été détaché.

Symboles portant un accent

t', v', x', m'_0, etc. grandeurs correspondantes à l'instant final de la période d'accélération
        (moment où la plus grande vitesse est atteinte)

Symboles portant plusieurs accents

a'', v''', $m'_0{}'$, etc. valeurs particulières

Symboles portant l'index f

t_f, t_{0f}, X_f, m_f, etc. grandeurs à la fin du voyage, lorsque le repos par rapport à l'observa-
        teur solaire est à nouveau atteint

Symboles divers

e   base des logarithmes népériens
π rapport entre une circonférence et son diamètre
a, b,   constantes auxiliaires dont la dimension est un temps inversé
ξ, ξ_0, η, η_0 coordonnées de la Fig. 2, dans laquelle ces symboles sont expliqués

I. La fusée classique

Imaginons une fusée qui parcourt une ligne droite et qui est libre de toute
action extérieure, telle que champs gravitatoires ou frottements à l'intérieur
d'une atmosphère.

Soit:

m la masse totale de la fusée à l'instant t.
v sa vitesse au même instant t.
$m + dm$ la masse à l'instant $t + dt$.
$(-dm)$ la masse expulsée pendant l'intervalle dt.
u la vitesse constante de la masse expulsée par rapport à la fusée, dirigée
        en sens contraire de v.

Le système matériel constitué par la fusée à l'instant t, possède une masse
m et une quantité de mouvement

$$m \, v \, . \tag{1}$$

Ce même système matériel, à l'instant $t + dt$, est divisé en deux parties, la masse restante de la fusée: $m + dm$ (plus petite que m parce que dm est négatif) dont la vitesse est $v + dv$, et la masse $(-dm)$ qui a été expulsée à l'arrière avec une vitesse relative u, et qui par conséquent possède une vitesse absolue $v - u$. La quantité de mouvement totale du système est alors:

$$(m + dm)\,(v + dv) + (v - u)\,(-dm)\,. \tag{2}$$

Les deux quantités de mouvements sont égales parce qu'aucune force extérieure n'agit sur le système.

Donc, en égalant (1) et (2) on obtient:

$$m\,dv + u\,dm = 0 \tag{3}$$

ou bien:

$$\frac{dv}{u} = -\frac{dm}{m}\,.$$

En intégrant cette expression et en supposant qu'à l'instant initial $t = 0$, on ait $m = M$ et $v = 0$, on obtiendra:

$$\frac{v}{u} = \text{lognep}\,\frac{M}{m}$$

et aussi:

$$\frac{M}{m} = e^{\frac{v}{u}}\,. \tag{4}$$

Cette formule nous donne le rapport entre les masses initiale et finale qui, avec l'expulsion à vitesse relative u, permettent d'atteindre la vitesse v de la fusée en partant du repos.

Elle nous fournit aussi la masse m, et par conséquent la masse expulsée par seconde, qui correspond à une loi des vitesses de la fusée, fixée d'avance.

II. Equations du mouvement du point de vue de relativité

Avec les hypothèses précédentes (absence de forces extérieures et mouvement en ligne droite) nous allons introduire dans les formules les corrections de relativité, lesquelles sont, naturellement, négligeables, si les vitesses v et u sont très petites par rapport à la vitesse de la lumière:

$$c = 3 \times 10^{10}\ \text{cm/sec.} \tag{5}$$

L'éjection des particules de matière avec la vitesse u demande une dépense d'énergie qui peut être obtenue par des réactions chimiques ou nucléaires, par l'accélération dans des champs électromagnétiques, etc.

Les réactions chimiques sont insuffisantes pour obtenir des valeurs élevées de u, et elles ne pourront pas être employées pratiquement pour des voyages interstéllaires. Cependant les raisonnements qui sont exposés ci-après et les formules auxquelles nous aboutirons, ne préjugent en rien la nature des dispositifs de propulsion et sont valables dans tous les cas, y compris le cas limite de la fusée photonique: $u = c$.

Il est seulement nécessaire que la source d'énergie ne soit pas extérieure à la fusée, mais qu'au contraire, elle soit inclue dans la masse instantanée considérée. (Fusée adiabatique.)

Nous suivrons le même raisonnement exposé ci-dessus pour la fusée classique, c'est-à-dire: la conservation de la quantité de mouvement du système matériel constitué par la fusée à un moment donné.

Le temps t à partir du moment initial, la masse m de la fusée à l'instant t, la vitesse correspondante v atteinte en partant du repos, et la distance x parcourue, sont mesurés par un observateur extérieur à la fusée, immobile par rapport au système solaire que nous supposons être notre point de départ.

Pour l'équipage de l'astronef, le temps écoulé depuis le départ (celui que nous appellerons "temps propre") est t_0, la masse de la fusée à cet instant m_0, et la vitesse constante avec laquelle sont éjectées en arrière les particules qui la font avancer u.

Comme il est bien connu d'après les principes de relativité, le rapport entre les intervalles de temps infiniments petits, pour l'observateur solaire et pour l'astronaute, est:

$$\frac{dt}{dt_0} = \frac{1}{\sqrt{1 - \frac{v^2}{c^2}}} \; . \tag{6}$$

Les particules éjectées par la fusée qui, d'après les lois de la cinématique classique, auraient, comme nous avons vu, une vitesse absolue $v - u$, possèdent maintenant pour l'observateur solaire une vitesse v_p qui d'après la loi de composition des vitesses de relativité, est:

$$v_p = \frac{v - u}{1 - \frac{uv}{c^2}} \; . \tag{7}$$

A l'instant t, la quantité de mouvement de la fusée pour cet observateur, est, comme auparavant: $m\,v$.

A l'instant $t + dt$ le système est fractionné: la masse de la fusée devient $(m + dm)$ et possède la vitesse $v + dv$, mais la masse éjectée n'est plus $(-dm)$, car l'énergie demandée par l'éjection a été obtenue par transformation d'une partie de la masse empruntée à la fusée.

En désignant par $(-dm_p)$ la masse que cet observateur solaire croit être éjectée et dont la vitesse est pour lui v_p, on obtiendra la quantité totale de mouvement du système en cet instant:

$$(m + dm) \cdot (v + dv) + v_p \, (-dm_p) \; .$$

Et la conservation de la quantité de mouvement de ce système, en absence de forces extérieures, exige l'égalité aux instants t et $t + dt$. D'où:

$$d\,(mv) + v_p \cdot (-dm_p) = 0 \; . \tag{8}$$

Pour l'équipage de la fusée, la masse de celle-ci à l'instant t_0 qui correspond à t, est m_0, et comme sa vitesse, pour l'observateur solaire, est v, il y a entre les masses le rapport dû à la relativité:

$$m = \frac{m_0}{\sqrt{1 - \frac{v^2}{c^2}}} \; . \tag{9}$$

L'intervalle dt de l'observateur solaire a pour correspondant dt_0 pour l'équipage. Pendant cet intervalle, la perte totale de masse du point de vue de l'équipage: $(-dm_0)$ doit être décomposée en deux parties. Une fraction de cette masse, que nous désignons par h, a été transformée en énergie et ce qui reste est la masse expulsée. C'est-à-dire:

$$\left. \begin{array}{l} h\,(-dm_0) \text{ masse transformée en énergie pendant l'intervalle } dt. \\ (1 - h)\,(-dm_0) \text{ masse expulsée pendant cet intervalle.} \end{array} \right\} \tag{10}$$

Par conséquent la masse expulsée pendant le temps dt est, pour l'observateur solaire, d'accord avec les lois de relativité:

$$(- d m_p) = \frac{(1-h)\,(- d\,m_0)}{\sqrt{1 - \dfrac{v_p^2}{c^2}}}\,. \tag{11}$$

En remplaçant (9) et (11) dans (8) on obtient finalement:

$$\left.\begin{array}{l} d\left[\dfrac{m_0\,v}{\sqrt{1 - \dfrac{v^2}{c^2}}}\right] = \dfrac{(1-h)\,v_p\,d\,m_0}{\sqrt{1 - \dfrac{v_p^2}{c^2}}} \\[3em] v_p = \dfrac{v - u}{1 - \dfrac{v\,u}{c^2}}\,; \quad \dfrac{dt}{dt_0} = \dfrac{1}{\sqrt{1 - \dfrac{v^2}{c^2}}} \end{array}\right\} \tag{12}$$

qui sont les équations du mouvement de la fusée. Je pense que ces équations sont connues, malgré que je ne les ai pas trouvées dans la bibliographie que j'ai été à même de consulter.

III. Equations énergétiques

Nous avons déjà dit que l'énergie nécessaire au jet de propulsion de la fusée était obtenue au moyen d'un moteur quelconque et que la masse perdue par la fusée pendant l'intervalle dt, devait être décomposée en une partie éjectée $(1-h) \cdot (- d m_0)$ et une autre partie transformée en énergie à l'intérieur du moteur: $h \cdot (- d m_0)$.

Par conséquent, l'énergie dépensée par le moteur pendant cet intervalle est, d'après la formule d'EINSTEIN:

$$dE = h\,(- d m_0)\,c^2\,. \tag{13}$$

Si le rendement total du moteur est ϱ, l'énergie utile fournie par le jet est:

$$\varrho\,dE$$

qui doit être égale à l'énergie cinétique acquise par la masse expulsée $(1-h) \cdot (- d m_0)$ et qui correspond à la vitesse u d'éjection. D'après l'expression de l'énergie cinétique de relativité:

$$\varrho\,dE = (1-h)\,(- d m_0)\left[\frac{1}{\sqrt{1 - \dfrac{u^2}{c^2}}} - 1\right] \cdot c^2\,. \tag{14}$$

Mais nous allons introduire une hypothèse qui simplifie énormément le calcul et permet d'arriver jusqu'à la fin. Nous supposerons qu'il n'y a pas de pertes, c'est-à-dire que toute l'énergie produite est utilisée dans le jet et que par conséquent $\varrho = 1$. Dans cette hypothèse, en portant (13) dans (14) on obtient:

$$1 - h = \sqrt{1 - \frac{u^2}{c^2}}\,. \tag{15}$$

Cette expression nous fournit une relation entre h et u et nous permet de calculer une de ces variables en fonction de l'autre. Puisque nous avons supposé que la vitesse u d'éjection est constante, h doit être aussi constante.

Les valeurs de u qui s'approchent de la vitesse de la lumière donnent des valeurs de h se rapprochant de l'unité.

Si la matière éjectée et celle que le moteur dépense sont indépendantes l'une de l'autre, il n'y a pas de limite théorique à la valeur de h. (Nous supposons résolus les problèmes pratiques de construction de l'engin, de résistance aux hautes températures, etc.)

Si la matière éjectée $(1 — h) (— dm_0)$ n'était autre chose que le résidu matériel d'une transformation nucléaire de la matière $(— dm_0)$, alors h serait égale au rendement de masse de cette transformation, qui dans les réactions aujourd'hui connues est très petit. Par exemple il est environ de 0,001 dans la fission de l'uranium 235, et de 0,007 dans la fusion théorique de 4 atomes d'hydrogène pour en former un d'hélium, ou dans celle du deuton avec le triton: $(D + T \rightarrow He^4 + n + + 17,6 \ Mev)$.

Mais, même avec cette hypothèse de l'existence d'un résidu, rien ne nous empêche de supposer h aussi élevé que nous voudrons dans un futur lointain, le futur des voyages interstellaires, seuls qui peuvent être affectés par les corrections de relativité qui font l'objet de cet étude, en arrivant même jusqu'à: $h = 1$ pour la fusée photonique.

Nous avons calculé (13) la consommation d'énergie de notre moteur pendant un intervalle dt. De cette expression (13) on déduit immédiatement la puissance instantanée:

$$W = \frac{dE}{dt_0} = h \, c^2 \, \frac{(— dm_0)}{dt_0} \, . \tag{16}$$

Et si le rapport de transformation h est constant et nous désignons par M la masse initiale de la fusée, nous obtiendrons l'énergie totale dépensée:

$$E = h c^2 \cdot (M — m_0) \, . \tag{17}$$

IV. La balance de l'énergie

Nous avons obtenu les équations du mouvement de la fusée par l'application du principe de conservation de la quantité de mouvement. Elles peuvent aussi être obtenues au moyen du principe de conservation de l'énergie, car nous avons supposé qu'il n'y a pas de pertes.

Pour un observateur solaire, l'énergie du système matériel constitué par la fusée, à l'instant t, est, d'après la formule donnée par la relativité:

$$\frac{m_0 \, c^2}{\sqrt{1 - \dfrac{v^2}{c^2}}} \, . \tag{18}$$

A l'instant $t + dt$, le système matériel est divisé en deux. La fusée proprement dite qui a une énergie, par rapport à cet observateur de:

$$\frac{(m_0 + dm_0) \, c^2}{\sqrt{1 - \dfrac{(v + dv)^2}{c^2}}} \, . \tag{19}$$

Et la matière expulsée pendant l'intervalle dt, dont la masse est $(1 — h) \cdot (— dm_0)$ pour l'équipage de la fusée, qui possède une énergie par rapport à ce même observateur solaire:

$$\frac{(1 — h) \, (— dm_0) \, c^2}{\sqrt{1 - \dfrac{v_p^2}{c^2}}} \, . \tag{20}$$

La conservation de l'énergie du système matériel éxige que l'expression (18) soit égale à la somme des expressions (19) et (20), d'où l'on déduit:

$$d\left[\frac{m_0}{\sqrt{1-\dfrac{v^2}{c^2}}}\right] = \frac{(1-h)\,dm_0}{\sqrt{1-\dfrac{v_p{}^2}{c^2}}}\,, \tag{21}$$

équation qui doit conduire au même résultat que le système (12), si nos calculs sont corrects.

V. Changements de variable et transformations

Les équations précédentes peuvent être simplifiées d'une façon extraordinaire au moyen de certains changements de variable, tellement naturels, que j'imagine qu'ils ont dû être employés déjà par d'autres auteurs, malgré que je n'en ai pas eu connaissance.

Soit a une nouvelle variable et β une constante telles que :

$$\left.\begin{aligned} v &= c \cdot \operatorname{th} a \\ u &= c \cdot \operatorname{th} \beta \end{aligned}\right\} \tag{22}$$

On obtient immédiatement :

$$\left.\begin{aligned} &\frac{v}{\sqrt{1-\dfrac{v^2}{c^2}}} = c \cdot \operatorname{sh} a \,; \qquad \frac{1}{\sqrt{1-\dfrac{v^2}{c^2}}} = \operatorname{ch} a \\[2mm] &\frac{1}{\sqrt{1-\dfrac{u^2}{c^2}}} = \operatorname{ch} \beta \\[2mm] &v_p = \frac{v-u}{1-\dfrac{u\,v}{c^2}} = c \cdot \operatorname{th}(a-\beta) \\[2mm] &\frac{v_p}{\sqrt{1-\dfrac{v_p{}^2}{c^2}}} = c \cdot \operatorname{sh}(a-\beta)\,; \qquad \frac{1}{\sqrt{1-\dfrac{v_p{}^2}{c^2}}} = \operatorname{ch}(a-\beta)\,. \end{aligned}\right\} \tag{23}$$

En remplaçant ces expressions dans les équations fondamentales du mouvement (12) et (21) on a :

$$d\,(m_0 \operatorname{sh} a) = (1-h) \operatorname{sh}(a-\beta)\,dm_0; \text{ (conservation de la quantité de mouvement)} \tag{24}$$

$$d\,(m_0 \operatorname{ch} a) = (1-h) \operatorname{ch}(a-\beta)\,dm_0; \text{ (conservation de l'énergie).} \tag{25}$$

La relation (15) est exprimée par :

$$1 - h = \frac{1}{\operatorname{ch} \beta}\,. \tag{26}$$

Le remplacement de h tirée de (26), dans l'éq. (24) ou dans l'éq. (25) conduit, en faisant les opérations nécessaires, à la même équation différentielle, qui est la suivante :

$$-\frac{dm_0}{m_0} = \operatorname{coth} \beta \cdot da\,. \tag{27}$$

Le fait que nous ayons obtenu cette même équation différentielle, qui d'ailleurs est extrêmement simple, en suivant deux procédés différents (premier et troisième principes de la dynamique des systèmes), nous fournit une vérification intéressante.

Quant au rapport des temps donné par l'éq. (12), il devient à présent :

$$\frac{dt}{dt_0} = \operatorname{ch} a\,. \tag{28}$$

VI. Equation du mouvement

L'intégration de l'éq. (27) est immédiate Si à l'instant initial $t = t_0 = 0$, la vitesse est nulle, c'est-à-dire $a = 0$, et la masse de la fusée est M, on a:

$$\frac{M}{m_0} = e^{\coth \beta \cdot a} . \qquad (29)$$

Tant que la vitesse v est très petite par rapport à la vitesse c de la lumière, a et th a sont des infiniments petits équivalents et l'exposant de (29) est:

$$\coth \beta \cdot a \approx \coth \beta \cdot \text{th } \alpha = \frac{\text{th } a}{\text{th } \beta} = \frac{v}{u}$$

et nous voyons alors que l'éq. (29) devient l'éq. (4) que nous avons obtenu pour la fusée classique.

Par contre, si v se rapproche de la vitesse de la lumière, a augmente indéfiniment, ce qui nous donne, d'après l'éq. (29) un rapport des masses qui tend vers l'infini.

Si c'est u qui tend vers c, c'est-à-dire si $h \to 1$, nous aurons à la limite la fusée photonique, pour laquelle $\frac{M}{m_0} = e^a$.

Nous pouvons aussi calculer au moyen de (27) l'équation de la période de freinage. Si le jet est inversé et dirigé dans le sens de v, il suffira de changer le signe de u, ou bien celui de β, dans (27). Si à l'instant t' de l'observateur solaire, équivalent à l'instant t_0' de l'équipage, la masse de la fusée est égale à m_0', et sa vitesse égale à v' ou (a') et c'est alors que le freinage commence, on a, à partir de cet instant:

$$\frac{m_0'}{m_0} = e^{\coth \beta (a' - a)} . \qquad (30)$$

En prenant (29) au moment où l'inversion commence on a:

$$\frac{M}{m_0'} = e^{\coth \beta \cdot a'} \qquad (31)$$

et si m_l désigne la masse finale correspondante au repos après freinage:

$$\frac{m_0'}{m_l} = e^{\coth \beta \cdot a'} . \qquad (32)$$

C'est-à-dire que, de même que pour la fusée classique, le rapport entre la masse initiale et celle qui correspond à une vitesse donnée, est égale au rapport entre cette dernière masse et la masse finale après freinage.

VII. La poussée et l'accélération sensible

Nous appellerons poussée la force qui agissant sur une fusée théorique de masse constante, égale à celle de la fusée réelle à l'instant t, produirait pendant l'intervalle dt la même variation de vitesse que subit la fusée réelle.

C'est-à-dire:

$$F = m_0 \frac{d}{dt} \left[\frac{v}{\sqrt{1 - \frac{v^2}{c^2}}} \right] = m_0 \cdot c \cdot \frac{d \text{ sh } a}{dt} = m_0 \cdot c \cdot \text{ch } a \frac{da}{dt} = m_0 c \frac{da}{dt_0} . \qquad (33)$$

Cette poussée est un invariant. C'est-à-dire: que la poussée est la même pour tous les observateurs se déplaçant les uns par rapport aux autres avec une vitesse constante.

Soit en effet un observateur qui se déplace à la vitesse v_1 par rapport à l'observateur solaire ($v_1 = c \cdot \text{th } a_1$).

Pour lui la vitesse de la fusée est: $v_r = \dfrac{v - v_1}{1 - \dfrac{v v_1}{c^2}} = c \cdot \text{th } (a - a_1)$;

$$F_1 = m_0 \frac{d}{dt_1} \cdot \frac{v_r}{\sqrt{1 - \dfrac{v_r{}^2}{c^2}}} = m_0 c \frac{d \text{ sh } (a - a_1)}{dt_1} = m_0 c \frac{d \text{ sh } (a - a_1)}{\text{ch } (a - a_1) \, dt_0} =$$

$$= \frac{m_0 c \, d(a - a_1)}{dt_0} = m_0 c \frac{da}{dt_0} = F .$$

Nous prendrons en particulier un observateur pour lequel $v_1 = v$, qui se déplace par conséquent à une vitesse constante et égale à celle de la fusée à l'instant propre t_0. A ce moment-là le temps du nouvel observateur est le même que celui de la fusée ($dt_1 = dt_0$) et par rapport à lui le système matériel m de la fusée est au repos.

Une fois l'intervalle dt_0 écoulé, les particules éjectées vers l'arrière avec une vitesse u, ont une quantité de mouvement:

$$(1 - h) (- dm_0) \frac{u}{\sqrt{1 - \dfrac{u^2}{c^2}}} = \frac{1}{\text{ch } \beta} (- dm_0) \cdot c \cdot \text{sh } \beta = - c \cdot \text{th } \beta \cdot dm_0$$

qui doit être la même que celle qui a été acquise en sens opposé par la fusée, par rapport au dit observateur. Si la vitesse de la fusée par rapport à ce dernier augmente de 0 à dw, cette quantité de mouvement est égale à:

$$\frac{(m_0 + dm_0)}{\sqrt{1 - \left(\dfrac{dw}{c}\right)^2}} \, dw = m_0 \, dw, \text{ en négligeant les infiniment petits, d'ordre plus élevé.}$$

Mais par ailleurs:

$$dw = \frac{(v + dv) - v}{1 - \dfrac{v \, (v + dv)}{c^2}} = \frac{dv}{1 - \dfrac{v^2}{c^2}} = c \cdot \text{ch}^2 \, a \cdot d \, \text{th } a = c \cdot d \, a$$

donc, l'égalité des quantités de mouvement nous conduit à:

$$- c \cdot \text{th } \beta \cdot dm_0 = m_0 c \, da \qquad (34)$$

qui est la même équation obtenue précédemment (27) et qui en fournit une vérification. Sous cette forme (34) le premier membre représente l'impulsion reçue pendant l'intervalle dt, et le deuxième membre donne l'effet de cette impulsion sur la fusée, laquelle est égale, comme il était à prévoir, à $F \, dt_0$, en prenant pour F la valeur donnée par (33).

On a donc:

$$F = \frac{c \cdot \text{th } \beta \, (- dm_0)}{dt_0} = m_0 c \frac{da}{dt_0} . \qquad (35)$$

Imaginons maintenant un voyageur à l'intérieur de la fusée, dont la masse propre est m_1, et qui dans l'intervalle dt passe de la vitesse v à la vitesse $v + dv$. A cet effet il est nécessaire que la partie de la fusée en contact avec le corps du voyageur (scl, fauteuil, etc.) exerce sur lui une force égale à:

$$m_1 c \frac{da}{dt_0} .$$

Par conséquent l'accélération de la gravitation apparente ou accélération "sensible", qui doit correspondre à cette force, est :

$$j = c \frac{da}{dt_0}. \tag{36}$$

Cette quantité ne dépend pas de l'observateur, pour les mêmes raisons ci-dessus exposées pour la poussée.

VIII. Loi des vitesses les plus favorables pour les fusées à équipage

Le programme des vitesses d'un voyage interstellaire en ligne droite, c'est-à-dire d'un voyage qui commencerait avec une vitesse nulle par rapport au système solaire et qui se terminerait avec une vitesse nulle également, peut être décomposé en trois périodes.

Une première période d'accélération pendant laquelle la vitesse croît de 0 à un maximum que nous désignerons par $\iota' = c \cdot \mathrm{th}\, \alpha'$.

Une deuxième période (qui pourrait ne pas se présenter) pendant laquelle la vitesse est constante et égale à v'.

Une troisième période de décélération pendant laquelle le jet est inversé et la vitesse décroît progressivement jusqu'à 0.

Le rapport entre les masses initiale M et finale m_f de la fusée, d'après les éqs. (31) et (32) est :

$$\frac{M}{m_f} = \left[\frac{M}{m_0'} \right]^2 = \mathrm{e}^{2\,a' \coth \beta}. \tag{37}$$

Par conséquent, le rapport des masses du début et de la fin du voyage, ne dépend pas de la loi des vitesses, mais seulement de la valeur maximum de α'.

Une fois que la vitesse maximum a été choisie (d'après le rapport des masses), il est évident que le voyage est d'autant plus court que la vitesse maximum est le plus rapidement atteinte pendant la première période et le plus rapidement annulée pendant la dernière période.

Mais d'autre part les accélérations ou les décélérations sont limitées par des considérations d'ordre physiologique.

Si nous évaluons à n fois la constante de gravitation terrestre : ng, la plus grande valeur qui soit admise pour la gravitation apparente, nous obtiendrons d'après (36) :

$$ng = c \frac{da}{dt_0}. \tag{38}$$

Et le temps du voyage est le plus court possible si nous appliquons cette loi à toute la période d'accélération (et de décélération).

Donc, pour la première période nous avons :

$$a = \frac{ng}{c} t_0; \quad a \leqslant a'. \tag{39}$$

IX. Equations de la periode d'accélération

La connaissance de α en fonction de t_0 permet de calculer les différentes grandeurs, visées ci-dessus.

Temps propre de la période totale d'accélération (39) :

$$t_0' = \frac{c\, a'}{ng} \tag{40}$$

Temps de l'observateur solaire (28):

$$t = \int_0^{t_0} \left(\text{ch} \, \frac{ng}{c} \, t_0 \right) dt_0 = \frac{c}{ng} \, \text{sh} \, \frac{ng}{c} \, t_0 = \frac{c}{ng} \, \text{sh} \, \alpha \; . \tag{41}$$

Temps correspondant à la période complète:

$$t' = \frac{c}{ng} \, \text{sh} \, \frac{ng}{c} \, t_0' = \frac{c}{ng} \, \text{sh} \, \alpha' \; . \tag{42}$$

Espace parcouru d'après l'observateur solaire:

$$x = \int_0^{t_0} v \, \frac{dt}{dt_0} \, dt_0 = c \int_0^{t_0} \text{th} \, \alpha \cdot \text{ch} \, \alpha \cdot dt_0 = c \int_0^{t_0} \left(\text{sh} \, \frac{ng}{c} \cdot t_0 \right) dt_0 \; ;$$

$$x = \frac{c^2}{ng} \left[\text{ch} \, \frac{ng \, t_0}{c} - 1 \right] = \frac{c^2}{ng} \, (\text{ch} \, \alpha - 1) \; . \tag{43}$$

Parcours total de la première période:

$$x' = \frac{c^2}{ng} \, [\text{ch} \, \alpha' - 1] \; . \tag{44}$$

Masse instantanée de la fusée (29):

$$m_0 = M \, e^{-\alpha \cdot \coth \beta} = M \, e^{-\frac{ng \, t_0}{c \cdot \text{th} \, \beta}} \; . \tag{45}$$

Puissance instantanée (16):

$$W = h \, c^2 \left[\frac{-d m_0}{dt_0} \right] = h \, c^2 \, m_0 \coth \beta \, \frac{da}{dt_0} = h \, c \, ng \coth \beta \cdot m_0 \; . \tag{46}$$

La puissance maximum est égale à la puissance initiale:

$$W_{max} = h \, c \, n \, g \coth \beta \cdot M \; . \tag{47}$$

Poussée (33):

$$F = m_0 \, c \, \frac{da}{dt_0} = m_0 \cdot ng \; ; \quad F_{max} \, (\text{initiale}) = M \, ng \; . \tag{48}$$

X. Le voyage en ligne droite

Le programme du voyage, d'après ce qui vient d'être dit, est déterminé par la connaissance des quatre données suivantes:

x_f parcours total du voyage.

$\dfrac{M}{m_f}$ rapport des masses initiale et finale.

h rapport entre la masse transformée en énergie et la somme de cette masse et de celle qui est expulsée.

n rapport de l'accélération sensible que l'équipage peut supporter à celle de la gravitation terrestre.

La valeur fondamentale a' est tirée de (37):

$$a' = \frac{\text{th} \, \beta}{2} \, \text{lognep} \, \frac{M}{m_f}$$

et l'on déduit de (26):

$$\text{th} \, \beta = \sqrt{h \, (2 - h)} \; . \tag{49}$$

D'où:

$$a' = \frac{1}{2} \sqrt{h\,(2-h)} \, \text{lognep} \, \frac{M}{m_j} \,. \tag{50}$$

Il est intéressant de remarquer que la relation entre la vitesse d'expulsion $u = c \cdot \text{th}\,\beta$, et le facteur de masse h, dépend de l'expression $\sqrt{h\,(2-h)}$, qui représente une circonférence, comme il est indiqué à la Fig. 1.

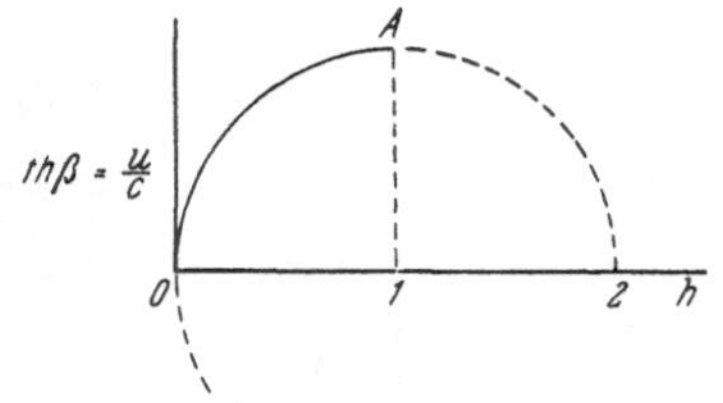

Il existe donc un maximum A aux environs duquel $\text{th}\,\beta$ est très peu sensible aux variations de h.

Par exemple, si $h = 0{,}6$ on a:

$$\text{th}\,\beta = \sqrt{0{,}6 \times 1{,}4} = 0{,}916 \tag{51}$$

Fig. 1

qui est à peu près égal à l'unité.

Une fois a' calculé et n connu, nous obtiendrons le temps propre $2\,t'_0$ nécessaire à l'accélération et à la décélération et le temps correspondant $2\,t'$ mesuré par l'observateur solaire, ainsi que le parcours partiel $2\,x'$, égal à la somme des parcours des deux périodes:

$$\left.\begin{aligned} 2t_0' &= \frac{2c}{ng}\,a'\,; \\[1mm] 2t' &= \frac{2c}{ng}\,\text{sh}\,a'\,; \\[1mm] 2x' &= \frac{2c^2}{ng}\,(\text{ch}\,a' - 1) = \frac{4c^2}{ng}\,\text{sh}^2\frac{a'}{2}\,. \end{aligned}\right\} \tag{52}$$

Le parcours total imposé doit normalement dépasser $2\,x'$:

$$x_j > 2x' \tag{53}$$

car il s'agit des grandes distances interstellaires. S'il n'en était pas ainsi, cela indiquerait que $\dfrac{M}{m_j}$, n ou h sont trop grands Nous pourrions alors nous passer de la période de vitesse constante et rajuster les valeurs indiquées de façon à ce que la dernière éq. (52) donne: $2x' = x_j$.

Mais, en général, il faudra prendre l'inégalité (53) et il y aura un parcours: $x_j - 2\,x'$ (par rapport à l'observateur solaire) à vitesse constante $c \cdot \text{th} \cdot a'$, pendant le temps (de cet observateur):

$$t_j - 2t' = \frac{x_j - 2x'}{c \cdot \text{th}\,a'} \tag{54}$$

auquel correspond un temps propre égal à:

$$t_{0j} - 2t_0' = \frac{t_j - 2t'}{\text{ch}\,a'} = \frac{x_j - 2x'}{c \cdot \text{sh}\,a'} \,. \tag{55}$$

Comme nous avons déjà dit, (47) et (48), la puissance et la poussée maximum sont celles du moment initial:

$$W_{max} = \frac{h\,c\,n\,g\,M}{\sqrt{h\,(2-h)}} \tag{56}$$

$$F_{max} = n\,g\,M \,. \tag{57}$$

Et l'énergie totale dépensée pendant le voyage est d'après (17):

$$E = h\,c^2 \cdot (M - m_j) \,. \tag{58}$$

XI. La fusée multiple

L'intégrale générale de l'éq. (27) nous dit que, si m_0'' et m_0''' désignent les masses de la fusée aux vitesses a'' et a''' correspondantes à une même période accélératoire :

$$\frac{m_0''}{m_0'''} = e^{\coth \beta \, (a''' - a'')} \tag{59}$$

rapport qui doit être inversé pour la période de freinage.

Supposons maintenant une fusée multiple à q étages. Le voyage est alors effectué en q étapes ou trajets partiels et un des corps auxiliaires est detaché à la fin de chacune des $q-1$ premières étapes.

D'après (59) lorsque le rapport des masses initiale et finale de chaque étape est constant et égal à r, on a :

$$r = e^{\coth \beta \cdot \Delta a} \tag{60}$$

c'est-à-dire que Δa est égal pour toutes les étapes.

Prenons maintenant q pair et décomposons chacune des périodes d'accélération et de décélération et $\frac{q}{2}$ étapes. On aura :

$$\Delta a = \frac{a'}{q} \cdot 2 \tag{61}$$

et en combinant (60) et (61) :

$$e^{2 \coth \beta \cdot a'} = r^q . \tag{62}$$

La comparaison de (62) et (37) nous montre que pour établir le programme du voyage d'une fusée multiple il suffit de remplacer, au paragraphe X, $\frac{M}{m_f}$ par r^q à la condition de négliger le temps nécessaire aux opérations de détachement des corps auxiliaires usagés.

Il n'est pas permis de faire ce remplacement pour les calculs énergétiques. En effet, à égalité des autres caractéristiques, l'énergie totale dépensée est en ce cas plus petite, parce que nous n'avons pas à accélérer et à freiner les corps succesifs détachés.

Supposons que la masse finale de chaque étape soit p fois supérieure à la masse initiale de l'étape suivante, à cause des détachements, et que p soit constant.

L'énergie totale dépensée est alors :

$$E = h c^2 \left[M - \frac{M}{r} + \frac{M}{rp} - \frac{M}{r^2 p} + \dots + \frac{M}{r^{q-1} p^{q-1}} - \frac{M}{r^q p^{q-1}} \right] =$$
$$= h c^2 M \frac{r-1}{r} \cdot \frac{1 - (r \, p)^{-q}}{1 - (r \, p)^{-1}} \tag{63}$$

et la masse finale est donnée par :

$$\text{Masse finale de la dernière étape} = \frac{M}{r^q \, p^{q-1}} . \tag{64}$$

XII. Applications numériques

Pour les applications numériques du programme d'un voyage interstellaire, il est commode de mesurer les temps en années et les distances en années-lumière.

Les mesures prises avec ces unités seront désignées par les mêmes lettres et indexes que précédemment, mais en employant des majuscules.

En tenant compte de:

$$x \text{ (cm)} = X \text{ (années-lumière)} \times 9 \times 10^{17}$$
$$t \text{ (sec)} = T \text{ (années)} \times 3 \times 10^{7}$$
$$g \qquad = 980 \text{ cm/sec}^2$$
$$c \qquad = 3 \times 10^{10} \text{ cm/sec}$$

nous aboutissons au groupe suivant de formules fondamentales:

$$\left.\begin{aligned}
a' &= \frac{q}{2}\sqrt{h\,(2-h)}\;\text{lognep } r \\[1mm]
2\,T_0' &= \frac{2{,}04}{n}\cdot a' \\[1mm]
2\,T' &= \frac{2{,}04}{n}\cdot \operatorname{sh} a' \\[1mm]
2\,X' &= \frac{2{,}04}{n}\,(\operatorname{ch} a' - 1) \leqslant X_{l}\,.
\end{aligned}\right\} \qquad (65)$$

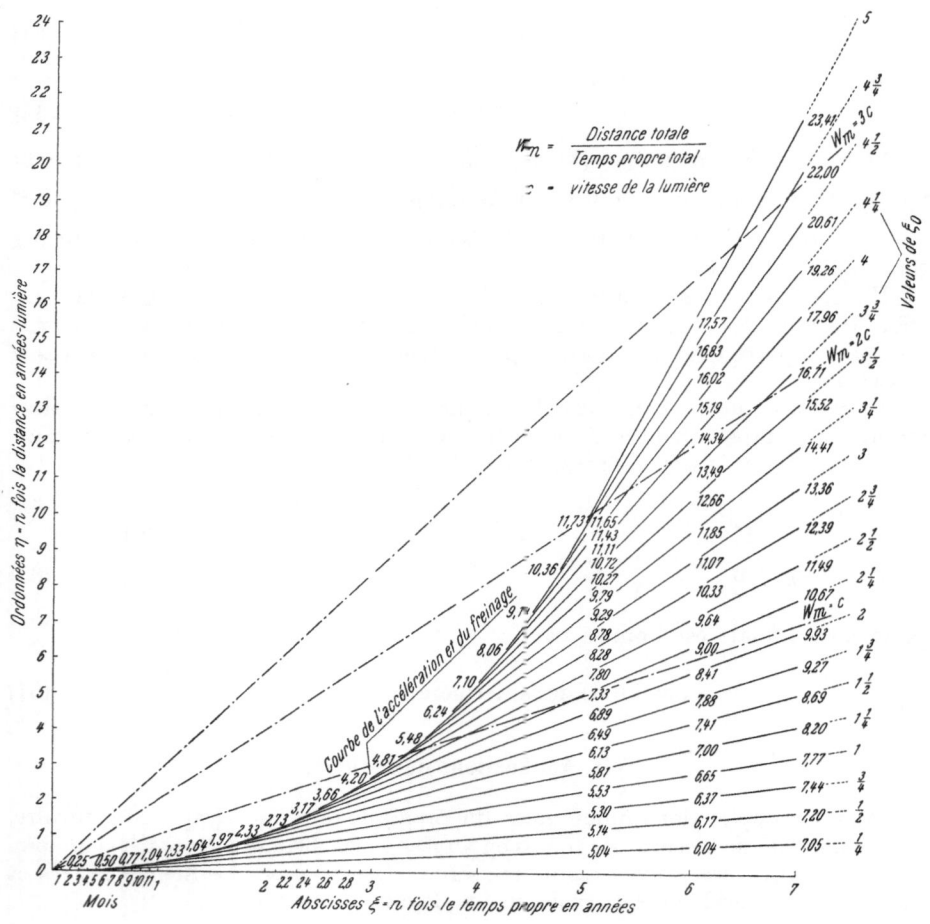

De (64) et (65) l'on tire les vitesses moyennes d'après le voyageur pour les périodes d'accélération et de freinage, ainsi que pour la période intermédiaire:

$$\left.\begin{array}{l} \dfrac{2\,X'}{2\,T_0{}'} = \dfrac{\operatorname{ch} a' - 1}{a'} \text{ fois la vitesse de la lumière.} \\[3mm] \dfrac{X_1 - 2\,X'}{T_{01} - 2\,T_0{}'} = \operatorname{sh} a' \text{ fois la vitesse de la lumière.} \end{array}\right\} \tag{67}$$

Il est intéressant de remarquer que ces vitesses moyennes sont indépendantes du nombre n choisi et que les temps d'accélération et de freinage ($2\,T'_0$), ainsi que les distances parcourues ($2\,X'$) sont inversement proportionnelles à n.

La Fig. 2 met en rapport les distances parcourues et le temps propre qui est celui de l'équipage de la fusée. Soient ξ les abscisses et η les ordonnées.

L'abscisse ξ représente le produit du nombre n définissant l'accélération constante choisie, par le nombre d'années-lumière ecoulées.

L'ordonnée η est égale au produit de n par la distance parcourue mesurée en années-lumière.

La courbe représentée par le graphique est donnée par les équations paramétriques:

$$\left.\begin{array}{l} \xi_0 = 2{,}04\,a' \\ \eta_0 = 2{,}04\,(\operatorname{ch} a' - 1)\,. \end{array}\right\} \tag{68}$$

Par conséquent, sur cette courbe, ξ_0 est le produit de n par le temps propre employé à l'accélération et au freinage, tandis que η_0 est le produit de n par la distance parcourue pendant ces deux phases.

Fig. 2. Voyage aux étoiles. Hypothèses: Mouvement rectiligne non soumis à des champs gravitatoires. La fusée part du repos et revient au repos.

1er Stade (Du repos à la vitesse maximum)	Accélération "sensible" constante: ng $(g = 980 \text{ cm/sec}^2)$ Durée: $\dfrac{\xi_0}{2\,n}$ (années du temps propre) Chemin parcouru: $\dfrac{\eta_0}{2\,n}$ (années-lumière)
2ème Stade (Peut être supprimé)	Vitesse constante Durée: $\dfrac{\xi - \xi_0}{n}$ (années du temps propre) Chemin partiel parcouru: $\dfrac{\eta - \eta_0}{n}$ (années-lumière)
3ème Stade (Depuis la vitesse maximum jusqu'au repos)	Accélération sensible constante: $-ng$ (négative) Durée: $\dfrac{\xi_0}{2\,n}$ (années du temps propre) Chemin partiel parcouru: $\dfrac{\eta_0}{2\,n}$ (années-lumière)

Temps propre total: $\dfrac{\xi}{n}$ (années)

Chemin total parcouru: $\dfrac{\eta}{n}$ (années-lumière)

Les nombres portés sur les lignes du graphique indiquent les temps multipliés par n, mesurés par l'observateur terrestre.

Ligne courbe (1er et 3ème stades): $\dfrac{\eta_0}{2{,}04} + 1 = \operatorname{ch} \dfrac{\xi_0}{2{,}04}$.

Lignes droites tangentes (2ème stade): $\eta - \eta_0 = (\xi - \xi_0)\,\operatorname{sh} \dfrac{\xi_0}{2{,}04}$.

Temps mesuré par l'observateur terrestre, multiplié par n:

$$\text{1er et 3ème stades: } 2{,}04\,\operatorname{sh} \dfrac{\xi_0}{2{,}04}$$

$$\text{2ème stade: } (\xi - \xi_0)\,\operatorname{ch} \dfrac{\xi_0}{2{,}04}$$

La tangente à la courbe au point (ξ_0, η_0) est donné par:

$$\eta - \eta_0 = \frac{d\,\eta_0}{d\,\xi_0}\,(\xi - \xi_0) = \text{sh}\ a'\ (\xi - \xi_0) \tag{69}$$

et comme sh a' est d'après (67) la vitesse constante de la phase comprise entre l'accélération et le freinage, la droite tangente (69) représente la relation entre ξ et η pendant cette période à vitesse constante.

En effet, il y a lieu de remarquer que:

$$\text{sh}\ a' = \frac{\eta - \eta_0}{\xi - \xi_0} = \frac{n \text{ fois l'accroissement de la distance}}{n \text{ fois l'accroissement du temps}} =$$
$$= \frac{\text{Accroissement de la distance}}{\text{Accroissement du temps}}\ .$$

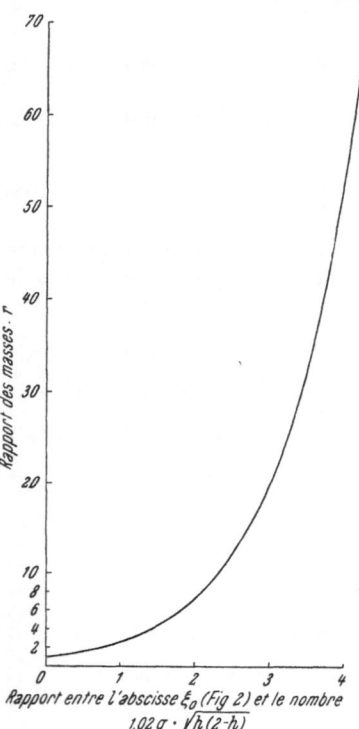

Par ailleurs, la valeur a' correspondante à la vitesse maximum, donnée par la première éq. (65), fournit une relation entre l'abscisse ξ_0 de la courbe et les caractéristiques de la fusée, qui est la suivante:

$$\xi_0 = 1{,}02\ q \cdot \sqrt{h\,(2-h)}\ \text{lognep}\ r\ . \tag{70}$$

Cette relation est représentée à la Fig. 3.

A la Fig. 4 sont indiquées la puissance W et la poussée F pour la période d'accélération des fusées à un seul étage. Les formules utilisées sont les (46) et (48). Ces formules exprimées en unités pratiques nous donnent:

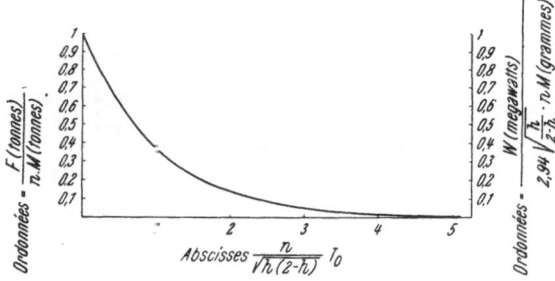

Fig. 3. $\xi_0 = 1{,}02\ q\sqrt{h\,(2-h)}$ lognep r.

q: nombre d'étages de la fusée; h: rapport entre la masse transformée en énergie et la perte totale de masse (masse expulsée et masse transformée en énergie); r: rapport des masses initiale et finale de la fusée, à chaque étage

Fig. 4. Poussée F et puissance W instantanées.

$$\frac{F\text{ (tonnes)}}{n \cdot M\text{ (tonnes)}} = \text{e}^{\displaystyle -\frac{0{,}98\ n\ T_0}{\sqrt{h\,(2-h)}}} =$$
$$= \frac{W\text{ (mégawatts)}}{2{,}94\ n\ \sqrt{\dfrac{h}{2-h}}\ M\text{ (grammes)}}$$

$$W\text{ (megawatts)} = 2{,}94 \times n \times M\text{ (grammes)} \times \sqrt{\frac{h}{2-h}} \times \text{e}^{\displaystyle -\frac{0{,}98\,\pi\,T_0}{\sqrt{h\,(2-h)}}} \tag{71}$$

$$F\text{ (tonnes)} = n \times M\text{ (tonnes)} \times \text{e}^{\displaystyle -\frac{0{,}98\,\pi\,T_0}{\sqrt{h\,(2-h)}}}\ . \tag{72}$$

Les Fig. 2, 3 et 4 mettent en rapport toutes les caractéristiques intéressant le voyage interstellaire.

La Fig. 2 est la plus compliquée et demande un commentaire spécial.

Chaque point (ξ, η), de la Fig. 2 se trouvant sur la courbe ou vers la droite, correspond à un voyage possible. La distance totale parcourue est $\frac{\eta}{n}$ et le temps propre employé $\frac{\xi}{n}$.

Lorsque le point se trouve sur la courbe elle même, il y aura seulement les deux phases d'accélération et de freinage et le temps nécessaire au voyage (n étant fixé) est le minimum possible.

Lorsque le point se trouve à droite de la courbe l'on pourra mener une tangente. Le point de contact de la tangente (ξ_0, η_0) nous fournira la distance parcourue pendant l'accélération et le freinage: $\frac{\eta_0}{n}$, et le temps nécessaire à ces deux phases $\frac{\xi_0}{n}$, et nous saurons aussi la relation qui doit exister entre les caractéristiques q, h, r de la fusée (70) au moyen de la Fig. 3, ainsi que l'espace $\frac{\eta - \eta_0}{n}$ et le temps $\frac{\xi - \xi_0}{n}$ correspondants à la phase de vitesse constante.

Des lignes droites à traits discontinus, portées aussi sur la Fig. 2, nous montrent les voyages pour lesquels la vitesse moyenne (par rapport au temps propre) est égale à la vitesse de la lumière ou à un multiple de cette dernière.

Les chiffres portés au milieu du graphique donnent les temps de l'observateur solaire.

Pour mettre d'avantage en lumière le maniement des graphiques, nous allons donner des exemples.

Supposons la distance à parcourir connue et qu'il s'agit d'en déduire les caractéristiques de la fusée adéquate au voyage.

L'on veut, par exemple, arriver à la région de l'Alfa du Centaure. Cette étoile se trouve à 4,3 années-lumière et l'on estime que l'objectif précis (dont la vitesse par rapport au soleil est négligée) est à 4,5 années-lumière:

$$X = 4,5 \ .$$

Il faut d'abord fixer le nombre n qui nous permet de calculer l'accélération "sensible" constante en fonction de la gravitation terrestre. Soit:

$$n = 1$$

qui correspond à l'accélération la plus commode à supporter pour l'organisme humain. On a:

$$\eta = n X = 4,5 \text{ années-lumière} \ .$$

L'ordonnée 4,5 de la Fig. 2 rencontre la courbe au point dont l'abcisse est:

$$\xi = 3,75 \text{ années}$$

à laquelle correspond pour le temps solaire:

$$n T = 6,24 \text{ années} \ .$$

Cela montre, que si nous décidons d'arriver le plus tôt possible au prix d'une plus grande dépense d'énergie, en réduisant à zéro la période de vitesse constante, c'est-à-dire: en accélérant la fusée pendant la moitié du trajet et en la freinant pendant l'autre moitié, le voyage durera 3 années et 9 mois d'après l'équipage, et 6,24 années d'après l'observateur terrestre.

Si au contraire, nous prétendons obtenir une économie d'énergie même si le temps employé augmente, nous chercherons sur la Fig. 2, à droite de la courbe, un point approprié sur l'ordonnée 4,5. Soit ce point celui d'intersection avec la tangente correspondante à $\xi = 1^1/_2$ années. L'abscisse de ce point est de 6 années, 4 mois et 18 jours et le temps solaire est obtenu par interpolation entre les chiffres consécutifs 7,41 et 8,69 entre lesquels se trouve le point choisi, ce qui donne 7,90 années.

Cette deuxième supposition nous conduit donc à 9 mois d'accélération et encore 9 mois de décélération (en tout $1^1/_2$ années), et à maintenir un régime à vitesse constante pendant 4 années, 10 mois et 18 jours; la durée totale du voyage est de 6 années, 4 mois et 18 jours, correspondant à $7,9 = 7$ années, 10 mois et 24 jours d'après l'observateur terrestre.

Pour fixer les caractéristiques de la fusée nous passons à la Fig. 3, qui met en relation le rapport des masses r et: ξ_0, h et q.

Soit $h = 0,6$ le rapport de conversion énergétique, qui conduit à une vitesse d'expulsion de:

$$\frac{u}{c} = \sqrt{h\,(2-h)} = 0,9165$$

soit: 91,6% de la vitesse de la lumière.

L'abscisse de la Fig. 3 est:

$$\frac{\xi}{1,02\,q\,\sqrt{h\,(2-h)}}$$

laquelle avec la première supposition devient: $\dfrac{3,75}{1,02 \cdot q \cdot 0,9165} = \dfrac{4}{q}$ et avec la deuxième supposition: $\dfrac{1,5}{1,02 \cdot q \cdot 0,9165} = \dfrac{1,6}{q}$.

Une fusée simple (à un seul étage) donnerait des abscisses 4 et 1,6 auxquels correspondraient des rapports de masse: $r = 53,7$ et $r = 4,5$ respectivement. Le deuxième rapport est acceptable. Il n'en est pas ainsi du premier, et nous supposerons alors une fusée multiple à deux étages, dont la nouvelle abscisse est égale à $\dfrac{4}{2} = 2$, qui correspond à $r = 6,7$; rapport qui peut être admis.

Finalement nous prendrons comme rapport p des masses de la fusée avant et après le détachement de la partie usagée: $p = 1,5$.

On obtient ainsi, avec la première supposition:

$$\begin{cases} h = 0,6 \\ r = 6,7 \\ q = 2 \\ p = 1,5 \ . \end{cases}$$

Et avec la deuxième:

$$\begin{cases} h = 0,6 \\ r = 4,5 \\ q = 1 \ . \end{cases}$$

Si dans les deux cas, la masse finale de la fusée, qui doit arriver à destination, était de 2 tonnes, on aurait:

1ère supposition

Masse initiale: $M = m_f \cdot r^q \cdot p^{q-1} = 2 \times 6{,}7^2 \times 1{,}5 = 134{,}67$ tonnes.

Energie totale dépensée: $E = h c^2 M \dfrac{r p - p}{r p - 1} [1 - (r\,p)^{-q}] = 6{,}8 \times 10^{28}$ ergs. $=$
$= 2{,}26 \times 10^8$ mégawatts-année.

Puissance maximum (initiale): $W_{max} = 2{,}94 n \sqrt{\dfrac{h}{2-h}}\, M = 2{,}59 \times 10^8$ megawatts.

Poussée maximum (initiale): $F_{max} = n\,M = 134{,}67$ tonnes.

Sur la Fig. 4 on peut observer la diminution rapide de W et F avec le temps.

2ème supposition

$M = 2 \times 4{,}5 = 9$ tonnes (ce qui est tout à fait raisonnable).

$E = 0{,}6 \times 9 \times 10^{20} \times 9 \times 10^6 \times \left(1 - \dfrac{1}{4{,}5}\right) = 3{,}78 \times 10^{27}$ ergs $= 1{,}26 \times 10^7$ mégawatts-année.

$W_{max} = 2{,}94 \sqrt{\dfrac{3}{7}} \times 9 \times 10^6 = 1{,}73 \times 10^7$ mégawatts.

$F_{max} = 9$ tonnes.

On trouve donc, que pour un même voyage: de 4,5 années-lumière, il faudrait des fusées de spécifications très différentes, selon que la durée envisagée est de 3,75 ou de 6,38 années. Les énergies à dépenser sont très considérables dans les deux cas. Elles peuvent être réduites à volonté en diminuant le facteur d'accélération $n = 1$ supposé, mais c'est au dépens de l'augmentation du temps employé.

Nous pourrions multiplier les exemples, qui sont particulièrement curieux lorsqu'on choisit des objectifs très éloignés, mais il suffit de l'exemple donné pour montrer la façon d'utiliser les graphiques.

XIII. Impulsion à puissance constante

Nous allons étudier maintenant, le cas d'une fusée impulsée par un moteur à puissance constante, c'est-à-dire nous supposons que la masse expulsée par seconde de temps propre, est constante.

Nous désignerons par μ_0 la perte totale constante de masse par seconde propre:

$$\mu_0 = -\frac{d m_0}{d t_0} = \text{const.} \tag{73}$$

D'après (16), la puissance constante du moteur, est:

$$W = h c^2 \mu_0 \tag{74}$$

et la quantité de masse matérielle expulsée par seconde:

$$(1 - h)\,\mu_0 = \frac{\mu_0}{\operatorname{ch}\beta} = \mu_0 \sqrt{1 - \frac{u^2}{c^2}}\;;$$

en combinant (73) avec la formule fondamentale (29) nous avons:

$$m_0 = M - \mu_0 t_0 = M\, e^{-a \cdot \coth \beta}. \tag{75}$$

D'où:

$$a = -\operatorname{th}\beta \cdot \operatorname{lognep}\left[1 - \frac{\mu_0}{M} t_0\right] \tag{76}$$

qui nous donne la loi de variation des vitesses de la fusée :

$$v = c \cdot \operatorname{th} a = c \frac{1 - \left[1 - \dfrac{\mu_0 t_0}{M}\right]^{\frac{2u}{c}}}{1 + \left[1 - \dfrac{\mu_0 t_0}{M}\right]^{\frac{2u}{c}}} . \qquad (77)$$

En tirant t_0 de (75) on a :

$$t_0 = \frac{M}{\mu_0} \left[1 - e^{-a \coth \beta}\right] . \qquad (78)$$

Et en prenant les différentielles :

$$dt_0 = \frac{M}{\mu_0} \coth \beta \cdot e^{-a \cdot \coth \beta} \cdot d a . \qquad (79)$$

D'où, à cause de (28) :

$$dt = dt_0 \cdot \operatorname{ch} a = \frac{M}{\mu_0} \coth \beta \cdot e^{-a \cdot \coth \beta} \cdot \operatorname{ch} a \cdot d a . \qquad (80)$$

Par conséquent, le temps solaire est :

$$t = \frac{M}{\mu_0} \coth \beta \cdot \int_0^a e^{-a \coth \beta} \operatorname{ch} a \cdot d a =$$

$$= \frac{M \coth \beta}{2 \mu_0} \int_0^a \left[e^{-a (\coth \beta + 1)} - e^{-a (\coth \beta - 1)}\right] d a =$$

$$= \frac{M \coth \beta}{2 \mu_0} \left[\frac{1 - e^{-a (\coth \beta + 1)}}{\coth \beta + 1} - \frac{1 - e^{-a (\coth \beta - 1)}}{\coth \beta - 1}\right] .$$

Cette expression simplifiée devient :

$$t = \frac{M}{\mu_0} \operatorname{ch} \beta \left[\operatorname{ch} \beta - \operatorname{ch} (a + \beta) e^{-a \coth \beta}\right] . \qquad (81)$$

Ce résultat peut être exprimé en fonction de t_0 au moyen de l'éq. (75) qui le met en rapport avec a. Si de plus nous remplaçons $\operatorname{th} \beta = \dfrac{u}{c}$, nous arrivons finalement à :

$$t = \frac{M}{\mu_0} \cdot \frac{1}{1 - \dfrac{u^2}{c^2}} \left[1 - \frac{c - u}{2c} \left[1 - \frac{M_0 t_0}{M}\right]^{\frac{c+u}{c}} - \frac{c + u}{2c} \left[1 - \frac{\mu_0 t_0}{M}\right]^{\frac{c-u}{c}}\right] . \qquad (82)$$

Voyons maintenant la distance parcourue. La différentielle, en tenant compte de (80), est :

$$d x = v \cdot d t = c \cdot \operatorname{th} a \frac{M}{\mu_0} \coth \beta \cdot e^{-a \coth \beta} \cdot \operatorname{ch} a \cdot d a =$$

$$= \frac{c M}{\mu_0} \coth \beta \cdot e^{-a \cdot \coth \beta} \cdot \operatorname{sh} a \cdot d a .$$

D'où :

$$x = \frac{c M}{2 \mu_0} \coth \beta \int_0^a \left[e^{-a (\coth \beta - 1)} - e^{-a (\coth \beta + 1)}\right] d a =$$

$$= \frac{c M}{2 \mu_0} \coth \beta \left[\frac{1 - e^{-a (\coth \beta - 1)}}{\coth \beta - 1} - \frac{1 - e^{-a (\coth \beta + 1)}}{\coth \beta + 1}\right] .$$

Et en simplifiant, on obtient finalement:

$$x = \frac{c \, M}{\mu_0} \, \text{ch} \, \beta \, [\text{sh} \, \beta - \text{sh} \, (\alpha + \beta) \cdot e^{-\alpha \cdot \coth \beta}]. \tag{83}$$

Cette expression est semblable à celle du n° (81) du temps solaire. Si nous voulons l'avoir en fonction du temps propre nous arriverons, par un procédé analogue, à:

$$x = \frac{M}{\mu_0} \cdot \frac{1}{1 - \frac{u^2}{c^2}} \left[u - \frac{c+u}{2} \left[1 - \frac{\mu_0 t_0}{M} \right]^{\frac{c-u}{c}} + \frac{c-u}{2} \left[1 - \frac{\mu_0 t_0}{M} \right]^{\frac{c+u}{c}} \right]. \tag{84}$$

Si r désigne le rapport des masses initiale et finale, les grandeurs précédentes auront les limites suivantes:

$$\left. \begin{aligned} t_{0\,max} &= \frac{r-1}{r} \cdot \frac{M}{\mu_0} \\[2mm] v_{max} &= c \, \frac{1 - r^{\frac{2u}{c}}}{1 + r^{\frac{2u}{c}}} \\[2mm] t_{max} &= \frac{M}{\mu_0} \cdot \frac{1}{1 - \frac{u^2}{c^2}} \left[1 - \frac{c-u}{2c} \, r^{\frac{c+u}{c}} - \frac{c+u}{2c} \, r^{\frac{c-u}{c}} \right] \\[2mm] x_{max} &= \frac{M}{\mu_0} \cdot \frac{1}{1 - \frac{u^2}{c^2}} \left[u + \frac{c-u}{2} \, r^{\frac{c+u}{c}} - \frac{c+u}{2} \, r^{\frac{c-u}{c}} \right] \end{aligned} \right\} \tag{85}$$

Quant à l'accélération "sensible", c'est-à-dire celle de la gravitation apparente à l'intérieur de la fusée, il suffit d'appliquer (36) en tenant compte de (78):

$$j = c \, \frac{d \, a}{d t_0} = c \cdot \text{th} \, \beta \cdot \frac{\mu_0}{M} \cdot e^{a \cdot \coth \beta} = \frac{u \, \mu_0}{m_0} = \frac{u \, \mu_0}{M - \mu_0 t_0}. \tag{86}$$

Par conséquent l'accélération "sensible" va en augmentant avec le temps propre en suivant une branche d'hyperbole, jusqu'à atteindre son maximum:

$$j_{max} = \frac{\mu_0}{M} \, r \, u = N \, g. \tag{87}$$

N étant le facteur choisi pour exprimer l'accélération maximum en fonction de la constante de gravitation terrestre, compte tenu de la résistance physiologique de l'équipage de la fusée. L'accélération initiale doit être r fois plus petite.

Enfin, la poussée (35) devient constante, comme il était à prévoir, étant donné la puissance du moteur supposée constante:

$$F = m_0 \, c \, \frac{d \, a}{d \, t_0} = u \, \mu_0. \tag{88}$$

XIV. Le paradoxe du temps propre limité

Jusqu'à présent nous avons considéré deux lois de vitesses correspondantes à une accélération sensible constante (c'est la loi la plus favorable pour une fusée à équipage) et à une puissance constante du moteur.

Parmi les lois en nombre infini que l'on pourrait envisager, il y en a qui ont des conséquences extrêmement curieuses. Nous allons examiner celles qui conduisent à une limite finie des temps propres, même si le parcours de la fusée est allongé indéfiniment.

De l'équation fondamentale (28) nous tirons:

$$t_0 = \int_0^t \frac{dt}{\operatorname{ch} a} = \int_0^a \left(\frac{dt}{da}\right) \cdot \frac{da}{\operatorname{ch} a} = \int_0^a \frac{da}{\left(\dfrac{da}{dt}\right) \operatorname{ch} a} . \qquad (89)$$

Si $\left(\dfrac{da}{dt}\right)$ était constante ou restait bornée pour toute valeur de t, on pourrait poser:

$$\left. \begin{aligned} a &\leqslant \frac{da}{dt} \leqslant b \\ at &\leqslant a \leqslant bt \end{aligned} \right\} \quad (a \text{ et } b \text{ étant des constantes positives}) \qquad (90)$$

et on obtiendrait:

$$t_0 \leqslant \int_0^a \frac{da}{a \cdot \operatorname{ch} a} \leqslant \frac{1}{a} \int_0^{bt} \frac{da}{\operatorname{ch} a} = \frac{2}{a} [\operatorname{arctg} e^a]_0^{bt} = \frac{2}{a} \left[\operatorname{arctg} e^{bt} - \frac{\pi}{4} \right]. \qquad (91)$$

Lorsque $t \to \infty$, $\operatorname{arctg} e^{bt} \to \dfrac{\pi}{2}$, donc:

$$\lim_{t \to \infty} t_0 \leqslant \frac{2}{a} \left[\frac{\pi}{2} - \frac{\pi}{4} \right] = \frac{\pi}{2a} . \qquad (92)$$

Quant à la distance parcourue:

$$x = \int_0^t c \cdot \operatorname{th} a \cdot dt = c \int_0^a \left(\frac{dt}{da}\right) \operatorname{th} a \cdot da \geqslant \frac{c}{t} \int_0^a \operatorname{th} a \cdot da \geqslant \frac{c}{b} \int_0^{at} \operatorname{th} a \cdot da =$$

$$= \frac{c}{b} [\text{lognep ch } a]_0^{at} = \frac{c}{b} \text{lognep ch } a\, t \qquad (93)$$

et, par conséquent, les distances croissent infiniment avec le temps t, tandis que le temps t_0 de l'équipage reste fini.

XV. Cas particulier à temps limité

Parmi les fonctions qui remplissent la condition (90) la plus simple est la suivante:

$$a = a\, t \ (a \text{ étant une constante}) . \qquad (94)$$

Et par conséquent:

$$\left. \begin{aligned} t_0 &= \frac{2}{a} \left[\operatorname{arctg} e^{at} - \frac{\pi}{4} \right] < \frac{\pi}{2a} \\ t_{0\,limite} &= \frac{\pi}{2a} \\ x &= \frac{c}{a} \text{lognep ch } c\, t . \end{aligned} \right\} \qquad (95)$$

En divisant les deux dernières expressions on obtient, pour l'équipage de la fusée:

$$\text{Vitesse moyenne} > \frac{2ax}{\pi} = \frac{2c}{\pi} \text{lognep ch } a\, t; \qquad (96)$$

A noter que la constante "a" ne figure pas au coefficient de cette expression. L'accélération sensible j d'après (36) devient:

$$j = c\,\frac{d\,a}{d\,t_0} = c \cdot \mathrm{ch}\,a\,\frac{d\,a}{d\,t} = c \cdot a \cdot \mathrm{ch}\,a\,t\,. \qquad\qquad (97)$$

Cette accélération croît indéfiniment avec le temps. Ce qui nous indique que l'accélération maximum est limitée uniquement en raison de l'effort auquel sont soumis les matériaux qui constituent la fusée. La limite est encore beaucoup plus petite si la fusée doit porter des voyageurs. Dans ce dernier cas elle est donnée par le maximum d'accélération que le corps humain est capable de supporter. Il faut noter, par contre, que les deux accélérations courantes vont en décroissant avec le temps et tendent vers zéro:

$$\frac{d\,v}{d\,t} = \frac{a\,c}{\mathrm{ch}^2\,a\,t}\;;\;\;\frac{d\,v}{d\,t_0} = \frac{a\,c}{\mathrm{ch}\,a\,t}\;;$$

à l'instant initial, les trois accélérations sont égales.

Supposons que l'accélération commune initiale soit $n\,g$ et que le maximum admissible, c'est-à-dire l'accélération sensible finale, soit $N\,g$:

$$\left.\begin{array}{l} c\,a = n\,g \\[2mm] \mathrm{ch}\,a\,t = \dfrac{N}{n} \end{array}\right\}\,. \qquad\qquad (98)$$

Alors:

$$\left.\begin{array}{l} a = \dfrac{n\,g}{c}\;; \\[3mm] a = \dfrac{n\,g}{c}\,t\;; \\[3mm] t_{0\,limite} = \dfrac{\pi\,c}{2\,n\,g}\;; \\[3mm] x = \dfrac{c^2}{n\,g}\,\mathrm{lognep}\,\dfrac{N}{n}\,. \end{array}\right\} \qquad\qquad (99)$$

Vitesse moyenne $> \dfrac{2\,c}{\pi}\,\mathrm{lognep}\,\dfrac{N}{n}\,.$

Et en exprimant les temps T en années et les distances X en années-lumière:

$$\left.\begin{array}{l} T_{0\,limite} = \dfrac{\pi \times 3 \times 10^{10}}{2\,n \times 980 \times 3 \times 10^7} = \dfrac{1{,}6}{n}\;(\text{années}) \\[4mm] X = \dfrac{9 \times 10^{20}}{n \times 980 \times 9 \times 10^{17}}\,\mathrm{lognep}\,\dfrac{N}{n} = \dfrac{1{,}02}{n}\,\mathrm{lognep}\,\dfrac{N}{n}\;(\text{années-lumière})\,. \end{array}\right\} \quad (100)$$

Ainsi, par exemple, avec une accélération initiale égale à celle de la gravitation terrestre $(n=1)$:

Temps limite: $1{,}6$ années $= 1$ année, 7 mois et 6 jours.

C'est-à-dire que cette fusée pourrait atteindre les nébuleuses les plus éloignées en un temps plus court que celui-ci, en supposant toujours que l'accélération progressive puisse être supportée.

En tenant compte des limitations imposées à l'accélération on obtient les parcours suivants:

$\{$ Fusée à équipage $(N=10)$; $X = 1{,}02\;\mathrm{lognep}\,10 = 2{,}34$ années-lumière.

$\phantom{\{}$ Fusée sans équipage $(N=200)$; $X = 1{,}02\;\mathrm{lognep}\,200 = 5{,}4$ années-lumière.

Quant à la loi des poussées, l'on déduit de (35):

$$F = m_0\, c\, \frac{d\,a}{d\,t_0} = m_0\, c\, a\, \frac{d\,t}{d\,t_0} = m_0\, c\, a\, \text{ch}\, \alpha\,.$$

Et:

$$\left.\begin{aligned}
&F = n\,g\,M\,\mathrm{e}^{-\,a\,\coth\beta}\,\text{ch}\,\alpha = \frac{n\,g\,M}{2}\,[\mathrm{e}^{-\,a\,(\coth\beta\,-\,1)} + \mathrm{e}^{-\,a\,(\coth\beta\,+\,1)}] \\
&a = \frac{n\,g}{c}\,t\,.
\end{aligned}\right\} \tag{101}$$

Il est à observer que la poussée diminue avec le temps. Sa valeur initiale maximum est:

$$F_{max} = n\,g\,M\,. \tag{102}$$

Quant à la puissance instantanée, d'après (16):

$$\begin{aligned}
W &= h\,c^2\left(\frac{-\,d\,m_0}{d\,t_0}\right) = h\,c^2\,M\,\frac{d\,(-\,\mathrm{e}^{-\,a\,\coth\beta})}{d\,t_0} = h\,c^2\,M\,\coth\beta\,\mathrm{e}^{-\,a\,\coth\beta}\,\frac{d\,a}{d\,t_0} = \\
&= \frac{h}{\sqrt{h\,(2-h)}}\,c^2\,M\,a\,\mathrm{e}^{-\,a\,\coth\beta}\cdot\text{ch}\,\alpha\,.
\end{aligned}$$

Soit:

$$W = \sqrt{\frac{h}{2-h}}\,c\,M\,n\,g\,\mathrm{e}^{-\,a\,\coth\beta}\cdot\text{ch}\,\alpha \tag{103}$$

laquelle, de même que la poussée, atteint sa plus grande valeur au moment initial:

$$W_{max} = \sqrt{\frac{h}{2-h}}\cdot n\,g\,c\,M\,. \tag{104}$$

Si nous exprimons la masse de la fusée en grammes et la puissance en mégawatts:

$$W_{max}\,(\text{megawatts}) = \frac{980 \times 3 \times 10^{10}}{10^{13}}\sqrt{\frac{h}{2-h}}\cdot n\,M\,(\text{grammes})$$

$$W_{max}\,(\text{megawatts}) = 2{,}94\sqrt{\frac{h}{2-h}}\cdot n\,M\,(\text{grammes}) \tag{105}$$

d'où, pour $n=1$ et un rapport de conversion de masse $h=0{,}6$ (qui suppose une vitesse d'éjection $u = c\sqrt{h\,(2-h)} = 0{,}916\,c$), il faudrait un moteur de 1,925 mégawatts par gramme de masse initiale de la fusée, ce qui représente un chiffre fantastique. Il est nécessaire de prendre des accélérations initiales (valeur de n) très petites, pour que la puissance soit admissible.

Enfin, si r désigne le rapport des masses initiale et finale pendant la période d'accélération que nous sommes en train de considérer:

$$\left.\begin{aligned}
&a_{max} = \text{th}\,\beta\,\text{lognep}\,r = \frac{u}{c}\,\text{lognep}\,r = \sqrt{h\,(2-h)}\,\text{lognep}\,r \\
&\frac{N}{n} = \text{ch}\,\alpha_{max} = \frac{r^{\sqrt{h\,(2-h)}} + r^{-\sqrt{h\,(2-h)}}}{2}\,,
\end{aligned}\right\} \tag{106}$$

relation qui complète les précédentes.

Les Fig. 5 et 6 fournissent les temps "propres" et les distances parcourues en fonction du temps de l'observateur terrestre (Fig. 5), et la puissance du moteur et la vitesse d'éjection en fonction du rapport h de conversion énergétique (Fig. 6).

N'importe quel voyage peut être calculé rapidement à l'aide de ces graphiques.

Supposons, en suivant l'exemple du paragraphe XII, que l'objectif soit à une distance de 4,5 années-lumière et que la moitié du voyage est réalisée en accélérant et l'autre moitié en freinant.

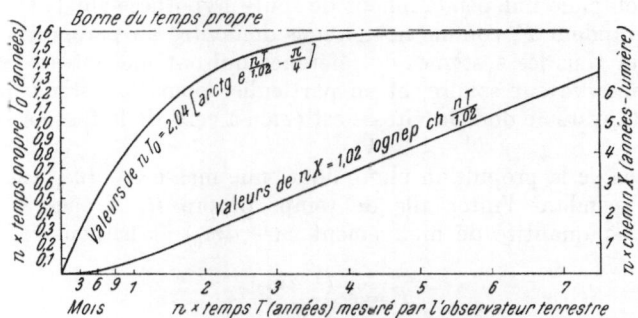

Fig. 5. Temps et chemin parcouru.

Paradoxe du temps borné: V (vitesse de la fusée) $= c \times \mathrm{th}\, \dfrac{nT}{1,02}$

Si l'accélération initiale est égale à celle de la gravitation terrestre: $n = 1$, nous chercherons sur la Fig. 5 l'abscisse de la courbe X qui correspond à l'ordonnée $X = 2,25$ années-lumière. Cette abscisse est $T = 2 + \dfrac{23}{24}$ et l'ordonnée correspondante de la courbe T_0, est $T_0 = 1,48$. Le voyage durerait $2\,T = 5$ années et 11 mois pour l'observateur terrestre et seulement 2,96 années pour l'équipage de la fusée. Pendant ce voyage, l'accélération sensible augmenterait jusqu'à atteindre sa valeur maximum au milieu du parcours:

$$j_{max} = n\,g\,\mathrm{ch}\,\frac{n\,T}{1,02} = g\,\mathrm{ch}\,2,89 = 9.02\,g$$

laquelle est dans la limite supportable.

Il est curieux d'observer (Fig. 5) comme on atteint rapidement des valeurs très prochaines du temps limite. Ainsi, pour le régime d'accé-

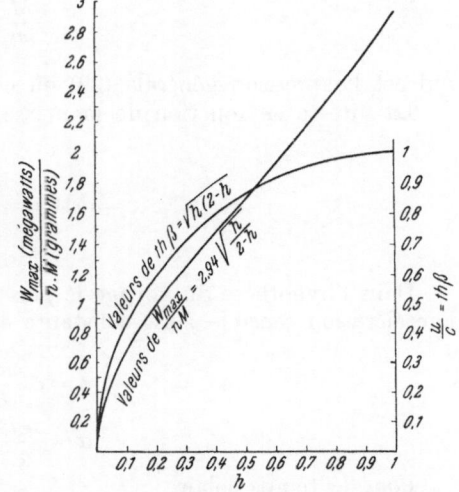

Fig. 6. Puissance maximum W et vitesse d'échappement ($u = c \cdot \mathrm{th}\,\beta$)

lération constante, et des distances: $n\,X = 5,5$ années-lumière, le temps propre est pratiquement le même: $n\,T_0 = 1,6$ années.

XVI. La fusée photonique

La fusée photonique est un cas particulier de l'étude générale précédente. Il s'agit de la transformation totale de la matière éjectée en énergie, c'est-à-dire, du cas où: $\hbar = 1$; $\mathrm{th}\,\beta = 1$.

Mais il peut être déduit directement, par exemple, de l'équation (38) qui fournit la poussée:

$$F = m_0\, c\, \frac{d\,\alpha}{d\,t_0}$$

et qui a été obtenue indépendamment de toute hypothèse sur le système impulseur. Cette grandeur F, comme nous avons démontré au paragraphe VII est un invariant pour tous les systèmes de référence qui ont une vitesse constante par rapport à l'observateur solaire, et en particulier pour le système propre, c'est-à-dire, pour le système dont la vitesse est égale à celle de la fusée à l'instant considéré.

Dans le cas de la propulsion photonique, une masse $(-\,dm_0)$ est transformée en radiation pendant l'intervalle de temps propre dt_0. L'énergie dégagée est $c^2 \cdot (-\,dm_0)$, la quantité de mouvement $c\,(-\,dm_0)$ et la poussée:

$$F = c\, \frac{(-\,dm_0)}{dt_0}\ .$$

En égalant les deux expressions que nous venons d'obtenir pour F, on a simplement:

$$m_0\, d\, \alpha = -\, dm_0\ . \tag{107}$$

Et en intégrant cette équation avec les hypothèses précédentes (vitesse initiale nulle et masse initiale M):

$$\frac{M}{m_0} = e^{\alpha} \tag{108}$$

qui est l'expression générale (29) où $\coth \beta = 1$.

La vitesse en fonction de la masse est donnée par:

$$v = c \cdot \mathrm{th}\, \alpha = c\, \frac{1 - \left[\dfrac{m_0}{M}\right]^2}{1 + \left[\dfrac{m_0}{M}\right]^2}\ . \tag{109}$$

Dans l'hypothèse du voyage le plus favorable (discutée au paragraphe VIII) l'accélération sensible j est constante et égale a n fois celle de la gravitation:

$$\left.\begin{aligned} j &= c\, \frac{d\, \alpha}{dt_0} = n\, g \\[2mm] \alpha &= \frac{n\, g}{c} \cdot t_0\ . \end{aligned}\right\} \tag{110}$$

Pour le temps solaire:

$$dt = \mathrm{ch}\, \alpha \times dt_0 = \frac{c}{n\, g} \cdot \mathrm{ch}\, \alpha\, d\, \alpha$$

$$t = \frac{c}{n\, g}\, \mathrm{sh}\, \alpha\ . \tag{111}$$

La distance parcourue est:

$$dx = v\, dt = c\, \mathrm{th}\, \alpha \cdot \frac{c}{n\, g}\, \mathrm{ch}\, \alpha \cdot d\, \alpha = \frac{c^2}{n\, g}\, \mathrm{sh}\, \alpha \cdot d\, \alpha$$

$$x = \frac{c^2}{n\, g}\, [\mathrm{ch}\, \alpha - 1]\ . \tag{112}$$

La poussée donnée par:

$$F = m_0\, c\, \frac{d\,a}{d\,t_0} = m_0\, n\, g = n\, g\, M\, \mathrm{e}^{-a} \tag{113}$$

est décroissante avec le temps et atteint sa plus grande valeur à l'instant initial:

$$F_{max} = n\, g\, M\,. \tag{114}$$

La puissance instantanée est:

$$W = c^2 \left(-\frac{d\,m_0}{d\,t_0} \right) = c^2\, m_0\, \frac{d\,a}{d\,t_0} = m_0\, n\, g\, c = n\, g\, c\, M\, \mathrm{e}^{-a} \tag{115}$$

dont la valeur maximum est la valeur initiale:

$$W_{max} = n\, g\, c\, M\,. \tag{116}$$

Et l'énergie totale dépensée est:

$$E = c^2\, (M - m_0) = c^2\, M\, (1 - \mathrm{e}^{-a})\,. \tag{117}$$

Toutes ces formules ont été obtenues de l'étude précédente, en posant $h = 1$ (th $\beta = 1$).

XVII. Comparaison des résultats de ce travail avec ceux obtenus par d'autres auteurs

La bibliographie sur le sujet traité, que nous avons pû consulter, est la suivante:

1. CH. MAUGUIN, Astronautique et Rélativité. C. r. acad. sci., Paris, **234**, 1004 (1952).
2. J. ACKERET, Alcanzará el hombre las profundidades del Cosmos ? Interavia, Genève, **11**, No. 12, 989 (1956).
3. E. SÄNGER, Die Erreichbarkeit der Fixsterne. Comptes rendus du VIIe Congrès International d'Astronautique, Rome 1956, p. 97. Rome: Associazione Italiana Razzi, 1956.
4. J. M. J. KOOY, On Relativistic Rocket Mechanics. Astronaut. Acta **4**, 31 (1958).

1. Prof. MAUGUIN

Il s'agit d'un travail de cinématique avec application des principes de relativité à l'étude d'un corps mobile doué d'une accélération propre constante, accélération qui est définie comme suit. Il emploi le vecteur $d\vec{\sigma}$ représentant l'élément de la ligne-univers dans l'espace à quatre dimensions de MINKOWSKI. Les carrés des modules des vecteurs à quatre dimensions représentant la vitesse et l'accélération propre sont égaux à:

$$\left| \frac{d\vec{\sigma}}{d\,\tau} \right|^2 = -\,c^2$$

$$\left| \frac{d^2\vec{\sigma}}{d\,\tau^2} \right|^2 = \gamma^2$$

où τ est le temps propre, c la vitesse de la lumière, γ l'accélération propre.

Il applique ces formules au cas du mouvement en ligne droite suivant l'axe des x et, après intégration, il obtient nos éqs. (41) et (43). C'est-à-dire que γ est équivalent à ce que nous avons appelé l'accélération sensible.

2. Prof. Ackeret

Il s'agit aussi d'un travail de cinématique pure appliquée au cas spécial de l'accélération sensible constante.

Il prend comme point de départ les relations de Lorentz-Fitz Gerald et les systèmes de référence constitués par la Terre et la fusée, en supposant cette dernière douée d'une vitesse constante v.

Il imagine ensuite un corps mobile ayant par rapport à ces deux systèmes des vitesses respectives u et u', reliées entre elles par la loi de relativité d'addition des vitesses. Il prend la dérivée par rapport au temps de l'expression qui en découle et, en supposant toujours v constant, il arrive à la formule:

$$\frac{du}{dt} = \frac{du'}{dt'} \cdot \frac{\left(1 - \frac{v^2}{c^2}\right)^{3/2}}{\left(1 + \frac{vu'}{c^2}\right)^3} \ .$$

Il suppose alors que u et v sont égales et par conséquent que $u' = 0$ tandis que $\frac{du'}{dt'}$ ne s'annule pas et représente donc une accélération, qu'il désigne par b, et qui est celle que nous avons dénommé accélération "sensible". Il suppose, ensuite, que v est variable et l'équation précédente devient:

$$\frac{dv}{dt} = b\left[1 - \frac{v^2}{c^2}\right]^{3/2} \ .$$

De cette formule il obtient t en fonction de v dans l'hypothèse de b constant, et il en tire les autres équations qui définissent le mouvement, et en particulier (41) et (43).

3. Prof. Sänger

C'est un travail sur les fusées photoniques. Les lois fondamentales: rapport des masses, temps, accélération, parcours, etc., censées être connues d'après un autre travail de l'auteur: ,,Zur Mechanik der Photonenstrahlantriebe", Mitteilungen Nr. 5 aus dem Forschungsinstitut für Physik der Strahlantriebe, Stuttgart. Nous n'avons pas pu consulter cet ouvrage.

Le travail dont nous faisons le commentaire a une grande généralité, mais il se rapporte exclusivement à la fusée photonique.

Les formules qu'il obtient sont exactement les nôtres.

Il y a lieu de faire remarquer l'idée de "vitesse propre" introduite par l'auteur, qui serait la fonction primitive de celle que nous avons appellée "accélération sensible"; ce serait par conséquent la seule vitesse que pourrait connaître un voyageur qui se trouverait isolé de l'extérieur.

Le travail se termine par un intéressant exposé sur la possibilité de vérifier expérimentalement la loi de relativité sur la dilatation du temps; cette vérification pourrait être faite au moyen de l'observation des mésons μ crées par les rayons cosmiques dans la haute atmosphère, et dont le parcours est d'une longueur environ 100 fois plus grande que celle qui semblerait possible d'après leur vie. L'augmentation est précisément égale à la dilatation du temps donnée par la relativité, correspondante à leur vitesse, laquelle peut être mesurée au moyen de l'énergie du méson.

4. Dr. Koοy

Le travail est très étendu et il envisage différents aspects du mouvement d'une fusée soumise aux lois de relativité.

L'auteur commence par calculer l'équilibre dynamique de la fusée en partant de la formule fondamentale de relativité:

$$K = m_0 \frac{d}{dt} \left[\frac{v}{\sqrt{1 - \dfrac{v^2}{c^2}}} \right]$$

dans laquelle K représente la force ou la poussée que fournirait le moteur au repos sur un banc d'essai, m_0 la masse instantanée d'après les voyageurs, t le temps de l'observateur terrestre et v la vitesse de la fusée.

D'autre part il identifie K avec l'augmentation par seconde de la quantité de mouvement des particules éjectées, par rapport à la fusée.

Ce point de vue est identique à celui que nous avons pris au chapitre VII de notre travail. Cependant il n'utilise pas notre changement de variable, si avantageux, $v = c$ th α; les expressions qu'il obtient deviennent de plus en plus compliquées, et lorsqu'il pousse les calculs plus en avant, pour le cas où la puissance est constante (développé par nous au chapitre XIII), il ne parvient pas à trouver les expressions des temps et des espaces en termes finis; il en laisse l'intégration indiquée en vue de l'évaluation par des méthodes numériques.

Il applique ensuite les formules, qu'il a établies précédemment, au cas de la fusée photonique à poussée constante; il parvient à trouver l'intégral du temps, sans essayer de calculer celle de l'espace.

Dans les deux cas il obtient le rapport des masses en fonction de la vitesse et il signale que la formule qu'il a trouvée est identique à celle obtenue par ACKERET dans l'étude intitulé: Zur Theorie der Raketen, Helvet. Physica Acta 19 (1946).

Nous ne sommes pas arrivés à voir cette étude. La formule en question est la même de la présente étude (29) quoique avec une apparence différente.

L'auteur fait ensuite une analyse détaillée et très intéressante des possibilités pratiques de propulsion au moyen d'ions accélérés par des champs électriques combinés avec l'énergie obtenue par la fusion nucléaire du deuton et du triton à l'état gazeux.

Cette étude minutieuse de la propulsion ionique est ensuite appliquée au problème général de la fusée à trajectoire courbe, soumise aux lois de relativité; il pose ce problème d'une façon générale en indiquant la méthode à suivre pour arriver à le résoudre. Il tire en conclusion: que les procédés de fusion nucléaire sont insuffisants pour obtenir des vitesses d'éjection très proches de celle de la lumière, et que même dans le cas d'une transformation totale de la matière en énergie, il faudrait des rapports de masse énormes pour explorer l'Univers pendant la très courte durée de la vie humaine. Et qu'il faudra peut être chercher à l'avenir la solution dans la gravitation "contrôlée".

Il fait ensuite une analyse de la dilatation des temps d'après le principe de relativité et il termine en exposant d'une façon brillante sa théorie sur le champ métrique général, à la création duquel l'Univers entier, considéré comme un ensemble de "quanta d'action" interdépendants, participerait.

A la fin du travail dont nous venons de faire le commentaire et parmi les références bibliographiques, se trouve un article de H. G. L. KRAUSE, Relativistische Raketenmechanik, Astronaut. Acta 2, 30 (1956); nous n'avons pas pu en prendre connaissance.

Résumé comparatif

Les renseignements, dont nous venons de faire le commentaire, nous ont procuré la satisfaction de vérifier nos formules donnant le rapport des masses en fonction de la vitesse, ainsi que celles correspondant à la poussée de l'accélération sensible constante, de la puissance constante et de la fusée photonique.

Dans ces ouvrages, nous n'avons rien trouvé de semblable à notre hypothèse générale qui fournit les vitesses et les énergies en fonction du rapport h de conversion énergétique; ni à notre théorie du temps limité (chapitre XIV et XV).

Nous regrettons de ne pas avoir pu obtenir des renseignements bibliographiques plus complets. Quand nous aurions eu le temps de nous les procurer, nous ne pensions pas à publier ce travail qui a été réalisé comme un simple exercice d'application des lois physiques de relativité et auquel des recherches bibliographiques aurait alors enlevé son charme.

Si nous osons maintenant le présenter au Xème Congrès International d'Astronautique tel qu'il avait été rédigé, c'est parce qu'il constitue un tout, comme les chaînons logiques d'un procédé très générale et très simple, et que nous avons voulu respecter son vrai caractère et sa spontanéité.

A Rocket for Manned Lunar Exploration[1]

By

M. W. Rosen[2] and F. C. Schwenk[2]

(With 15 Figures)

(Received July 27, 1959)

Abstract — Zusammenfassung — Résumé

A Rocket for Manned Lunar Exploration. One of the significant human accomplishments of the next decade will be the manned exploration of the moon. Previously, the uncharted regions of the earth, the Arctic and Anarctic, the Amazon and Himalayas challenged the skill and fortitude of explorers. But these regions cannot long retain their status—the new frontier lies beyond the confines of our planet —on the nearest sizeable aggregation of matter in space—the moon.

Significantly, man's exploration has been paced by his technical progress. The discovery of America was made possible by ships and sails of sufficient size and by advances, however crude, in the art of navigation. Oxygen masks made possible the conquest of Everest, and rockets—the exploration of the upper atmosphere.

The exploration of the moon is within view today. If it may be assumed that Project Mercury in the U.S.A. and similar efforts by the U.S.S.R. will establish that man can exist for limited periods of time in space, then a trip to the moon requires mainly the design, construction and proving of a large rocket vehicle.

In one concept of a manned lunar vehicle the entire mission, the trip to the moon and the return, is staged on the earth's surface. A highly competitive technique, one favored by many engineers, is to stage the lunar mission by refueling in a low earth orbit. This would permit the use of a smaller launching vehicle but would require development of orbital rendezvous techniques. In any case, a vehicle of the larger type will be needed for lunar as well as other exploratory missions.

This paper presents a parametric study of vehicle scale for the direct flight manned lunar mission. The main parameter is the take-off thrust which is influenced by many factors; principally the propellants in the several stages and the flight trajectory. A close choice exists in the second stage where conventional and high energy propellants are compared. The size of the final stage and hence the entire vehicle is governed mainly by the method of approach to the earth's surface, whether it is elliptic, parabolic or hyperbolic. The various methods are applied to an illustrative vehicle configuration.

Reliability will be a major factor in the success of any manned lunar flight. While no formula is proposed for improving component reliability, certain operational procedures can be used to advantage in enhancing the probability of a successful round trip to the moon.

Monderforschung mittels einer bemannten Rakete. Eine der bedeutendsten menschlichen Errungenschaften der nächsten Dekade wird die „bemannte" Erforschung des Mondes sein. Früher forderten die nichtkartierten Gegenden der Erde, die Arktis

---

[1] The opinions expressed in this paper are those of the authors and do not necessarily represent the views of the National Aeronautics and Space Administration.

[2] National Aeronautics and Space Administration, Washington 25, D.C., U.S.A.

und die Antarktis, der Amazonas und das Himalajagebirge, die Tüchtigkeit und Tapferkeit der Forschungsreisenden heraus. Diese Regionen können jedoch nicht länger ihren Status bewahren. Die neue Grenze liegt jenseits der Begrenzung unseres Planeten, auf der nächsten großen Ansammlung von Materie im Weltraum — dem Mond.

Bezeichnenderweise wurde die Erforschungstätigkeit des Menschen durch seinen technischen Fortschritt geleitet. Die Entdeckung von Amerika wurde durch Schiffe und Segel von ausreichender Größe und durch, wenngleich rohe, Fortschritte in der Navigationskunst möglich gemacht. Sauerstoffmasken ermöglichten die Eroberung des Everest und Raketen die Erkundung der Hohen Atmosphäre.

Die Erforschung des Mondes liegt heute in Sicht. Wenn man annehmen darf, daß das „Project Mercury" in den USA und ähnliche Bestrebungen in der UdSSR es zustandebringen werden, daß der Mensch während begrenzter Zeiträume im Weltraum existieren kann, dann erfordert eine Reise zum Mond hauptsächlich die Planung, Konstruktion und Erprobung eines großen Raketenfahrzeuges.

Nach der einen Vorstellung von einem bemannten Mondschiff ist das ganze Unternehmen, die Fahrt zum Mond und zurück, von der Erdoberfläche aus auszuführen. In starkem Wettbewerb damit steht die Idee, die vielen Ingenieuren vorteilhaft scheint, daß das Mondunternehmen mit Hilfe des Auftankens mit Brennstoff in einer niedrigen Kreisbahn um die Erde gestartet werden solle. Dies würde zwar die Verwendung eines kleineren Startfahrzeuges gestatten, aber die Schaffung einer Technik für das Zusammentreffen in den Umlaufbahnen erfordern. Auf alle Fälle wird ein Fahrzeug der größeren Art sowohl für die Erforschung des Mondes wie auch für andere Forschungsunternehmen benötigt werden.

Die vorliegende Arbeit gibt eine parametrische Studie des Fahrzeugmaßstabes für das „Unternehmen Mond" einer Mannschaft im Direktflug. Der Hauptparameter ist der Startschub, der durch viele Faktoren beeinflußt wird, hauptsächlich durch die Treibstoffe in den verschiedenen Stufen und durch die Flugbahn. Eine begrenzte Wahl bietet sich für die zweite Stufe, wo konventionelle und hochenergetische Treibstoffe verglichen werden. Die Größe der Endstufe und daher auch das ganze Fahrzeug wird vor allem durch die Methode der Annäherung an die Erdoberfläche beherrscht, ob diese nämlich in elliptischer, parabolischer oder hyperbolischer Bahn erfolgt. Die verschiedenen Methoden werden auf eine beispielmäßige Gestaltung des Fahrzeuges angewendet.

Verläßlichkeit wird ein Hauptfaktor für den Erfolg jedes „bemannten" Mondfluges sein. Ohne daß eine Methode für die Erhöhung der Verläßlichkeit der verschiedenen Komponenten vorgeschlagen wird, können bestimmte Methoden vorteilhaft benützt werden, um die Wahrscheinlichkeit einer erfolgreichen Reise zum Mond zu vergrößern.

Une fusée d'exploration lunaire. Un des exploits marquants de la prochaine décade sera l'exploration de la Lune par un équipage. Les régions inconnues de la Terre: l'Arctique et l'Antarctique, l'Amazone et l'Himalaya ont en leur temps défié l'habileté et le courage de l'homme. La nouvelle frontière se situe en dehors des limites de notre planète, sur l'agrégat de matière le plus proche: la Lune.

L'exploration de l'homme a été conditionnée par son progrès technique. La découverte de l'Amérique a été rendue possible par les bateaux et les voiles de dimensions suffisantes et par le progrès même élémentaire dans l'art de la navigation. Les masques à oxygène ont rendu la conquête de l'Everest possible et les fusées celle de la haute atmosphère.

L'exploration de la Lune est proche. On peut admettre que le projet Mercury aux Etats-Unis et des efforts semblables en U.R.S.S. prouveront que l'homme peut subsister un certain temps dans l'espace. Un voyage à la Lune ne demandera plus alors que la conception, la construction et l'essai d'une fusée de grandes dimensions.

Dans une des conceptions toute la mission aller-retour est prévue depuis la surface de la Terre. Une conception rivale, qui a la faveur de nombreux ingénieurs, prévoit un ravitaillement sur une orbite proche. Elle permettrait l'utilisation d'un véhicule plus petit mais nécessite la mise au point de techniques de rendez-vous orbitaux.

De toute façon un véhicule de grandes dimensions sera nécessaire pour les missions d'exploration lunaire et autres.

L'article présente une étude paramétrique de l'échelle que doit avoir un véhicule à mission lunaire directe. Le paramètre principal est la poussée au décollage influencée par de nombreux facteurs dont: la quantité d'ergols dans les différents étages et le type de trajectoire. Un choix difficile se présente pour le second étage quand on compare l'utilisation d'ergols conventionnels avec les ergols à haute énergie. Les dimensions du dernier étage et par suite tout le véhicule sont gouvernés par la méthode d'approche de la surface terrestre: elliptique, parabolique ou hyperbolique. Les différentes possibilités sont comparées sur une configuration exemplative.

La sécurité de fonctionnement sera un des facteurs principaux pour assurer le succès d'une telle mission. Sans examiner la façon d'améliorer la sécurité de fonctionnement des éléments séparés, on montre que certaines procédures peuvent être utilisées avec avantage pour accroître les probabilités de succès d'un voyage autour de la Lune.

Introduction

When one views the history of exploration he finds the dominant role played by man. Men, many of them, explored the coasts and interior of America. Fewer numbers endured Arctic cold and coped with other physical hazards to reach previously inaccessible regions of the earth. Hence, it is not surprising that exploration is linked in history with the names of the men who accomplished it: COLUMBUS, BALBOA, PEARY, AMUNDSEN, to name a few.

Only in our time has it seemed important to support or defend this point of view. We have sought reasons to justify sending men into space to explore the moon and the planets. Because they are merely expressions of a desire, most of these sought-after reasons are unconvincing, such as the vapid reason, "because the moon and the planets are there."

Indeed, there are many who maintain that most of space exploration should be done with instruments, that men should be sent only after years of unmanned examination—and for some ill-defined reason centered mainly around national prestige.

Moreover, a distinction is being drawn between manned exploration and scientific exploration. Not that sending a man into space is unscientific but perhaps not scientific enough. After all, a man cannot see ultraviolet light or sense magnetic fields, nor can he detect cosmic rays. These things are done by instruments, and if the instruments can transmit their findings back to earth, why do we need a man in space? Since an instrument can fail, we make it redundant. If it needs adjusting, we make it self-adjusting. Certainly it can be built to withstand a greater range of temperature, pressure, acceleration and radiation than the sensitive body of man.

But we have a tendency to look only at one side of this picture. Because our knowledge of distant celestial bodies is so meagre, we tend to magnify the importance of the simple data that can be most readily obtained by instruments. We overlook that, if an instrument can do one or several things, there are thousands indeed millions, of things it cannot do. To put it bluntly, no instrument or array of instruments exists that can duplicate the sensing capabilities of a man. When to this is added man's capability to record, remember, interpret and discriminate, we see how paltry are the powers of the most sophisticated mechanical substitute.

If this line of reasoning is accepted, there remains the question of timing and the argument runs, "Instruments first—men later." Many scientists see years of instrumented exploration required before a first manned mission. The standard program is now quite familiar. First comes a close approach or hard impact in which measurements are made of the magnetic field, the local radiation and perhaps a few photographs are taken. Then some vehicles that orbit around the moon doing extensive reconnaissance. Finally instrumented packages are set down on the moon (so-called soft landing) to closely examine the lunar surface. At first the packages are stationary but later they are mobile. A decade or more of intensive engineering development is envisioned to make possible this type of exploration, and we are even now designing the rockets that can carry the freight.

It is argued here that we would learn much more at an earlier date by a bold and immediate approach to manned lunar exploration. Moreover, that instruments should be used mainly for a certain type of reconnaissance—to provide the information necessary to attempt a manned lunar landing. The early attempts will not be without risk of failure and probable loss of life. How could it be otherwise. Exploration implies risk and manned exploration implies risk of life. The names of those who failed are numerous though not always well remembered. SCOTT reached the South Pole one month after AMUNDSEN, but died with four of his men on the return trip. NUNGESSER and COLI took off from Paris and were lost in the Atlantic twelve days before LINDBERGH left New York.

The argument that we cannot afford the risk of human life to explore the moon is historically unsound—moreover it is economically unsound. The attempt to duplicate with instruments what could be accomplished by a few men on the moon would be immeasurably more expensive.

In this paper we examine the type of vehicle required for manned lunar exploration by the simplest operational method, a direct flight to the moon and a direct return to the earth. First the direct flight method is compared with orbital rendezvous. Then, various factors influencing the size of the vehicle are examined. Finally a typical vehicle and its employment are described.

Direct Flight or Orbital Rendezvous

There are two important approaches among many for achieving a manned lunar landing. One approach is the direct-flight method that presupposes the development of a very large vehicle containing the complete capability for the mission. The other, orbital rendezvous, employs a smaller booster and involves accumulating in an earth orbit the required mass of hardware and fuel for the escape from orbit, landing on the moon, and returning to earth. Each method has many supporters among rocket engineers. Although this paper describes a direct-flight vehicle, the rendezvous method is a worthy contender for providing the earliest capability for a manned lunar landing.

As far as booster availability is concerned, the orbital rendezvous method leads the direct-flight approach since the smaller vehicle will be available earlier. However, booster availability is only a small part of the mission picture. Techniques must be developed for orbital rendezvous—an operation that poses many problems. If we consider launching from a nonequatorial base, then accurate timing of the launching is required to establish coplanar orbits; otherwise, a plane change is required in the rendezvous maneuver. Plane changes are costly in payload and require added developments in guidance. Possibly, the rendezvous method requires an equatorial launch site—a vast undertaking—to rid the method of some of its complications and the strict requirement on launch time. Consequently, these factors may delay the orbital technique to a time long after that required for booster availability.

Another important facet is the number of vehicles required for the rendezvous method. To build up the capability in orbit for just one lunar vehicle, eight successful flights of a booster vehicle that uses eight engines to produce over one million pounds of thrust at launch are required. In addition, a crew of men would be needed in orbit to perform the tasks of assembly, transferral of fuel and vehicle check out. Surely, the operation would be performed at least one time to provide an unmanned test flight to the moon before a man is sent. If we include the need for a spare lunar vehicle, the result is that at least twenty-four launchings of the one-million-pound booster must occur for the sole purpose of the manned lunar landing.

Admittedly, the development of a large vehicle with direct-flight capability will be costly and require a long time to reach operational status. In fact, the time required may be too long to satisfy those who wish to see a man on the moon as soon as possible. We believe, however, that space exploration of all types will require the development of the larger vehicle that has direct flight to the moon capability. For example, if a manned lunar landing is achieved first by the rendezvous method, the supply of a lunar base will be accomplished more readily by a direct-flight vehicle.

Direct-Flight Vehicle — Design Factors

Having set our sights on a direct-flight vehicle, we wish to examine some of the factors that affect its design and, ultimately, to describe a vehicle suitable for a round trip to the moon.

First, we must define the mission. A $2^1/_2$-day flight from earth to moon is chosen. A shorter trip time minimizes effects of errors in burnout velocity, but demands more total impulse. The first three stages accelerate the payload and remaining stages to an inertial velocity of 36,000 feet per second. After coasting to the vicinity of the moon, the fourth stage lowers the remainder of the vehicle to a landing on the moon. At the time of departure, the fifth stage propels the vehicle toward the earth. After $2^1/_2$ days, the payload approaches the earth. Here there is a choice; a sixth stage of propulsion can be employed to slow the payload to orbital speed, or the vehicle can enter the earth's atmosphere at hyperbolic velocity. We shall delay discussion of this choice until later, but assume for the moment that hyperbolic re-entry can be tolerated as we discuss some other factors related to the direct flight vehicle.

One of the major concerns is the selection of propellants for the various stages. High-energy propellants, liquid oxygen and liquid hydrogen, are most desirable to achieve the mission with the least vehicle gross weight. Naturally, this propellant combination can only be used if the necessary engines are available and if the techniques for handling liquid hydrogen are developed. We believe that both these conditions can be met in the smaller stages. Consequently, high-energy propellants were chosen for the third and fourth stages of the vehicle.

For a return capsule weight of between 8000 and 9000 pounds we can show that the vehicle at lift-off must weigh more than 4 million pounds. A sea-level thrust rating of over 6 million pounds is, therefore, a necessity. NASA is presently developing a rocket engine capable of producing $1^1/_2$ million pounds of thrust with liquid oxygen and kerosene propellants. A cluster of several of these engines is therefore the logical choice for first-stage propulsion—a choice that also specifies liquid oxygen and kerosene as first-stage propellants.

Now we must decide on the propellants for the second stage. Fig. 1 compares the payload variations with earth take-off thrust for two cases. In one case, high energy propellants are used in the second stage; in the other case, liquid oxygen and kerosene are the second-stage propellants and an altitude version of the 1.5

million pound thrust engine serves as second-stage propulsion. (High energy third and fourth stages were assumed and the first three stages were optimized by the method described in [1].) The performance advantage afforded by the high-energy propellants is obvious—a 6 million pound thrust first stage with high-energy propellant in the second stage can provide the same payload capability as a vehicle having a take-off thrust of 9 million pounds and lox-kerosene in the

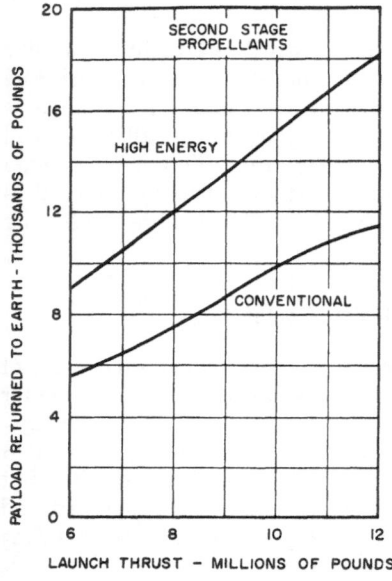

Fig. 1. Effect of the choice of second stage propellants on the variation of payload with launch thrust

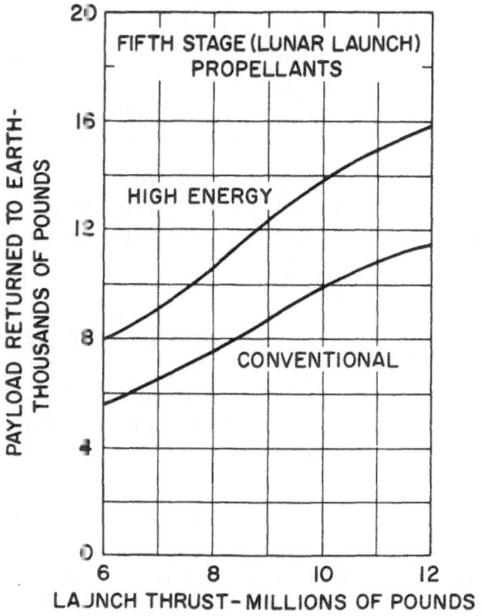

Fig. 2. Effect of the choice of fifth stage propellants on the variation of payload with launch thrust

second stage. However, the calculations also show that a second-stage thrust level of 2.4 million pounds is required in the vehicle that uses the high-energy propellants. Such a thrust level in a high-energy engine may not be available for a long time. Hence, our choice at this time is the conventional lox-kerosene second stage using one large engine.

Possibly, liquid hydrogen could be stored long enough on the surface of the moon to allow its use for launching from the lunar surface. Fig. 2 shows an impressive increase in payload (or reduction in first-stage thrust for the same payload) if high-energy propellants are employed in the lunar launch. In this situation, the term "conventional propellants" refers to those that are liquids at normal temperatures and pressures such as nitrogen-tetroxide and hydrazine. A fundamental question confronts us—Which propellants can be stored in the vehicle tanks on the moon whose surface temperature [2] varies, in the extreme, from — 150° C to 134° C ? There is a strong possibility that, with careful vehicle design and proper shielding against thermal radiation, high-energy propellants (liquid oxygen and liquid hydrogen) can be stored as well as any in the lunar environment. However, this is an area that has not yet been explored, and we chose at this time the more conservative propellant combination.

As mentioned → previously, we have assumed that the return vehicle, or manned capsule, enters the atmosphere at hyperbolic velocities. Of course, a powered sixth stage could be employed to first slow the vehicle to orbital speed;

thereupon the landing would be similar to that of NASA's Project Mercury. Fig. 3 shows what a retro-to-earth orbit costs. For a capsule payload weighing 6000 pounds, 24 million pounds of thrust at launch (16 engines) is required if we must provide propulsion to place the capsule in orbit on the return trip. (Actually, a capsule weight of 8000 pounds is desired.) Clearly, retro-to-orbit is a costly maneuver and its use would require a vehicle so large as to make the task of a manned lunar landing too ponderous unless we considered nuclear or electrical propulsion schemes The same conclusion applies to the orbital rendezvous method using the one-million-pound booster. Whereas 8 or 9 launchings are required if hyperbolic reentry is employed, approximately 24 firings are required to place into orbit the necessary mass of material if the returning vehicle must be decelerated to orbital speed.

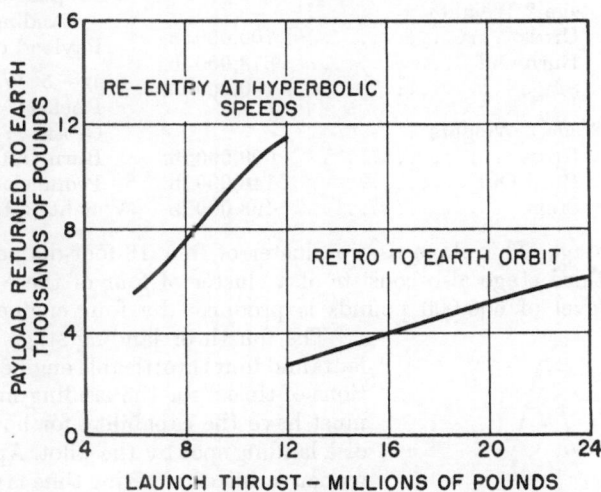

Fig. 3. Effect of re-entry method on the variation of payload with launch thrust

Fig. 3, therefore, presents the reasons for assuming re-entry at hyperbolic velocity. It also emphasizes the need for research and development to provide this capability irrespective of direct or rendezvous approaches to a manned lunar mission.

CHAPMAN of NASA in two papers [3, 4] describes the hyperbolic re-entry phenomena and corridors. He shows that the heating rates and heat absorbed are several times as great as in orbital decay of a non-lifting body, and that guidance requirements are severe as far as path angle accuracy is concerned. CHAPMAN states that the tolerance on flight path angle for proper entry into the atmosphere is approximately 1 minute of arc at distances of 10 to 100 earth radii from the earth. These are formidable problems, but considering how ballistic missile re-entry was solved once the problem could be stated, one expects that hyperbolic re-entry will yield to a similar treatment.

Typical Direct-Flight Vehicle—Vehicle Characteristics

In describing a typical direct-flight vehicle, our purpose is to summarize the previous discussions on the various design factors. Fig. 4 shows an outline drawing of the typical direct-flight vehicle and Table I presents vehicle weights. The vehicle stands about 220 feet high and the first stage is 48 feet in diameter. The conical portion at the top contains the landing or fourth stage, the take-off or fifth stage, and the manned capsule or payload. Upon return to the earth, the payload will weigh 8000 pounds including men, equipment, capsule, guidance and control, and parachute. Two or three men will constitute the crew.

Six engines each of 1.5 million pounds of thrust power the first stage. Liquid oxygen and kerosene are carried in a cluster of 7 tanks each one 16 feet diameter. One altitude version of the 1.5 million-pound-thrust engine propels the second

Table I. *Weight Breakdown for Typical Direct Flight Vehicle*

Stage 1 Weights		Stage 4 Weights (Landing Rocket)	
Launch	6,700,000 lb.	Gross	102,000 lb.
Burn-Out	2,000,000 lb.	Burn-Out (on Moon) ...	49,100 lb.
Stage	5,000,000 lb.	Propulsion, Tanks, Landing Gear	13,100 lb.
Stage 2 Weights		Payload on Moon	36,000 lb.
Gross	1,700,000 lb.	Stage 5 Weights (Return Rocket)	
Burn-Out	678,000 lb.		
Stage	1,100,000 lb.	Gross	36,000 lb.
Stage 3 Weights		Burn-Out	13,700 lb.
Gross	600,000 lb.	Propulsion and Tanks ..	3,800 lb.
Burn-Out	146,000 lb.	Weight Returned to Earth	8,000 lb.
Stage	498,000 lb.		

stage. This stage uses a cluster of four 16-foot-diameter tanks. The high-energy third stage also consists of a cluster of four of these 16-foot tanks and a thrust level of 600,000 pounds is produced by four engines.

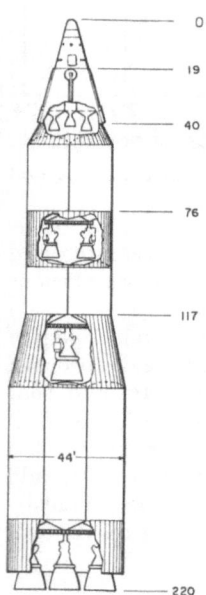

Fig. 4. Outline drawing of direct flight vehicle

The fourth or landing stage utilizes high-energy propellants and four throttleable engines provide the required variations of thrust for the landing maneuver. The landing stage must have the capability for hovering to allow final choice of a landing spot by the pilot. Approximately one minute of maneuvering or hovering time is provided. Retracted landing legs appear on the side of the fourth stage. When extended for landing, the legs span a distance of 40 feet for purposes of stability.

The fifth stage is placed in a cylindrical tube that pierces the tankage of the landing stage. At take-off from the moon, the fifth stage slides out of the landing vehicle on rollers. We chose this arrangement because it presents a vehicle with a low center of gravity which will reduce any tendency for the vehicle to topple on the surface of the moon. In addition, the propellant tanks of the spent landing stage which surround the fifth stage serve as meteor bumpers and shielding against thermal radiation. Furthermore, no landing loads are transmitted through the return stage, thus minimizing the danger of a rough landing.

The manned capsule is an enlarged version of the one used in Project Mercury. It is a truncated cone, with a maximum diameter of 12 feet and a height of 14 feet. Inside the capsule, two levels are provided. The lower level contains contoured couches for the crew, controls, communications, and a folding air-lock for use on the moon. The upper level contains food, power supply, exploration gear, and work space. The outer surface of the capsule is covered with ablative material for insulation against and removal of heat generated during atmospheric re-entry.

Guidance Systems

Guidance system requirements normally are divided into three phases: initial, mid-course, and terminal. For this mission, we must provide these three functions for both the moon-bound and the earth-bound trips. In addition, we should consider the pilot's capabilities in performing major guidance tasks or in monitoring an automatic system. At present, the latter is most reasonable, since we believe that an unmanned return vehicle, a spare so to speak, should be placed on the moon prior to the manned flights to provide an escape route should the manned vehicle be damaged upon landing.

The initial guidance phase from launch to earth-escape can be accomplished with sufficient accuracy by inertial systems now under development. Mid-course guidance by means of earth-based radio can direct the vehicle to an accuracy of 50 miles for a lunar impact trajectory. The terminal phase involves the final approach to the moon and the lunar-landing. These maneuvers require vehicle-contained guidance; however, lunar-based radio beacons will assist. A combination radar-optical system will sense altitude and velocity components relative to the lunar surface. In all but the initial guidance phase (during launching), the pilot can effectively monitor and override the automatic system if necessary. During the mid-course phase, in particular, the pilot can make optical observations of the lunar disk for distance and path angle measurements. The pilot will also be very effective in the final phase of the landing on the moon.

Launching from the surface of the moon will be guided by an inertial system that is aligned and calibrated by the pilot on optical sightings of stars and earth. The proper re-entry corridor in the atmosphere is reached by a combination of optical sightings from the vehicle and earth-based radio signals. During re-entry, the lift of the capsule is utilized to modify the trajectory such that the vehicle follows a prescribed deceleration program and lands within the recovery area. The first phase of the re-entry maneuver will utilize vehicle-contained guidance monitored from the earth. After the initial slow down to orbital speeds, earth-based radar in the landing area will control the vehicle.

Description of the Flight

Let us dismiss for a few moments considerations of time and space and imagine that we are on a Pacific Island some five to ten years in the future. The latest of a series of large rockets stands erect in the launching area. Only a few men can be seen working on the rocket in contrast to the hundreds that used to crowd the launch areas of the late fifties. For we have learned along with increasing size to make our rockets less complicated and more reliable. No battery of speakers blares out the count. Instead each worker has a small transceiver attached to his helmet through which he receives the count and communicates with the block-house. Finally the 300-foot-high gantry rolls away and the rocket is left standing alone poised for its launching. The six giant motors ignite in pairs while the rocket is held fast to the launch stand. Finally the umbilical cables drop away and the rocket rises with the roar of nine million pounds of thrust (Fig. 5). The light of the exhaust illuminates the entire island. The rocket rises vertically for ten seconds and then tilts slightly to the east. It continues to burn for 135 seconds to an altitude of 35 miles. Then it cuts off and separates to be recovered for later use. The second stage ignites immediately (Fig. 6) and burns for 177 seconds accelerating to a speed of 15,800 feet per second. Finally the third stage fires (Fig. 7) along a path almost parallel to the earth's surface, but at an altitude of about 150 miles. After third stage burnout the cone-shaped vehicle coasts silently through cislunar space for 60 hours. As it approaches the moon (Fig. 8) the vehicle starts to turn under the influence of control jets to orient itself for the descent to the lunar surface. The four braking rockets are now firing (Fig. 9), maneuvering the vehicle toward its selected landing area. The landing struts extending from the side of the cone span 40 feet. The cone settles down slowly (Fig. 10) and comes to rest on the moon.

Fig. 5. Launching, gross weight — 6,700,000 lbs., thrust — 9,000,000 lbs.

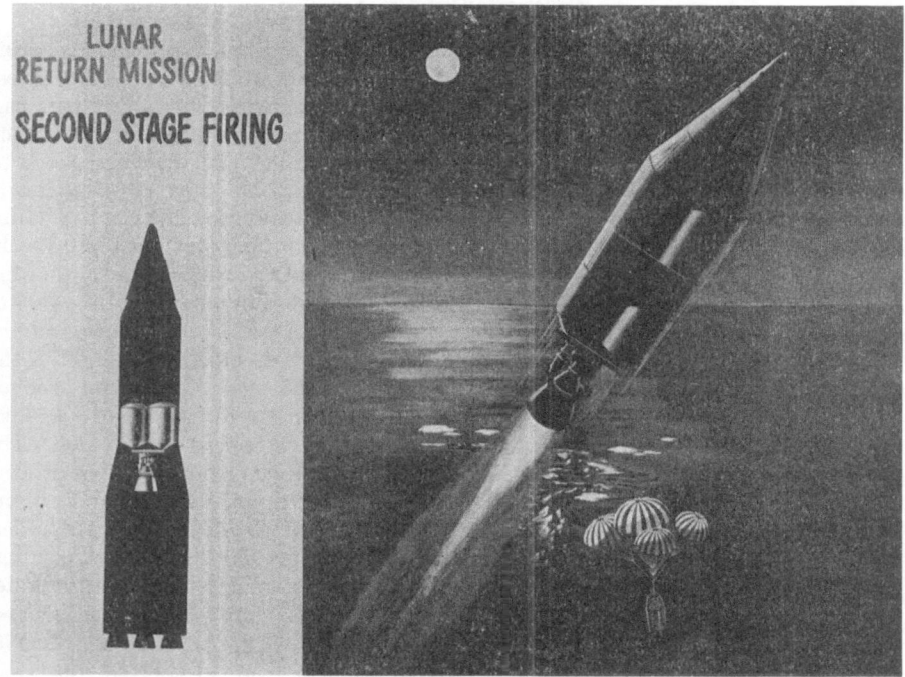

Fig. 6. Second stage firing, jettisoned first stage at bottom center

Fig. 7. Third stage firing, jettisoned second stage at bottom center

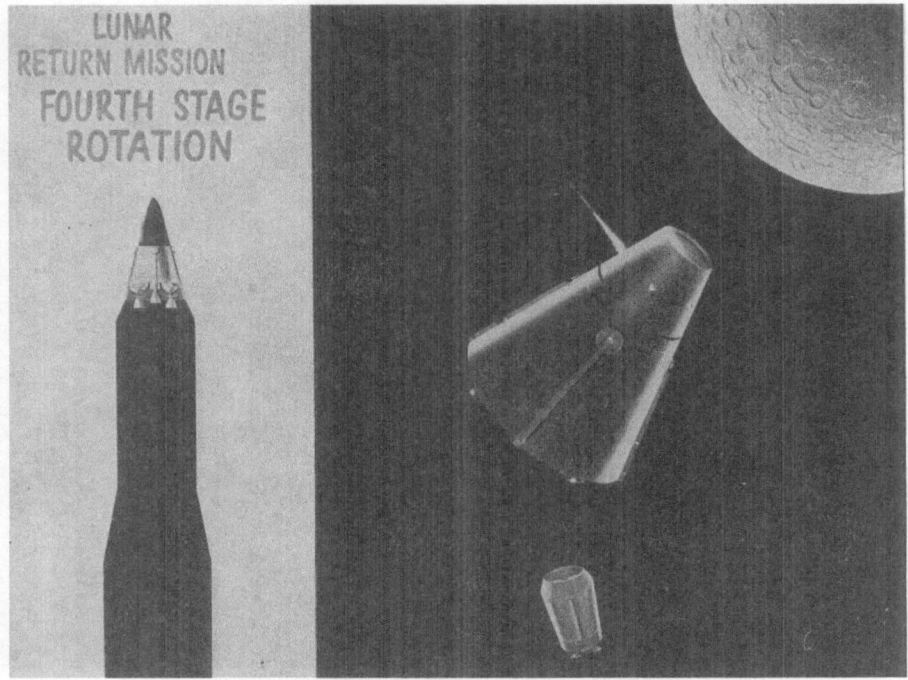

Fig. 8. Approach to moon vehicle rotating under jet control

Fig. 9. Descent to lunar surface, braking rockets firing

Fig. 10. Landing on the moon. Radio beacon is at far right

Fig. 11. Exploring the moon. Spare return vehicle in background

Fig. 12. Fifth stage taking off from moon

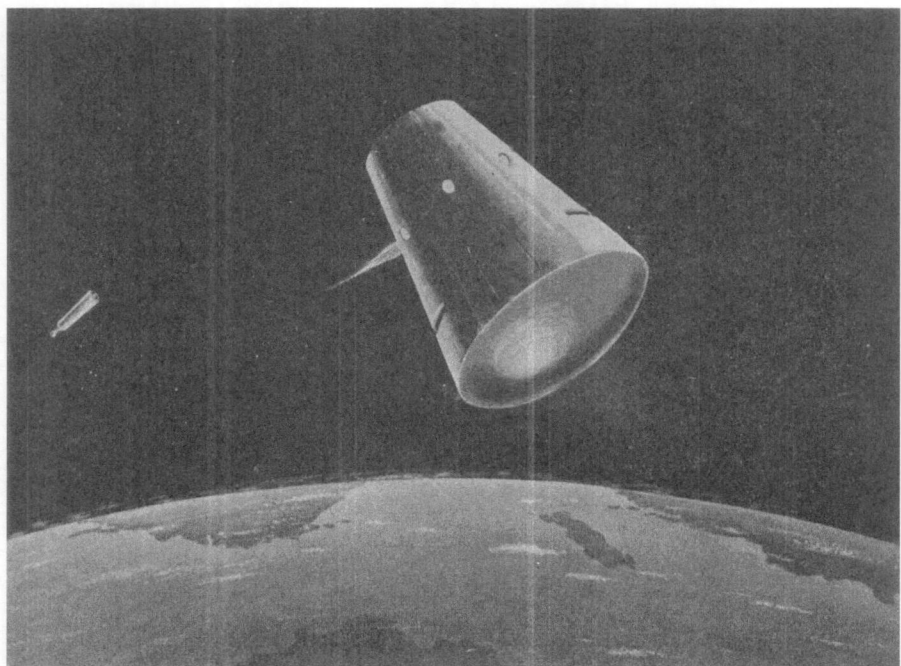

Fig. 13. Return capsule orienting for re-entry. Spent fifth stage at left

Fig. 14. Capsule re-enters earth's atmosphere

Fig. 15. Return to earth and recovery at sea

As the two occupants emerge (Fig. 11), they see 500 yards away an exact duplicate of the vehicle that brought them to the moon. This spare return vehicle had been sent up one month earlier, had landed on the moon, checked itself out and radioed its state of readiness to earth. Farther away is the radio beacon sent to the moon a year earlier on a smaller rocket to mark the landing area.

How the two men occupy themselves during their 12 days on the moon can be better described by those who have for years speculated about the lunar crust.

When they are ready to depart, the men re-enter the capsule and fire the fifth stage (Fig. 12) which uses the fourth stage as a launching stand. The final stage burns for 220 seconds. Then starts the long 60-hour return trip during which a few precisely timed corrective blasts put the cone in the correct corridor for re-entry to the earth's atmosphere. Then the fifth-stage motor is discarded (Fig. 13), and the cone begins its descent with careful control of its angle of attack. The cone approaches the earth (Fig. 14), its ablative surface glowing from the heat of re-entry. At 30,000 feet a large parachute is deployed which slowly lowers the capsule to the ocean (Fig. 15).

If, at first glance, the preceding account appears fanciful, it is because our thinking has not caught up with the engineering advances of the last few years. What has been presented here is based on a preliminary design study of the type conducted by many agencies to assess the feasibility of a vehicle design. All of the engines are either being developed or programmed to be developed in the next few years. No new or exotic fuels are required. Indeed, our calculations reflect the sober degree of conservatism that should characterize a preliminary study. We believe feasibility has been shown. There remains now the intriguing task of doing the job.

References

1. H. H. Hall and E. D. Zambelli, On the Optimization of Multistage Rockets. Jet Propulsion **28**, 463 (1958).
2. R. B. Baldwin, The Face of the Moon. Chicago: University Press, 1949.
3. D. R. Chapman, An Approximate Analytical Method for Studying Entry into Planetary Atmospheres. N.A.C.A. T.N. 4276, 1958.
4. D. R. Chapman, On the Corridor and Associated Trajectory Accuracy for Entry of Manned Spacecraft Into Planetary Atmospheres. Proceedings of the Xth International Astronautical Congress, London 1959, p. 254. Wien: Springer, 1960.

Differential Expressions for Low-Eccentricity Geocentric Orbits[1]

By

Samuel Herrick[2,3], Louis G. Walters[2] and C. Geoffrey Hilton[2,3]

(With 2 Figures)

(Received June 18, 1959)

Abstract — Zusammenfassung — Résumé

Differential Expressions for Low-Eccentricity Geocentric Orbits. Differential relationships between observed quantities, especially range and range rate, and selected orbit parameters are developed for low-eccentricity geocentric orbits. These relationships serve in (1) the mathematical correction of approximate orbits into better approximations, (2) the physical correction of an orbit to satisfy objectives such as interception, and (3) the determination of uncertainties in the orbit from uncertainties in the observations, and vice versa for uncertainties in ephemerides.

Attention is given to the selection of suitable parameters for "selectivity", "linearity," "efficiency," and "compatibility." In the present experiment, for example, the semi-major axis, a, is chosen because of its selective relationship to progressive changes in the residuals with time; on the other hand, components of initial position and velocity, if used for reference parameters, would all be tied in to such progressive changes. The argument of perigee, ω, may change erratically in the differential correction of orbits of low eccentricity, e, and so is rejected on the score of linearity; $e \cos \omega$ and $e \sin \omega$ are used as parameters to satisfy this criterion. Efficiency has to do, of course, with computational procedures and the elimination of functions whose evaluation consumes too much computer time; thus two of the three components of $\underline{W}$, the unit vector normal to the orbit plane, are used in place of i and Ω, the inclination and the longitude of the node. although the differentials of $\underline{W}$, are expressed in terms of those of i and Ω. Compatibility has to do both with the usefulness of the adopted parameters in related computation programs (e.g. in the "variation of parameters") and with consistency between parameters; thus M_0, the initial mean anomaly, is rejected as a parameter because it is associated with

[1] This research was sponsored by the National Aeronautics and Space Administration under contract NAS 1-204, titled "A Range Planning Study for Project Mercury," by Boeing Airplane Company P. O. 7-400120-7600, titled "Orbit and Related Studies for Advanced Operational Weapon Systems," and by the United States Air Force under contract AF 49 (638)-498, monitored at the University of California, Los Angeles, by the Air Force Office of Scientific Research of the Air Research and Development Command.

[2] Aeronutronic, a Division of Ford Motor Company, Newport Beach, California, U.S.A. (Publication No. U-456, June 5, 1959).

[3] University of California, Los Angeles, California, U.S.A. (Contribution to Astrodynamics, No. 2, June 5, 1959).

an ill-defined perigee independently specified by ω; the parameter $U_0 = M_0 + \omega$ is adopted to insure compatibility with ω. Similarly, if i were small, so that both ω and Ω were poorly determined, ω would be replaced by $\Pi = \omega + \Omega$ and U_0 by $L_0 = = M_0 + \Pi$.

The development of the differential expressions is divided into three parts. The first of these is concerned with the relationships between the differentials of the components of position and velocity in the orbit plane and the differentials of a, $e \cos \omega$, $e \sin \omega$, and U_0. Herein the differentials expressions are particularized to very small or zero eccentricities. The second part extends to general components of position and velocity by taking the differentials of i and Ω into account. The third part relates the differentials of geocentric position and velocity to differentials in observed coordinates. Emphasis is placed upon range and range rate (or topocentric distance and radial velocity) because these are judged to be the quantities most accurately observed in the problem for which the study was undertaken ("Project Mercury" or "Man-in-Space").

The study is arranged for later comparison with other techniques now under development.

Differentialgleichungen für geozentrische Umlaufbahnen geringer Exzentrizität.
In der vorliegenden Arbeit werden Differential-Beziehungen zwischen beobachteten Größen, insbesondere Entfernung und Entfernungsänderung, und ausgewählten Bahnparametern für geozentrische Bahnen geringer Exzentrizität entwickelt. Diese Beziehungen sind von Nutzen: 1. für die mathematische Korrektion angenäherter Bahnen zu noch besseren Annäherungen; 2. für die physikalische Korrektion einer Umlaufbahn, um Absichten wie etwa das Einfangen zu verwirklichen; 3. für die Bestimmung von Unsicherheiten in den Umlaufbahnen aus Unsicherheiten in den Beobachtungen sowie umgekehrt von Unsicherheiten in den Ephemeriden.

Berücksichtigt wird auch die Auswahl geeigneter Parameter für „Selektivität", „Linearität", „Leistungsfähigkeit" und „Vereinbarkeit". Bei dem vorliegenden Versuch beispielsweise wird die große Halbachse a wegen der selektiven Beziehung zu fortschreitenden Veränderungen in der Zeitdifferenz gewählt; anderseits würden Komponenten der Anfangslage und -geschwindigkeit im Fall der Benützung als Bezugsparameter in solche progressive Änderungen verwickelt werden. Das Argument ω des Perigäums kann regellos bei der Differentialkorrektur der Bahnen geringer Exzentrizität e schwanken und wird deshalb auf Grund der Linearität herausgeworfen. $e \cos \omega$ und $e \sin \omega$ werden als Parameter benützt, um dieses Kriterium zu erfüllen. Die Zweckmäßigkeit ist eine Sache der Rechenverfahren und der Eliminierung von Funktionen, deren Auswertung zu viel Zeit für die Rechenmaschine benötigen würde. So werden zwei der drei Komponenten von $\underline{W}$, dem zur Bahnebene normal stehenden Einheitsvektor, an Stelle der Neigung i und der Länge des Knotens Ω verwendet, obwohl die Differentiale von $\underline{W}$ als diejenigen von i und Ω ausgedrückt werden. Die Verträglichkeit steht im Zusammenhang sowohl mit der Nützlichkeit der gewählten Parameter in den betreffenden Rechenprogrammen (z. B. in der „Variation der Parameter") wie auch mit der Einheitlichkeit zwischen den Parametern; so wird M_0, die anfängliche mittlere Anomalie, als Parameter verworfen, weil sie mit einem schlecht definierten Perigäum zusammenhängt, das unabhängig durch ω bestimmt ist. Der Parameter $U_0 = M_0 + \omega$ wird gewählt, um die Vereinbarkeit mit ω zu sichern. Wenn i klein wäre, so daß sowohl ω wie Ω schlecht bestimmt wären, würde in ähnlicher Weise ω durch $\Pi = \omega + \Omega$ und U_0 durch $L_0 = M_0 + \Pi$ ersetzt werden.

Die Entwicklung der Differentialausdrücke wird in drei Teile geteilt. Der erste davon befaßt sich mit den Beziehungen zwischen den Differentialen der Komponenten der Lage und Geschwindigkeit in der Bahnebene und den Differentialen von a, $e \cos \omega$, $e \sin \omega$ und U_0. Hiebei werden die Differentialausdrücke ausführlich für sehr kleine Exzentrizitäten oder solche der Größe null angegeben. Der zweite Teil umfaßt die allgemeinen Komponenten der Lage und Geschwindigkeit, indem die Differentiale von i und Ω in Rechnung gesetzt werden. Der dritte Teil verbindet die Differentiale der geozentrischen Position und Geschwindigkeit mit den Differentialen der beobachteten Koordinaten. Das Schwergewicht wird auf die Entfernung und Entfernungsänderung (oder topozentrischen Abstand und Radialgeschwindigkeit) gelegt, weil diese als jene Größen aufgefaßt werden, welche am genauesten bei dem Problem beobachtbar sind, für welches die Studie ausgeführt wurde („Project Mercury" oder „Mensch im Weltraum").

Die Studie wurde so angelegt, daß sie sich zu späterem Vergleich mit den anderen jetzt in Entwicklung befindlichen Techniken eignet.

Expressions différentielles pour les orbites géocentriques de faible excentricité.
On développe des relations différentielles entre grandeurs observées, spécialement la distance et sa variation, et certains paramètres choisis des orbites géocentriques de faible excentricité. Leur utilité réside: (1) dans la correction mathématique des approximations à l'orbite réelle; (2) dans la modification physique d'une orbite aux fins d'interception, par exemple; (3) dans l'estimation des erreurs sur l'orbite à partir des erreurs d'observation et réciproquement en ce qui concerne la précision des éphémérides.

Le choix des paramètres s'inspire de considérations sur la "sélectivité", la "linéarité", "l'efficacité" et la "compatibilité". Le choix du demi-grand-axe a par exemple est justifié par sa relation sélective vis-à-vis des changements progressifs des différences du temps; au contraire les éléments de position et de vitesse initiales sont des paramètres qui sont tous impliqués dans de tels changements. L'argument ω du périgée est susceptible de changements erratiques dans les corrections différentielles d'orbites de faible excentricité e; il est en conséquence rejeté comme paramètre en vertu de la clause de linéarité. Pour satisfaire ce critère $e \cos \omega$ et $e \sin \omega$ sont adoptés comme paramètres. L'efficacité s'entend évidemment en relation avec les méthodes de calcul; elle conduit à éliminer les fonctions d'évaluation trop longue. Ainsi deux des trois composantes de la normale $\underline{W}$ au plan de l'orbite remplacent l'inclinaison i et la longitude du node Ω; quoique les différentielles de ces grandeurs soient en relation directe. La compatibilité se juge d'après l'utilité des paramètres dans des programmes de calcul voisins (par exemple, dans la méthode de variation des paramètres) et dans le conditionnement des paramètres entre-eux. Ainsi l'anomalie moyenne initiale M_0 est éliminée parce que liée à un périgée mal défini tel que indépendamment spécifié par ω. Le paramètre $U_0 = M_0 + \omega$ est adopté par compatibilité avec ω. De même, si i était petit, ce qui rendrait ω et Ω mal définis on remplacerait ω par $\Pi = \omega + \Omega$ et U_0 par $L_0 = M_0 + \Pi$.

Les développements sont divisés en trois parties. La première établit les relations entre les différentielles des composantes de position et de vitesse dans le plan orbital et les différentielles des grandeurs a, $e \cos \omega$, $e \sin \omega$ et U_0. Les expressions différentielles sont spécialisées pour des excentricités petites ou nulle. La seconde partie fait l'extension aux composantes générales de position et de vitesse en tenant compte des différentielles de i et de Ω. La troisième relie les différentielles de position géocentrique et de vitesse à celles dans les coordonnées d'observation. L'accent est mis sur la distance topocentrique et la vitesse radiale car ces grandeurs sont les plus précises mesurées dans le problème qui a initié l'étude (le projet "Mercury").

La présentation est prévue pour faciliter une comparaison ultérieure avec d'autres techniques en cours de développement.

I. Introduction

In the planning and evaluation of a tracking complex for space vehicles, two functions are served by differential expressions relating uncertainties, errors, or corrections in selected orbital elements to corresponding uncertainties, errors, or corrections in the observed coordinates. These functions are:

(1) The translation of orbit uncertainties into observed quantities, or vice versa, for an evaluation of ability to correct an orbit from given observations, or for an evaluation of the accuracy of an ephemeris based upon these observations.

(2) The translation of residuals in observed coordinates into differential corrections to the orbital elements (or parameters).

The differential expressions are linear equations relating differentials in orbit parameters to corresponding differentials in observed coordinates and they assume a familiar form:

$$\Delta O_1 = A_1 \Delta p_1 + A_2 \Delta p_2 + \cdots$$
$$\Delta O_2 = B_1 \Delta p_1 + B_2 \Delta p_2 + \cdots$$

The p_i's denote orbit parameters, O_j's the observed coordinates, and the A's, B's, etc., the partial derivatives of observed quantities with respect to the parameters[1]. These partial derivatives may be determined analytically, as in this paper, or may be determined by variant trajectory calculations wherein the parameters are varied, one by one, and the resulting changes in the calculated coordinates are noted. The latter technique is preferable or even necessary when the orbit is strongly perturbed from the two-body representation, as in re-entry and, to a lesser extent, in translunar trajectory studies. During the normal orbit periods of the Project Mercury capsule and of other satellites employed for communication and reconnaissance, the analytical expressions derived from two-body formulas should describe adequately the differential effects for the functions stated above, and, for the low-eccentricity case at least, may exhibit superior computational efficiency to the method of variant trajectory calculation.

The analysis presented here is divided into two steps because of the need for data intermediate between the parameters and the ephemeris. These steps are:

(1) The development of differential expressions relating the parameters to rectangular components of position and velocity, the latter of which are necessary to the evaluation of orbital and re-entry flight conditions; and

(2) The extension of these differential expressions to the observational framework employed in the range. This is carried out for radar-type instruments where the emphasis is on very accurate range and range-rate data.

These steps are outlined in sections III, IV and V. Additional sections are devoted to the choice of parameters and to related computational procedures.

II. Selection of Orbit Parameters

The orbit parameters that are to define the two-body orbit are selected with the following criteria in mind:

(1) Selectivity in the determination, from observations, of which of the parameters are in error;

(2) Linearity in relationships between cause and effect, as in the linear equations employed in differential correction procedures;

(3) Computational efficiency in the representation of the position of the object, i.e., reduction of the computational burden;

---

[1] Where the word, "parameter," is employed in the following discussions, it will refer to the set of six two-body orbit parameters selected to describe the orbit, and perhaps also to selected physical constants entering into the equations of motion.

(4) Compatibility or usefulness of the parameters in other portions of the orbit determination problem.

Six independent parameters are required to define the instantaneous orbit; familiar sets of these include: position and velocity at the epoch; the set a, e, M_0 or T, Ω i, ω; the set a, e, M_0 or T, $\underline{P}$, $\underline{Q}$, in which the six vector-components are reduced by the condition-equations to three independent quantities. Careful consideration of the special problems posed by low-eccentricity orbits suggests the adoption of the following set of six parameters[1]:

a, $e \sin \omega$, $e \cos \omega$, U_0, and two of the three components of $\underline{W}$.

This selection is the result of extensive analysis and experience with heliocentric planetary orbit determination. The means by which this choice circumvents some of the difficulties encountered with the more conventional elements or injection position and velocity is discussed below.

The selectivity criterion is most readily demonstrated by noting that residuals determined from a comparison of the assumed orbit with observations of the object are strongly influenced by the validity of the orbital period, determined directly from semi-major axis a. An orbit determined by injection position and velocity has a period dependent upon both velocity and altitude and, consequently, variations in any of these parameters will show the progressive change in residuals characteristic of incorrect orbital period. The use of the semi-major axis itself, on the other hand, provides a direct cause-and-effect link to this important property demonstrated by the residuals.

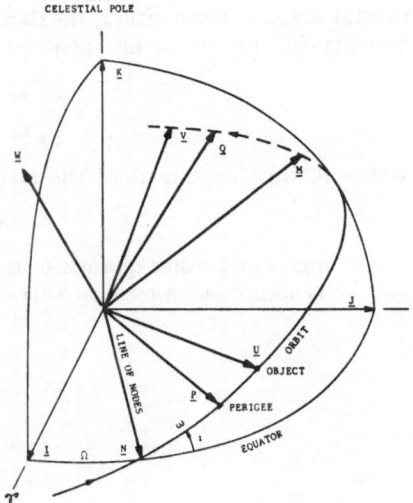

Fig. 1. Projection of orbit on celestial sphere, with orientation unit vectors and angles displayed

The problem of linearity is demonstrated by the use of e and ω or e and $\pi = \Omega + \omega$ as parameters in low-eccentricity orbits. The eccentricity enters into the equation as a coefficient of $\Delta \omega$ or $\Delta \pi$, and the effect of a small correction may accordingly be a very large shift in the position of perigee in an orbit, nearly circular, in which the perigee is not actually important. In fact, the mechanism for introducing a perigee into an initially circular orbit is indeterminate unless the perigee is coupled with the eccentricity. The use of $e \sin \omega$ and $e \cos \omega$, or of $e \sin \pi$ and $e \cos \pi$, avoids this difficulty.

The computational efficiency and compatibility criteria are coupled to the total problem of orbit representation and correction. Aeronutronic research into this problem, and specifically into a special perturbations approach employing the efficient variation-of-parameters technique for orbit integration, shows great promise. The formulation employed in the variation-of-parameters orbit integration technique employs the vector $\underline{a}$, whose components in the instantaneous orbit plane are $e \sin \omega$ and $e \cos \omega$; thus selection of these as parameters strengthens the compatibility of the representation and correction procedures.

---

[1] The nomenclature employed here is defined in Fig. 1.

III. Differential Expressions for Position and Velocity Components in the Orbit Plane[1]

This section is concerned with the derivation of the differential relationships between the adopted elements and the position and velocity components in the orbit plane. These derivations are conveniently separable from the broader problem in that these quantities depend only on four of the six parameters, i.e., a, U_0, $e \cos \omega$, and $e \sin \omega$, and the analysis involves only scalar two-body formulas. The remaining differential relationships are derived in section IV from the expressions for the vectors defining the orientation of the orbit plane.

The whole of the analysis will employ the $\underline{N}$, $\underline{M}$ and $\underline{W}$ unit vectors to specify the orientation and the chosen axes in the orbit plane. Such a reference system is preferable to one identified with perigee, which is poorly determined for very low eccentricity. The $\underline{M}$ and $\underline{N}$ vectors lie in the orbit plane, with $\underline{N}$ directed to the node as shown in Fig. 1. The sense of the vector $\underline{W}$ is determined by that of the orbital angular momentum, northerly for direct (eastward) motion. Position and velocity in this plane are denoted by the components x_N and y_N, defined by

$$x_N = r \cos u = \underline{N} \cdot \underline{r} \tag{3.1}$$

$$y_N = r \sin u = \underline{M} \cdot \underline{r} \tag{3.2}$$

where u, the "argument of the latitude," is given by

$$u = v + \omega . \tag{3.3}$$

These angles and other quantities used in this section are defined in Fig. 1. Other two-body expressions used in this development include:

$$n = k' \sqrt{\mu}\, a^{-\frac{3}{2}} \tag{3.4}$$

$$U = U_0 + n\,(t - t_0) \tag{3.5}$$

$$E - e \sin E = M = U - \omega \tag{3.6}$$

$$r = a\,(1 - e \cos E) \tag{3.7}$$

$$r \cos v = a\,(\cos E - e) \tag{3.8}$$

$$r \sin v = a \sqrt{1 - e^2}\, \sin E . \tag{3.9}$$

Reference to the glossary will clarify the meanings of these quantities.

The differential expressions for x_N and y_N are obtained by differentiating (3.1) and (3.2)

$$\Delta x_N = \Delta r \cos u - r \Delta u \sin u \tag{3.10}$$

$$\Delta y_N = \Delta r \sin u + r \Delta u \cos u . \tag{3.11}$$

The determination of Δr follows from (3.7). The determination of $r \Delta u$ is made in terms of its components $r \Delta v$ and $r \Delta \omega$, the first of these from the derivatives of (3.8) and (3.9) and the latter from the derivatives of the parameters $e \cos \omega$ and $e \sin \omega$. These parameters will be denoted a_{xN} and a_{yN}, respectively, in subsequent developments. They are the components, referred to axes determined by $\underline{N}$ and $\underline{M}$, of a vector $\underline{a}$ directed to perigee with magnitude e.

The procedure outlined above is relatively straightforward and will be summarized in this section. Starting with the parameters a_{xN} and a_{yN}

[1] Much of the material in this and subsequent sections has been developed in conjunction with Herrick's forthcoming book "Astrodynamics," to be published by Van Nostrand, Princeton, N. J., in 1960.

$$\Delta a_{zN} = \Delta (e \cos \omega) = \Delta e \cos \omega - e \Delta \omega \sin \omega$$
$$\Delta a_{yN} = \Delta (e \sin \omega) = \Delta e \sin \omega + e \Delta \omega \cos \omega.$$

Expressions for Δe and $e \Delta \omega$ follow directly

$$\Delta e = \Delta a_{zN} \cos \omega + \Delta a_{yN} \sin \omega \tag{3.12}$$

$$e \Delta \omega = - \Delta a_{zN} \sin \omega + \Delta a_{y} \cos \omega. \tag{3.13}$$

Next, the differential of (3.7) yields Δr

$$\Delta r = r \frac{\Delta a}{a} + ae \Delta E \sin E - a \Delta e \cos E. \tag{3.14}$$

The term ΔE follows from (3.4), (3.5), and (3.6), for

$$\frac{\Delta n}{n} = - \frac{3}{2} \frac{\Delta a}{a}$$

$$\Delta U = \Delta U_0 + (t - t_0) \Delta n = \Delta U_0 - \frac{3}{2} (U - U_0) \frac{\Delta a}{a}$$

and finally

$$\Delta E = \frac{a}{r} \Delta U_0 - \frac{a}{r} \Delta \omega - \frac{3}{2} (U - U_0) \frac{a}{r} \frac{\Delta a}{a} + \frac{a}{r} \sin E \Delta e.$$

Substitution for ΔE in (3.14) leads to

$$\Delta r = \frac{\Delta a}{a} \left[r - \frac{3}{2} (U - U_0) \frac{a^2}{r} e \sin E \right] + \Delta U_0 \left[\frac{a^2}{r} e \sin E \right] +$$
$$+ \Delta \omega \left[- \frac{a^2}{r} e \sin E \right] + \Delta e \left[-a \cos E + \frac{a^2}{r} e \sin^2 E \right].$$

Further substitution for $\Delta \omega$ and Δe from (3.12) and (3.13) leads to the form

$$\Delta r = R_u \Delta U_0 + R_a \frac{\Delta a}{a} + R_{zN} \Delta a_{zN} + R_{yN} \Delta a_{yN}. \tag{3.15}$$

The R's are the partial derivates of r with respect to the indicated parameters, and are tabulated below:

$$R_u = \frac{a^2 e}{r} \sin E$$

$$R_a = r - \frac{3}{2} (U - U_0) R_u$$

$$R_{zN} = \frac{a^2}{r} [e \cos \omega - \cos (E + \omega)]$$

$$R_{yN} = \frac{a^2}{r} [e \sin \omega - \sin (E + \omega)].$$

Alternative expressions may be derived for use in special circumstances, but the present form is well suited to the low-eccentricity orbits of the present analysis.

For the partial derivatives entering into $r \Delta u$, one obtains the differentials of $r \Delta v$ from ecs. (3.8) and (3.9) as follows:

$$\Delta r \cos v - r \Delta v \sin v = \Delta a (\cos E - e) - a \Delta e - a \sin E \Delta E$$

$$\Delta r \sin v + r \Delta v \cos v = \Delta a \sqrt{1 - e^2} \sin E - \frac{a e \sin E}{\sqrt{1 - e^2}} \Delta e + a \sqrt{1 - e^2} \cos E \Delta E$$

from which we obtain

$$r \Delta v = a \sqrt{1 - e^2} \Delta E + \frac{a \sin E}{\sqrt{1 - e^2}} \Delta e.$$

Following the substitution pattern established above for Δe and ΔE leads to

$$r\,\Delta v = V_u\,\Delta U_0 + V_a\frac{\Delta a}{a} + V_{zN}\,\Delta a_{zN} + V_{yN}\,\Delta a_{yN} \tag{3.16}$$

where the coefficients or partial derivatives are

$$V_u = \frac{a^2}{r}\sqrt{1-e^2}$$

$$V_a = -\frac{3}{2}\,(U-U_0)\,V_u$$

$$V_{zN} = \frac{a^2}{r}\left[\sqrt{1-e^2}\,\sin E \cos \omega \left(1+\frac{r}{p}\right) + \frac{\sqrt{1-e^2}}{e}\,\sin \omega\right]$$

$$V_{yN} = \frac{a^2}{r}\left[\sqrt{1-e^2}\,\sin E \sin \omega \left(1+\frac{r}{p}\right) - \frac{\sqrt{1-e^2}}{e}\,\cos \omega\right].$$

The instability of these derivatives for low-eccentricity is a reflection of the indeterminate nature of perigee. By reserving the specialization of low-eccentricity to $r\Delta u$ rather than to $r\,\Delta v$ or $r\Delta\omega$, this situation can be avoided.

The $r\Delta\omega$ component of $r\Delta u$ follows directly from (3.13). In combination with (3.16) there results

$$r\,\Delta u = r\,\Delta v + r\,\Delta \omega$$
$$= U_u\,\Delta U_0 + U_a\frac{\Delta a}{a} + U_{zN}\,\Delta a_{zN} + U_{yN}\,\Delta a_{yN} \tag{3.17}$$

with the coefficients

$$U_u = V_u = \frac{a^2}{r}\sqrt{1-e^2}$$

$$U_a = V_a = -\frac{3}{2}\,(U-U_0)\,U_u$$

$$U_{zN} = \frac{a^2}{r}\left\{2\sin (E+\omega) + \sin E \cos \omega \left[\sqrt{1-e^2}\left(1+\frac{r}{p}\right)-2\right]\right.$$
$$\left. - e \sin \omega \left[\frac{1}{1+\sqrt{1-e^2}} + \cos^2 E\right]\right\}$$

$$U_{yN} = \frac{a^2}{r}\left\{-2\cos (E+\omega) + \sin E \sin \omega \left[\sqrt{1-e^2}\left(1+\frac{r}{p}\right)-2\right]\right.$$
$$\left. + e \cos \omega \left[\frac{1}{1+\sqrt{1-e^2}} + \cos^2 E\right]\right\}.$$

Up to this point no restrictions to low-eccentricity have been imposed, but considerable simplification is achieved by substituting the following special values into the R's and U's:

$$e \cong 0$$
$$r \cong p \cong a$$
$$U = M + \omega \cong E + \omega \cong v + \omega = u \cong U_0 + n\,(t-t_0)$$
$$\dot{U} = \dot{M} \cong \dot{E} \cong \dot{v} = \dot{u} \cong \sqrt{\mu}/a^{3/2}.$$

The partial derivatives then assume the simple forms,

$$R_u = 0 \qquad\qquad\qquad U_u = a$$
$$R_a = a \qquad\qquad\qquad U_a = -\frac{3}{2}\,(u-U_0)\,a$$
$$R_{zN} = -a \cos u \qquad\qquad U_{zN} = 2a \sin u$$
$$R_{yN} = -a \sin u \qquad\qquad U_{yN} = -2a \cos u.$$

Table I. *Coefficients Appearing in Differential Correction Formulae*
(Valid for $e \approx 0$)

Coefficient of:	ΔU_0	$\dfrac{\Delta a}{a}$	Δa_{zN}	Δa_{yN}
Δx_N	$-a \sin u$	$a\left[\cos u + \dfrac{3}{2}(u-U_0)\sin u\right]$	$-a(1+\sin^2 u)$	$a \sin u \cos u$
Δy_N	$a \cos u$	$a\left[\sin u - \dfrac{3}{2}(u-U_0)\cos u\right]$	$a \sin u \cos u$	$-a(1+\cos^2 u)$
$\Delta \dot x_N$	$-a\dot u \cos u$	$a\dot u\left[\dfrac{1}{2}\sin u + \dfrac{3}{2}(u-U_0)\cos u\right]$	$-2a\dot u \sin u \cos u$	$a\dot u(\cos^2 u - \sin^2 u)$
$\Delta \dot y_N$	$-a\dot u \sin u$	$a\dot u\left[-\dfrac{1}{2}\cos u + \dfrac{3}{2}(u-U_0)\sin u\right]$	$a\dot u(\cos^2 u - \sin^2 u)$	$2a\dot u \sin u \cos u$

in Differential Expression for

The coefficients in the Δa_{zN} and Δa_{yN} columns may also be expressed effectively in terms of $\cos 2u$ and $\sin 2u$.

By means of (3.10) and (3.11), the differentials in x_N and y_N then follow the pattern

$$\Delta x_N = \Delta r \cos u - r \Delta u \sin u$$
$$= X_u \Delta U_0 + X_a \frac{\Delta a}{a} + X_{zN}\Delta a_{zN} + X_{yN}\Delta a_{yN} \tag{3.18}$$

where

$$X_u = R_u \cos u - U_u \sin u = -a \sin u$$
$$X_a = R_a \cos u - U_a \sin u$$
$$= a \cos u + \frac{3}{2}(u - U_0)\sin u$$

et cetera.

The remaining coefficients, and those for Δy_N as well, are tabulated in the first two rows of Table I. The third and fourth row in Table I are the coefficients in the differential expressions for $\dot{x}_N$ and $\dot{y}_N$, derived from

$$\dot{x}_N = -\sqrt{\frac{\mu}{p}} \,(\sin u + e \sin \omega) \tag{3.19}$$

$$\dot{y}_N = \sqrt{\frac{\mu}{p}} \,(\cos u + e \cos \omega). \tag{3.20}$$

The procedure for the velocity terms parallels that for the displacement terms.

IV. Differential Relationships Extended to the Orientation of the Orbit Plane

The orientation of the orbit plane is defined by the longitude of the node, Ω, and the inclination, i, or alternatively by the unit vector $\underline{W}$ shown in Fig. 1. The latter definition is convenient for manipulation; and the required differential relationships between Ω and i, on the one hand, and $\underline{W}$, on the other hand, will be set forth first.

The unit vector $\underline{W}$ has the components W_x, W_y, W_z measured in the inertial equatorial reference frame defined by the unit vectors $\underline{I}$, $\underline{J}$, and $\underline{K}$. These components are:

$$\begin{aligned}
W_x &= \underline{I} \cdot \underline{W} = \sin \Omega \sin i \\
W_y &= \underline{J} \cdot \underline{W} = -\cos \Omega \sin i \\
W_z &= \underline{K} \cdot \underline{W} = \cos i \,.
\end{aligned} \tag{4.1}$$

Differentiation yields

$$\begin{aligned}
\Delta W_x &= \cos \Omega \sin i \,\Delta \Omega + \sin \Omega \cos i \,\Delta i \\
\Delta W_y &= \sin \Omega \sin i \,\Delta \Omega - \cos \Omega \cos i \,\Delta i \\
\Delta W_z &= -\sin i \,\Delta i \,.
\end{aligned}$$

The coefficients herein of $\Delta \Omega \sin i$ and of $-\Delta i$ are the components of the unit vectors $\underline{N}$ and $\underline{M}$, which form a right-handed set with $\underline{W}$ and may be determined directly from its components:

$$\begin{aligned}
N_x &= \cos \Omega = -W_y/M_z \\
N_y &= \sin \Omega = +W_x/M_z \\
N_z &= 0
\end{aligned} \tag{4.2}$$

$$\begin{aligned}
M_x &= -\sin \Omega \cos i = -N_y W_z \\
M_y &= \cos \Omega \cos i = +N_x W_z \\
M_z &= \sin i = +\sqrt{W_x^2 + W_y^2}
\end{aligned} \tag{4.3}$$

so that we may write, considering also the differentials of eq. (4.2)

$$\begin{aligned}
\Delta \underline{W} &= \underline{N} \,\Delta \Omega \sin i - \underline{M} \,\Delta i \\
\Delta \underline{N} &= \underline{M} \,\Delta \Omega \cos i - \underline{W} \,\Delta \Omega \sin i \\
\Delta \underline{M} &= \underline{W} \,\Delta i - \underline{N} \angle \Omega \cos i \,.
\end{aligned} \tag{4.4}$$

In addition, the orthogonality of $\underline{N}$, $\underline{M}$, and $\underline{W}$ leads to

$$\begin{aligned}
\underline{N} \cdot \Delta \underline{W} &= -\underline{W} \cdot \Delta \underline{N} = \Delta \Omega \sin i \\
\underline{M} \cdot \Delta \underline{N} &= -\underline{N} \cdot \Delta \underline{M} = \Delta \Omega \cos i \\
\underline{W} \cdot \Delta \underline{M} &= -\underline{M} \cdot \Delta \underline{W} = \Delta i \,.
\end{aligned} \tag{4.5}$$

These expressions will be used in the following development.

The differential relationships between[1] topocentric position and the parameters follow directly from

$$\varrho = \underline{N}\, x_N + \underline{M}\, y_N + \underline{R} \tag{4.6}$$

$$\Delta \varrho = \underline{N}\, \Delta x_N + \underline{M}\, \Delta y_N + x_N\, \Delta \underline{N} + y_N\, \Delta \underline{M} \tag{4.7}$$

where Δx_N and Δy_N were derived in section III as functions of the four corrections, ΔU_0, Δa, Δa_{xN}, and Δa_{yN}. Since the components of $\Delta \underline{N}$ and $\Delta \underline{M}$ are expressed by eqs. (4.4) in terms of the two corrections, $\Delta \Omega$ and Δi, it is evident that eq. (4.7) involves the six parameter-corrections on the one side and the three components of the residual on the other. In a differential correction scheme, then, each determinable component of $\Delta \varrho$ gives us one equation toward the determination of the six unknown corrections to the parameters. A minimum of six such equations, whether from six components of position and velocity (see below) observed at one time, or from selected components observed at different times, is necessary of course to the completion of the differential corrections.

If all three components of $\Delta \varrho$ are known, we may use

$$
\begin{aligned}
\underline{N} \cdot \Delta \varrho &= \Delta x_N - y_N \Delta \Omega \cos i \\
\underline{M} \cdot \Delta \varrho &= \Delta y_N + x_N \Delta \Omega \cos i \\
\underline{W} \cdot \Delta \varrho &= - x_N \Delta \Omega \sin i + y_N \Delta i \,.
\end{aligned}
\tag{4.8}
$$

Similarly, the differential relationships for the velocity components follow from

$$\dot{\varrho} = \dot{x}_N \underline{N} + \dot{y}_N \underline{M} \tag{4.9}$$

$$\Delta \dot{\varrho} = \Delta \dot{x}_N \underline{N} + \Delta \dot{y}_N \underline{M} + \dot{x}_N \Delta \underline{N} + \dot{y}_N \Delta \underline{M} \tag{4.10}$$

and if all three components of $\Delta \dot{\varrho}$ are known,

$$
\begin{aligned}
\underline{N} \cdot \Delta \dot{\varrho} &= \Delta \dot{x}_N - \dot{y}_N \Delta \Omega \cos i \\
\underline{M} \cdot \Delta \dot{\varrho} &= \Delta \dot{y}_N + \dot{x}_N \Delta \Omega \cos i \\
\underline{W} \cdot \Delta \dot{\varrho} &= - \dot{x}_N \Delta \Omega \sin i + \dot{y}_N \Delta i \,.
\end{aligned}
\tag{4.11}
$$

A 6×6 matrix relating differential quantities to orbit parameters may be derived from (4.8) and (4.11), and from the entries in Table I. This is presented in Table II. It has the same limitation to $e \approx 0$ as does Table I. It should be noted, however, that a direct solution is possible in these circumstances and that the inversion of a 6×6 matrix, however simple, is unnecessary.

The expressions in (4.7) and (4.10) lead directly to orbit correction formulas for range and range-rate instruments, as outlined in the following section.

V. Orbit Correction Formulas for Range and Range-Rate Tracking Instruments

This section translates the differential expressions for position and velocity in terms of the six orbit parameters, as derived in the previous section, into orbit correction formulas based on measurements of topocentric range and range-rate. These formulas have special application to single-site electronic instruments where great emphasis is placed on range and range-rate accuracy. The available

---

[1] See eq. (5.1).

angle observations, by contrast, are less precise by at least an order of magnitude below the ranging data and usually serve little purpose in an orbit correction program.

The observation framework is shown in Fig. 2, where the following vectors are identified:

$$\underline{R} = \text{position-vector of geocenter with respect to observer (topocenter)}$$
$$\underline{r} = r\,\underline{U} = \text{position-vector of object referred to geocenter}$$
$$\underline{\varrho} = \varrho\,\underline{L} = \text{position-vector of object referred to topocenter.}$$

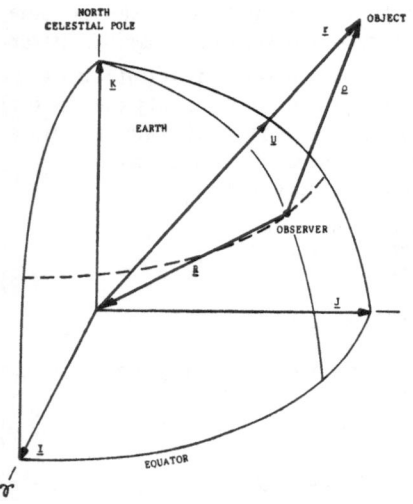

Fig. 2. Observational framework

We shall employ the following relationships between these vectors and associated ones:

$$\underline{\varrho} = \varrho\,\underline{L} = \underline{r} + \underline{R} = r\,\underline{U} + \underline{R} \quad (5.1)$$

$$\underline{U} = \underline{N}\cos u + \underline{M}\sin u \quad (5.2)$$

$$\underline{V} = -\underline{N}\sin u + \underline{M}\cos u . \quad (5.3)$$

Like $\underline{N}$, $\underline{M}$, $\underline{W}$, the vectors $\underline{U}$, $\underline{V}$, $\underline{W}$ make up a right-handed orthogonal system.

To suppress $\underline{L}$ and its dependence upon angles, we form the dot product of $\underline{\varrho}$ and $\Delta\underline{\varrho}$, the latter from eq. (4.7).

$$\varrho\,\Delta\varrho = \underline{\varrho} \cdot \Delta\underline{\varrho}$$
$$= (\underline{\varrho} \cdot \underline{N})\,\Delta x_N + (\underline{\varrho} \cdot \underline{M})\,\Delta y_N +$$
$$- x_N\,(\underline{\varrho} \cdot \Delta\underline{N}) + y_N\,(\underline{\varrho} \cdot \Delta\underline{M}) . \quad (5.4)$$

If we now particularize to the circular (or very nearly circular) orbit, with $x_N = a\cos u$, $y_N = a\sin u$, with the coefficients for Δx_N and Δy_N from Table I, and with $\Delta\underline{N}$ and $\Delta\underline{M}$ expressed in terms of $\Delta\Omega$ and Δi by eqs. (4.4), we find that

$$\varrho\,\Delta\varrho = a\underline{\varrho} \cdot [\underline{M}\cos u - \underline{N}\sin u]\,\Delta U_0$$
$$+ a\underline{\varrho} \cdot [\underline{N}\cos u + \underline{M}\sin u - \frac{3}{2}(u - U_0)(\underline{M}\cos u - \underline{N}\sin u)]\frac{\Delta a}{a}$$
$$+ a\underline{\varrho} \cdot [\underline{M}\sin u\cos u - \underline{N}(1 + \sin^2 u)]\,\Delta a_{zN}$$
$$+ a\underline{\varrho} \cdot [\underline{N}\sin u\cos u - \underline{M}(1 + \cos^2 u)]\,\Delta a_{yN}$$
$$+ a\underline{\varrho} \cdot [(\underline{M}\cos u - \underline{N}\sin u)\,\Delta\Omega\cos i - \underline{W}\cos u\,\Delta\Omega\sin i + \underline{W}\sin u\,\Delta i] .$$

Introducing eqs. (5.2) and (5.3), and (5.1) so far as advantageous, noting that $\underline{U} \cdot \underline{U} = 1$, $\underline{U} \cdot \underline{V} = \underline{U} \cdot \underline{W} = 0$, we find

$$\varrho\,\Delta\varrho = a\,(\underline{R} \cdot \underline{V})\,\Delta U_0$$
$$+ a\,[a + \underline{R} \cdot \underline{U} - \frac{3}{2}(u - U_0)(\underline{R} \cdot \underline{V})]\frac{\Delta a}{a}$$
$$+ a\,[(\underline{R} \cdot \underline{V})\sin u - \underline{\varrho} \cdot \underline{N}\,\Delta a_{zN}$$
$$+ a\,[-(\underline{R} \cdot \underline{V})\cos u - \underline{\varrho} \cdot \underline{M}]\,\Delta a_{yN}$$
$$+ a\,[(\underline{R} \cdot \underline{V})\cos i - (\underline{R} \cdot \underline{W})\cos u\sin i]\,\Delta\Omega$$
$$+ a\,[(\underline{R} \cdot \underline{W})\sin u]\,\Delta i . \qquad (5.5)$$

Table II

	ΔU_0	Δa	Δa_{zN}	Δa_{yN}	$\Delta\Omega$	Δi
$N\cdot\Delta\underline{\varrho}$	$-a\sin u$	$a\cos u + \frac{3}{2}(u-U_0)\,a\sin u$	$-a(1+\sin^2 u)$	$a\sin u\cos u$	$-a\sin u\cos i$	0
$M\cdot\Delta\underline{\varrho}$	$a\cos u$	$a\sin u - \frac{3}{2}(u-U_0)\,a\cos u$	$a\sin u\cos u$	$-a(1+\cos^2 u)$	$+a\cos u\cos i$	0
$W\cdot\Delta\underline{\varrho}$	0	0	0	0	$-a\cos u\sin i$	$+a\sin u$
$N\cdot\Delta\underline{\dot\varrho}$	$-a\dot u\cos u$	$\frac{1}{2}a\dot u\sin u + \frac{3}{2}a\dot u(u-U_0)\cos u$	$2a\dot u\sin u\cos u$	$a\dot u(\cos^2 u-\sin^2 u)$	$-a\dot u\cos u\cos i$	0
$M\cdot\Delta\underline{\dot\varrho}$	$-a\dot u\sin u$	$-\frac{1}{2}a\dot u\cos u + \frac{3}{2}a\dot u(u-U_0)\sin u$	$a\dot u(\cos^2 u-\sin^2 u)$	$2a\dot u\sin u\cos u$	$-a\dot u\sin u\cos i$	0
$W\cdot\Delta\underline{\dot\varrho}$	0	0	0	0	$+a\dot u\sin u\sin i$	$+a\dot u\cos u$

Alternative forms applicable to particular computational problems may easily be derived; further study is warranted to optimize computational compatibility with other aspects of orbit correction.

At this point it is worthwhile to note how angular data, rather than range data, might be employed. If the product $\varrho \times \Delta \varrho$ is formed, rather than the dot product, there results

$$\varrho \times \Delta \varrho = \varrho \underline{L} \times (\Delta \varrho \underline{L} + \varrho \Delta \underline{L}) = \varrho^2 \underline{L} \times \Delta \underline{L} . \tag{5.6}$$

In this expression, the residuals in right ascension and declination, or azimuth and altitude, can be expressed directly as components of $\Delta \underline{L}$. An extensive development of this approach to angular residuals, and of other approaches, will be found in Herrick's *Astrodynamics*, when published.

The orbit correction formula for range-rate observations follows from

$$\Delta (\varrho \cdot \dot{\varrho}) = \Delta (\varrho \dot{\varrho})$$

$$\varrho \cdot \Delta \dot{\varrho} + \dot{\varrho} \cdot \Delta \varrho = \varrho \Delta \dot{\varrho} + \dot{\varrho} \Delta \varrho$$

$$\therefore \varrho \Delta \dot{\varrho} = \varrho \cdot \Delta \dot{\varrho} + \dot{\varrho} \cdot \Delta \varrho - \dot{\varrho} \Delta \varrho$$

$$= \frac{d}{d\tau} (\varrho \cdot \Delta \varrho) - \frac{\dot{\varrho}}{\varrho} (\varrho \Delta \varrho) .$$

In this form, the expression given in eq. (5.5) is employed for the right-hand terms. No effort is made ro reduce $\varrho \Delta \dot{\varrho}$ further for the purpose of this present treatment.

References

Good modern treatments of the differential correction, and bibliographic data, are to be found in:

1. J. Bauschinger, Die Bahnbestimmung der Himmelskörper, 2nd ed. Leipzig: Engelmann, 1928.
2. R. T. Crawford, Determination of Orbits of Comets and Asteroids. New York: McGraw-Hill, 1930.
3. J. G. Porter, The Differential Correction of Orbits. Monthly Notices, Roy. Astronom. Soc. **109**, 409-420 (1949).

Line-of-Sight Criteria for Interplanetary Navigation

By

T. L. Connors[1], W. K. Huggett[1] and A. C. Lawson, Jr.[1]

(With 10 Figures)

(Received August 27, 1959)

Abstract — Zusammenfassung — Résumé

Line-of-Sight Criteria for Interplanetary Navigation. Instead of the usual techniques proposed for interplanetary navigation, which require the explicit computation of an orbit or orbits to be followed, it is possible to navigate along a trajectory implicitly defined by a line through the space vehicle and a convenient reference, such as the Sun or a point on the celestial sphere. Studies of two particular line-of-sight criteria show them to possess distinct advantages over more conventional methods. In one method, the line-of-sight from the vehicle to the Sun must be held colinear with the line-of-sight from the vehicle to the destination planet; in the other, the vehicle must remain on a line through the destination planet and a fixed point on the celestial sphere. Both methods reduce the amount of measuring and computing equipment to be carried by the vehicle, and generally simplify the entire mid-course navigational process. An example of a feasible Earth-to-Mars orbit generated implicitly by the first technique requires only slightly more energy than that required for the HOHMANN ellipse. Two examples of the second technique illustrate its applicability. Line-of-sight techniques are suitable only for space vehicles propelled by continuous low-thrust motors. The maximum continuous acceleration required for the sample case is less than 60 micro *g*'s.

Sichtlinien-Kriterien für interplanetare Navigation. Anstatt die gewöhnliche Technik für interplanetare Navigation zu verwenden, die die explizite Berechnung der Bahn oder Bahnen benötigt, denen gefolgt werden soll, kann man einer Geschoßbahn entlang fahren, die implizit durch eine Linie definiert ist, die durch das Raumschiff und eine geeignete Bezugsstelle, wie die Sonne oder einen anderen Himmelskörper, geht. Studien zweier Sichtlinien-Kriterien zeigen, daß sie deutliche Vorteile gegen die konventionellen Methoden haben. Bei der ersten Methode muß die Sichtlinie vom Raumschiff zur Sonne ko-linear mit der Sichtlinie vom Fahrzeug zum Bestimmungs-Planeten verlaufen. Bei der zweiten Methode muß das Raumschiff in einer Linie mit dem Planeten und einem festgesetzten Himmelskörper bleiben. Beide Methoden reduzieren die Anzahl von Meß- und Berechnungsausrüstung, die im Raumschiff mitgetragen werden muß, und vereinfachen im allgemeinen den ganzen Navigationsprozeß. Ein Beispiel einer möglichen Erde-Mars-Bahn nach der ersten Methode zeigt, daß nur ein wenig mehr Energie wie für die HOHMANN-Ellipse benötigt wird. Zwei Beispiele nach der zweiten Methode zeigen ihre Anwendbarkeit. Sichtlinien-Techniken sind nur für Raumschiffe anwendbar, die mit kontinuierlichen Motoren mit geringem Schub angetrieben sind. Die höchste Dauerbeschleunigung, die für den Beispielsfall benötigt wird, ist geringer als 60 Mikro-*g*.

[1] Willow Run Laboratories, The University of Michigan, Ann Arbor, Michigan, U.S.A.

Critères de ligne de visée pour navigation interplanétaire. Au lieu des techniques usuelles impliquant un calcul explicite de l'orbite suivie ou à suivre, il est possible de naviguer le long d'une trajectoire définie implicitement par une ligne de visée entre l'astronef et une référence adéquate telle que le soleil ou un autre point de la sphère céleste. Deux critères particuliers possèdent des avantages bien définis sur les méthodes conventionnelles. Dans le premier les lignes de visée du véhicule à la planète de destination et au soleil sont maintenues colinéaires, dans le second le véhicule reste aligné sur une droite joignant la planète de destination et un point fixe de la sphère céleste. Les deux méthodes réduisent le volume de l'équipement de mesure et de calcul à emporter et simplifient l'ensemble du processus intermédiaire de navigation. Une orbite Terre-Mars générée par la première technique requiert une énergie à peine supérieure à celle d'une ellipse de Hohmann. Deux exemples de la seconde technique illustrent ses possibilités d'application. Ces techniques ne sont applicables qu'aux véhicules à faible propulsion continue. L'accélération continue requise dans le cas traité est inférieure à 60 micro-g.

The choice of a principle of navigation for interplanetary space travel is usually guided by the search for a minimum-fuel orbit, or for a minimum-fuel correction for navigational errors. Must it be so? From the navigator's viewpoint, space flight can be simplified considerably through the use of what we call here line-of-sight criteria. Other proposed navigational schemes require a space vehicle

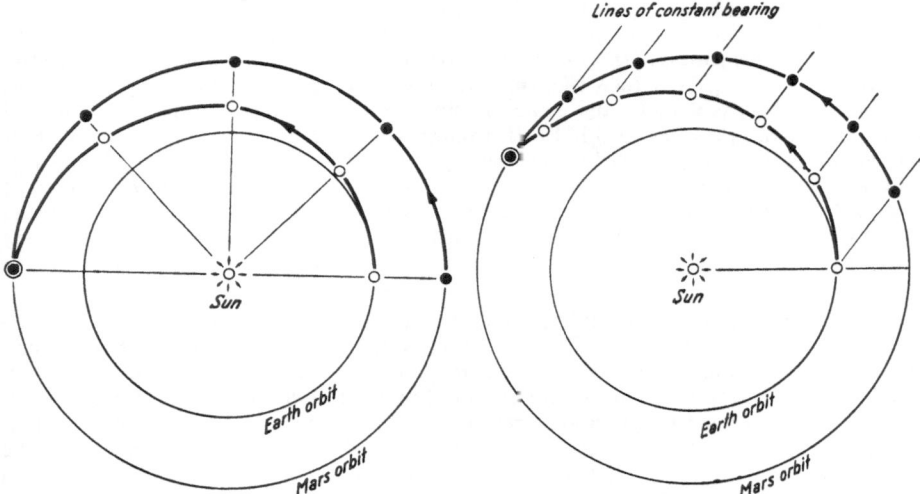

Fig. 1. Constant-angular-velocity trajectory. ○ Vehicle positions, ● Mars positions   Fig. 2. Constant-bearing trajectory. ○ Vehicle positions, ● Mars positions

to follow a precomputed trajectory, or to compute a new trajectory in flight. In contrast, line-of-sight techniques require only that the vehicle remain on a specific line in space.

For example, the line between the Sun and Mars rotates around the Sun as Mars rotates—a vehicle which remained on this line, and which supported an outwardly directed velocity component, would eventually reach Mars (Fig. 1). The condition on this vehicle would be that its angular velocity be always equal to Mars' angular velocity, which for purposes of this discussion we consider constant—and therefore call this the constant-angular-velocity technique.

Or, for another example, suppose that a space vehicle applies thrust to insure that the apparent longitude of Mars is constant; in other words, that the position of Mars as seen from the vehicle remains fixed against the stellar background (Fig. 2). With this condition, and the restriction that the vehicle always approach Mars, interception is again assured.

A space vehicle using either of these navigational criteria could navigate from Earth to Mars, or with obvious modifications from any planet to any other planet, with only the simplest of navigational equipment. The vehicle needs never determine its own position or velocity. Exact time is of little consequence. No planetary or vehicle orbits need be computed. No tables of the positions of celestial bodies are required. The only equipment needed is tracking equipment, to establish the line-of-sight from the vehicle to Mars and the line-of-sight from the vehicle to the Sun or to a star, and thrust control equipment, to apply a corrective thrust when the two lines of sight drifted out of coincidence. Specified permissible directions and magnitudes of thrust would insure that a component of the vehicle velocity was always directed toward Mars.

These minimal requirements for measurements and computation mean less equipment, less fuel, more payload, and more over-all reliability. Furthermore, suitable equipment for tracking stars and planets exists today, when equipment for measuring celestial distances and space ship velocities needs further development. The simplification of the navigational process and the availability of equipment constitute the primary advantages of line-of-sight techniques.

The drawbacks are two: Paths generated by line-of-sight criteria require in general somewhat more fuel than optimum-fuel trajectories, and they require a continuous, low-thrust powerplant which is not yet available. By way of compensation for these drawbacks, the reduction of the amount of equipment carried not only saves the fuel required to propel the equipment, but saves secondary power, such as that required to operate computers. And finally, ion and other low thrust powerplants are being developed now.

The preliminary investigation of two line-of-sight techniques presented here shows that they do not require unreasonably large power plants. For this preliminary evaluation, we are interested only in feasibility—the practical utility of the results will be to point out a direction for further research. The solutions presented apply to coplanar, circular, planetary orbits and to constant-angular-velocity planets.

Sun-Mars Line

The first technique requires the vehicle to maintain an angular velocity equal to Mars angular velocity. One further restriction serves the dual purpose of insuring that the vehicle continuously approaches Mars and simplifying the mathematical derivation. The restriction is: the applied thrust is assumed perpendicular to the line between the sun and the vehicle. There exists, therefore, no radial thrust and the radial acceleration is given by the sum of the Sun's attraction and the centrifugal reaction, with Mars' attraction neglected.

$$\frac{d^2 r_v}{dt^2} = r_v \left(\frac{d\theta_v}{dt}\right)^2 - \frac{K_s^2}{r_v^2} \tag{1}$$

in which

r_v = distance from the Sun to the vehicle,
θ_v = angular position of the vehicle, and
K_s^2 = solar gravitational constant.

Integrating the radial acceleration, and applying the condition that the radial acceleration is zero when the vehicle is at the same distance from the Sun as Mars, yields an expression for the radial velocity as a function of radial distance, which is plotted in Fig. 3.

When $r_v = r_m$, $\dfrac{dr_v}{dt} = 0$

$$\frac{dr_v}{dt} = \sqrt{\left(\frac{d\theta_v}{dt}\right)^2 (r_v{}^2 - r_m{}^2) + 2K_s{}^2 \left(\frac{1}{r_v} - \frac{1}{r_m}\right)} \tag{2}$$

in which

r_m = distance from the Sun to Mars (assumed constant).

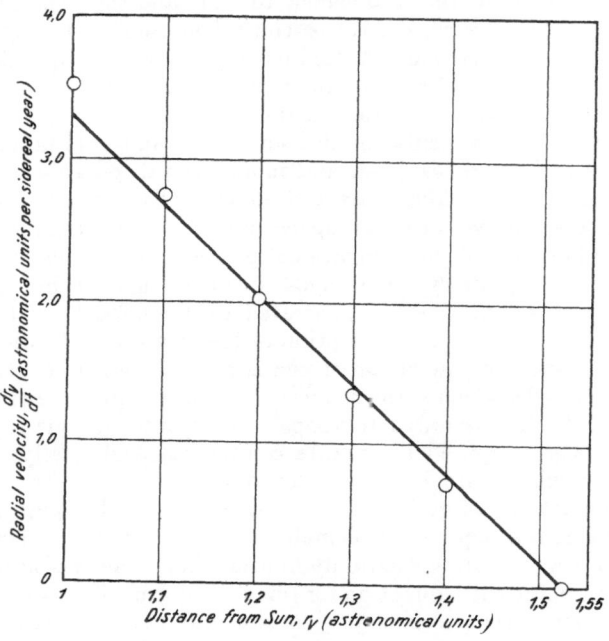

Fig. 3. Radial velocity vs distance from Sun. Constant-angular-velocity technique

The radial velocity, $\dfrac{dr_v}{dt}$, turns out to be an almost linear function of radius, and can be so approximated:

$$\frac{dr_v}{dt} \approx \frac{r_m - r_v}{0.159} \frac{\text{astronomical units}}{\text{sidereal year}}. \tag{3}$$

The thrust required to maintain the vehicle on the line connecting Mars and the Sun is calculated from the transverse force equation with constant angular velocity,

$$A_t = \frac{F}{M_v} = \frac{1}{r_v} \frac{d}{dt}\left(r_v{}^2 \frac{d\theta}{dt}\right)$$

in which

A_t = applied acceleration

$\dfrac{d\theta_v}{dt}$ = vehicle's angular velocity = Mars angular velocity =

= 3.34 radians per sidereal year

$$A_t = \left(\frac{1}{r_v}\right)\left(2\,r_v\,\frac{dr_v}{dt}\right) \text{ (3 34)}$$

$$A_t = 6.68\,\frac{dr_v}{dt}\;\frac{\text{astronomical units}}{\text{(sidereal year)}^2} \tag{4}$$

The applied acceleration is a linear function of the radial velocity, $\frac{dr_v}{dt}$. Thus, the linear graph of radial velocity, with a charge of units, serves as a graph of the applied acceleration for the trip (Fig. 4). The acceleration required decreases as the vehicle approaches Mars; at the start of the trip it is rather high.

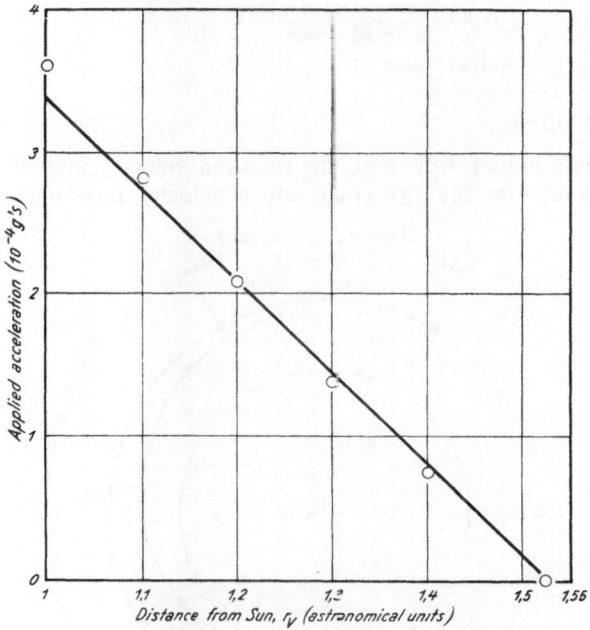

Fig. 4. Applied acceleration vs distance from Sun. Constant-angular-velocity technique

A method for overcoming the excessive acceleration required early in the trip is to start on a free-fall elliptical orbit, chosen to produce an intersection with the line-of-sight trajectory at some time during the trip. If the initial guidance is somewhat inaccurate and the initial ellipse is not exactly as planned, no great harm is done—the vehicle will make its corrective maneuvers when it reaches line-of-sight coincidence. At that time it will begin to apply the corrective thrusts necessary to maintain the line-of-sight coincidence. One possible initial ellipse is tangent to the Earth orbit and intersects the line-of-sight trajectory some 150 degrees after takeoff at a distance from the sun of approximately 1.43 astronomical units [1] (Fig. 5). At 1.43 astronomical units, the acceleration required to maintain the line-of-sight trajectory is 5.63×10^{-5} g's, which is well within the capabilities of proposed continuous-thrust powerplants. The acceleration required diminishes from that value to zero at interception.

Another useful measure of power requirements is the total velocity increment, defined as the time integral of the applied acceleration:

$$\Delta V = \int \frac{T}{M_v}\,d\,t \tag{5}$$

in which

ΔV is the velocity increment,
T is thrust, and
M_v is the mass of the vehicle

or, from Eq. (4),

$$\Delta V = 6.68 \int_{r_s}^{r_m} \frac{dr_v}{dt}\, dt$$

$$\Delta V = 6.68\, (r_m - r_s)\ \frac{\text{astronomical units}}{\text{sidereal year}}$$

$$\Delta V = 3.5\ \frac{\text{astronomical units}}{\text{sidereal year}}$$

$$\Delta V = 16.6\ \text{km/sec.}$$

For the complete line-of-sight trip, the required velocity increment is 16.6 kilometers per second—for the two stage trip a velocity increment of 3.04 km/sec

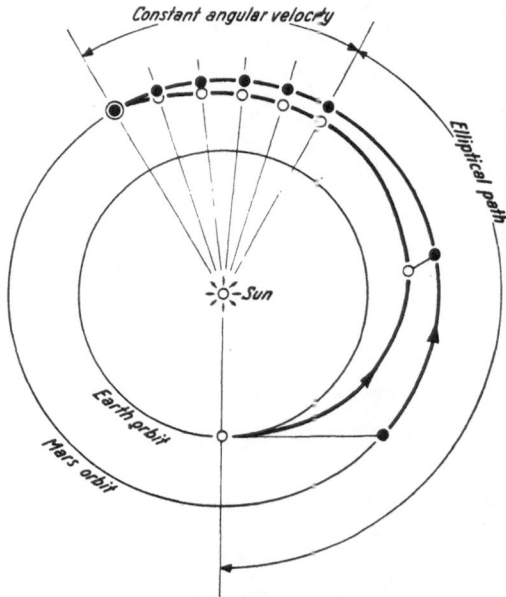

Fig. 5. Combined trip—elliptical and constant-angular velocity trajectories. ○ Vehicle positions,
● Mars positions

suffices to put the vehicle into its heliocentric elliptical orbit and only 2.92 km/sec
are required to get the vehicle to within one million kilometers of Mars, a total
of about 6 km/sec for the trip[1].

The constant-angular-velocity case is thus within the limits of feasible total
velocity increment and feasible maximum acceleration.

---

[1] Significant advantages accrue to a trip which ends at the point-of-no-force, the
point at which the gravitational attraction of the sun is equal to the sum of the gravitational attraction of Mars and the centrifugal reaction of the space vehicle. This point
is approximately one million kilometers from Mars. The subject is treated more fully
for this trajectory in [1].

Constant Bearing

Consider now the constant-bearing method, which is similar to the method used by nautical navigators to determine whether they are on a collision course with another ship. A ship whose bearing from your ship remains constant will eventually collide with you, an event less attractive to seagoing navigators than to astronauts, who seek as their goal a near collision with their destination.

The suitable choice of coordinates helps to provide a qualitative description of a constant-bearing trajectory (Fig. 6). Select a polar coordinate system such

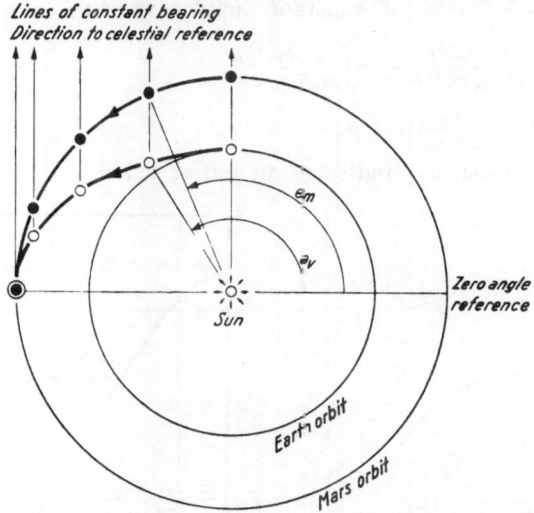

Fig. 6. Geometry of constant-bearing trajectory. ○ Vehicle positions, ● Mars positions

that the line of constant bearing is 90 degrees from zero reference. The starting and ending points of the journey can be selected arbitrarily by selecting starting and ending positions of Mars. The trajectory selected cannot pass through the zero degree or 180 degree points unless the vehicle travels outside Mars orbit, for the constant-bearing line lies completely outside Mars orbit at those points. The general condition implied by the geometry of this technique is that the cosines of the angles of the vehicle and Mars be in the inverse ratio of their radial distances from the Sun.

$$r_v \cos \theta_m = r_v \cos \theta_m$$

in which

r refers to distance from the sun
θ refers to angle from the reference
and the subscripts $_v$ and $_m$ refer to the vehicle and Mars respectively.

$$r_v = r_m \frac{\cos \theta_m}{\cos \theta_v}. \tag{6}$$

In the constant-bearing case, the derivation is considerably simplified if the vehicle is restricted to radial thrusts only; with this assumption the angular momentum of the vehicle

$$h_v = r_v^2 \frac{d\theta_v}{dt} = \text{constant} \tag{7}$$

is constant and the substitution of the geometric condition into the angular momentum equation and integration produce an expression for the vehicle's angular position in terms of two parameters: Mars' angular position at either the start or the end of the voyage, and the vehicle's constant angular momentum.

$$r_m^2 \frac{\cos^2 \theta_m}{\cos^2 \theta_v} \frac{d\theta_v}{dt} = h_v \tag{8}$$

$$\frac{d\theta_v}{\cos^2 \theta_v} = \frac{h_v \, \omega_m \, dt}{r_m^2 \, \omega_m \cos^2 \theta_m}$$

in which $\omega_m = \dfrac{d\theta_m}{dt} =$ Mars' constant angular velocity

$$\int \frac{d\theta_v}{\cos^2 \theta_v} = \int \frac{h_v \, \omega_m \, dt}{r_m^2 \, \omega_m \cos^2 \theta_m} = \int \frac{h_v \, d\theta_m}{r_m^2 \, \omega_m \cos^2 \theta_m}$$

$$\tan \theta_v - \tan \theta_{vo} = \frac{h_v}{r_m^2 \, \omega_m} [\tan \theta_m - \tan \theta_{mo}] \tag{9}$$

in which the subscript $_o$ indicates an initial value.

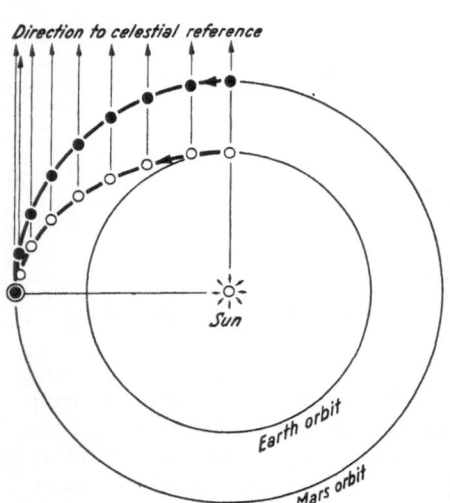

Fig. 7. Longest trajectory with constant angular momentum. ○ Vehicle positions, ● Mars positions

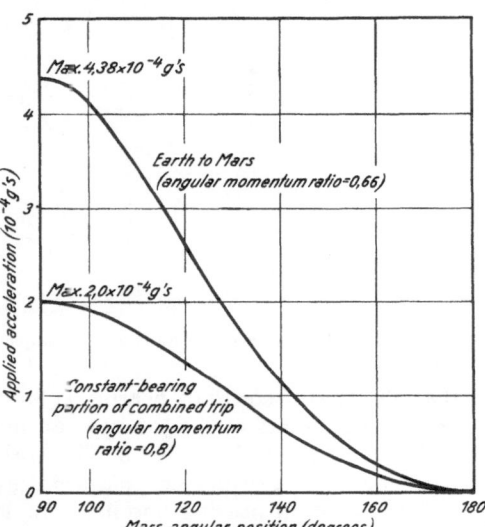

Fig. 3. Applied acceleration vs Mars angular positions. Constant-bearing trajectory

A parameter which turns out to be very useful to describe constant-bearing orbits is the ratio of the angular momentum of the vehicle to that of Mars, here called the ratio R.

$$R = \frac{h_v}{r_m^2 \, \omega_m} = \frac{\tan \theta_v - \tan \theta_{vo}}{\tan \theta_m - \tan \theta_{mo}}. \tag{10}$$

A single orbits is defined by two parameters out of three, R and the angular position of Mars at takeoff or at interception. R is, of course, a constant.

A vehicle following the line-of-sight criterion with constant angular momentum will trail Mars in angle in the first quadrant, be exactly at Mars angle at 90° and lead Mars' in the second quadrant (Fig. 6). (The vehicle never reaches the third or fourth quadrants without going outside Mars' orbit.) Thus the angular velocity of the vehicle is slightly greater than Mars' angular velocity just before 90 degrees and slightly less just after 90 degrees. The vehicle cannot pass through the 90-

degree position if its angular momentum (and thus R) remains constant. The longest interesting orbit is therefore the orbit from 90 degress to 180 degrees,

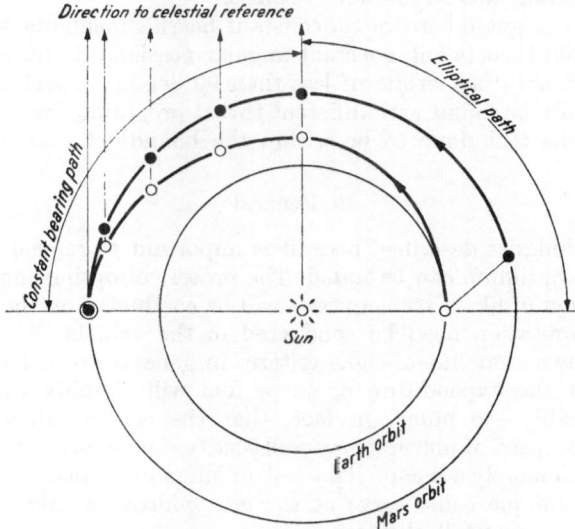

Fig. 9. Combined trip—elliptical and constant—bearing trajectories. ○ Vehicle positions, ● Mars positions

which is completely specified by these two end points (Fig. 7). The ratio R turns out to be the ratio of the distances of the vehicle and Mars from the Sun at take-off—in this case the ratio of the distances of Earth and Mars—about $^2/_3$.

In the appendix, an expression for the applied acceleration is derived; the plot of Fig. 8 shows this acceleration versus time. In common with the constant-angular-velocity technique, the constant-bearing technique requires high acceleration during the beginning of the trip— 438 micro g's maximum acceleration. The same trick can be played with the constant-bearing technique as with the constant-angular-velocity technique however—takeoff on a free-fall ellipse with power applied at line-of-sight coincidence. For example, a trip which began at the 90-degree point but instead of at the Earth's orbit at a point halfway between Earth's and Mars'

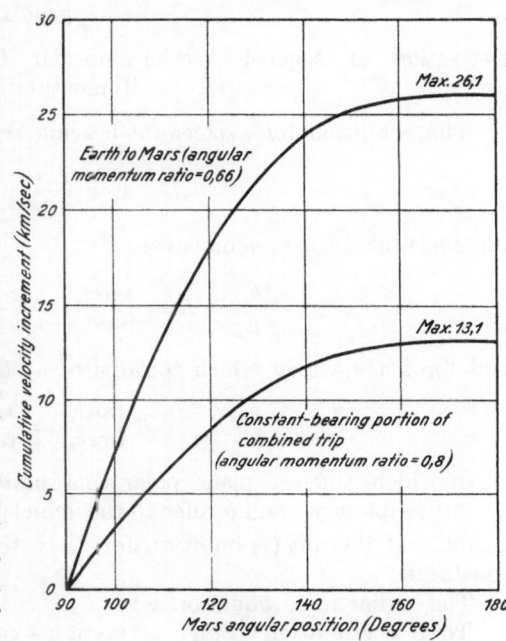

Fig. 10. Cumulative velocity increment vs Mars angular position. Constant-bearing trajectory

orbits could be achieved by a takeoff at zero degrees on a free-fall ellipse (Fig. 9). At ninety degrees, and for the remaining portion of the trip, R would be $^4/_5$ and acceleration required would drop by about a factor of two. Maximum acceleration still occurs at the beginning of the powered portion, but is here only 200 micro g's. Velocity increment for both cases is shown in Fig. 10. The power requirements are now approaching more reasonable values

The orbits investigated here for the constant-bearing technique were artificially restricted to radial thrusts and constant angular acceleration for simplicity—the restriction allows only fast orbits of less than 90 degrees travel. More optimum orbits can certainly be found with different thrust programs, but this preliminary analysis shows the technique to be within the bounds of reason.

In General

For both techniques described here, it is important to remember that thrust criteria closer to optimum can be found. The process of optimizing—even by numerical integration of likely trajectories—will be conducted on the ground before takeoff; no computation need be conducted in the vehicle.

We have shown that line-of-sight criteria in general are not completely unreasonable; that the expenditure of extra fuel will simplify the navigational process considerably—so much, in fact, that the compensation in weight of equipment and in speed of journey may well justify the method; that some further restriction beyond merely a line-of-sight restriction must be placed on a trajectory; and that the technique could serve as the late midcourse guidance for a flight which started as a free-fall ellipse.

Appendix

Derivation of Applied Acceleration for Constant-Bearing, Constant-Angular-Momentum Case

The condition for a constant-bearing trajectory is:

$$r_v = \frac{r_m \cos \theta_m}{\cos \theta_v} ; \tag{6}$$

for constant angular momentum:

$$\frac{r_v^2 \dot{\theta}_v}{r_m^2 \dot{\theta}_m} = R = \frac{\tan \theta_v - \tan \theta_{vo}}{\tan \theta_m - \tan \theta_{mo}} \quad (R \text{ constant}); \tag{10}$$

and for a trajectory which terminates at $\theta_v = \theta_m = 180°$:

$$R \frac{\tan \theta_v}{\tan \theta_m} = \frac{\tan \theta_{vo}}{\tan \theta_{mo}} \tag{A-1}$$

in which r, θ are plane polar coordinates.

Subscripts v, m, and o refer to the vehicle, Mars, and original position respectively, and the dot ($\cdot$) notation for derivatives is used. Note that $\dot{\theta}_m$ and R are constants.

The radial force equation is:

Total acceleration = Sun's attraction + centrifugal + applied
reaction   thrust

$$m_v \ddot{r}_v = - m_v \frac{GM_s}{r_v^2} + m_v r_v \dot{\theta}_v^2 + m_v A_t$$

which yields

$$A_t = \ddot{r}_v + \frac{GM_s}{r_v^2} - r_v \dot{\theta}_v^2 \tag{A-2}$$

in which G is the gravitational constant and M_s is the mass of the Sun. The total vehicle acceleration, $\ddot{r}_v$, is derived from Eqs. (6) and (A-1)

$$\dot{r}_v = r_m \left[\tan \theta_v \frac{\cos \theta_m}{\cos \theta_v} \dot{\theta}_v - \frac{\sin \theta_m}{\cos \theta_v} \dot{\theta}_m \right]$$

$$\dot{\theta}_v = R \frac{\cos^2 \theta_v}{\cos^2 \theta_m} \dot{\theta}_m \tag{A-3}$$

$$\dot{r}_v = r_m \dot{\theta}_m \left(R \frac{\sin \theta_v}{\cos \theta_m} - \frac{\sin \theta_m}{\cos \theta_v} \right)$$

$$\ddot{r}_v = r_m \dot{\theta}_m{}^2 \left(R^2 \frac{\cos^3 \theta_v}{\cos^3 \theta_m} - \frac{\cos \theta_m}{\cos \theta_v} \right)$$

$$\ddot{r}_v = r_m \dot{\theta}_m{}^2 \left(R^2 \frac{r_m{}^3}{r_v{}^3} - \frac{r_v}{r_m} \right). \tag{A-4}$$

The assumption of constant angular velocity in a circular orbit for Mars yields:

$$r_m \dot{\theta}_m{}^2 = \frac{G M_s}{r_m{}^2} . \tag{A-5}$$

(A-3), (A-4), and (A-5) substituted into Eq. (A-2) yield:

$$A_t = \frac{G M_s}{r_m{}^2} \left(\frac{r_m{}^2}{r_v{}^2} - \frac{r_m}{r_v} \right) \tag{A-6}$$

which is the required expression.

Fig. 8 is a graph of Eq. (A-6) and Fig. 10 a graph of the integral of Eq. (A-6) obtained by graphical integration.

Reference

1. W. R. DE HART, Navigational Techniques for Interplanetary Space Flight. Report 2752-15-F. University of Michigan, Willow Run Laboratories, May 1959.

Etude théorique et expérimentale des mélanges propergoliques

Mesure des rendements de combustion

Par

Lucien Reingold[1]

(Reçu le 8 juin 1959)

Résumé — Zusammenfassung — Abstract

Etude théorique et expérimentale des mélanges propergoliques. Mesure des rendements de combustion. On étudie les réactions de combustion en utilisant les notions de dilution relative pour le mélange carburant-comburant et de fractions relatives pour le comburant et le combustible incomplètement utilisés.

On montre que le rendement de combustion peut alors s'exprimer en fonction de deux nombres:

1. La fraction relative de combustion inefficace θ_O;-
2. La fraction relative de comburant inutilisé θ_A.

Le rendement de combustion, tel qu'il est défini théoriquement, s'exprime sous la forme:

$$r_O = f\left[\theta_{O1}\left(\theta_A - \theta_e\right)\right] .$$

Les deux nombres θ_A et θ_O sont déterminés au moyen de mesures expérimentales classiques: ce sont les mesures des concentrations relatives dans les gaz brûlés, du CO, du CO_2 et de l'oxygène. On peut ainsi distinguer séparément les conditions de fonctionnement:

1. Du dispositif d'injection du combustible, caractérisé par la valeur de θ_O;
2. Du dispositif d'introduction du comburant, caractérisé par la valeur θ_A.

Cette méthode de recherche théorique et expérimentale permet de déterminer les conséquences d'une modification de l'un d'eux sur l'autre des deux dispositifs constituant un foyer.

Messung des Verbrennungswirkungsgrades propergolischer Gemische. Die Verbrennungsreaktionen werden untersucht, indem man die Begriffe der relativen Verdünnung für das Gemisch aus Brennstoff und Verbrennungsstoff und der Verhältnisanteile für den Verbrennungsstoff und den unvollständig ausgenutzten Brennstoff benützt.

Man kann zeigen, daß sich der Verbrennungswirkungsgrad ausdrücken läßt als Funktion zweier Zahlenwerte, nämlich:

1. des Bruchteiles der unvollständigen Verbrennung θ_O;
2. des Bruchteiles von überschüssigem Verbrennungsstoff θ_A.

Der Verbrennungswirkungsgrad läßt sich gemäß seiner theoretischen Definition dann in folgender Form schreiben:

$$r_O = f\left[\theta_{O1}\left(\theta_A - \theta_O\right)\right].$$

---

[1] Ingénieur — Docteur — E.C.P.; Attaché aux services techniques de l'Armée française. 9, Boulevard Péreire, Paris XVIIe, France.

Die beiden Werte θ_A und θ_C werden mittels klassischer experimenteller Messungen bestimmt, nämlich der Messungen der Gaskonzentrationen im Verbrennungsgemisch, des CO, des CO_2 und des Sauerstoffs. Man kann also, getrennt voneinander, den Einfluß der beiden folgenden Vorrichtungen [auf die Güte der Verbrennung) unterscheiden:

1. der Einspritzanordnung des Brennstoffs, die durch den Wert von θ_C erfaßt wird;

2. der Vorrichtung für die Einführung des Verbrennungsstoffes, die durch den Wert von θ_A charakterisiert wird.

Diese gemischt theoretisch-experimentelle Untersuchungsmethode gestattet es, die Folgen einer Abänderung einer dieser beiden Anordnungen, die zusammen eine Brennkammer bestimmen, auf die andere zu erkennen.

Measurement of Combustion Performance of Propellant Mixtures. A study is made of combustion reactions utilizing the relative dilution conception for the fuel-oxidizer mixture and of relative fractions of the incompletely used fuel and oxidizer.

It is shown that combustion can then be expressed as a function of two fractions:
1. the relative fraction of combustion inefficiency θ_C;
2. the relative fraction of unused oxidizer θ_A.

Combustion performance, such as it is theoretically defined, is expressed in the form

$$r_C = f \left[\theta_{C1} \left(\theta_A - \theta_C \right) \right].$$

The two terms θ_A and θ_C are determined by means of classical experimental measurements: these are measurements of the relative concentrations of CO, of CO_2 and of O_2 in the burned gases. One can thus separately distinguish the functioning conditions:
1. of the fuel injector, characterized by the value of θ_C;
2. of the oxidizer injector, characterized by the value of θ_A.

This method of theoretical and experimental research permits the consequences of modifying one in terms of the other of the two injectors constituting a chamber to be determined.

Résumé des notations et définitions utilisées dans le mémoire

1. Le comburant: $O_2 + \lambda N_2$.
2. Le combustible: $C_n H_m$.
3. Une mole de combustible brûlé parfaitement en mélange théorique donne $\mathfrak{M}$ moles de gaz de combustion à haute température ou $\mathfrak{M}^*$ moles de gaz secs refroidis à 15° C.
4. La concentration relative du CO_2 dans les gaz brûlés refroidis et secs d'une combustion stoéchiométrique est $a_{CO_2}{}^*$.
5. Une mole d'hydrocarbure en mélange pauvre avec dilution relative σ donne $(\mathfrak{M}^*)_p$ moles de gaz de combustion secs en combustion complète.

Elle donne $(\mathfrak{M}^*)_p{}^i$ moles de gaz de combustion secs en combustion quelconque avec:

$(1 - k_{CO_2})$ de CO_2 et k_{CO_2} de CO
$(1 - k_{H_2O})$ de H_2O et k_{H_2O} de H_2
θ_C de combustible inefficace.
θ_A d'air théorique non utilisé.

6. Une mole d'hydrocarbure en mélange riche avec dilution relative $-\varrho$ donne $(\mathfrak{M}^*)_r$ moles de gaz de combustion secs en combustion complète.

Elle donne $(\mathfrak{M})^*_r{}^i$ moles de gaz de combustion secs en combustion quelconque avec:

$1 - k_{CO_2}$ de CO_2 et k_{CO_2} de CO
$1 - k_{H_2O}$ de H_2O et k_{H_2O} de H_2
θ_C de combustible inefficace.
θ_A d'air théorique non utilisé.

7. Une mole d'un combustible complexe $C_n H_m O_p N_r$ en mélange théorique avec le comburant $O_2 + \lambda N_2$ donne en combustion complète $\mathfrak{M}_o{}^*$ moles de gaz de combustion secs refroidis à $15°$ C.

8. La concentration relative du CO_2 dans les gaz brûlés refroidis et secs d'une combustion stoéchiométrique de $C_n H_m O_p N_r$ est $\alpha_{CO_2}{}^*$.

9. Combustion en mélange pauvre pour une mole de combustible complexe:
$(\mathfrak{M}_o{}^*)_p$ moles de gaz de combustion refroidis et secs en combustion complète.
$(\mathfrak{M}_o{}^*)_p{}^i$ moles de gaz de combustion refroidis et secs en combustion incomplète.

10. Combustion en mélange riche pour une mole de combustible complexe:
$(\mathfrak{M}_o{}^*)_r$ moles de gaz de combustion refroidis et secs en combustion complète.
$(\mathfrak{M}_o{}^*)_r{}^i$ moles de gaz de combustion refroidis et secs en combustion incomplète.

11. Les dilutions relatives σ, ϱ et les fractions relatives k_{CO_2}, k_{H_2O}, θ_o et θ_A ont les mêmes significations que pour les hydrocarbures $C_n H_m$.

12. Un mélange en proportion quelconque pour une combustion propergolique est symbolisée d'une manière générale:

$$K_1 . C_n H_m O_p N_r + K_2 [(\mathfrak{R})_\nu (O_2)_\omega (N_2)_\mu] ;$$

dans cette réaction $(\mathfrak{R})_\nu (O_2)_\omega (N_2)_\mu$ est l'oxydant où $\mathfrak{R}$ est un symbole général représentant soit un radical d'hydrocarbure, soit tout autre corps chimique susceptible d'être oxydé ou non.

Exemples d'oxydants: a) H_2O_2 $(\mathfrak{R})_\nu = H_2$
$\omega = 1$
$\mu = 0$

b) NO_3H $(\mathfrak{R})_\nu = H$
$\omega = 1,5$
$\mu = 0,5$

Introduction

Pour exploiter les mesures expérimentales classiques, nous utilisons une méthode de calcul fondée essentiellement sur la notion de *dilution relative* que nous rappelons ci-dessous.

Pour un combustible (C), la réaction stœchiométrique avec le comburant $O_2 + \lambda N_2$ est symbolisée:

$$(C) + A (O_2 + \lambda N_2) = A_1 \cdot CO_2 + A_2 H_2O + A \lambda N_2 .$$

La dilution relative a pour un mélange en proportion quelconque de ce combustible (C) et du comburant $O_2 + \lambda N_2$ est définie par la réaction:

$$(C) + (1+a) A (O_2 + \lambda N_2) = A_1' CO_2 + \ldots + A_2 H_2O + \ldots + A_3' O_2 + (1+a) A \lambda \cdot N_2.$$

Si a est positif, la combustion a lieu en mélange pauvre.
Si a est négatif, la combustion a lieu en mélange riche.

Dans ce qui suit, tous les calculs théoriques d'exploitation des mesures expérimentales classiques, sont effectués en utilisant la dilution relative dont la valeur absolue est symbolisée σ dans le cas des mélanges pauvres et ϱ dans le cas des mélanges riches.

Par analogie, nous utilisons les notions de *fraction relative* d'imbrûlés (k_{CO_2} et k_{H_2O}), de combustible inefficace (θ_o) et de comburant inutilisé (θ_A).

Ces différentes dilutions ou fractions relatives peuvent être calculées à partir de mesures expérimentales classiques (concentrations relatives de CO_2, de CO et d'oxygène des gaz de combustion). Ces mesures ne nécessitent pas d'exposé particulier.

La méthode de calcul permettant l'exploitation de ces mesures classiques fait l'objet du présent mémoire.

Elle permet notamment de déterminer sans mesure de débit, la dilution relative du mélange carburant-comburant des combustions complètes.

Elle permet de plus de calculer la fraction k_{ω_2} de combustible incomplètement oxydée et transformée en CO, ainsi que la fraction k_{H_2O} incomplètement oxydée et transformée en H_2.

Enfin, elle permet de calculer les fractions de combustible inefficace θ_C et de comburant inutilisé θ_A à partir desquelles se calcule le rendement de combustion global $\eta_C = \dfrac{Q_R}{Q_{Th}}$, rapport de la quantité d'énergie effectivement transformée, à la quantité d'énergie totale théoriquement transformable en énergie thermique utilisable.

Pour faciliter l'exposé de ce mode de calcul et des raisonnements qu'il nécessite, nous avons un peu insisté sur ceux-ci dans le chapitre I concernant les cas simples de combustions parfaites d'hydrocarbures en mélange pauvre et en mélange riche.

Le chapitre II concerne l'étude des foyers utilisant un hydrocarbure en mélange pauvre (cas des turbo-réacteurs et des stato-réacteurs).

Le chapitre III est consacré à l'étude des foyers utilisant un hydrocarbure en mélange riche (cas des stato-réacteurs et des fusées).

Dans le chapitre IV, réservé en cas général des combustions propergoliques, on montre comment les calculs théoriques peuvent être effectués en utilisant les calculs théoriques précédemment effectués pour les études d'hydrocarbures.

I. Etude des combustions parfaites d'hydrocarbures

A. Combustion parfaite d'hydrocarbure en mélange pauvre

a) Données théoriques

1. Une combustion parfaite en mélange théorique d'un combustible $C_n H_m$ avec un comburant $O_2 + \lambda N_2$ se symbolise:

$$C_n H_m + \left(n + \frac{m}{4}\right)(O_2 + \lambda N_2) = n\, CO_2 + \frac{m}{2}\, H_2O + \lambda\left(n + \frac{m}{4}\right) N_2\,.$$

2. La combustion d'une mole de combustibles donne $\mathfrak{M}$ moles de gaz de combustion:

$$\mathfrak{M} = n + \frac{m}{2} + \lambda\left(n + \frac{m}{4}\right)$$

d'où les concentrations relatives

$$a_{CO_2} = \frac{n}{\mathfrak{M}} \qquad a_{H_2O} = \frac{1}{2}\frac{m}{\mathfrak{M}}$$

pour le gaz carbonique et la vapeur d'eau. D'où la relation simple existant entre ces deux concentrations relatives.

$$a_{CO_2}(1 + \lambda) + a_{H_2O}\left(1 + \frac{\lambda}{2}\right) = 1\,. \tag{1}$$

3. Dans le cas des mesures expérimentales, on opère sur *gaz secs*. On a, dans ces conditions, les relations suivantes correspondant aux données définies ci-dessus:

$$\mathfrak{M}^* = n + \lambda\left(n + \frac{m}{4}\right)$$

$$a_{CO_2}^* = \frac{n}{\mathfrak{M}^*}$$

et la relation (1) devient:

$$a_{CO_2}* (1 + \lambda) + \frac{1}{2} \lambda \cdot \Delta_{H_2O} = 1 \text{ en posant } \Delta_{H_2O} = \frac{1}{2} \frac{m}{\mathfrak{M}*}. \tag{2}$$

Cette égalité est l'équation caractéristique de l'hydrocarbure $C_n H_m$ pour le comburant $O_2 + \lambda N_2$. De ce qui précède se déduisent deux autres égalités qui seront utilisées ultérieurement:

$$a_{CO_2}* + \frac{1}{2} \Delta_{H_2O} = \frac{1 - a_{CO_2}*}{\lambda} \tag{3}$$

$$a_{CO_2} = a_{CO_2}* \frac{1}{1 + \Delta_{H_2O}}. \tag{4}$$

b) Etude préalable d'un hydrocarbure par combustion parfaite en excès d'air

1. On réalise une combustion parfaite en proportion quelconque de comburant et de combustible en excès de comburant suffisant pour que la température des gaz de combustion soit de l'ordre de 1200° K (foyer type chambre de turbo-réacteur) de telle sorte que les phénomènes de dissociation soient négligeables.

La réaction de combustion a la forme générale:

$$C_n H_m + (1 + \sigma) \left(n + \frac{m}{4}\right) (O_2 + \lambda N_2) = n CO_2 + \left[\frac{m}{2} H_2O\right] + \sigma \left(n + \frac{m}{4}\right) O_2 +$$
$$+ (1 + \sigma) \left(n + \frac{m}{4}\right) \lambda N_2.$$

La vapeur d'eau est symbolisée entre crochets, pour mémoire, les mesures expérimentales étant effectuées sur *gaz secs* à 15° C.

2. Pour une mole de combustible brûlé, on a $(\mathfrak{M}*)_p$ moles de gaz de combustion secs, tels que:

$$(\mathfrak{M}*)_p = \mathfrak{M}* + \sigma (1 + \lambda) \left(n + \frac{m}{4}\right). \tag{5}$$

3. Sur les gaz refroidis et secs, en mesure avec des appareils classiques et couramment utilisés dans les laboratoires spécialisés:

α) La concentration relative en CO_2 $(a_{CO_2}*)_p = a$
β) La concentration relative en O_2 $(a_{O_2}*)_p = b$

4. D'après les définitions et rappels théoriques initiaux:

$$(a_{CO_2}*)_p = \frac{n}{\mathfrak{M}* + \sigma (1 + \lambda) \left(n + \frac{m}{4}\right)} \tag{6}$$

$$(a_{O_2}*)_p = \frac{\sigma \left(n + \frac{m}{4}\right)}{\mathfrak{M}* + \sigma (1 + \lambda) \left(n + \frac{m}{4}\right)}. \tag{7}$$

5. Les relations (6) et (7) utilisées en se servant également des relations (2), (3) et (4) permettent d'obtenir le système (8) ci-dessous de 2 équations à 2 inconnues $a_{CO_2}*$ et σ.

$$a = \frac{\lambda a_{CO_2}*}{\lambda + \sigma (1 + \lambda) (1 - a_{CO_2}*)}$$
$$b = \frac{\sigma (1 - a_{CO_2}*)}{\lambda + \sigma (1 + \lambda) (1 - a_{CO_2}*)} \tag{8}$$

$a_{CO_2}*$ et σ étant ainsi déterminées, on calcule Δ_{H_2O} d'après la relation (3).

6. Si l'on remarque, d'après les données théoriques, que

$$\mathfrak{M} - \mathfrak{M}^* = \frac{m}{2}$$

il est facile, en se servant des définitions de a_{CO_2} et de $a_{CO_2}^*$ de voir que:

$$\mathfrak{M} - \mathfrak{M}^* = \frac{m}{2} = n\left(\frac{1}{a_{CO_2}} - \frac{1}{a_{CO_2}^*}\right)$$

qui s'écrit finalement en utilisant l'égalité (4):

$$\frac{m}{n} = 2\frac{\Delta_{H_2O}}{a_{CO_2}^*} \,. \tag{9}$$

7. Il suffit d'avoir la masse molaire du combustible utilisé pour déterminer les valeurs numériques de m et n d'après la relation (9).

8. L'expérimentation consiste d'une part à déterminer les valeurs mesurées a de $(a_{CO_2}^*)_p$ et b de $(a_{O_2}^*)_p$ ainsi que la masse molaire du combustible utilisé.

9. La résolution du système (8) permet de déterminer les valeurs de $a_{CO_2}^*$ et de σ.

On trouve:

$$a_{CO_2}^* = \frac{a}{1 - b(1 + \lambda)} \qquad \sigma = \frac{b\lambda}{1 - b(1 + \lambda) - a} \,.$$

On a ainsi déterminé, sans mesure de débit, la *dilution relative* σ du comburant lors de la combustion.

10. La relation (3) donne la valeur de (Δ_{H_2O}); ce dernier résultat permet en définitive de calculer la valeur du rapport $\frac{m}{n}$ en utilisant l'égalité (9). Tous calculs faits la valeur de ce rapport est donnée par la relation:

$$\frac{m}{n} = \frac{4}{\lambda}\frac{1 - (a + b)(1 + \lambda)}{a} \,. \tag{10}$$

Connaissant la valeur de ce rapport et celle de la masse molaire, on calcule les valeurs respectives de m et de n.

Soit $M_{C_nH_m}$ la masse molaire du combustible étudié.

$$M_{C_nH_m} = 12\,n + m \,.$$

En utilisant cette égalité et la relation (10), on a, tous calculs faits:

$$n = \frac{M_{C_nH_m}}{4}\frac{a\lambda}{1 + a(2\lambda - 1) - b(1 + \lambda)} \tag{11}$$

$$m = M_{C_nH_m} \cdot \frac{1 - (a - b)(1 + \lambda)}{1 + a(2\lambda - 1) - b(1 + \lambda)} \,. \tag{12}$$

B. Combustion parfaite d'hydrocarbure en mélange riche

a) Données théoriques

1. La réaction de combustion parfaite d'un hydrocarbure en mélange riche est symbolisée:

$$C_nH_m + (1 - \varrho)\left(n + \frac{m}{4}\right)(O_2 + \lambda N_2) = n(1 - k_{CO_2})\,CO_2 + n\,k_{CO_2}\,CO +$$

$$+ \left[\frac{m}{2}(1 - k_{H_2O})\,H_2O\right] + \frac{m}{2}\,k_{H_2O}\,H_2 + (1 - \varrho)\left(n + \frac{m}{4}\right)\lambda\,N_2 \,.$$

ϱ est la valeur positive de la dilution relative.

k_{CO_2} et k_{H_2O} représentent respectivement les fractions relatives d'imbrûlés gazeux des systèmes carbone et hydrogène.

La vapeur d'eau est symbolisée entre crochets, pour mémoire, les mesures expérimentales étant effectuées sur *gaz secs* à 15° C.

On admet que la combustion ne donne lieu a aucun dépôt de carbone.

2. Pour une mole de combustible consommé, on a $(\mathfrak{M}*)_r$ moles de gaz de combustion secs:

$$(\mathfrak{M}*)_r = \mathfrak{M}* + \frac{m}{2} k_{H_2O} - \varrho \lambda \left(n + \frac{m}{4}\right) \tag{13}$$

en utilisant les données et les notations précisemment définis.

b) Conditions expérimentales

1. Sur les gaz de combustion refroidis et secs, on mesure avec des appareils classiques et couramment utilisés dans les laboratoires spécialisés:

α) La concentration relative en CO_2 $(a_{CO_2}*)_r = a_1$

β) La concentration relative en CO $(a_{CO}*)_r = g$

2. On a ainsi immédiatement la valeur de k_{CO_2}:

$$k_{CO_2} = \frac{g}{a_1 + g} . \tag{14}$$

3. En se référant aux définitions et aux égalités (2) et (3) du chapitre I, un calcul simple permet de voir que:

$$g = \frac{a_{CO_2}* \dfrac{g}{a_1 - g}}{1 + k_{H_2O} \cdot \Delta_{H_2O} - \varrho \lambda \left(a_{CO_2}* + \dfrac{1}{2} \Delta_{H_2O}\right)}$$

d'où l'on tire:

$$\varrho (1 - a_{CO_2}*) = 1 + k_{H_2O} \cdot \Delta_{H_2O} - \frac{a_{CO_2}*}{a_1 + g} . \tag{15}$$

4. Posons:

$$h = \frac{m}{2} \frac{k_{H_2O}}{(\mathfrak{M}*)_r} . \tag{16}$$

En opérant comme ci-dessus, on obtient:

$$h = \frac{\Delta_{H_2O} \cdot k_{H_2O}}{1 + k_{H_2O} \cdot \Delta_{H_2O} - \varrho \lambda \left(a_{CO_2}* + \dfrac{1}{2} \Delta_{H_2O}\right)}$$

d'où l'on tire:

$$\varrho (1 - a_{CO_2}*) = 1 - k_{H_2O} \Delta_{H_2O} \frac{1 - h}{h} . \tag{17}$$

Des relations (15) et (17), il résulte:

$$k_{H_2O} \cdot \Delta_{H_2O} = h \frac{a_{CO_2}*}{a_1 + g} . \tag{18}$$

Les relations (14) et (18) permettent de calculer:

$$\frac{k_{H_2O}}{k_{CO_2}} = h \frac{a_{CO_2}*}{\Delta_{H_2O}} . \tag{19}$$

5. Tenant compte de l'égalité du nombre d'atomes d'oxygène dans les corps réagissants et les corps réactants, on obtient, tous calculs faits:

$$a_{CO_2}{}^* \cdot k_{CO_2} + \Delta_{H_2O} \cdot k_{H_2O} = 2\,\varrho\left(a_{CO_2}{}^* + \frac{1}{2}\,\Delta_{H_2O}\right). \qquad (20)$$

Les relations (19) et (20) permettent de calculer k_{H_2O} et h en tenant compte des relations (17) et (18). On trouve tous calculs faits:

$$h = \frac{2\,(a_1 + g) - a_{CO_2}{}^*\,(2 + g\lambda)}{a_{CO_2}{}^*\,(\lambda - 2)} \qquad (21)$$

$$k_{H_2O} = \frac{\lambda}{2}\,\frac{2\,(a_1 + g) - a_{CO_2}{}^*\,(2 + g\lambda)}{(a_1 + g)\,(\lambda - 2)\,[1 - a_{CO_2}{}^*\,(1 + \lambda)]}\ . \qquad (22)$$

6. Finalement, en portant dans l'égalité (15) la valeur de k_{H_2O} donné par l'égalité (22), on a

$$\varrho = \frac{\lambda}{\lambda - 2}\,\frac{a_1 + g - a_{CO_2}{}^*\,(1 + g)}{(a_1 + g)\,(1 - a_{CO_2}{}^*)}\ . \qquad (23)$$

Ainsi, connaissant $a_{CO_2}{}^*$ et Δ_{H_2O}, calculés d'après les expérimentations préalables décrites précédemment et en effectuant les mesures expérimentales des concentrations relatives d'anhydride carbonique et d'oxyde de carbone de la combustion en mélange riche, on peut déterminer les conditions exactes dans lesquelles cette dernière a été effectuée en calculant ϱ par la relation (23), k_{CO_2} et k_{H_2O} par les relations (14) et (22) ainsi que la concentration d'hydrogène imbrûlé h par la relation (21).

7. Il faut remarquer qu'au cas où un recoupement expérimental pourrait être effectué en mesurant h, on pourrait déterminer directement

$$k_{H_2O} = \frac{1}{2}\,\frac{h}{a_1 + g}\,\frac{\lambda\,a_{CO_2}{}^*}{1 - a_{CO_2}{}^*\,(1 + \lambda)}$$

$$\varrho = \frac{1}{1 - a_{CO_2}{}^*}\left(1 - a_{CO_2}{}^*\,\frac{1 - h}{a_1 + g}\right).$$

8. Pour terminer on voit que, si un combustible de masse molaire connue, a été étudié d'abord en excès d'air pour déterminer sa formule élémentaire ainsi que $a_{CO_2}{}^*$ et Δ_{H_2O}, on peut étudier complètement toute combustion en mélange riche de ce combustible, en calculant la dilution relative et les fractions relatives d'imbrûlés gazeux de carbone et d'hydrogène de cette combustion si elle est parfaite.

II. Etude d'une combustion incomplète d'hydrocarbure en mélange pauvre

La combustion incomplète d'hydrocarbure en mélange pauvre correspond au cas des chambres de combustion de turbo-réacteur et de stato-réacteur.

a) Données théoriques

1. La réaction de combustion incomplète d'un hydrocarbure en mélange pauvre est symbolisée:

$$C_n H_m + (1 + \sigma)\left(n + \frac{m}{4}\right)(O_2 + \lambda\,N_2) = (1 - \theta_c)\,n\,[(1 - k_{CO_2})\,CO_2 + k_{CO_2}\,CO] +$$

$$+ (1 - \theta_c)\,\frac{m}{2}\,[(1 - k_{H_2O})\,H_2O + k_{H_2O}\,H_2] + \left(n + \frac{m}{4}\right)(\theta_A + \sigma)\,O_2 +$$

$$+ \left(n + \frac{m}{4}\right)(1 + \sigma)\,\lambda\,N_2 + n\,\theta_c\,[CH_r]\,.$$

σ est la valeur de la dilution relative.

k_{CO_2} et k_{H_2O} représentent respectivement les fractions relatives d'imbrûlés gazeux des systèmes carbone et hydrogène, comme précédemment.

La vapeur d'eau, soulignée en traits interrompus, est indiquée pour mémoire, les mesures expérimentales étant effectuées sur gaz secs à 15° C.

θ_σ est la *fraction relative* de combustible non efficace, c'est à-dire qui n'a pas été effectivement consommée et transformée en CO_2 ou CO et en H_2O ou H_2.

θ_A est la *fraction relative* d'air théorique non utilisé par suite de la combustion incomplète.

Comme précédemment, au chapitre II, on admet que la combustion imparfaite n'a donné lieu à aucun dépôt de carbone. Les divers imbrûlés gazeux sont bloqués sous le symbole $[CH_\nu]$.

2. Pour une mole de combustible consommé, on a $(\mathfrak{M}^*)_p{}^i$ moles de gaz de combustion secs:

$$(\mathfrak{M}^*)_p{}^i = (\mathfrak{M}^*)_p + \theta_A \left(n + \frac{m}{4}\right) \tag{24}$$

à noter de plus que dans ces $(\mathfrak{M}^*)_p{}^i$ moles, on a

$$\left.\begin{array}{l} (1 - \theta_\sigma)\, n\, (1 - k_{CO_2})\ \text{moles de } CO_2 \\[2mm] (1 - \theta_\sigma)\, n\, k_{CO_2}\ \text{moles de CO} \\[2mm] n\, \theta_\sigma\ \text{moles d'imbrûlés gazeux} \end{array}\right\} \tag{25}$$

Leur total est n (d'où la valeur de $(\mathfrak{M}^*)_p{}^i$ ci-dessus) mais leurs concentrations relatives respectives interviendront dans les calculs ultérieurs.

b) Conditions expérimentales

1. Sur les gaz de combustion, refroidis et secs, on mesure avec des appareils classiques et couramment utilisés dans les laboratoires spécialisés:
α) La concentration relative en CO_2 $(a_{CO_2}{}^*)$ réel $= A$
β) La concentration relative en CO $(a_{CO}{}^*)$ réel $= G$
γ) La concentration relative en O_2 $(a_{O_2}{}^*)$ réel $= B$

2. Comme précédemment, on a de suite la valeur de k_{CO_2}:

$$k_{CO_2} = \frac{G}{A + G} . \tag{26}$$

3. En se référant aux définitions et aux égalités (2) et (3) du chapitre I et en faisant des calculs analogues à ceux effectués au chapitre II pour obtenir les égalités (15) et (20), on obtient les relations finales récapitulées ci-dessous:

$$A = \frac{a_{CO_2}(1 - \theta_\sigma)}{1 + \sigma\,(1 + \lambda)\left(a_{CO_2}{}^* + \frac{1}{2}\Delta_{H_2O}\right) + \theta_A\left(a_{CO_2}{}^* + \frac{1}{2}\Delta_{H_2O}\right)} \tag{27}$$

$$B = \frac{\left(a_{CO_2}{}^* + \frac{1}{2}\Delta_{H_2O}\right)(\sigma + \theta_A)}{1 + \sigma\,(1 + \lambda)\left(a_{CO_2}{}^* + \frac{1}{2}\Delta_{H_2O}\right) + \theta_A\left(a_{CO_2}{}^* + \frac{1}{2}\Delta_{H_2O}\right)} \tag{28}$$

$$G = \frac{a_{CO_2}{}^* k_{CO_2}(1 - \theta_\sigma)}{1 + \sigma\,(1 + \lambda)\left(a_{CO_2}{}^* + \frac{1}{2}\Delta_{H_2O}\right) + \theta_A\left(a_{CO_2}{}^* + \frac{1}{2}\Delta_{H_2O}\right)} \tag{29}$$

$$a_{CO_2}{}^* k_{CO_2} + \Delta_{H_2O}\, k_{H_2O} = 2\left(a_{CO_2}{}^* + \frac{1}{2}\Delta_{H_2O}\right)\frac{\theta_A - \theta_\nu}{1 - \theta_\sigma} . \tag{30}$$

De l'examen des relations (27) à (30), il résulte que le calcul de σ par ce système d'équation n'est plus possible du fait de la présence simultanée de $(1 - \theta_c)$ aux numérateurs des égalités (27) et (29).

On est donc amené, pour ces combustions incomplètes en mélange pauvre à mesurer expérimentalement σ par des mesures convenables de débits du combustible et du comburant qui permettent de calculer la dilution relative σ si l'on connait leurs formules élémentaires (documentation classique à partir de l'analyse immédiate et de mesures cryométriques de masses molaires).

On opère les calculs comme indiqué ci-après, la valeur de σ étant considérée comme une donnée:

Des éq. (27), 28) et (29), on tire la valeur de $\dfrac{B}{A + G}$ qui donne une relation entre θ_A et $(1 - \theta_c)$.

Cette relation permet de calculer θ_A au moyen de la relation (27) et $(1 - \theta_c)$ par l'expression de $\dfrac{B}{A + G}$.

La valeur de k_{H_2O} est calculée au moyen de l'éq. (30). On trouve:

$$k_{H_2O} = 2 \frac{a_{CO_2}{}^* + \dfrac{1}{2} \Delta_{H_2O}}{\Delta_{H_2O}} \cdot \frac{\theta_A - \theta_c}{1 - \theta_c} - \frac{a_{CO_2}{}^*}{\Delta_{H_2O}} \cdot \frac{G}{A + G} \cdot \tag{31}$$

Remarque. Les égalités (27), (28) et (29) sont écrites en explicitant les différents termes du dénominateur afin de permettre de reconstituer les calculs si nécessaire.

En fait, ces égalités peuvent s'écrire d'une manière plus simple notamment en mettant en facteur la somme $a_{CO_2}{}^* + \dfrac{1}{2} \Delta_{H_2O}$ et en la remplaçant par sa valeur $\dfrac{1 - a_{CO_2}{}^*}{\lambda}$ tirée de l'éq. (2) caractéristique des hydrocarbures (voir chapitre I).

c) Exploitation des études théoriques et expérimentales

1. Ainsi que le montre l'exposé qui précède, les mesures expérimentales classiques de concentrations de CO_2, CO et O_2 permettent de déterminer les valeurs de k_{CO_2}, k_{H_2O} ainsi que θ_c et θ_A.

En ce qui concerne l'utilisation de ces deux derniers résultats, remarquons qu'ils permettent de déterminer la valeur du rendement de combustion r_c.

Celui-ci se définit de la façon suivante:

$$r_c = \frac{Q_R}{Q_{Th}}$$

en appelant Q_{Th} la chaleur de réaction molaire du combustible et Q_R la chaleur réellement dégagée par la réaction symbolisée au début du présent chapitre.

Q_R se calcule d'après la relation précitée; on a:

$$Q_R = Q_{Th} - n\, k_{CO_2} (1 - \theta_c)\, Q_{CO} - \frac{m}{2} (1 - \theta_c)\, k_{H_2O}\, Q_{H_2} - n\, \theta_c\, Q_{[CH_p]}$$

où Q_{CO} est la chaleur de réaction molaire de l'oxyde de carbone.

Q_{H_2} est la chaleur de réaction molaire de l'hydrogène.

$Q_{[CH_p]}$ est la chaleur de réaction molaire du méthane.

En faisant le quotient $\dfrac{Q_R}{Q_{Th}}$ et en remplaçant k_{CO_2} et k_{H_2O} par leurs valeurs, on trouve, tous calculs faits:

$$r\,c = 1 - A_0\,(1 - \theta_c) - A_1\,(\theta_A - \theta_c) - A_2\,\theta_c \tag{32}$$

dans l'expression (32):

$$A_0 = a_{CO_2}{}^* \left[\frac{Q_{CO}}{(Q_F)_0} - \frac{Q_{CH_2}}{(Q_F)_0} \right] \frac{G}{A+G}$$

$$A_1 = 2 \left(a_{CO_2}{}^* + \frac{1}{2} \Delta_{H_2O} \right) \frac{Q_{H_2}}{(Q_F)_0}$$

$$A_2 = a_{CO_2}{}^* \frac{Q_{[CH_\nu]}}{(Q_F)_0}$$

$(Q_F)_0$ est l'enthalpie d'une mole de gaz de combustion théorique.

Remarque. Dans ce cas, on peut recouper les calculs et l'expérimentation en calculant le rendement suivant l'expression simplifiée connue:

$$r_c = \frac{(a_{CO_2}{}^*) \ \text{réel}}{(a_{CO_2}{}^*) \ \text{théorique}} \ .$$

III. Etude d'une combustion quelconque d'hydrocarbure en mélange riche

La combustion quelconque d'hydrocarbure en mélange riche correspond au cas de foyers de stato-réacteurs et de fusées.

a) Données théoriques

1. La réaction de combustion imparfaite d'un hydrocarbure en mélange riche est symbolisée:

$$C_n H_m + (1 - \varrho) \left(n + \frac{m}{4} \right) (O_2 + \lambda N_2) = n (1 - \theta_c) [(1 - k_{CO_2}) CO_2 + k_{CO_2} CO] +$$

$$+ \frac{m}{2} (1 - \theta_c) [(1 - k_{H_2O}) \underline{\ H_2O} + k_{H_2O} H_2] + \theta_A \left(n + \frac{m}{4} \right) O_2 + n \theta_c [CH_\nu] +$$

$$+ (1 - \varrho) \left(n + \frac{m}{4} \right) \lambda N_2 \ .$$

ϱ est la valeur positive de la dilution relative du comburant.

k_{CO_2} et k_{H_2O} représentent respectivement les fractions relatives d'imbrûlés gazeux des systèmes carbone et hydrogène comme précédemment.

La vapeur d'eau, soulignée en traits interrompus, est indiquée pour mémoire, les mesures expérimentales étant effectuées sur gaz secs à 15° C.

θ_c et θ_A sont comme précédemment et respectivement les fractions relatives de combustible inefficace et d'air theorique non utilisé.

Les divers inbrûlés gazeux sont symbolisés [CH$_\nu$] et l'on admet qu'il n'y a pas de dépôt de carbone.

2. Des calculs, analogues à ceux effectués au chapitre précédent, donnent:

$$(\mathfrak{M}^*)_r{}' = (\mathfrak{M}^*)_r + \theta_A \left(n + \frac{m}{4} \right) - \theta_c \frac{m}{2} k_{H_2O} \ . \tag{33}$$

Comme dans le cas des combustions incomplètes en mélange pauvre, on a θ_c qui s'introduit dans les relations telles que (33); de plus ici, nous avons également θ_A dans cette expression.

Il en résulte que la valeur de la dilution relative du mélange utilisé ne peut pas être calculée: pour les études des combustions incomplètes en mélange riche, il faut mesurer expérimentalement les débits de combustible et de comburant pour obtenir la valeur de ϱ.

3. Dans ces conditions, les calculs d'exploitation des mesures permettent de déterminer les valeurs de k_{CO_2}, k_{H_2O}, θ_c et θ_A comme exposé-ci-après.

b) Conditions expérimentales

1. On mesure les débits de combustible et de comburant pour calculer $-\varrho$ à partir de leurs formules élémentaires connues. Puis on mesure, sur les gaz de combustion, refroidis et secs:

α) La concentration relative du $CO_2 = A$

β) La concentration relative du $CO = G$

γ) La concentration relative d'oxygène $= B$.

2. Les résultats des mesures des concentrations relatives donnent de suite:

$$k_{CO_2} = \frac{G}{A+G} \tag{34}$$

et en exploitant par des calculs déjà exposés les concentrations mesurées G et B, on a encore:

$$\frac{(\mathfrak{M}^*)_r^i}{\mathfrak{M}^*} = 1 - (\varrho\,\lambda - \theta_A)\,\frac{1 - a_{CO_2}{}^*}{\lambda} + k_{H_2O}\,\Delta_{H_2O}\,(1 - \theta_c) \tag{35}$$

l'expression (35) étant utilisée pour établir les deux relations:

$$A + G = \frac{a_{CO_2}{}^* \, (1 - \theta_c)}{1 - (\varrho\,\lambda - \theta_A)\,\dfrac{1 - a_{CO_2}{}^*}{\lambda} + k_{H_2O}\,\Delta_{H_2O}\,(1 - \theta_c)} \tag{36}$$

$$\frac{B}{A+G} = \frac{\theta_A}{1 - \theta_c} \cdot \frac{1 - a_{CO_2}{}^*}{\lambda\,a_{CO_2}{}^*}. \tag{37}$$

On a enfin l'équation résultant de la conservation du nombre d'atomes d'oxygène dans les deux membres de la réaction de combustion:

$$a_{CO_2}{}^*\,k_{CO_2} + k_{H_2O}\cdot\Delta_{H_2O} = 2\,\frac{1 - a_{CO_2}{}^*}{\lambda}\cdot\frac{\varrho - (\theta_c - \theta_A)}{1 - \theta_c}. \tag{38}$$

3. ϱ étant calculé à partir des mesures de débit, les équations (36), (37) et (38) constituent un système de 3 équations à 3 inconnues $1 - \theta_c$, θ_A et k_{H_2O} dont la résolution est classique. La valeur de k_{CO_2} est donnée directement par l'éq. (34).

c) Exploitation des études théoriques et expérimentales

Nous calculons le rendement r_c de combustion en mélange riche en fonction de θ_c et de θ_A, comme dans le chapitre précédent, en partant de la définition théorique.

$$r_c = \frac{Q_R}{Q_{Th}}.$$

Le calcul de Q_R donne comme précédemment:

$$Q_R = Q_{Th} - n\,(1 - \theta_c)\,k_{CO_2}\,Q_{CO} - \frac{m}{2}\,(1 - \theta_c)\,k_{H_2O}\,Q_{H_2} - n\,\theta_c\,Q_{[CH_\nu]}. \tag{39}$$

Tous calculs faits on trouve pour r_c:

$$r_c = 1 - A_0\,(1 - \theta_c) - A_1\,[\varrho - (\theta_c - \theta_A)] - A_2\cdot\theta_c \tag{40}$$

dans cette relation :

$$A_0 = a_{CO_2}{}^* \left[\frac{Q_{CO}}{(Q_F)_0} - \frac{Q_{E_2}}{(Q_F)_0} \right] \frac{G}{A+G}$$

$$A_1 = 2 \left(a_{CO_2}{}^* + \frac{1}{2} \Delta_{H_2O} \right) \frac{Q_{H_2}}{(Q_F)_0}$$

$$A_2 = a_{CO_2}{}^* \frac{Q_{[CH_\nu]}}{(Q_F)_0} .$$

On constate que l'expression (40) de la valeur r_σ est la même que celle (32) de r_σ du chapitre précédent, à cela près que dans l'expression (40) intervient la dilution relative. Les valeurs des coëfficients A_0, A_1 et A_2 sont les mêmes que dans le cas des combustions en mélange pauvre.

IV. Etude théorique et expérimentale des mélanges propergoliques

Un combustible complexe est symbolisé dans ce qui suit $C_n H_m O_p N_r$.

Un comburant complexe est symbolisé $(\mathfrak{R})_\nu (O_2)_\omega (N_2)_\mu$.

On distingue le premier du second par le fait que ce dernier est le corps oxydant : la quantité d'oxygène par mole du corps dit combustible est inférieure à la quantité d'oxygène par mole du corps dit comburant.

$\mathfrak{R}$ est un symbole général représentant soit un radical d'hydrocarbure, soit tout autre corps chimique susceptible d'être oxydé ou non.

Un mélange en proportion quelconque du combustible et du comburant est dit *mélange propergolique*.

Dans ce chapitre, on se propose de présenter les principes d'une étude des réactions de mélanges propergoliques fondée sur la méthode théorique et expérimentale exposée et exploitée au cours des trois chapitres précédents pour l'étude des combustions d'hydrocarbures avec le comburant $O_2 + \lambda N_2$.

Le présent chapitre comporte deux parties :

Dans la première, on se limite à l'étude des combustions d'un combustible complexe $C_n H_m O_p H_r$ avec le comburant $O_2 + \lambda N_2$.

Dans la seconde, on étudie la réaction propergolique générale.

A. Etude théorique et expérimentale de la combustion d'un combustible complexe

a) Données théoriques

1. Une combustion complète en mélange théorique d'un combustible $C_n H_m O_p N_r$ avec un comburant $O_2 + \lambda N_2$ se symbolise :

$$C_n H_m O_p N_r + \left(n + \frac{m}{4} - \frac{p}{2} \right) (O_2 + \lambda N_2) = n\, CO_2 + \left[\frac{m}{2}\, H_2O \right] +$$

$$+ \left(n + \frac{m}{4} - \frac{p}{2} \right) \lambda N_2 + \frac{r}{2} N_2 .$$

2. La combustion d'une mole de combustible donne $\mathfrak{M}_\sigma{}^*$ moles de gaz de combustion secs :

$$\mathfrak{M}_\sigma{}^* = n + \left(n + \frac{m}{4} - \frac{p}{2} \right) \lambda + \frac{r}{2}$$

d'où la concentration relative $a_{CO_2}{}^* = \dfrac{n}{\mathfrak{M}_\sigma{}^*}$ du gaz carbonique ; ce qui donne la relation caractéristique du combustible :

$$a_{\mathrm{CO_2}}{}^* \, (1 + \lambda) + \frac{1}{2} \, \lambda \cdot \delta_{\mathrm{H_2O}} - \lambda \varDelta_{\mathrm{O_2}}{}^* + \varDelta_{\mathrm{N_2}}{}^* = 1 \tag{41}$$

en symbolisant comme précédemment $\delta_{\mathrm{H_2O}} = \dfrac{m}{2 \, \mathfrak{M}_c{}^*}$

et $\varDelta_{\mathrm{O_2}}{}^* = \dfrac{p}{2 \, \mathfrak{M}_c{}^*}$ $\qquad \varDelta_{\mathrm{N_2}}{}^* = \dfrac{r}{2 \, \mathfrak{M}_c{}^*}$

posons: $\zeta^* = \lambda \varDelta_{\mathrm{O_2}}{}^* - \varDelta_{\mathrm{N_2}}{}^*$

on voit que l'éq. (41) peut se mettre sous la forme:

$$a_{\mathrm{CO_2}}{}^* \, (1 + \lambda) + \frac{1}{2} \, \lambda \, \delta_{\mathrm{H_2O}} = 1 + \zeta^* . \tag{42}$$

Cette équation caractéristique des combustibles $C_n H_m O_p N_r$ correspond à l'équation caractéristique (2) des hydrocarbures $C_n H_m$ (chapitre I).

3. Cette éq. (42) peut se mettre sous la forme:

$$(1 + \lambda) \, \frac{a_{\mathrm{CO_2}}{}^*}{1 + \zeta^*} + \frac{1}{2} \, \lambda \, \frac{\delta_{\mathrm{H_2O}}}{1 + \zeta^*} = 1 . \tag{43}$$

En comparant cette éq. (43) à l'éq. (2) du chapitre I, on voit que:

$$\frac{a_{\mathrm{CO_2}}{}^*}{1 + \zeta^*} = a_{\mathrm{CO_2}}{}^* \quad \text{et} \quad \frac{\delta_{\mathrm{H_2O}}}{1 + \zeta^*} = \varDelta_{\mathrm{H_2O}} . \tag{44 et 44 bis}$$

4. D'après cette remarque, on voit que l'étude de la combustion d'un combustible complexe $C_n H_m P_p N_r$ peut être ramenée à celle d'un hydrocarbure $C_n H_m$, en utilisant les calculs exposés aux chapitres précédents:

Connaissant la formule élémentaire[1] du combustible complexe, on calcule $a_{\mathrm{CO_2}}{}^*$, $\delta_{\mathrm{H_2O}}$ et ζ^*. On utilise les relations établies aux chapitres II et III, et y remplaçant $a_{\mathrm{CO_2}}{}^*$ et $\varDelta_{\mathrm{H_2O}}$ par leurs valeurs (44) et (44 bis) ci-dessus.

De plus, pour les dilutions relatives du comburant (σ ou $- \varrho$), il faut les exprimer pour l'hydrocarbure en fonction de la dilution relative réelle du combustible complexe:

Un calcul simple donne:

$$\sigma_{C_n H_n} = \sigma - \frac{(1 + \sigma) \, \varDelta_{\mathrm{O_2}}{}^*}{a_{\mathrm{CO_2}}{}^* + \dfrac{1}{2} \, \delta_{\mathrm{H_2O}}} \tag{45}$$

$$\varrho_{C_n H_m} = \varrho + \frac{(1 - \varrho) \, \varDelta_{\mathrm{O_2}}{}^*}{a_{\mathrm{CO_2}}{}^* + \dfrac{1}{2} \, \delta_{\mathrm{H_2O}}} . \tag{45 bis}$$

b) Combustion quelconque d'un combustible complexe en mélange pauvre

La réaction de combustion pour la dilution σ, est symbolisée:

$$C_n H_m O_p N_r + (1 + \sigma) \left(n + \frac{m}{4} - \frac{p}{2} \right) (O_2 + \lambda N_2) = (1 - \theta_\sigma) \, n \, [(1 - k_{\mathrm{CO_2}}) \, CO_2 + k_{\mathrm{CO_2}} CO] +$$
$$+ (1 - \theta) \, \frac{m}{2} \, [(1 - k_{\mathrm{H_2O}}) \, H_2O + k_{\mathrm{H_2O}} H_2] + \left(n + \frac{m}{4} - \frac{p}{2} \right) (\theta_A + \sigma) \, O_2 +$$
$$+ \left(n + \frac{m}{4} - \frac{p}{2} \right) (1 + \sigma) \, \lambda N_2 + \frac{r}{2} \, N_2 + n \, \theta_\sigma \, [CH_\nu] .$$

---

[1] Cette formule élémentaire s'établit expérimentalement au moyen:
1. d'une analyse immédiate.
2. d'une mesure cryométrique de la masse molaire.

On utilise les éq. (26) à (30) du chapitre II et on trouve après la substitution précitée:

$$k_{CO_2} = \frac{G}{A + G} \tag{46}$$

$$A = \frac{a_{CO_2}{}^* (1 - \theta_C)}{1 + \zeta^* + \sigma_{C_nH_m}(1 + \lambda)\left(a_{CO_2}{}^* + \frac{1}{2}\delta_{H_2O}\right) + \theta_A\left(a_{CO_2}{}^* + \frac{1}{2}\delta_{H_2O}\right)} \tag{47}$$

$$B = \frac{\left(a_{CO_2}{}^* + \frac{1}{2}\delta_{H_2O}\right)(\sigma_{C_nH_m} + \theta_A)}{1 + \zeta^* + \sigma_{C_nH_m}(1 + \lambda)\left(a_{CO_2}{}^* + \frac{1}{2}\delta_{H_2O}\right) + \theta_A\left(a_{CO_2}{}^* + \frac{1}{2}\delta_{H_2O}\right)} \tag{48}$$

$$G = \frac{a_{CO_2}{}^* k_{CO_2}(1 + \theta_c)}{1 + \zeta^* + \sigma_{C_nH_m}(1 + \lambda)\left(a_{CO_2}{}^* + \frac{1}{2}\delta_{H_2O}\right) + \theta_A\left(a_{CO_2}{}^* + \frac{1}{2}\delta_{H_2O}\right)} \tag{49}$$

$$a_{CO_2}{}^* k_{CO_2} + \delta_{H_2O} \cdot k_{H_2O} = 2\left(a_{CO_2}{}^* + \frac{1}{2}\delta_{H_2O}\right)\frac{\theta_A - \theta_c}{1 - \theta_c}. \tag{50}$$

Les équations ci-dessus permettent de calculer les valeurs de k_{CO_2}, k_{H_2O} ainsi que celles de θ_A et θ_c en fonction des mesures expérimentales de concentrations relatives de CO_2, CO et O_2 dans les gaz de combustion d'une combustion quelconque en mélange pauvre d'un combustible complexe $C_nH_mO_pN_r$, après avoir également mesuré expérimentalement la valeur de la dilution relative σ, comme il a été exposé au chapitre II pour les combustions imparfaites en mélange pauvre.

La valeur du rendement de combustion est finalement:

$$r_c = 1 - A_0(1 - \theta_c) - A_1(\theta_A - \theta_c) - A_2\theta_c \tag{51}$$

dans laquelle:

$$A_0 = \frac{a_{CO_2}{}^*}{1 + \zeta^*}\left[\frac{Q_{CO}}{(Q_F)_0} - \frac{Q_{H_2}}{(Q_F)_0}\right]\frac{G}{A + G} \tag{52}$$

$$A_1 = \frac{2}{1 + \zeta^*}\left(a_{CO_2}{}^* + \frac{1}{2}\delta_{H_2O}\right)\frac{Q_{H_2}}{(Q_F)_0} \tag{53}$$

$$A_2 = \frac{a_{CO_2}{}^*}{1 + \zeta^*} \cdot \frac{Q_{[CH_v]}}{(Q_F)_0}. \tag{54}$$

$(Q_F)_0$ est l'enthalpie d'une mole de gaz de combustion pour une mole de combustible $C_nH_mO_pN_r$ consommé.

c) Combustion incomplète d'un combustible complexe en mélange riche

1. La réaction de combustion, pour la dilution $-\varrho$, est symbolisée:

$$C_nH_mO_pN_r + (1-\varrho)\left(n + \frac{m}{4} - \frac{p}{2}\right)(O_2 + \lambda N_2) = (1-\theta_c)\,n\,[(1-k_{CO_2})\,CO_2 + k_{CO_2}CO] +$$

$$+ (1-\theta_c)\frac{m}{4}[(1-k_{H_2O})\,H_2O + k_{H_2O}H_2] + \theta_A\left(n + \frac{m}{4} - \frac{p}{2}\right)O_2 +$$

$$+ (1-\varrho)\left(n + \frac{m}{4} - \frac{p}{2}\right)\lambda N_2 + \frac{r}{2}N_2 + \theta_c\,n\,[CH_v].$$

2. Rappelons que dans le cas des mélanges riches, la dilution $-\varrho$ est déterminée expérimentalement en plus des mesures des concentrations relatives de CO_2 (A), de CO (G) et d'oxygène (B).

3. On utilise les éq. (35) à (39) établies au chapitre précédent et on y fait les substitutions

$$\varrho_{C_nH_m} = \varrho + \frac{(1-\varrho)\,\Delta_{O_2}^{\;*}}{a_{CO_2}^{\;*} + \frac{1}{2}\,\delta_{H_2O}}\,, \quad \frac{a_{CO_2}^{\;*}}{1+\zeta^*} = a_{CO_2}^{\;*}\,, \quad \frac{\delta_{H_2O}}{1+\zeta^*} = \Delta_{H_2O}\,.$$

On obtient ainsi les équations nécessaires aux calculs de k_{CO_2}, k_{H_2O}, θ_A et θ_σ (ce calcul s'effectue comme indiqué au chapître précédent).

$$k_{CO_2} = \frac{G}{A+G} \tag{55}$$

$$A+G = \frac{a_{CO_2}\,(1-\theta_\sigma)}{1+\zeta^* - (\varrho_{C_nH_m}\lambda - \theta_A)\,\dfrac{1+\zeta^* - a_{CO_2}^{\;*}}{\lambda} + k_{H_2O}\delta_{H_2O}\,(1-\theta_\sigma)} \tag{56}$$

$$\frac{B}{A+G} = \frac{\theta_A}{1-\theta_\sigma} \cdot \frac{1+\zeta^* - a_{CO_2}^{\;*}}{a\,\lambda_{CO_2}^{\;*}} \tag{57}$$

$$a_{CO_2}^{\;*} \cdot k_{CO_2} + \delta_{H_2O} \cdot k_{H_2O} = 2\,\frac{1+\zeta^* - a_{CO_2}^{\;*}}{\lambda} \cdot \frac{\varrho_{C_nH_m} - (\theta_\sigma - \theta_A)}{1-\theta_\sigma}\,. \tag{58}$$

Les relations ci-dessus permettent, comme précédemment, de calculer:
les fractions relatives k_{CO_2} et k_{H_2O}
les fractions relatives θ_σ et θ_A
en fonction de la dilution relative — ϱ et des concentrations relatives mesurées expérimentalement du CO_2, du CO et de l'oxygène dans les gaz de combustion du combustible complexe, en mélange riche.

Le rendement de combustion se calcule comme précédemment:

$$\eta_\sigma = 1 - A_0\,(1-\theta_\sigma) - A_1\,[\varrho_{C_nH_m} - (\theta_\sigma - \theta_A)] - A_2\,\theta_\sigma\,. \tag{59}$$

Dans l'égalité (59), A_0, A_1 et A_2 ont les mêmes valeurs que celles des relations (52), (53) et (54).

B. Etude théorique et expérimentale des réactions propergoliques

Données théoriques

1. La forme générale d'un mélange propergolique est symbolisée par:

$$K_1 \cdot C_nH_mO_pN_r + K_2\,[(\Re)_\nu\,(O_2)_\omega\,(N_2)_\mu]\,. \tag{60}$$

Le corps oxydant dans ce mélange est celui symbolisé entre crochets.
$\Re$ est un radical ou un corps susceptible d'être oxydé ou non. S'il est oxydable, en appliquant le principe de conservation de l'énergie sous forme de principe de l'état initial et de l'état final, nous pouvons écrire le mélange précédent sous la forme:

$$K_1 \cdot C_nH_mO_pN_r + K_2\,[R + (O_2)_{\omega - \psi} + \mu\,N_2] \tag{61}$$

en supposant effectuée la combustion préliminaire du radical combustible de l'oxydant[1].

On a: $R = \psi \cdot O_2 + (\Re)_\nu$.

---

[1] Du point de vue énergétique, il y aura lieu de tenir compte de l'énergie thermique résultant de cette combustion.

La formule du mélange (64) peut s'écrire sous la forme:

$$K_1 \cdot C_n H_m O_p N_2 + K_2 (\omega - \psi) \left[O_2 + \frac{\mu N_2 + R}{\omega - \psi} \right].$$

On peut finalement symboliser un mélange propergolique sous la forme:

$$C_n H_m O_p N_2 + \frac{K_2}{K_1} (\omega - \psi) \left[O_2 + \frac{\mu N_2 + R}{\omega - \psi} \right]. \tag{62}$$

2. Cette forme (62) d'un mélange propergolique en proportion quelconque est analogue à celle que nous avons étudié pour la combustion d'un combustible complexe avec un comburant $O_2 + \lambda N_2$.

Ici le comburant est $O_2 + \dfrac{\mu N_2 + R}{\omega - \psi}$

c'est-à-dire que le présent λ est λ_p tel que

$$\lambda_p = \frac{\mu + R}{\omega - \psi} \tag{63}$$

d'autre part: $\dfrac{K_2}{K_1} (\omega - \psi) = (1 + a) \left(n + \dfrac{m}{4} - \dfrac{p}{2} \right).$

a) Si $\dfrac{K_2}{K_1} (\omega - \psi) = n + \dfrac{m}{4} - \dfrac{p}{2}$ le mélange propergolique est dit en proportion théorique. Le rapport $\dfrac{K_2}{K_1}$ a une valeur γ_0 telle que

$$\gamma_0 = \frac{n + \dfrac{m}{4} - \dfrac{p}{2}}{\omega - \psi}.$$

b) Si $\dfrac{K_1}{K_2} \dfrac{\omega - \psi}{n + \dfrac{m}{4} - \dfrac{p}{2}} > 1$ le mélange propergolique est qualifié de mélange pauvre. Dans ce cas $\dfrac{K_2}{K_1}$ a une valeur γ_p telle que

$$\gamma_p > \frac{n + \dfrac{m}{4} - \dfrac{p}{2}}{\omega - \psi}$$

et

$$1 + \sigma = \frac{\gamma_p (\omega - \psi)}{n + \dfrac{m}{4} - \dfrac{p}{2}}. \tag{64}$$

c) Si $\dfrac{K_2}{K_1} \dfrac{\omega - \gamma}{n + \dfrac{m}{4} - \dfrac{p}{2}} < 1$ le mélange propergolique est qualifié de mélange riche. Dans ce cas $\dfrac{K_2}{K_1}$ a une valeur γ_2 telle que

$$\gamma_2 < \frac{n + \dfrac{m}{4} - \dfrac{p}{2}}{\omega - \psi}$$

et

$$1 - \varrho = \frac{\gamma_r (\omega - \psi)}{n + \dfrac{m}{4} - \dfrac{p}{2}}. \tag{65}$$

Remarque. Tout ce qui précède suppose connues les formules élémentaires de chacun des 2 constituants des mélanges.

3. D'après ce qui précède, le mélange propergolique (65) peut finalement se mettre sous la forme:

$$C_n H_m O_p N_r + (1 + \sigma)\left(n + \frac{m}{4} - \frac{p}{2}\right)(O_2 + \lambda_p \, \Re_2)$$

$$\text{ou} \quad C_n H_m O_p N_r + (1 - \varrho)\left(n + \frac{m}{4} - \frac{p}{2}\right)(O_2' + \lambda_p \, \Re_2)$$

dans ces deux expressions: λp est défini par l'égalité (63)
σ est défini par l'égalité (64)
ϱ est défini par l'égalité (65)
et l'on constate que l'étude d'un mélange propergolique sous sa forme générale peut être ramenée à celle de la combustion d'un combustible complexe, laquelle, comme nous l'avons montrée, peut elle-même être ramenée à celle d'un hydrocarbure.

Conclusion

Quelque soit la forme complexe générale d'un mélange réactionnel susceptible de dégager de l'énergie thermique utilisable, les résultats de la réaction peuvent être étudiés complètement en effectuant trois mesures expérimentales classiques dans les gaz de combustion:

$A =$ mesure de la concentration relative du CO_2
$G =$ mesure de la concentration relative du CO
$B =$ mesure de la concentration relative de l'oxygène.

Dans le cas des études de combustions incomplètes en mélange pauvre, (dilution relative σ) ou en mélange riche (dilution relative $- \varrho$), on mesure également les débits de combustible et de comburant pour déterminer respectivement les valeurs de σ et de $- \varrho$, données expérimentales nécessaires à l'exploitation des calculs théoriques.

Dans ces conditions, on peut calculer:
1. La fraction k_{CO_2} du système carbone incomplètement oxydée.
2. La fraction k_{H_2O} du système hydrogène incomplètement oxydée.
3. La fraction θ_σ de combustible inefficace.
4. La fraction θ_A de comburant inutilisé.
5. Le rendement de combustion global en fonction de θ_A et de θ_σ.

D'une façon générale, le processus expérimental d'étude est le suivant:
à la sortie du foyer, on dérive, par un dispositif convenable, un faible débit de gaz chauds à la sortie d'un mélangeur destiné à homogénéiser les gaz de combustion pour avoir des mesures statistiques.

Les gaz dérivés sont refroidis et séchés avant de passer aux appareils de mesures.

Ceux-ci peuvent être avantageusement:
a) des analyseurs à absorption Infra-Rouge pour le CO_2 et le CO. On a intérêt à faire des dérivations distinctes pour séparer les mesures.
b) un analyseur à variation de la perméabilité magnétique pour l'oxygène.

Funzioni del rapporto di massa in campo relativistico

Formulazione generale per qualunque velocità di trasferimento e di getto

Enrico Ostinelli [1]

(Ricevuto il 22 luglio 1959)

Riassunto — Zusammenfassung — Abstract — Résumé

Funzioni del rapporto di massa in campo relativistico. Formulazione generale per qualunque velocità di trasferimento e di getto. L'autore osserva preliminarmente che le esigenze dell'astronautica intersiderale sono tali da restituire al rapporto di massa quella preminente importanza che con i sistemi di propulsione jonica e fotonica esso ha perduto nei progetti di trasferte interplanetarie.

Procede quindi alla determinazione della misura delle principali grandezze di interesse astronautico in campo relativistico, che esprime secondo semplici funzioni del rapporto di massa.

L'autore generalizza le formule già presentate dal Crocco per il caso di una velocità di getto eguale alla velocità della luce (propulsione fotonica) rendendole valide per qualunque velocità di getto.

Le formule presentate concernono:

1. la misura del rapporto di massa in campo relativistico in funzione della velocità di trasferimento teorica e della velocità di getto;
2. la formula del "tempo proprio" dell'astronauta in funzione del rapporto di massa;
3. la formula dello spazio percorso in funzione della stessa grandezza;
4. la formula del tempo relativo all'osservatore immobile rispetto al punto di partenza;
5. le formule di alcuni rapporti di particolare interesse astronautico.

Funktionen des Massenverhältnisses im relativistischen Feld. Allgemeine Formulierung für beliebige Reise- und Ausströmgeschwindigkeiten. Der Verfasser weist zuerst darauf hin, daß das Massenverhältnis nicht allzu bedeutungsvoll ist, wenn interplanetarische Fahrten im Sonnensystem mittels Ionen- oder Photonenantriebes geplant werden, daß es jedoch zweifellos als wichtigster Faktor betrachtet werden muß, wenn es sich um interstellare Räume handelt.

Hierauf gibt der Verfasser die zur Astronautik im relativistischen Feld gehörenden Grundgrößen an und drückt sie als einfache Funktionen des Massenverhältnisses aus.

Unter Benützung der von General Crocco erhaltenen Formeln erweitert der Verfasser schließlich deren Gültigkeit für jede Ausströmgeschwindigkeit bis zur Lichtgeschwindigkeit (Photonenantrieb).

Die vom Verfasser eingeführten Formeln betreffen:

1. Das Massenverhältnis im relativistischen Feld im Zusammenhang mit der theoretischen Reise- und Ausströmgeschwindigkeit.
2. die Formel der „Eigenzeit" des Raumfahrers im Zusammenhang mit dem Massenverhältnis;
3. die Formel des durchfahrenen Raumes im Zusammenhang mit dem Massenverhältnis;

---

[1] Consigliere della Associazione per le Scienze Astronautiche, Via Banchi Vecchi, 61, Roma, Italia.

4. die Zeitformel bezogen auf den Beobachter, der sich relativ zum Startort in Ruhe befindet;

5. Formeln betreffend die Werte gewisser Verhältnisse von besonderem Interesse für die Astronautik.

Functions of Mass Ratio in the Relativistic Field. General Formulation for Whatever Transfer and Jet Velocity. The author points out, firstly, that, while the mass ratio is not so important when planning interplanetary travel within the solar system by means of ionic or photonic propulsion, it is to be undoubtedly considered the most important factor, when dealing with interstellar spaces.

Next, he establishes the principal quantities of concern in atronautics in the relativistic field and expresses them in simple mass ratio functions.

Finally, utilizing the formulae obtained by Gen. Crocco, he extends their validity to any jet velocity up to light velocity (photonic propulsion).

The formulae introduced by the author concern:

1. Mass ratio in the relativistic field, in relation to the theoretical transfer velocity and jet velocity;

2. Formula of the "proper time" of the space-traveller in relation to the mass ratio;

3. Formula of the covered space, in relation to the mass ratio;

4. Formula of time referred to the observer motionless relating to the starting point;

5. Formulae concerning some ratio values of particular interest in astronautics.

Fonctions du rapport de masse dans le champ relativiste. Formulation pour vitesses de transfert et d'éjection arbitraires. Le rapport de masse, moins important en cas de propulsion ionique ou photonique pour les voyages interplanétaires, est certainement le facteur le plus décisif pour les transports intersidéraux.

Les données principales sont établies dans le champ relativiste et exprimées en fonctions simples du rapport de masse.

La validité des formules du Gen. Crocco est étendue aux vitesses d'éjection atteignant celle de la lumière (propulsion photonique).

Les formules introduites par l'auteur sont relatives:

1. au rapport de masse dans le champ relativiste en relation avec les vitesses théoriques de transfert et d'éjection,

2. au temps propre du navigateur en fonction du rapport de masse,

3. au rayon d'action en fonction du rapport de masse,

4. au temps de l'observateur lié au point de départ,

5. à certains rapports intéressants dans le domaine de l'astronautique.

Con riferimento a quanto esposto nella memoria presentata al IX Congreso I.A.F. di Amsterdam, concernente le prestazioni-limite ammissibili nelle astronavi a propulsione jonica sfruttanti l'energia di fissione o di fusione nucleare, l'A. osserva che nei progetti di trasferte intersiderali si ripropone in tutta la sua gravità il problema del rapporto di massa che le alte velocità di getto teoricamente consentite dalla propulsione jonica permettono di trascurare nei progetti di trasferte interplanetarie.

Ravvisa quindi l'opportunità di presentare una formulazione generale delle principali grandezze di interesse astronautico in funzione del rapporto di massa, valide sia in campo classico che relativistico per qualunque velocità di trasferimento e di getto. Osserva che analoga formulazione, valida nel solo caso di propulsione fotonica, venne elaborata recentemente dal Crocco [1].

Ne consegue che le formule presentate in questa memoria si identificano con quelle del Crocco qualora si ponga la velocità di getto $u = c$. Per velocità subrelativistiche, invece, esse riconducono ai risultati delle formule classiche Newtoniane.

E' appena il caso di osservare che la realizzazione di astronavi per le quali possa doversi rendere necessario il ricorso alle correzioni relativistiche non presuppone solamente la possibilità di una conversione in energia di frazioni di massa considerevolmente maggiori di quella convertita con la fusione nucleare del deuterio, bensì anche la possibilità tecnica di utilizzare nella propulsione l'energia liberata con un rendimento ed una spinta specifica accettabili.

I. Determinazione del rapporto di massa in campo relativistico

$M_0 =$ massa alla partenza
$M =$ massa all'arrivo
$v =$ velocità raggiunta al termine della propulsione in assenza di attriti e di campi gravitazionali
$u =$ velocità di getto.

Sia un sistema S' animato da una velocità uniforme v rispetto al sistema S in istato di quiete.

Sia altresì un mobile in condizione di moto incipiente rispetto ad S', dotato di accelerazione costante a, ottenuta per mezzo di un getto di velocità u (eguale o minore di c). Se M è la massa attuale di quiete (rispetto ad S') del mobile in oggetto, avremo:

$$dM = -\, M\, \frac{dv'}{u} \tag{1}$$

dove dv' è l'incremento di velocità (ed altresì la velocità) valutata in S' all'istante $(t + dt)$ del sistema S, o, se si vuole, all'istante $(t' + dt')$ del tempo proprio.

Per il teorema della somma delle velocità di EINSTEIN:

$$v + dv = \frac{v + dv'}{1 + \dfrac{v\, dv'}{c^2}} \tag{2}$$

dove $v + dv$ è la velocità assoluta del mobile al tempo $(t + dt)$ rispetto ad S.
Si ricava allora

$$dv = \left(1 - \frac{v^2}{c^2}\right) dv' \tag{3}$$

od anche, dato che su S' si ha

$$dv' = a\, dt'$$

$$\frac{dv}{1 - \dfrac{v^2}{c^2}} = a\, dt'\ .$$

Sarà allora

$$t' = \frac{1}{a} \int_0^v \frac{dv}{1 - \dfrac{v^2}{c^2}} = \frac{c}{2a} \ln \frac{1 + \dfrac{v}{c}}{1 - \dfrac{v}{c}}\ . \tag{4}$$

Ma

$$\frac{dM}{M} = -\, \frac{a\, dt'}{u}$$

$$\ln \frac{M_0}{M} = \frac{a}{u}\, t' = \frac{c}{2u} \ln \frac{1 + \dfrac{v}{c}}{1 - \dfrac{v}{c}}$$

e perciò:

$$R = \frac{M_0}{M} = \left(\frac{c+v}{c-v}\right)^{\frac{c}{2u}} \tag{5}$$

che, risolta rispetto a v, dà:

$$\frac{v}{c} = \frac{R^{\frac{c}{2u}} - 1}{R^{\frac{c}{2u}} + 1}. \tag{6}$$

La formula (6) rappresenta la più generale espressione di v in funzione del rapporto di massa.

Nel caso della propulsione fotonica, per $u = c$, per semplice sostituzione in detta formula (6), riotterremo la nota formula che, nella formulazione adottata dal CROCCO (op. cit. pag. 11)

$$\frac{v}{c} = \frac{R^2 - 1}{R^2 + 1} \; ; \quad v = c \, \frac{R^2 - 1}{R^2 + 1} \tag{7}$$

è applicabile al solo caso-limite esaminato.

II. Formula del tempo proprio in funzione di *R*

Assumendo opportune ipotesi semplificatrici ($c = 1$ ed accelerazione del mobile unitaria) dalle note:

$$\begin{cases} t' = \dfrac{1}{2} \ln \dfrac{1+v}{1-v} \\ v = \dfrac{R^{2u}-1}{R^{2u}+1} \end{cases}$$

sostituendo avremo:

$$t' = \frac{1}{2} \log \frac{1 + \dfrac{R^{2u}-1}{R^{2u}+1}}{1 - \dfrac{R^{2u}-1}{R^{2u}+1}} =$$

$$= \frac{1}{2} \ln \frac{R^{2u} + 1 + R^{2u} - 1}{R^{2u} + 1 - R^{2u} + 1} =$$

$$= \frac{1}{2} \ln R^{2u} = \ln R^u \tag{8}$$

che, per accelerazioni diverse da 1, dovrà essere divisa per il valore dell'accelerazione in questione.

Nel caso-limite di propulsione fotonica ($u = c = 1$) sarà

$$t' = 1/2 \ln R^2 = \ln R \,.$$

Osservazioni al capo II

La formula

$$t' = \ln R$$

richiama l'identica formula (classica) legante il tempo t e la velocità teorica ottenibile v al rapporto di massa:

$$t = \frac{v}{u} = \ln R \text{ (per accelerazioni unitarie)}.$$

Si osserva altresì che, rinunciando alla semplificazione di comodo $c = 1$, la (8) dovrà assumere l'espressione:

$$t' = \ln R^u = u \ln R \tag{9}$$

e che infine, per accelerazioni a, diverse da 1, avremo:

$$t' = \frac{u}{a} \ln R . \tag{10}$$

III. Formula relativistica dello spazio percorso in funzione della velocità e del rapporto di massa

Detto Sp lo spazio percorso, t il tempo rispetto al sistema S (terra-stella) e t' il tempo proprio (del viaggiatore) sarà:

$$d\,Sp = v\,dt = v\,\frac{dt'}{\sqrt{1 - \dfrac{v^2}{c^2}}}$$

$$= \frac{1}{a}\,\frac{v\,dv}{\left(1 - \dfrac{v^2}{c^2}\right)^{\frac{3}{2}}} \tag{11}$$

da cui:

$$Sp = \frac{c^2}{a}\left(\frac{1}{\sqrt{1 - \dfrac{v^2}{c^2}}} - 1\right) . \tag{12}$$

Sostituendo ora a v il suo valore in funzione di R potremo ottenere:

$$Sp = \frac{c^2}{a}\,\frac{\left(R^{\frac{u}{c}} - 1\right)^2}{2\,R^{\frac{u}{c}}} . \tag{13}$$

IV. Determinazione della relazione tra R e il tempo relativo all'osservatore immobile rispetto al sistema di partenza

Dalla:

$$dt = \frac{dt'}{\sqrt{1 - \dfrac{v^2}{c^2}}} = \frac{1}{a}\,\frac{dv}{\left(1 - \dfrac{v^2}{c^2}\right)^{\frac{3}{2}}}$$

integrando avremo

$$t = \frac{v}{a}\,\frac{1}{\sqrt{1 - \dfrac{v^2}{c^2}}} \tag{14}$$

nella quale, sostituendo v in funzione di R otterremo:

$$t = \frac{c}{a}\,\frac{R^{\frac{2u}{c}} - 1}{2\,R^{\frac{u}{c}}} . \tag{15}$$

V. a) Determinazione del rapporto Sp/t (velocità media)

Dalle formule esposte agevolmente si deduce:

$$\frac{Sp}{t} = c \ \frac{R^{\frac{u}{c}} - 1}{R^{\frac{u}{c}} + 1} < c \,. \tag{16}$$

V. b) Determinazione del rapporto Sp/t'

Analogamente:

$$\frac{Sp}{t'} = \frac{c^2}{u} \cdot \frac{\left(R^{\frac{u}{c}} - 1 \right)^2}{2\,R^{u/c}\ln R} \ \text{(velocità media fisiologica)} \tag{17}$$

che non ha limite superiore teorico.

È interessante osservare che le formule esposte in questo paragrafo sono indipendenti dal valore dell'accelerazione a.

Letteratura

1. A. Crocco, Cosmonautica. Possibilità e limiti. Civiltà delle macchine, 1957, n. 5-6.

Gravitational-Magnetic Origin of Sunspots and Related Phenomena

By

Louis Gold[1] [2]

(With 1 Figure)

(Received August 27, 1959)

Abstract — Zusammenfassung — Résumé

Gravitational-Magnetic Origin of Sunspots and Related Phenomena. A wholly satisfactory theory for the origin of sunspots has evidently not as yet been proposed. Contemporary approaches are generally founded upon the conviction that they originate somehow from within the solar core. It is now theorized that the anisotropic streaming of the solar atmosphere induced by the perturbing gravitational action of the planetary system gives rise to a corresponding anisotropic magnetic field which in turn produces variations in particle pressure and temperature. Unlike the older tidal action theory based upon the marked correlation of sunspot character and the planetary configuration, the present theory, founded upon a gravitational-magnetic driving force, affords a physical explanation within the framework of magnetohydrodynamics. Thus, the interrelation between sunspot and coronal pattern periodicity, along with the localization of spots within the equatorial region of the sun, can be understood. The observed non-uniform rotation of the sun may also be accounted for in this light; the fact that rotation measurements based upon sunspot movement and Doppler shift agree quite closely suggests that the spots are highly disturbed regions in the solar atmosphere Perhaps the achievement of properly designed solar probes will permit viewing the sun from sufficiently varied attitudes to gain new information that may elucidate its structure.

Gravitations-magnetischer Ursprung der Sonnenflecken und verwandter Phänomene. Bis jetzt wurde noch keine vollauf befriedigende Theorie des Ursprunges der Sonnenflecken vorgeschlagen. Gegenwärtige Versuche dazu beruhen allgemein auf der Überzeugung, daß die Sonnenflecken irgendwie aus dem Sonnenkern entspringen. Es wird die Ansicht vertreten, daß die anisotrope Strömung der Sonnenatmosphäre, die durch die störende Gravitationswirkung des Planetensystems erzeugt wird, Ursache der Entstehung eines entsprechenden anisotropen magnetischen Feldes ist, das seinerseits Schwankungen des Partikeldruckes und der Temperatur bewirkt. Anders als die ältere Gezeitenwirkungstheorie, die sich auf die merkliche Wechselbeziehung des Charakters der Sonnenflecken und der Stellung der Planeten stützte, gründet sich die hier vertretene Theorie auf eine gravitations-magnetische Antriebskraft und bietet eine physikalische Erklärung innerhalb des Rahmens der Magnetohydrodynamik. So kann der Zusammenhang zwischen Sonnenflecken und Periodizität der Erscheinungsform der Korona, zusammen mit dem Auftreten der Flecken innerhalb der Äquatorzone der Sonne, verstanden werden. Die beobachtete ungleichmäßige Rotation der

[1] Radiation, Inc., Research Division, Orlando, Florida, U.S.A.
[2] Present address: Project Matterhorn, James Forrestal Research Center, Princeton University, P.O. Box 451, Princeton, New Jersey, U.S.A.

Sonne kann auch von diesem Gesichtspunkt aus erklärt werden. Die Tatsache, daß solche Rotationsmessungen, die auf der Bewegung der Sonnenflecken beruhen, und diejenigen, welche die Dopplerverschiebung benützen, recht gut übereinstimmen, läßt darauf schließen, daß die Sonnenflecken stark gestörte Regionen der Sonnenatmosphäre sind. Möglicherweise wird die Verwirklichung geeignet konstruierter Sonnen-Sondenfahrzeuge es gestatten, die Sonne aus genügend verschiedenen Lagen zu studieren, um neue Kenntnisse zu gewinnen, die ihre Struktur aufklären können.

Origine gravitationnelle et magnétique des taches solaires et phénomènes alliés.
Il est clair qu'une théorie complètement satisfaisante des taches solaires n'a pas encore été proposée. Les tentatives actuelles sont généralement fondées sur la conviction qu'elles trouvent leur origine dans le noyau solaire. On suppose que l'anisotropie dans l'écoulement superficiel de l'atmosphère solaire, induite par les perturbations gravitationnelles d'un système planétaire, provoquent une anisotropie correspondante du champ magnétique qui induit à son tour des variations de pression et de température particulaires. Contrairement à l'ancienne théorie des marées basée sur la correlation marquée entre la distribution des taches et la configuration des planètes, la théorie actuelle, fondée sur une force de caractère gravitationnel et magnétique, fournit une explication physique dans le cadre de la magnétodynamique des fluides. Les relations entre la périodicité des taches et celle de la structure coronaire peuvent maintenant se comprendre, aussi bien que la localisation des taches dans la région équatoriale. La rotation non uniforme du soleil, telle qu'observée, peut aussi s'expliquer; le fait que les mesures de rotation basées sur le mouvement des taches et sur l'effet Doppler concordent bien suggère que les taches sont des zones fortement perturbées de l'atmosphère solaire. Peut-être l'utilisation de sondes solaires adéquatement équipées permettra-t-elle l'observation sous des angles suffisamment variés pour fournir de nouveaux renseignements susceptibles de nous éclairer sur la structure solaire.

I. Introduction

It is becoming more evident that the earth, particularly in the upper regions of the atmosphere, is subject to the streaming of ionized matter from the sun's corona—the so-called solar wind [1]. While it is recognized that the coronal gas likely has velocities in excess of 100 km/sec there has been no apparent attempt to relate this circumstance with the sunspot phenomenon [2]. The present report suggests that the solar wind is created by the pertubing gravitational fields of the planetary system which thus establish a magnetic field configuration that ultimately culminates in the production of sunspot patterns.

In this manner, one is compelled to resurrect the discredited theory which sought to show the correlation between the sunspot cycle and the planetary configuration [3, 4]. A study of Cowling's appraisal of sunspot theories will reveal that none of the possible explanations can account for the key facts which are

(1) The approximate 11-year cycle.

(2) Correlation of coronal shape with the sunspot cycle.

(3) Confinement of spots within a narrow belt parallel to the solar equator.

(4) Essential repetitive history of the spot evolution during each cycle.

Cowling, like others, rejects the thought that the spots can originate anywhere but in the photosphere and/or the solar core itself [5]. Outright disavowal of any connection between planetary motion and sunspots pervades the more modern astronomical literature [6]. The preponderant view has it that sunspots originate from within the solar core by some kind of convective instability [7].

In this report the consequences of the perturbing gravitational field of the planets on the solar atmosphere will be elaborated in some detail to at least suggest how sunspots are generated. Furthermore, a possible explanation of the supposed non-uniform rotation of the sun will be offered which may cast serious doubt on mechanisms for sunspots based upon solar rotation. Finally, some recommendations on improved experiments for securing solar data involving space vehicles will be made.

Solar observations from the earth have been much too restrictive. SCHWARZ-SCHILD has recently demonstrated that information of great significance may be obtained by measurements beyond the earth s atmosphere [8]. What is most sorely needed is the capability of viewing the sun from a number of aspects simultaneously; some possibilities in this connection will be offered. It is never-theless amazing the wealth of information that has been gained about the structure of the sun and solar activity from earth-bound observation stations. With the establishment of space stations and the general expansion of the space program, a multifold amplification of solar knowledge may be anticipated.

II. Some Basic Features of Sunspots

It may be well to summarize the principal facts about sunspots and related solar activity which must ultimately be accounted for by a comprehensive theory of the origin and nature of sunspots. Reference has already been made to certain characteristics of the sunspot motif not truly embodied in contemporary theories; these being the 11-year periodicity, spot confinement within the equatorial zone, etc.

Spot coolness and magnetic field have been the cornerstones of several prev-alent theories. RUSSEL [9] and ROSSELAND [10] have attempted to explain the somewhat lower temperature of sunspots in terms of adiabatic cooling; but SIEDENTOPF [11] has contended that the associated stable flow really accounts for the overall granulation of the solar disc. The large magnetic field ($\sim 10^3$ gauss) has not been satisfactorily accounted for—yet BIERMANN [12] and others [13, 14] have evolved theories of spot coolness premised on the inhibiting action of the magnetic field on thermal convection.

A vast amount of work has been performed in connection with the Doppler shift. The Evershed effect provides some insight as to the proper motion of spots, i.e. the Doppler displacement for standard spectral lines of certain elements allows evaluation of gas movement within the spot. While an overall outflow of gas with velocities in the range of several km/sec appears to be the case, detailed investigation [15] has revealed a remarkable correlation between velocity and line intensity; a velocity profile seems evident. Later research [16] uncovered a considerable variance in the outward velocity with an upper bound in the neigh-borhood of 6 km/sec.

Positional Doppler studies allow further inferences about the gas flow patterns that manifest themselves in sunspots. Typical information is contained in Table I. The interpretation rendered for these data supposes a radial outflow from spots in horizontal directions.

Table I

	Locale Description	Nature of Doppler Shift
1.	Edge of sunspot toward center of disc	Toward violet
2.	Edge of sunspot toward limb	Toward red
3.	Spot center to edge	Increasing shift
4.	Spot displaced away from center of disc	Increasing shift

Doppler shifts recorded at the solar limbs more or less support deductions made about the solar rotation from direct observation of spot motion across the solar disc. Since a number of sunspot theories have been premised on the non-uniform rotation of the sun, it is important to appraise this phenomenon for whatever possible rôle it may play. Indeed, a novel explanation of solar rotation will be proposed later.

Whether one is concerned with Doppler effects in the spots or at the solar limbs, it is essential to allow for the interference of the earth's atmosphere [17]. Thus, a re-examination of the Doppler measurements indicated should be carried out in connection with the space program. To allow for possible solar atmospheric effects, it is of particular interest to determine the Doppler displacements from a number of vantage points which could be provided by space vehicles and observation stations established on the moon or some of the planets. It is unlikely that interpretations of the Doppler data are on firm ground without these further checks.

That sunspots do not occur in the photosphere and so reside in the upper solar atmosphere is evidenced by the interpenetration of spot groups. This fact has been ignored and may be crucial in the theory to be described.

III. The Question of Solar Rotation

The suspicion persists that the mystery of the non-uniform solar rotation somehow bears on the question of sunspots. As a basis for presenting a possible explanation of the observations relating to solar rotation, which may lay to rest sunspot theories so characterized [18], a brief review of the current status is first given.

It has long been known that spots near the equator of the sun appear to move more rapidly than those at the higher latitudes. In fact an empirical relation which closely represents the behavior of the angular velocity of rotation per day is described by [19]

$$\omega = 14.38° - 2.77 \sin^2 \Phi \qquad (1)$$

where Φ denotes the number of degrees beyond the equatorial belt. Spectroscopic observations of Doppler displacement at the solar limbs essentially support eq. (1), although differences among various researchers have been noted; local motions in the solar atmosphere have been blamed for such discrepancies [20]. A systematic increase in the Doppler shift has also been detected at higher altitudes in the solar atmosphere [21].

These basic facts, as well as some others to be pointed out, can be accounted for by adopting the view that the solar disc is essentially stationary and only the solar

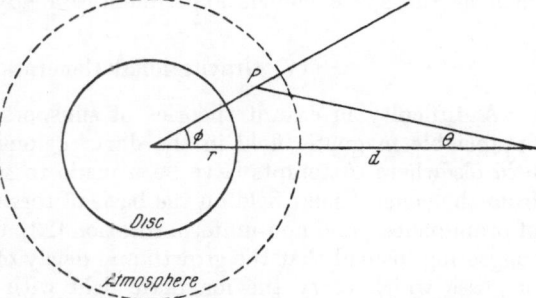

Fig. 1. Diagram for the latitude and height effect on Doppler-observations based upon uniformly rotating solar atmosphere

atmosphere rotates about it[1]—somewhat in the manner of the moons of saturn. Thus, the non-uniform rotation of the sun is illusory and it is not strange

---

[1] KIEPENHAHN has presented a more sophisticated model for the rotating solar atmosphere (Liége Conference on Stellar Evolution, summer 1959).

to find little or no flattening of the polar regions as well as little or no general magnetic field.

Next, referring to Fig. 1, the latitude effect of diminishing apparent solar rotation becomes intelligible. Then, too, the enhancement of the Doppler shift at higher regions in the solar atmosphere must follow. Indeed, quantitative formulation of the Doppler velocity is now possible. It is seen from the figure

$$V_{observed} = V_{equator} \cos \theta \, . \tag{2}$$

To secure the latitude relation θ must be expressed in terms of Φ—the altitude effect enters from $R = r + h$, where h is the distance above the photosphere.

Thus, from the trigonometric properties

$$\frac{R}{\sin \theta} = \frac{d}{\sin (\Phi + \theta)} \tag{3}$$

or

$$\frac{r + h}{d} = \frac{\sin \theta}{\sin (\Phi + \theta)} \, . \tag{3}$$

Some rather obvious manipulation leads to the desired finding

$$\cos \theta = \frac{\dfrac{d}{r + h} - \cos \Phi}{\left(1 + \dfrac{d^2}{(r + h)^2} - \dfrac{2d}{r + h} \cos \Phi\right)^{1/2}} \tag{4}$$

whence [since $v_{equator} = \omega \, (r + h)$]:

$$v_{observed} = \frac{\omega \, [d - (r + h) \cos \Phi]}{1 + \left(\dfrac{d^2}{(r + h)^2} - \dfrac{2d}{r + h} \cos \Phi\right)^{1/2}} \, . \tag{5}$$

If the rotation model is inherently correct, the lack of correspondence between eq. (1) and eq. (5) suggests that spurious effects enter into the observed latitude behavior of the Doppler shift. This may be in the nature of solar atmospheric effects and the averaging process that enters in the location of the center of gravity of the emitting spectral source. Clarification of this point must await future studies such as the space vehicle experiments already indicated.

IV. Gravitational Generation of Sunspot

A difficulty in extant theories of sunspots has been the explanation of an appreciable magnetic field in the dark regions and the relative absence of the field elsewhere. Attempts have been made to show the creation of sunspot fields from the general solar field on the basis of torsional oscillations [22], compression of prominences, and non-uniform rotation [23]. COWLING finds these unacceptable, suggesting instead that the growth and decay of the field is caused by the motion of gases which carry the magnetic field with them.

With full knowledge of his sentiment that any theory beginning with the sun's upper atmosphere must be suspect because there is not enough matter to hold back radiation or enough pressure to offset magnetic pressure, the mechanism of gravitational perturbation in the upper regions of the solar atmosphere is nevertheless argued to be responsible for the generation of the magnetic fields. This does not conflict with the observation of the interpenetration of spots which actually supports the position that they are not in the photosphere and so reside in the more tenuous regions of the solar atmosphere.

The planetary system, immersed in the outermost regions of the corona, produces an anisotropic streaming of gas. The effective gravitational field in the neighborhood of a planet is sufficiently altered to establish a channeling of the escaping gas—the so-called solar wind. It is this oriented movement of the solar atmosphere which gives rise to the magnetic fields. The resultant cooling of the gas in these channels then comes about by adiabatic expansion and/or the inhibited thermal convection at the base of the spot near the photosphere due to the magnetic field.

ALFVÉN [18] has offered the following argument for the lowered temperature in the spot. The requirement of hydrostatic equilibrium means that

$$\left(P_{particle} + \frac{\mu\,H^2}{8\,\pi}\right)_{spot} = P_{exterior} \tag{6}$$

where $P_{particle}$ is the particle pressure within the spot and $\mu\,H^2/8\,\pi$ is the magnetic pressure; eq. (6) merely states the balance of spot and exterior solar atmosphere pressure ($P_{exterior}$). Consequently, the temperature of the spot T_{spot} must be less than outside, i.e.

$$T_{spot} < T_{exterior} \,. \tag{7}$$

A rigorous calculation of the magnetic field associated with the streaming ionized gas is difficult. However, a plausibility determination may be performed in similar manner to that offered by CHANDRASEKHAR and FERMI for spiral arms in galaxies [24]. Thus, with the magnetic energy at least equal to or greater than the kinetic energy,

$$\frac{H^2}{8\,\pi} \geq \frac{1}{2}\,\varrho\,v^2 \,. \tag{8}$$

Supposing particle density $\varrho = 10^{-14}$ gm/cc and a velocity $v = 100 - 1,000$ km/sec, the magnetic field must range somewhere near 10^3 gauss. This is only a rough estimate and is within reasonable range of actual spot fields.

It is incomprehensible that the correlation between planetary periods and sunspot activity could have been so thoroughly discredited when many earlier workers were impressed with the remarkable association! This is all the more incredible in the light of no present-day theories which offer any explanation for the periodic motif of sunspots.

These disturbed regions in the solar atmosphere only superficially resemble storms in our terrestrial atmosphere. The hydrodynamic considerations are vastly different because of the dominant magneto-ionic contributions in the former medium. The prediction is put forth that a complete peripheral inspection of the sun will reveal all stages of sunspot patterns; a solar probe would be invaluable for this purpose. The related coronal structure could simultaneously be recorded and very likely the familiar tongue patterns characteristic of sunspot maximum and minimum will appear about 180° out of phase.

V. The Future of Sunspot Investigations

The gravitational-magnetic origin of sunspots that has been posed here can be subjected to a critical experiment as indicated above. The simultaneous recording of the sunspot and coronal features from many observation points should become a major objective of the IGY program. Properly executed experiments may also provide information on the nucleation and growth of spots and so deepen our understanding of solar activity.

While it certainly appears worthwhile to re-examine the diverse positional Doppler shifts that have been described to ascertain the rôle of the attitude of observation; theoretical work, somewhat in the nature of the Doppler effect in a medium with both a density and velocity profile, ought to prove illuminating[1]. This study is now in progress and will be reported on the future.

Acknowledgement

The author is indebted to R. P. HAVILAND for a helpful discussion and referral to G. M. CLEMENCE who provided a clue to the astronomical literature on the suppressed theory of sunspot and planetary correlation.

References

1. E. N. PARKER, Astrophysic. J. **60**, 171 (1955).
2. R. K. LANDSHOFF, The Plasma in a Magnetic Field. Stanford: University Press, 1958.
3. W. LUBY, Astronom. J. **40**, 101 (1930).
4. A. SCHUSTER, Proc. Roy. Soc. London **85**, 309 (1911).
5. G. P. KUIPER, The Sun, p. 578. Chicago: University Press, 1954.
6. Ibid. p. 333.
7. T. G. COWLING, Magnetohydrodynamics. New York-London: Interscience, 1957.
8. M. and B. SCHWARZSCHILD, Sci. American **200**, 52 (1959).
9. H. N. RUSSELL, Astrophysic. J. **54**, 293 (1921).
10. S. ROSSELAND, Astrophysic. J. **63**, 536 (1926).
11. H. SIEDENTOPF, Astronom. Nachr. **255**, 157 (1935).
12. L. BIERMANN, Vjschr. Astronom. Ges. 76 (1941).
13. F. HOYLE, Recent Researches in Solar Physics. Cambridge: University Press, 1949.
14. W. B. THOMPSON, Philos. Mag. **42**, 1417 (1951).
15. C. F. ST. JOHN, Astrophysic. J. **37**, 322 (1913); **38**, 341 (1913).
16. G. ABETTI, Handbuch der Astrophysik, Vol. 4. Berlin: Springer, 1929.
17. R. E. DE LURY, J. Roy. Astronom. Soc. Canada **33**, 345 (1939); also see H. D. BABCOCK, Trans. Internat. Astronom. Union **6**, 65 (1938).
18. H. ALFVÉN, Tellus **8**, 274 (1956).
19. H. W. NEWTON and M. L. NUNN, Monthly Notices **111**, 413 (1951).
20. Ref. [5], p. 21.
21. W. S. ADAMS, Astrophysic. J. **31**, 30 (1910).
22. C. WALÉN, On the Vibratory Rotation of the Sun. Stockholm: H. Lindstahls Bokhandel.
23. D. MENZEL, Nature **166**, 31 (1950).
24. S. CHANDRASEKHAR and E. FERMI, Astrophysic. J. **118**, 113 (1953).

[1] Paper presented at IV Ionization Phenomena in Gases Congress, Uppsala, Aug. 14-17 (1959): L. GOLD, Doppler Phenomenon in Radiant Gaseous Atmospheres with Velocity and Density Fields (to be published).

Global Aspects of the Exploration of Space[1]

By

Hugh L. Dryden[2]

(Received August 27, 1959)

Introduction

It is an honor and privilege for me to participate in this Congress, a meeting of persons from all the world devoted to the purpose of bringing to accomplishment that which has so long been only a vision of a few gifted men, the exploration of space by mankind. To explore space to gain knowledge of the physical universe in which he lives; to explore space as a demonstration of his mastery of advanced technology; to open space to his own travel to satisfy his desire to see and experience for himself; to explore applications of space technology to improve worldwide communications and weather forecasting—all these aims reflect as in a mirror the desires of men everywhere.

Older Methods of Space Exploration

The interest of man in outer space began long ago among uncivilized peoples to whom the face of the sky was clock and almanac; the celestial bodies, objects of worship. Exploration was at first by visual observation, later aided by armillary spheres and quadrants, and still later by more precise measuring instruments, telescopes and spectroscopes. The information obtained was that borne by the light that was transmitted from the distant celestial object through the atmosphere to the observing instrument on the ground. In recent years the light waves have been supplemented by radio waves as carriers of information from the stars and planets.

Men of many nations have contributed through the centuries to the exploration of space by the methods of astronomy. The history of advances in astronomical knowledge and technique includes the records of Chinese, Babylonians, Greeks, Arabians, and of nearly every nation of the modern world. International co-operation was early recognized as essential and beneficial; the countless number of the stars and the vastness of space present mankind with a truly global task.

The picture of the universe obtained by the astronomers early stirred the imagination of men to speculate about the existence of life elsewhere in the universe, about means of communication with distant stars, and in the last centuries about the possibility of the travel of man to the moon and planets. Some sought to apply the science and engineering of their day to describe the vehicles to be used. For example, JULES VERNE published in 1865 in "From the Earth to

[1] Address before the Tenth International Astronautical Congress, London, England, August 31, 1959.

[2] Deputy Administrator, National Aeronautics and Space Administration, 1520 H Street, N.W., Washington 25, D.C., U.S.A.

the Moon" a description of a gun-launched projectile carrying passengers to orbit the moon. Today we have taken the first steps to bring this inspired vision to reality. The exploration of space by unmanned vehicles carrying scientific apparatus began on October, 4, 1957; exploration by man will follow in due course.

Space Exploration Today

The present stage of development of vehicles for space exploration corresponds to some degree to that of the airplane in 1905. We find it as difficult to predict the pace and scope of developments in space exploration now as it was then to foresee the amazing developments in performance and utility of the airplane which have since come to pass. But there are important differences. In 1905 it was possible for an individual to learn all that had been discovered about aeronautics and aircraft design. Airplanes were designed by individual inventors or engineers. This situation soon changed. Today there is no single designer of a modern airplane; it represents the work of many specialists organized into an effective team and its design is based on the contributions of many men of past and present generations. The development is supported by an extensive research program in the basic aeronautical sciences. In fact, the development of a modern supersonic airplane is a task commensurate with the technical and financial resources of a whole nation rather than with those of a single private company.

I do not foresee any projects in space exploration similar to Lindbergh's flight across the ocean. For the present only a few nations are able to undertake a comprehensive program of space exploration. There is every indication that the task may soon develop beyond the resources of any single nation. The incentives to international co-operation thus transcend those which led to world-wide collaboration in the exploration of space by astronomical methods. Even at the present stage the global aspects of space exploration demand some degree of international co-operation. I will discuss some of these questions by reference to the program objectives of the National Aeronautics and Space Administration, a newly established United States governmental agency with the mission of expanding human knowledge of phenomena in the atmosphere and space; the development, improvement, and operation of space vehicles; and the application of aeronautical and space science and technology to the conduct of peaceful activities within and outside the atmosphere. So far as can be learned, the programs in other countries include all or some activities of a similar character to those to be described, and the global aspects and opportunities for international co-operation are similar.

Space Science Program

One of the principal objectives of current space activity is the study of the space environment by the conduct of scientific experiments using apparatus carried by sounding rockets, man-made earth satellites, man-made planets, and deep space probes. In the United States we have used the term "space science" as a shorthand expression for experiments in physics, chemistry, bioscience, astronomy, astrophysics, and geophysics which employ instruments transported into the high atmosphere and outer space.

High altitude rocket research began at least as early as 1945. As a part of the program of the International Geophysical Year many hundreds of sounding rockets were fired by Australian, Canadian, French, Japanese, Soviet, United Kingdom and United States scientists in a co-ordinated global program. The US and USSR together have successfully launched 12 satellites to date, have sent two space probes to become man-made planets of the sun, and have projected two additional probes to distances of 63,000 and 71,000 miles from the earth.

As I have stated elsewhere: "Just as the fullest development of space science involves the whole spectrum of scientific disciplines, so also does it require the interest, support, and participation of the whole world. In the mechanics of conducting a space research program there is need for international co-operation. The tracking of satellites and space probes and the collection of data from their radio signals provide examples of cases where such co-operation is essential. In scientific research itself there are also many areas in which international co-operation is essential to the fullest realization of potential scientific gains. Joint efforts in the investigation of the ionosphere and the fundamentals of radio propagation through the upper atmosphere are required to obtain the world-wide coverage that alone can provide a complete picture."

"But most of all, space research needs to draw upon an entire world for its ideas. Those ingenious insights into the real meaning behind a set of observed facts that lead to real advances in the understanding of our universe are not the prerogative of a single nation or group but come from every quarter of the world where men are seriously occupied with scientific research. So vast is the challenge of space reserach and so great is the promise to mankind in the way of increased knowledge and ultimate benefits that the world can not afford to neglect or slight the opportunities that lie before it."

The NASA program for the future, and presumably those of other countries, looks to the orderly development of increased capabilities which permit the more and more detailed study of the space within our solar system and of the moon and planets. Already space probes have been sent to distances beyond the moon and the miss distance will decrease as development of vehicles and guidance proceeds. In due course artificial satellites of the moon will be established, followed by hard and soft landings of scientific equipment on the lunar surface. In similar fashion the nearer planets Mars and Venus will be studied when the requisite vehicle systems are available.

Applications of Earth Satellites

The NASA objectives include the investigation of the uses of earth satellites to perform more efficiently and effectively some tasks which are now carried out by other means and to perform other tasks which cannot be done at all with present means. The applications which seem most promising at present are to weather observation, analysis, and forecasting on a global scale; to the improvement of long distance radio communication; to the study of the size and shape of the earth and of the distribution of land masses and water; and to all-weather global navigation. It is believed that such applications brought to successful fruition will further the welfare of mankind everywhere.

More accurate weather forecasts have tremendous economic implications for people everywhere in their homes, on the farms, and at their work. Large monetary savings would accrue to many industries, for example, food processing, hydro-electric power, and public utilities.

At present attempts are made to study global weather by balloon soundings at a limited number of points on the land areas and from a few airplanes and ships. Properly instrumented satellites enable fast global observation of hurricanes, tornadoes, cloud heights and type, presence of precipitation, thunderstorms, incoming and reflected solar radiation, and perhaps temperatures at various levels.

Meteorology is already a recognized field for international cooperation as an inherently global activity of great significance to human welfare. Many nations can co-operate in the collection, interpretation and application of the large amount of data receivable from meteorological satellites.

NASA has an active project under way for a simple form of communication satellite which can be used by any nation or person having the necessary ground equipment without interference with any other user. It is well known that the moon may be used as a reflector of radio and radar signals if very powerful transmitters and very sensitive receivers are used on the ground and the geometrical relations are correct. Satellites can provide smaller moons nearer the earth which require less transmitter power and less expensive equipment. According to some studies a passive satellite system may prove economically competitive with ocean cable for transatlantic communication. Television or other wide band transmission would be possible over this system.

The basic data needed to determine the practicality of such a system will be determined by launching a large inflatable sphere 100 ft. in diameter and made of aluminized mylar plastic. A strong radio signal beamed to strike the satellite will be reflected over a wide angle and can be received as a much weaker signal by a highly sensitive receiving antenna pointing toward the satellite.

Manned Exploration of Space

The NASA program objectives presumably like those of other countries include the orderly development of means for the manned exploration of space. En route to the long-range objective of manned exploration of the solar system are the temporary ballistic flight of man into space and return (already accomplished with animals), orbital flight for one or a few circuits in the simplest vehicle below the level of the Great Radiation Belt, manned flight in advanced maneuverable vehicles, in larger satellites carrying several men, in permanent manned orbiting space laboratories, manned flight to the vicinity of the moon and return to earth, and manned landing on the moon and return.

NASA's present project in this field, Project Mercury, has been repeatedly described in the international public and technical press. Its successful completion requires the co-operation of several countries in permitting the installation of portable tracking radars, communication stations, and telemetry receiving stations at suitable points along the intended course. Such negotiations are under way.

Even the first steps in the manned exploration of space are very expensive as may be inferred from the estimated cost of Project Mercury of $ 300 million or more. The resources required for the advanced missions may well demand a world-wide collaboration and serve to give a true measure of man's response to the challenge to discover and explore the new frontier of our day.

Vehicle Development Program

An early task of NASA was the planning of a program of rocket and vehicle development which would provide all the desired missions with a minimum number of new rockets and new vehicles. As in other countries our present vehicles are assembled from rockets developed in the ballistic missile program and available smaller rockets. For increased thrust, two new developments have been started in the United States—(1) a cluster of existing engines to give an early capability of about one and one-quarter million pounds thrust, and (2) a new single-chamber rocket of one to one and one-half million pounds thrust, which can be clustered to give six million pounds thrust or more.

In addition to these first stage boosters, several upper stage rockets are under development, including some using high energy fuels. In addition nuclear rockets are being developed by the Atomic Energy Commission and NASA along with the general application of nuclear energy for auxiliary power in space vehicles.

Of particular interest to other countries may be the Scout vehicle under development by NASA. This is a four-stage solid propellant satellite vehicle to carry about 150 pounds into a 300 mile orbit. It will be much more economical than existing vehicles, hopefully to cost not more than $ 500,000 per firing. It will satisfy many, though not all needs of our scientific program. We expect to use this vehicle, if its development is a success, in early international co-operative programs.

The NASA program was reported in considerable detail in the June 22, 1959 issue of Aviation Week, the account being based on testimony before Congressional Committees in defense of the budget estimates for the period July 1, 1959 to June 30, 1960. The total estimate for the aeronautical and space activities of NASA amounted to approximately $ 500 million.

There is of course much research essential to activities of the type described but carried out on the ground with the aid of suitable equipment for simulating various aspects of the vehicle or space environment. As in many branches of science and technology, contributions will come from a large number of countries by groups without direct access to space vehicles.

Areas of International Co-operation

Throughout the discussion repeated reference has been made to the desirability of international co-operation arising from the global nature of space exploration. The desirable types of activity, it seems to me, are exchanges of scientific and technical information and data, exchanges of scientists, co-ordinated programs of observation and experimentation, and co-operative programs of space exploration.

Exchange of information in its usual form consists of the exchange of publications and the holding of international scientific meetings. In the space activities initiated during the IGY it was found desirable to exchange information on the planning of experiments, to give prompt notice of launchings, early information on orbits, and such other data as would permit participation of others in observations of scientific value.

It has been remarked earlier that space science is not a new scientific discipline but comprises the use of new tools of experimentation by trained scientists in physics, geophysics, astronomy and similar established fields. The exchange of scientists between countries permits a more rapid transfer of the new techniques than can be accomplished by publications or presentation of papers. NASA has established a few fellowships available to scientists of other countries and has provided research opportunities to a few guest scientists. Exchange of scientists in addition to providing training in new techniques may also be used for substantive participation of senior scientists in co-operative programs.

It is obviously desirable that national programs in the space field be co-ordinated to avoid undesired duplication and to provide the enhanced increase in knowledge that comes from co-ordinated efforts. This co-ordination was well done under the non-governmental international committee for the IGY (CSAGI) and we look forward to the early establishment on a more permanent basis of the Committee on Space Research to continue co-ordination of basic scientific research in the space field. There is need for co-ordination in program planning, and in the execution of certain programs. Activities in the tracking of satellites and in the reception of telemetered data, in research on the upper atmosphere and ionosphere by means of sounding rockets launched simultaneously in various parts of the world, in investigation of the ionosphere by observation of radio signals from satellites, and in laboratory and theoretical research in areas supporting space activities are examples of program areas in which international co-ordination would be most productive.

The ultimate step in international co-operation is joint participation in a single program with participation of scientists of two or more countries in the design of experiments and in the preparation of payloads for rockets, satellites, and space probes. Discussions are under way between NASA scientists and their colleagues from other countries with the view of beginning activities of this type.

In view of the large expense of space activities and the consequent support by governments in each country, international co-operation in space activities can ultimately succeed only when supported by appropriate action on a governmental level. Thus the activity of the non-governmental international scientific groups must be supplemented by suitable action at the intergovernmental level.

As most of you know the first steps toward a basis for the furtherance of international co-operation in the peaceful uses of outer space at the governmental level have been taken by the United Nations. By resolution 1348 (XIII) of December 13, 1958, the General Assembly established an *Ad Hoc* Committee on the Peaceful Uses of Outer Space. This Committee met during the period May 6 to June 25, 1959 and prepared a report under date of July 14, 1959, Document A/4141, which will be submitted at the fall meeting of the General Assembly. This report suggests certain general functions and tasks that might appropriately be undertaken within the framework of the United Nations at the present time. The suggestions in the report represent a hopeful first step toward fuller international co-operation. Many of these tasks require consideration at the governmental level and it is suggested that the General Assembly "may wish to consider the establishment of an Assembly committee, composed of representatives of Member States and having such membership as the Assembly may decide, to perform these functions, to report to the General Assembly and to make recommendations as appropriate." One of the urgent tasks is "to study practicable and feasible measures for facilitating international co-operation." The report discusses international co-operation in the conduct of space activities in considerable detail and the areas of space activity in which international co-operation should be strengthened. Requirements for mutual agreements in such matters as co-ordination of radio frequencies for use in association with space vehicles are noted as well as opportunities for international co-operation in joint projects.

Conclusion

When one considers the vast distances of the solar system—93 million miles to the sun; 26 million miles to Venus, the nearest planet; 3,680 million miles to Pluto—and when one catalogues the problems to be solved and the new knowledge needed in almost every branch of science and technology from magneto-hydrodynamics to cosmology, from materials to biology and psychology, the magnitude of the task before us becomes apparent. It is a task that challenges the peoples of the earth as a whole. There is room for co-operation of men of many skills and of nations large and small.

Many of you will regard me as ultra conservative to have limited our objective to the exploration of the solar system. This Congress deals with the subject of astronautics—travel to the stars. Is then the travel of man to the stars a futile dream ?

You remember the verse:

> The world will last when gone are we
> Without a trace of thee or me
> Before we came there was no void,
> And when we're gone the same 't will be.

I wonder. Since the invention of writing the thoughts, the knowledge and the influence of men who lived long ago have been available to us and our work will be available to future generations. Each age builds on the shoulders of the past. Who then dares to limit the horizons of the physical universe to be ultimately explored by man ? The exploration of space has begun; it is our task to advance it as much as we can. Who knows where it will end ?

Present Status of the Earth's Radiation Belt[1]

By

S. F. Singer[2]

(Received August 27, 1959)

Abstract — Zusammenfassung — Résumé

Present Status of the Earth's Radiation Belt. The present available observations on the geomagnetically trapped radiation are reviewed. The inner belt contains a large flux of high energy penetrating protons while the outer belt consists mainly of low energy electrons. The inner belt, therefore, is of considerable importance from a spacemedical point of view since shielding presents a real problem. We discuss again the possible use of "sweeper satellites" to depress the intensity of the inner belt. We discuss, in more detail, the creation of artificial radiation belts by means of electrons injected from a rocket or satellite-borne accelerator. Such artificial radiation belts can be of great value for geophysical studies.

Gegenwärtiger Stand der Ansichten über den Strahlungsgürtel der Erde. Die derzeit verfügbaren Beobachtungen über die geomagnetisch „eingefangene" Strahlung werden kritisch betrachtet. Der innere Strahlungsgürtel enthält einen intensiven Fluß energiereicher, durchdringender Protonen, während der äußere Gürtel hauptsächlich aus Elektronen geringer Energie besteht. Infolgedessen ist der innere Gürtel vom raumfahrtmedizinischen Gesichtspunkt von wesentlicher Bedeutung, da die Abschirmung ein ernstes Problem darstellt. Der Verfasser erörtert nochmals die mögliche Verwendung von „Reinfeger-Satelliten", um die Intensität des inneren Gürtels zu unterdrücken. Eingehender wird außerdem die Schaffung künstlicher Strahlungsgürtel mit Hilfe von Elektronen diskutiert, die von einer Rakete oder einem satellitengetragenen Beschleuniger eingeschossen wurden. Derartige künstliche Strahlungsgürtel können für geophysikalische Studien von großem Wert sein.

Etat actuel des investigations concernant les ceintures de rayonnement de la Terre. Les données d'observation acquises sur le rayonnement capturé par géomagnétisme sont recensées. La ceinture intérieure contient un flux important de protons de haute énergie, tandis que la ceinture extérieure est composée principalement d'électrons de faible énergie. D'un point de vue médical la ceinture intérieure est importante en raison des difficultés de protection. La possibilité d'utiliser des satellites de balayage est à nouveau discutée. La création d'une ceinture artificielle par injection d'électrons à partir d'accélérateurs de particules fuséo-portés est envisagée de façon plus détaillée. Les ceintures artificielles peuvent être d'un grand intérêt pour les études géophysiques.

[1] University of Maryland, Technical Report No. 155. — This research was supported in part by the United States Air Force under Contract AF 49 (638)-530 monitored by the Air Force Office of Scientific Research of the Air Research and Development Command and by the Electronics Research Directorate of the Air Force Cambridge Research Center, Air Research and Development Command, under Contract No. AF 19 (604) 5575.

[2] Professor at the University of Maryland, Physics Department, College Park, Maryland, U.S.A.

From theoretical considerations one expects to find two belts of trapped radiation around the earth. The following table, first presented by the writer in November 1958 at the San Antonio Symposium on the Physics and Medicine of Space (New York: J. Wiley and Sons, 1960) lists the principal characteristics. These predictions have now been well supported by observations in satellites and space probes.

Table I. *Properties of the Earth's Radiation Belts*

	Hard (cosmic ray)[1]	Soft (Auroral)[2]
Origin	From the decay of cosmic ray albedo neutrons which come out of the earth's atmosphere	From Solar Corpuscular streams with subsequent acceleration near the earth
Nature	Protons between 10 and 400 Mev	Electrons and protons of less than 1 Mev
Shielding	Difficult to absorb and shield	Easily absorbed
Location	Equator and low latitudes Close to the earth	Auroral latitudes Far from the earth
Time Dependence ...	Constant in time	Variable, increases when sun is active

[1] S. F. Singer, Trapped Albedo Theory of the Radiation Belt. Physic. Rev. Letters 1, 181 (1958).

[2] S. F. Singer, Trapped Orbits in the Earth's Dipole Field. Bull. Amer. Physic. Soc. 1, 229 (1956); A New Model of Magnetic Storms and Aurorae. Trans. Amer. Geophysic. Union 38, 175 (1957).

Status of Observations

The observations of the trapped radiation around the earth can be briefly summarized. There appear to be two belts produced most likely by two distinct sources of trapped particles: (1) an outer belt (of solar origin) whose particles are quite "soft" (nonpenetrating); and (2) an inner belt (probably of cosmic ray origin) whose outstanding feature is the presence of a large flux of high energy, penetrating protons. The existence of two regions was clearly shown by the space probes, Pioneer III and the Russian Cosmic Rocket (reports by Van Allen et al. and by Vernov et al. in December 1958, January 1959).

Concerning the nature and composition of the radiation the following data are available and were discussed at a conference in Moscow in July 1959. Russian workers made observations starting with Sputnik II. In Sputnik III they began to use scintillation crystals under very thin absorbers and under shields of 1 g/cm² thick.

Outer Belt

The identification of the particles of the outer belt as soft electrons was first given by Krassovsky et al. on the basis of Sputnik III experiments. Vernov and Chudakov were able to measure the spectrum of these electrons in the cosmic rocket from the pulse height distribution in the scintillation crystal. At very low energies ($E = 10$—20 Kev) the spectrum follows a law E^{-8}; between 20 and 100 Kev the spectrum is more like E^{-5} (in the center of the region), and perhaps like E^{-3} at the fringe of the region. The average energy near the maximum (at about 3 earth radii) is 25 Kev and the flux is $\sim 10^{11}$ particles/cm² sec.

Inner Belt

For the inner radiation zone VERNOV and CHUDAKOV find that their ionization data are best explained by protons of average energy 100 Mev. Similarly, VAN ALLEN finds that the penetrating particles of the inner zone are most likely protons of energy greater than 40 Mev. Their flux is $\sim 2 \times 10^4$/cm² sec, while the flux of electrons of energy greater than 600 Kev is $\sim 10^7$/cm² sec. ster. But the most direct indication for the nature of the particles of the inner radiation belt comes from the recent experiments of FREDEN and WHITE. In nuclear emulsions exposed at 1,200 km they can identify the penetrating radiation as protons, with energies ranging up to 700 Mev.

Space Medical Implications

The report given at the IAF Congress in Amsterdam (August 1958) is still applicable[1]. The danger comes only from the penetrating particles produced by neutron decay; hence only the inner belt in the altitude region 1000 to about 4000 km up to magnetic latitude 30° presents any problem. The radiation level reaches a maximum of a few roentgens per hour. As discussed previously, shielding is difficult.

Artificial Modification of the Inner Belt

Sweeping: One may take advantage of the excellent trapping properties of the earth's field in two different ways. If the lifetime is as long as one calculates based on atmospheric scattering only, then it may be possible to substantially decrease the intensity level in the inner belt by providing additional absorbing bodies, e.g., large satellites. The use of such "sweeper satellites" was first suggested at the International Astronautical Congress in Amsterdam in 1958. It appears now, on the basis of the observations, that this method is indeed feasible. In all likelihood, however, the operation of artificial satellites for other purposes besides sweeping, for example the operation of communication satellites and meteorological satellites, will serve to depress the intensity of the inner radiation belt provided the number of such operating satellites is large.

Injection: However, before carrying out any large-scale experiments, one would like to study the trapping properties of the field experimentally. For this purpose, it would be most interesting to inject particles into the magnetic field directly at high altitudes and trap them. Electrons are better suited for injection than protons since they are much easier to generate and since their radius of curvature is very small. For this reason, they become more insensitive to magnetic field fluctuations and to questions of constancy of the magnetic moment. For example, a 2 Mev electron has a radius of curvature at an altitude of 1000 km above the equator, which is only $1/3$ km as compared to 70 km for a 100 Mev proton. The injection is best carried out by means of a linear accelerator which is carried aloft in a satellite; even a high altitude rocket may be adequate.

---

[1] S. F. SINGER, Scientific Problems in Cislunar Space and their Exploration with Rocket Vehicles. Astronaut. Acta 5, 116 (1959); Proceedings of the IXth International Astronautical Congress, Amsterdam 1958, p. 904. Wien: Springer, 1959.

It turns out that an injection time of only a few minutes is quite sufficient to give an increase of the electron intensity observable over the natural radiation belt intensity.

The way in which such an experiment might work is as follows: The accelerator should be as small and as light as possible. This suggests a linear accelerator of the type which uses a klystron as an electron source and also as a means of supplying energy to a wave guide which accelerates the electrons. The instrument can be constructed to be extremely light-weight since the vacuum system can of course be dispensed with. This leaves only the electron gun, the tube structure, focusing magnet, and wave guide. By supplying a small magnet the electron beam can be doubled up and the whole accelerator can be folded into a smaller space. The power supply requirements can be estimated reasonably as about 50 kw for a few minutes. Applying a normal weight factor for batteries, we derive a total payload weight of about 500 pounds which is surprisingly small compared to what one might expect from laboratory types of electron accelerators.

The mean output current of such an accelerator would be about 500 micro-amperes giving therefore 3×10^{15} electrons per second. If we inject over an altitude interval of 10 km, then the volume into which the electrons are released and diffuse into is of the order of 10^{24} cm³. Its shape resembles a "napkin ring". The mean injection rate thus becomes 3×10^{-9} electrons per cm³ per second. For simplicity we may assume that the injection is isotropic, i.e., the satellite or rocket is tumbling so that electrons are injected into all directions (it should be noticed here that it is not necessary to provide a well-defined electron beam, and this again saves considerable weight in the construction of the accelerator).

Hence, about one-half of the electrons remain trapped. Since the concentration of particles in the natural radiation belt is of the order of 10^{-7} per cm³, we can double the radiation intensity after about 100 seconds of operation.

After building up the intensity in this manner, it will decay away slowly to its original value after the accelerator is turned off. The trapping volume will change from a napkin ring into a ring of more nearly circular cross-section. The main point of the experiment now is to study the rate and manner in which the radiation intensity falls off after the injection.

After the electrons are released, we would then check on the artificial radiation belt, perhaps once a day, by means of simple sounding rockets which traverse the region of the belt and determine the profile of electron intensity with altitude. From a comparison with the theoretically predicted variation one may then expect to study the trapping properties of the earth's magnetic field, therefore the mean lifetime of the electrons, and thereby the density of the atmosphere at various altitude levels.

For a more detailed account of the theory, see Appendix of paper in J. Astronaut. Sci. (Amer. Astronaut. Soc.) 6, 1 (1959).

For a more technical review of geomagnetically trapped radiation forming radiation belts, see: Progress in Cosmic Ray Physics (J. G. WILSON, editor), Vol. 6. Amsterdam: No. Holland Publ. Co., 1960.

Observations on Small Primates in Space Flight

By

Ashton Graybiel[1], **Joseph H. McNinch**[2] and **Robert H. Holmes**[3]

(With 5 Figures)

(Received August 27, 1959)

Abstract — Zusammenfassung — Résumé

Observations on Small Primates in Space Flight. Two biological packages were sent 480 kilometers into space in the nose cones of Jupiter-C Missiles. These experiments required close cooperation between the biologists and the engineers. The first flight with one squirrel monkey was launched in December 1958, but was not recovered. The second flight with two animals, one female Rhesus monkey ("Able") and one squirrel monkey ("Baker") was launched and recovered in May 1959. Flight data were obtained by fourteen telemetered channels plus a spaceborne tape recorder.

The primary objective of the experiment was to recover the animals in good condition. Secondary objectives were to verify technical aspects of the biopackages and to prove the telemetry for physiological measurement applications.

Electrocardiogram records of the monkeys were obtained. The first flight monkey's heart rate fluctuated continually in a saw-tooth fashion, which was attributed to shivering, environmental noise and vibration. Transitory steep rises were noted at missile lift-off, and near cut-off. The records of the other monkeys showed quite different responses. The heart beat rate for Baker was extraordinarily stable although relatively fast. Variations are attributable to endovascular reflexes during major flight events. From the records of the electrocardiograms, no severe injury to the pericardium or heart wall was evidenced. Respiration measurements showed that changes in rate were associated with major events of the flight. Individual variations provided different patterns of these changes. Changes in respiration rates were well within physiological ranges for Able and Baker. The body and ambient temperatures of the animals remained relatively unaffected by the large variations of the nose cone skin temperatures. After the recovery the animals were in good condition and Baker has remained in good health since the flight.

The report recommends that the period between placing the passenger in the vehicle and take-off should be reduced.

Beobachtungen an kleinen Primaten während des Raumfluges. Es wird über zwei Raketenaufstiege (Dezember 1958 und Mai 1959) berichtet, bei denen Äffchen als Passagiere verwendet wurden. Zur Verwirklichung dieser Versuche waren sehr schwierige Vorbereitungen notwendig, die nur durch enge Zusammenarbeit zwischen Biologen und Ingenieuren zum Erfolg geführt werden konnten. Dabei zeigte sich, daß man sich vorläufig auf eine engere Auswahl von Informationen beschränken muß, um nicht den Gesamterfolg zu gefährden. Die verhältnismäßig größte Gefahr besteht für die tierischen Passagiere während der langen Vorbereitungszeit, die vor dem Start unvermeidlich ist. Für den Transport in den Raum wurden Nasenkegel

---

[1] Captain USN (M.C.).

[2] Brigadier General USA (M.C.).

[3] Colonel USA (M.C.).

von Jupiter-C-Fernlenkgeschossen verwendet. Dabei wurden Höhen von 480 km erreicht.

Bei dem zweiten der erwähnten Versuche, an dem zwei Affen („Baker" und „Able") teilnahmen, wurden insgesamt 14 elektrische Signale für die Indikation des physiologischen Zustandes und der „Umwelt" übertragen. Das Verhalten der Tiere wurde mit Kameras beobachtet. Bei der Temperaturmessung der Umgebung und der Körper der Passagiere zeigten sich trotz großer Schwankungen der Außentemperatur an der Oberfläche des Nasenkegels der Rakete verhältnismäßig geringe Veränderungen. Die Aufnahme der Herzschläge zeigte bei beiden Tieren im wesentlichen Zusammenhänge mit den äußeren Ereignissen des Fluges. Auch die Meßkurven der Atmungsgeschwindigkeit wiesen auf solche Zusammenhänge hin. Bei der Befreiung der Tiere aus der Kapsel befanden sich beide in gutem Zustand.

Observations sur des petits primates en vol spatial. Un rapport sur deux ascensions en fusées (décembre 1958 et mai 1959) au cours desquelles de petits singes ont été emportés comme passagers. La préparation de ces recherches a été longue et a requis la plus étroite collaboration entre ingénieurs et biologistes. Il est établi qu'il faut temporairement limiter l'étendue des recherches possibles. La période la plus aléatoire est celle, inévitable, durant laquelle les passagers sont préparés pour subir les conditions du vol. Des cones de pénétration de fusées Jupiter-C ont été utilisés comme habitacles et des altitudes de 480 km atteintes.

Au cours de la seconde de ces recherches, à laquelle deux singes "Baker" et "Able" ont participé, 14 signaux ont été transmis pour les conditions ambiantes et les indications physiologiques. Le comportement des animaux a été observé par cameras. Les mesures de température ambiante et physiologique ont montré des oscillations relativement faibles comparées aux variations de température de paroi. L'activité cardiologique des deux animaux a été en rapport étroit avec les conditions extérieures du vol. Il en a été de même pour les activités respiratoires. A leur libération de la capsule les animaux se trouvaient en bonne condition.

In presenting a report on the experiments in which three monkeys were sent 480 kilometers up into space, we are acting as representatives of a large group of scientists and of men with great technical skill whose combined effort made these experiments possible. Someone has said that the greatest hurdle facing anyone who would carry out an experiment in space is to arrange for transportation. In the first flight which took place December 13, 1958, and the second flight which took place May 28, 1959, the transportation was by courtesy of the U.S. Army Ballistic Missile Agency on a non-interference basis. Both flights represented joint Army-Navy cooperation between the medical departments, and the second flight was carried out under the sponsorship of the National Aeronautics and Space Administration.

First Test Flight

The first test flight took place in December, 1958. In this test, a squirrel monkey weighing 350 grams (designated "Old Reliable" and hereinafter referred to as O.R.) was transported in the nose cone of a Jupiter C missile a distance of about 1300 nautical miles in less than a quarter of an hour. Unfortunately, the vehicle was not recovered, which might be said to emphasize the factor of reliability in this mode of travel. Up until some time during re-entry, the condition of the animal, based on the telemetered flight data, was good. In other words, this monkey after being placed in the nose cone was not over-stressed as a result of the countdown period of seven hours, the acceleration of about 12 G during liftoff, the sudden transition to the subgravity state at cutoff, the period of near zero gravity till spinup, and the initial period of re-entry. The responses of the animal during this flight will be mentioned later in connection with the similar experiment in the second test flight. At this time, it will suffice to make two generalizations: first, the success achieved inspired much confidence in the bio-technical aspects of the experimental procedure. This included the design and construction of the tiny capsule with its equipment, which provided both a life-support environment and equipment for telemetering essential physiological and environmental data during flight. Secondly, we also acquired a fear of the waiting period between placing the capsule in the nose cone and lift off. This is best emphasized by the fact that by a close decision, one of two identical biopacks was used in this flight. The monkey in the rejected biopack died before the missile left the pad, despite the fact that he was removed from the capsule as soon as the point was past when he might have served as a substitute.

In turning now to a brief description of the second test flight, we cannot help but feel that we have not emphasized sufficiently the over-all effort of many persons of knowledge and skill in the disciplines they represented over a period of about six months.

Second Test Flight

In this experiment, two monkeys were recovered in good condition after a flight in all major respects resembling the first flight save for the fact that re-entry and pickup were successful. The first, a female Rhesus monkey, born in

Fig. 1

St. Louis, Mo., designated as "Able", weighed 2.72 kg. During the training period, she was exposed to the environmental living space similar to the capsule in which she was to ride (Fig. 1). In this capsule, provision was made for telemetering 14 electrical signals to serve as indicators of the capsule environment and the physiological condition of the animal. Additionaly, cameras were provided to obtain pictures and movies during the flight. Some of these signals were not utilized or failed. At the last minute, a shock and avoidance test was canceled since her poor performance resulted in interference with other signals. The electrocardiogram was lost because of a poor connection and failure of transducer, movement of the animal, and other factors resulted in the loss. Because of the position of the biopack in the nose cone, the countdown period lasted nearly three days. The launch crew felt that the biggest hurdle had been passed when the moment of launch arrived with the animal in good condition.

The second passenger was a squirrel monkey, designated "Baker," born in South America and weighing 300 grams. She was selected more for her temperamental qualifications than for the physical requirements which were met by a number of monkeys in the colony. Fig. 2 shows a monkey prepared for flight in a capsule which was similar to the one used in the first test flight but containing a few improvements. The launch crew for Baker also felt that the anxious period was over at lift-off, 6 and $1/_2$ hours after the animal had been placed in the nose cone.

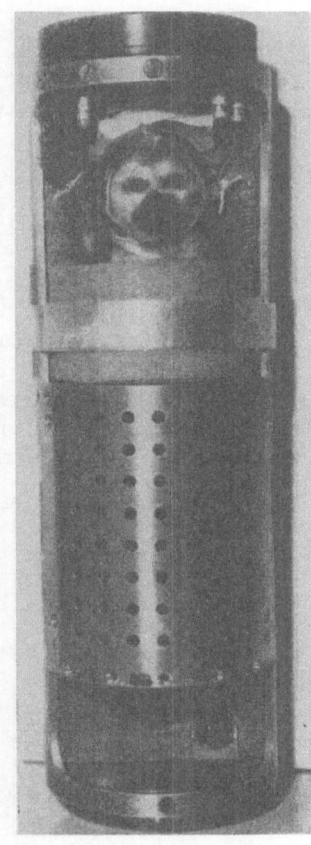

Fig. 2

Neither animal was given any premedication prior to liftoff, although Able was fed by means of an intra-peritoneal tube. Both animals were monitored during flight, and it was readily apparent that they were withstanding the stresses without difficulty. At pickup, both were in good condition and neither required any medical attention. Their survival, in good condition, constituted not only a primary objective of the experiment but offered the best testimony to the reliability of the life support system. There is time to comment only briefly on a few of the experimental findings. It is necessary to re-emphasize that the sophistication in this experiment lay in the technical aspects of the biopackages and the excellent telemetry and not in the physiological measurements *per se*. It is also worth emphasizing that only a few measurements were made because of our concern lest we overstress the animals. In our opinion, the restraint exercised by the investigators in this particular was commendable.

Recordings of body and ambient temperatures during flight were not remarkable, except that they varied but little, this despite the anticipated wide and extreme variations in the temperature of the outside surface of the nose cone.

Fig. 3 shows the variations in heart rate recorded in all three experiments. The differences in the patterns are striking. In the O.R. experiment the changes in the heart rate appear to be occurring continually in a saw-tooth fashion, fluctuating between 225 and 325 beats per minute. Subsequent analysis of the ECG recording indicated that the pattern of these data is largely the result of artifact, possibly a combination of shivering movements and environmental noise and vibration. The ECG tracing indicated a pre-launch rate between 220 and 225. At lift-off there was a transitory acceleration in the rate to 339 which quickly dropped to 240; thereafter during the boost phase the rate again increased gradually to a rate between 300 and 320. Just before cut-off there was a fall in the rate to the pre-launch values followed by a sharp rise to 360. This rise was very brief, and thereafter the rate varied without pattern, and at a level slightly above the pre-launch values. The sudden but very brief acceleration at lift-off and cut-off is believed to be the result of startle.

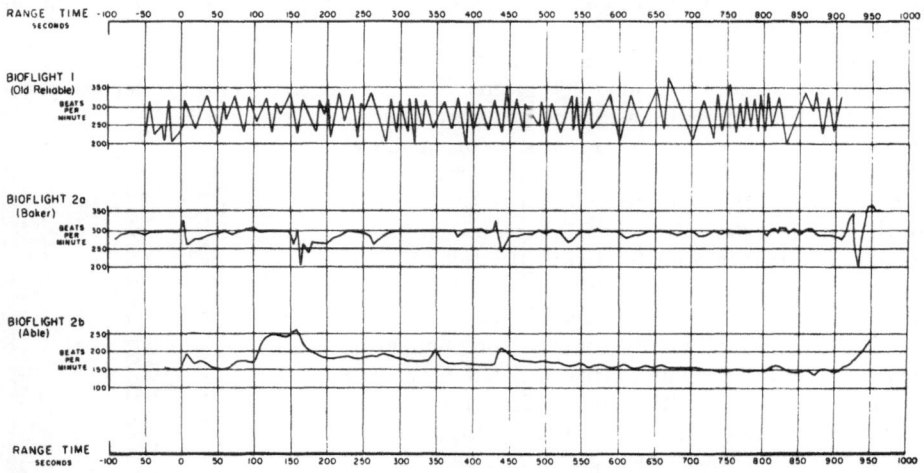

Fig. 3. Heart rate of monkeys during Jupiter missile flights

The changes in the heart rate of Able and Baker clearly indicate the major events throughout the course of the flight. Although the responses are quite different, the extraordinarily stable base line in the case of Baker as well as the

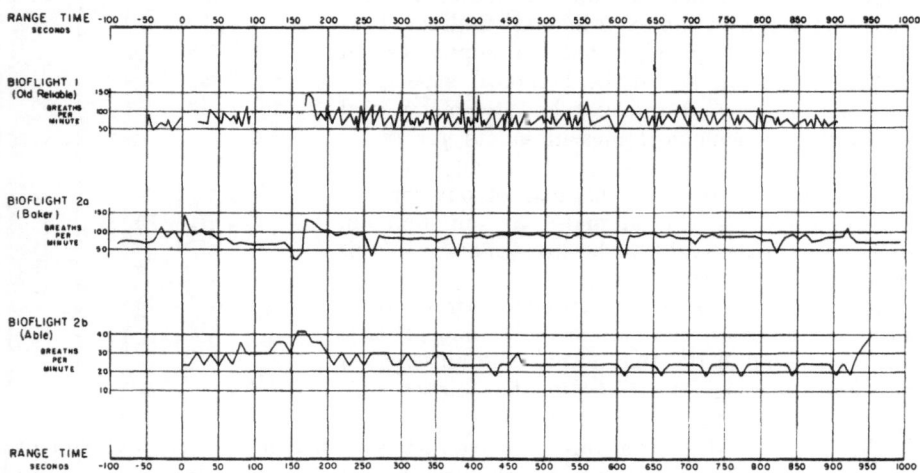

Fig. 4. Respiration of monkeys during Jupiter missile flights

relatively fast heart rate suggest not only the presence of a slight constant stress but also the freedom from alarm. If such was the case, then the variations observed are mainly the reflections of the play of the endovascular reflexes during the major events. The transition period from burnout to free flight is the most interesting, and investigators have speculated on the effects from what might be termed a "rebound stress".

Baker responded with the slowest heart rate observed during the flight and with a very slow return to base line. Able responded with a sharp fall in heart rate which did not reach the base line. A review of the findings of other investigators indicates even greater variations in response at this critical point in ballistic flight, and it is evident that not all of these variations are due solely to individual differences in response to the same stimulus; other factors must be involved as well.

In Fig. 4 are shown the measurements obtained of respiratory rate in all three animals. With O.R., just as in the case of the heart rate, the respirations appeared to fluctuate in saw-tooth fashion, principally within the range of 60

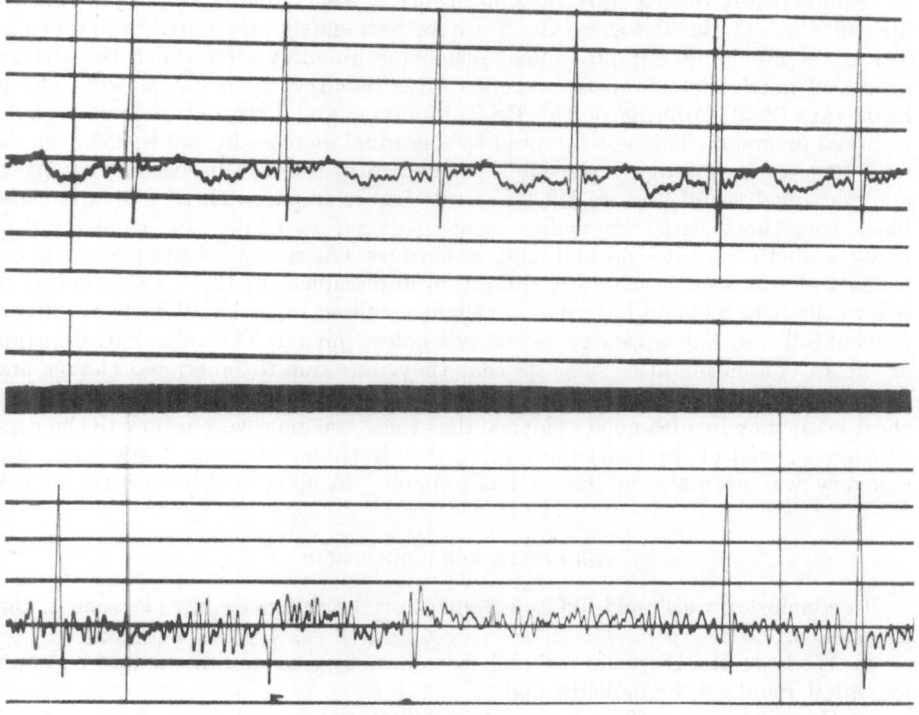

Fig. 5

to 110. It is believed that variations reflecting major events of the flight did take place but were marked by artifact. In the second test flight (Able and Baker), changes in respiratory rate reflect the major events throughout flight with the exception of spinup, although the pattern of these changes is not the same for both animals. Able responded with a moderate gradual rise during the boost phase followed by a steeper decline at cutoff, whereas Baker showed a sharp initial rise followed by a gradual fall with a second sharp increase at the time of cutoff. On re-entry, Able showed a considerable rise in respiratory rate whereas Baker exhibited slight changes. In summary, the changes in respiratory rates observed in Able and Baker were well within the physiological ranges for these animals.

Electrocardiograms were obtained only on O.R. and Baker during flight. A bipolar chest lead was used with electrodes implanted under the skin and in apposition to the anterior surface of the heart. Technically the records were good except when distorted by artifact. This was most prominent during the re-entry phase as the result of vibration and electro-magnetic phenomena due to ionization.

In the case of O.R., only slight alterations were seen in addition to changes in heart rate and none had pathological significance. The alterations were greater in the case of Baker and most notably during the re-entry period. A preliminary analysis has been made of the record obtained during the 98 seconds before impact. A space-borne tape recorder was used. The spinup had occurred prior to the onset of the recording and re-entry occurred 36 seconds after the onset. The rhythm was normal during the first 23 seconds, the rate varying between 272 and 300. At that point in time, two interpolated ventricular beats occurred in successive cycles but the basic rhythm was unaltered. This was followed 9 seconds later (4 seconds before re-entry) by the appearance of artifact sufficient to prevent the identification of the P waves; the T waves became slightly lower in amplitude. There was momentary improvement just before re-entry after which the artifact produced much distortion. Six seconds after re-entry there was slowing of the heart rate (225), lowering of the RS-T junction, and lowering of T with slight terminal inversion. This was followed by a gradual increase in rate to 332 over the next 13 seconds. Twenty-two seconds after re-entry the rhythm became abnormal and remained so off and on till 8 seconds before impact. There was sino-atrial block sometimes with ventricular escape, premature beats with compensatory pause sometimes with normal and sometimes bizarre QRS complexes. Gross artifact at times made close interpretation impossible, but there was lowering of S-T to the base line and lowering and slight terminal inversion of T. These altera-tions of S-T and T disappeared 14 seconds before impact. The very short duration of the ECG abnormalities suggests that they were due to functional change and disorder. Although the ECG findings were confined to those obtained in a single chest lead, they furnish good evidence that there was no severe injury to the peri-cardium or wall of the heart underlying the electrode. The record obtained after recovery was normal, and Baker has remained in good health since the flight[1].

Summary and Conclusions

In conclusion we should like to comment on the things we have learned in this experience.

1. We have the bio-technical skill necessary to provide a life support system for small monkeys in ballistic flights.

2. Carrying out such experiments requires much time and effort, and the amount of data obtained seems small in comparison. The natural desire to obtain as much information as possible must be tempered by the fact that everything may be lost if the attempt is over-ambitious.

3. These experiments were a new experience for us in cooperative effort. A large number of persons were involved, and the problem of coordinating all the efforts in time and space toward a highly specific goal was not easy. Cooperation between biologist and engineer is as important as their respective skills.

4. We acquired a great respect for the engineers whose standard of perfec-tion and ingenuity in devising new or adapting old equipment was outstanding.

5. With regard to the future, it is obvious that we need to add performance tests to our battery, but not at the expense of giving up indicators of physiological performance.

6. Of the many suggestions which have come in for improving the procedure in experiments of this kind, the greatest gain would come from reducing the period between placing the passenger in the vehicle and takeoff.

---

[1] True as this goes to press (March 1960)

Acknowledgement

The authors wish to make grateful acknowledgement to the many individuals who made these tests possible and contributed to their success. The list of such persons is long, and includes but is not limited to the following:

Major General JOHN B. MEDARIS, Brigadier General JOHN A. BARCLAY, Mr. GEORGE PEDIGO, Mr. WALLACE KISTLER, Mr. JAMES PCWELL, Mr. SANDFORD DOWNS, Mr. JOHN AVERY and Private First Class RUSSELL LEAGUE of the US Army Ordnance Missile Command.

Doctor DIETRICH BEISCHER, Mr. W. CARROLL HIXSON, Doctor JORMA NIVEN, and Doctor DONALD E. STULLEN, of the US Naval School of Aviation Medicine, and Captain NORMAN BARR, Bureau of Medicine and Surgery, US Navy.

Colonel CHARLES W. HILL, MSC, Lt Colonel RICHARD TAYLOR, M.C., Major GERALD CHAMPLIN, M.C., Major JOSEPH BRADY, M.C., Captain EDWARD S. WIL-BARGER, MSC, Doctor THOMAS R. A. DAVIS, Doctor SIEGFRIED GERATHEWOHL, and Captain WILLIAM S. AUGERSON, M.C., and Doctor ARTHUR RIOPELLE of the US Army Medical Research and Development Command.

Doctor T. KEITH GLENNAN, NASA.

Ionospheric Scintillations of Satellite Signals

By

H. P. Hutchinson[1] and P. R. Arendt[1]

(With 9 Figures)

(Received June 29, 1959)

Abstract — Zusammenfassung — Résumé

Ionospheric Scintillations of Satellite Signals. The scintillation of satellite-emitted radio signals has been observed using two different techniques, namely, doppler-shift frequency measurements and radio direction finding. Variations from a smooth doppler-shift curve obtained during individual orbits give a measure of frequency scintillation and thus of the roughness of the ionospheric path between the satellite and the observer. As expected, these variations are a function of frequency. Examples of the range of scintillations are given; both as a function of frequency and as correlated with an artificial ionospheric disturbance (*Argus*). Variations in direction-finder bearings are presented and their rate of change is discussed in relation to the roughness of the ionosphere. Both techniques are applicable to measurement of the hyperfine structure of the ionosphere. Tentative indices of ionospheric inhomogeneities are presented; namely, scintillation quality index from doppler measurements and probability of angular deviation from direction-finding technique.

Ionosphärische Szintillationen von Satellitensignalen. Es wurde die Szintillation von Radiosignalen, die von Satelliten ausgesandt wurden, unter Benützung von zwei verschiedenen Methoden beobachtet: nämlich durch Messung der Frequenz der Doppler-Verschiebung und durch Radio-Richtungsfindung. Aus einer glatten Doppler-Verschiebungskurve während individueller Umläufe erhaltene Variationen geben ein Maß der Frequenzszintillation und damit der Unregelmäßigkeit des Weges der Signale in der Ionosphäre zwischen Satellit und Beobachter. Wie erwartet, sind diese Variationen eine Funktion der Frequenz. Beispiele des Bereiches der Szintillationen werden beigebracht, sowohl als Funktion der Frequenz wie in Beziehung zu einer künstlichen Störung der Ionosphäre (*Argus*). Es werden auch Variationen der Bestimmungen mit dem Richtungsfinder aufgezeigt und die Geschwindigkeit ihrer Änderung in Beziehung zur Rauhigkeit der Ionosphäre erörtert. Beide Methoden können auf die Messung der Hyperfeinstruktur der Ionosphäre angewendet werden. Versuchsweise werden Indizes der ionosphärischen Inhomogenitäten angegeben, und zwar der Szintillationsqualitätsindex nach der Doppler-Methode einerseits, die Wahrscheinlichkeit der Winkelabweichung auf Grund der Richtungsfindungsmethode anderseits.

Scintillation ionosphérique des signaux de satellites. La scintillation des signaux radio émis par satellite a été observée par deux techniques différentes: la mesure du déplacement de fréquence par effet Doppler et la détection d'orientation. Les écarts à une courbe de déplacement régulière de la fréquence par effet Doppler, obtenus au cours de passages individuels, donnent une mesure de la scintillation de

---

[1] U.S. Army Signal Research and Development Laboratory, Fort Monmouth, New Jersey, U.S.A.

fréquence et donc de l'irrégularité du trajet ionosphérique entre le satellite et l'observateur. Comme il fallait s'y attendre ces écarts sont fonctions de la fréquence. On donne des exemples d'amplitudes de scintillation, aussi bien en fonction de la fréquence qu'en relation avec des perturbations ionosphériques artificielles (*Argus*). Les variations dans la position angulaire sont présentées et leurs changements sont discutés en relation avec l'irrégularité ionosphérique. Les deux techniques sont applicables à la mesure de la structure hyperfine de l'ionosphere. Des indices approximatifs d'inhomogénéité de l ionosphère sont déduits de la qualité de scintillation dans l'effet Doppler et de la probabilité de déviation angulaire dans le repérage d'orientation.

The short-time variations of satellite-emitted radio signals have been observed by using two techniques, namely doppler-shift frequency measurements and radio-direction-finding measurements. These variations are most interesting to observe and it is believed that they afford a good measure of the roughness or inhomogeneity of the ionosphere. As one expects, these variations are more noticeable at the lower satellite-transmitted radio frequencies, namely 20 and 40 mc's than at the higher ones. However, as you will see, these scintillations cannot be considered negligible at 108 mc/s, which is the frequency most commonly used for determining the position location of previous U.S. satellites.

The doppler shift that is observed is primarily a function of the radial velocity of the satellite with respect to the observer and the emitted frequency. However,

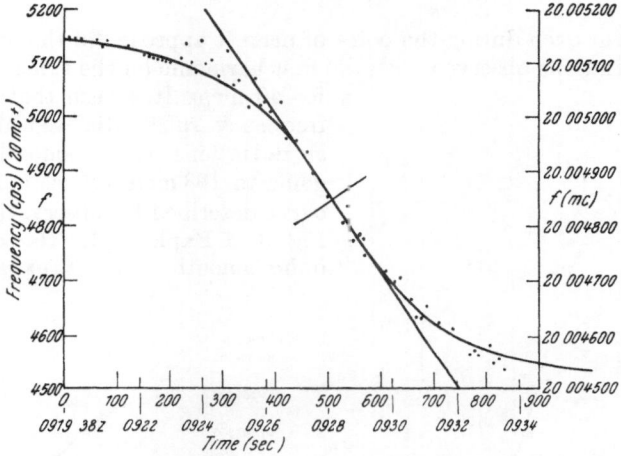

Fig. 1. Satellite, 1958 δ; orbit # 2294; Time reference 09 19:38 G.M.T.; Zero frequency 20.004847 mc/s; Zero time 09 28:05 G.M.T.

for frequencies in the region between 20 and a few hundred mc/s, the effects of the wave velocity of propagation through the ionosphere as a function of the refractive index are noticeable. Therefore, the observed doppler shift is further affected by changes in refractive index or ionization along the path between the satellite's position and that of the observer. Also, the doppler shift is affected by the intensity of ionization at the satellite. The degree to which the observed variations are attributable either to changes in ionization at the satellite when the satellite is passing through the ionosphere or to changes in the integral of electron density along the path, is a suitable subject for further study.

These frequency measurements were taken with an instrumental accuracy of ± 7 cycles in frequency measurement and at 1 second time intervals based on simultaneous reception of WWV standard time transmissions. In Fig. 1 are shown the deviations of the observed doppler shift recordings of Sputnik III from a smooth curve drawn through the observed points. It is seen that substantial

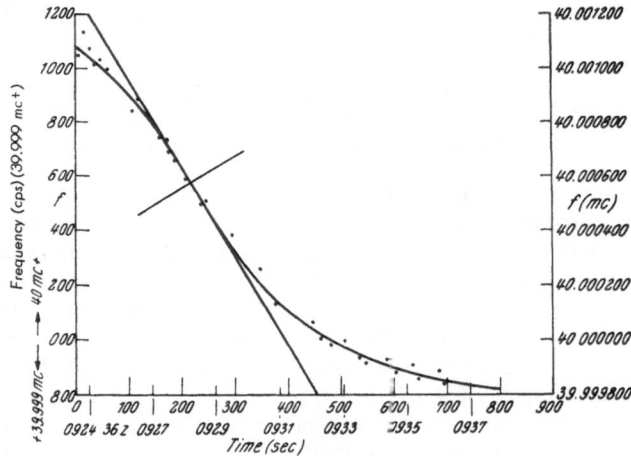

Fig. 2. Satellite, 1958 δ; orbit # 2294; Time reference 09 24:36 G.M.T; Zero frequency 40.000575 mc/s; Zero time 09 28:13 G.M.T.

variations occur even during the point of nearest approach of the satellite. Fig. 2 shows the variations observed with the first harmonic on the same orbit, namely for 40 mc/s. It is seen that even at this frequency substantial scintillations occur as deviations from a smooth curve. Now going to 108 mc/s we see a fairly smooth curve described by observations shown in Fig. 3 of Explorer I. The center part is quite smooth but, as expected from a

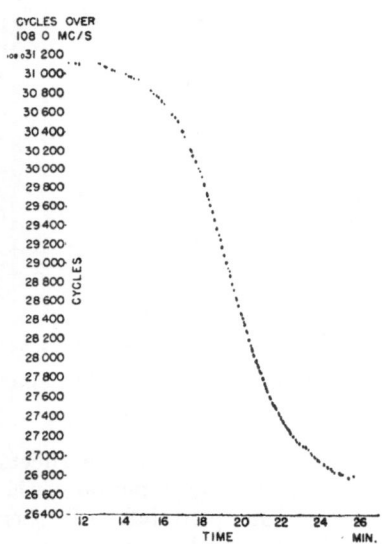

Fig. 3. Satellite, 1958 α; orbit # 97, February 8, 1958; Time reference 18 11:00 E.S.T.

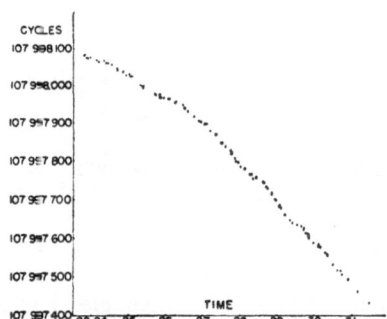

Fig. 4. Satellite, 1958 β; orbit # 128, March 28, 1958; Time reference 22 24:00 E.S.T.; Frequency deviations in the order of 20 to 25 cycles

longer ionospheric path at the outer ends of the curve, these ends show some deviations from a smooth curve. This orbit has an apogee of 2500 km and a perigee of 360 km. Thus, for a good part of the time it lies within the ionosphere and the ionospheric path from the observer to the satellite is somewhat less than the total thickness of the ionosphere in the direction of the satellite.

Such is not the case for the orbit of Vanguard I, which at all times is well above the maximum of the normal F region. Its orbit has an apogee of 3960 km and a perigee of 660 km. There are shown in Fig. 4 frequency deviations of the order of 20 to 25 cycles observed during a particular orbit of Vanguard I. This orbit occurred during a time when the ionospheric quality figure was 7 (good).

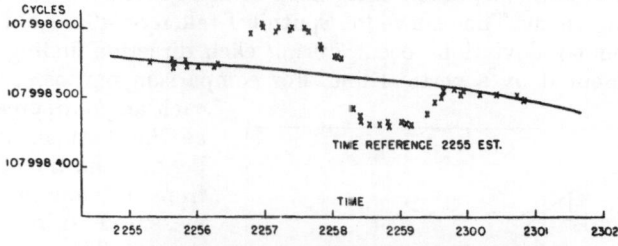

Fig. 5. Satellite, 1958 β; orbit #138, March 29, 1958

Subsequently, on the next day, there occurred the largest short-period deviation, which we have observed on the 108 mc/s signals (Fig. 5). This anomaly also was taken from an orbit of Vanguard I. It appears to be of greater magnitude than that might be expected from variations in the integrated electron density of the iono-sphere. The quality index was "good" again. Since the orbit of this satellite is well above the normal F region and at the time of this observation the satellite was at an altitude of approximately 3900 km, that is, close to its apogee, a possibility exists that the occurrence may have been caused by the satellite encountering a cloud of charged particles above the ionosphere. However, one must also consider the possibility of an ionized cloud coming between the satellite's position and the observer for the three minute time duration of the anomaly by some disturbance in the F region of the ionosphere.

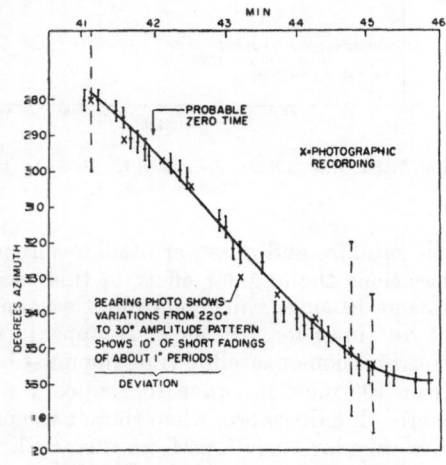

Fig. 6. Satellite, 1957 a; October 18, 1957; 21 41 E.S.T.; 40 mc/s

From this and the previous illustrations, it is clear that ionospheric frequency scintillations of substantial magnitudes do occur with satellite signals. This observation coincides with the so called "color" scintillations of the radio star signals which have been cited by GORDON LITTLE [1].

For all the above figures, the smoothed curves were drawn by hand through the plotted data points and thus no measure is obtained of the systematic errors caused by effects of the earth's rotation and of the curvature of the satellites' path. The comparison of these variations of actual doppler-shift observations from a calculated doppler-shift curve has been discussed elsewhere [2] and time does not allow its inclusion in this paper.

Let us now consider the use of direction-finding azimuthal bearings with regard to studies on ionospheric scintillations. These bearings may be used to obtain Bearing versus Time curves as the satellite signals come within range of the observer. Since one may regard the satellite velocity as varying smoothly over the orbital sector under observation, the angular deviation of successive bearings from a smooth curve is indicative of ionospheric refractive effects. Possibly an even more useful measure of short-period ionospheric disturbances or scintillations is to tabulate in time sequence all the deviations of each set of bearing measurements from a smooth curve and then obtain successive differences between these deviations. A series of data was taken using both of these methods. There is shown in Fig. 6 a Bearing versus Time curve for Sputnik I taken on 40 mc/s. It is seen that substantial angular deviations occur during each direction-finding observation, which is represented by a vertical line. For comparison purposes, the center of each angular spread was taken as the best measure of that bearing and then the variations from a smooth curve where measured from these center points. The average bearing spread observed is of the order of 6—7 degrees. More precise observations are gained by a photographic method. Now turning to Fig. 7 we see the Bearing versus Time curve obtained on a recording on 108 mc/s of Explorer I satellite. Here again deviations from the smooth curve are obtained. Thus, these angular variations indicate that the actual path of major energy transfer between the satellite and observer oscillates about the average ray path. Furthermore, on occasions the angular effect of this oscillation can be as much at 108 mc/s as it was at 40 mc/s. Thus one cannot assume that these angular effects are neglectible at all times for 108 mc/s. It appears that frequencies for undisturbed satellite transmission or satellite tracking must be raised by at least one order of magnitude from 108 mc/s in order to reduce the angular beam spread by a substantial factor. Furthermore, when these two special sets of data are compared for successive angular variations from a smooth Bearing versus Time curve, the following results are obtained:

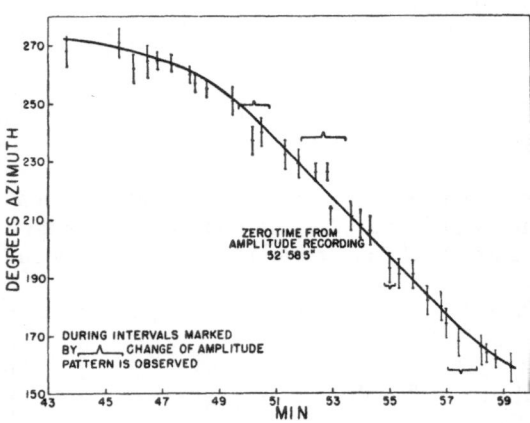

Fig. 7. Satellite, 1958a, February 13, 1958; 16 43 E.S.T.;
108.00 mc/s

Table I. *Roughness Distribution*

40 mc/s Sputnik I		108 mc/s Explorer I	
Angle (deg)	p	Angle (deg)	p
1	0.37	1	0.61
2	0.12	2	0.39
3	0.04	3	0.35
4	0.02	4	0.32
5	0.00	5	0.21
		6	0.11
		7	0.07
		8	0.00

p = Probability that the difference between successive angular variations from a smooth curve exceeds a given angle.

For the Sputnik I orbit, the maximum rate of change of the direction-finder bearings was 70° per minute, while for the Explorer I orbit it was 20° per minute. Also, the Explorer I path length was considerably greater than that for the Sputnik I. The longer path length for the Explorer supports a greater variability in the integral of electron density over the path, since it is well known that Explorer traveled outside the normal F region whereas Sputnik I was within the F region, except for a few occasions when it was below the F region. Thus this preliminary comparison gives some clues as to what we may expect when we have a satellite orbiting outside the ionosphere and transmitting several frequencies in the 40—
—1000 mc/s region. Such transmissions are vitally necessary to obtain better comparative data for studying the degrading effects of the ionosphere and upper atmosphere upon signals passing through to or from the earth's surface. Meanwhile, those interested in designing earth to satellite communications systems would do well to include in their systems means for overcoming the degrading effects.

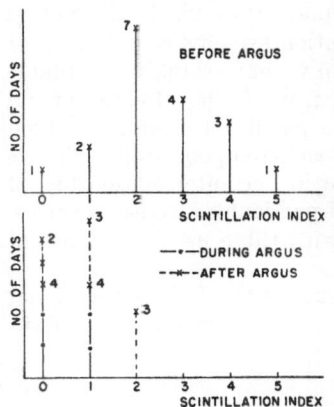

Fig. 8. Distribution of scintillation quality index

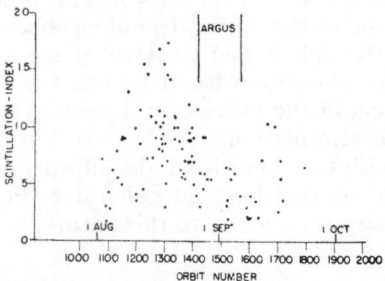

Fig. 9. Satellite 1958δ, Sputnik III

From the above illustrations it is seen that both frequency deviations taken on doppler-shift measurements and direction-finder bearing-deviations can be used as a measure of ionospheric roughness. With the recent public announcement of the *Argus* high-altitude nuclear detonations and the CHRISTOFILOS effect, our doppler shift recordings were studied to see whether any systematic effect was discernable. As a preliminary measurement, the records of eighty-nine orbits were examined visually and ranked on a qualitative grading basis from 0 to 5 with the number zero denoting no sincillations and the number five maximum scintillations. We call this a scintillation quality index. These observations were all on Sputnik III orbits and taken at 20 mc/s, which is our most sensitive frequency for that purpose. In Fig. 8 are shown histograms of these distributions taken before, during and after the three *Argus* high-altitude detonations. It is clearly seen that systematic changes occurred which initially showed a strong reduction in scintillation of the doppler-shift signals and even several days later indicated a much lower index than for the period before the detonations. Meanwhile quantitative measurements were initiated using the measured areas generated by the difference between a smoothed curve through the data and the curve generated by the data points. For each orbit this area was measured in arbitrary units and the resultant number was divided by the time (in minutes) of the observation. The results of this procedure are shown in Fig. 9. On this figure data to the left of the first vertical line cover the period of 18 days prior to the first detonation and to the right, the period of 16 days afterwards. It is clear that in this latter period the scintillations of these Sputnik III signals were systematically less than prior to the event.

The median valves of both these visual and quantitative scintillations measurements are in good agreement. One further question is under investigation. This is a study on the correlation of this scintillation index with suitable indices of solar activity. If it is found that there is a positive correlation between an appropriate solar index and our scintillation index at times before and considerably after the *Argus* operation period, then such results would prove our present belief that the effects shown on these last slides are primarily caused by the *Argus* detonations. Further work is needed to develop the correlation, which seems to be indicated by the data which has already been reduced.

In conclusion, we wish to indicate our belief that these techniques using doppler shift and direction-finding may prove to be most useful tools in further studies in the ionosphere and outer space. Furthermore, they yield information of direct interest to designers of satellite communication systems as a large part of their problem is to overcome such variations. When we have satellites available in orbits outside the ionosphere and transmitting from each satellite several signals in the 40—1,000 mc/s region, it will then be possible to obtain a much better appreciation of the degrading effects of the ionosphere upon satellite signals passing through it and to develop more knowledge in the interrelationship of measurements made with many other techniques. Of particular interest is such a comparison of the previously described techniques with those used in studies of radio star scintillation.

We wish to acknowledge the efforts of our colleagues at the U.S. Army Signal Research and Development Laboratory in collecting the large amount of raw data so necessary for a study of this nature.

References

1. G. LITTLE, J. Geophysic. Res. **59**, 152 (1954).
2. 1959 I.R.E. National Convention Record, Vol. 5: H. P. HUTCHINSON, Satellite Radio Emission, Part II: Slant Range at Nearest Approach.

Ion Rocket Efficiency Studies

By

Guntis Kuskevics[1]

(With 6 Figures)

(Received August 27, 1959)

Abstract — Zusammenfassung — Résumé

Ion Rocket Efficiency Studies. The energy losses for a general ion rocket engine using a surface ionization ion source are outlined. The surface ionization efficiency is calculated for all alkali metals on tungsten and for Cs on various base metals assuming no adsorption. The adsorption of Cs on tungsten is analyzed by curves showing work function. The rate of energy loss per unit of current is shown to decrease with temperature. The total major loss which is associated with emission and acceleration can be minimized for an optimum design. An extensive bibliography of surface ionization is included.

Studien über die Leistung von Ionenraketen. Es werden die Energieverluste für den Motor einer Ionenrakete, die eine Oberflächenionisationsquelle benutzt, auseinandergesetzt. Die Leistung der Oberflächenionisation für alle Alkalimetalle auf Wolfram und für Cäsium auf verschiedenen Grundmetallen unter der Annahme des Stattfindens keiner Adsorption wird berechnet. Die Adsorption von Cäsium auf Wolfram wird durch Diagramme illustriert. Das Ausmaß des Energieverlustes je Stromeinheit sinkt, wie sich zeigen läßt, mit der Temperatur. Der hauptsächliche Verlust, der mit Emission und Beschleunigung verbunden ist, kann für eine optimale Konstruktion auf ein Mindestmaß gesenkt werden. Ein ausführliches Literaturverzeichnis über Oberflächenionisation ist beigeschlossen.

Etudes sur le rendement des fusées ioniques. Les pertes d'énergie accompagnant les sources d'ionisation par surface dans un moteur ionique sont indiquées. Le rendement en ionisation de surface est calculé pour les couples formés par le tungstène avec les métaux alcalins et le Césium avec différents métaux de base, dans l'hypothèse où il n'y a pas d'adsorption. L'adsorption du Césium par le tungstène est montrée sur des graphiques illustrant la fonction de forces. Le taux de perte d'énergie par unité de courant décroit avec la température. La perte principale, associée à l'émission et l'accélération peut être rendue minimum par un dessin adéquat. L'article comprend une bibliographie étendue sur l'ionisation de surface.

Ion Rocket Components

The conceptual design of ion rocket propulsion systems has been described by many previous authors over the last decade [1—22]. In general, the system includes some source of energy, probably either solar or nuclear, conversion equipment to change heat into electrical energy, which may be a turbine with generators or a thermionic converter, and the ion rocket engine. Parts of the ion

---

[1] Flight Propulsion Laboratory Department, General Electric Company, Cincinnati 15, Ohio, U.S.A.

rocket engine have been described [23—31]. The ion engine consists of a surface ionization source, an accelerating and focussing electrode system and a space charge neutralization arrangement. The surface ionization source [32—35] incorporates propellant storage tanks, a feed mechanism and a hot ionization surface with its heater power supplies. The accelerator electrode system would utilize PIERCE's geometry to get high current densities. Some designs are based on just acceleration, while some others utilize acceleration followed by deceleration [25, 31]. The space charge neutralizer is an electron gun injecting an equal number of electrons into the beam, so that an electrically neutral plasma would leave the ion rocket.

General Energy Loss Analysis

This analysis considers only the ion rocket engine or ion accelerator exclusive of the power source, converters, radiators, etc. We assume further that the ion accelerator utilizes a surface ionization source, any acceleration geometry and charge neutralization. For this system we can classify the energy losses as follows.
1. Propellant vaporization, transfer and control losses
   a) Heat of vaporization
   b) Loss due to pumping or pressurizing
   c) Loss due to valving or temperature control
2. Ion emitter loss
   a) Loss of propellant as neutrals
   b) Thermal radiation from emitter
   c) Conduction and low temperature radiation loss
   d) Ionization energy
3. Acceleration loss
   a) Beam interception
   b) Thermal radiation and conduction from heater
   c) Losses due to sputtered material
4. Neutralization loss
   a) Thermal radiation from electron filaments
   b) Electrons accelerated back into the emitter
   c) Recombination radiation
5. Other losses
   a) Ohmic losses in the leads
   b) Auxiliary electrode losses.

Many of these losses are insignificant for the particular case we are considering. In this paper we will concentrate on the analysis of the emission and acceleration losses and their minimization.

Surface Ionization Theory

The utilization of propellant mass in an ion rocket depends upon the ability of the ion source to convert neutral atoms into ions. The surface ionization process is considered to be the most efficient. The efficiency of this process is expressed by the ionization coefficient β defined as the ratio of the number of ions emitted per unit surface area per unit time n_p to the number of particles incident per unit area per unit time n_i.

$$\beta = \frac{n_p}{n_i} = (1 - r_i) \left\{ 1 + \frac{g_a}{g_p} \frac{(1 - r_a)}{(1 - r_p)} \exp\left[\frac{e(V_i - \varphi)}{KT}\right] \right\}^{-1}$$

where V_i is the ionization potential and φ the surface work function, r's are reflection coefficients. In the ion rocket we desire values of β close to 1.0. This means we must select a base metal of high φ and propellant of low V_i.

The elements having the lowest ionization potentials are the alkali metals, while refractory metals in general have high work functions. For a given φ increased temperature decreases the ionization efficiency so that we should operate at low temperature. However, adsorption at low temperature reduces φ to a value below that of alkali metals which is not only much lower than φ of base metal but also lower than V_i of the alkali metal. Hence, high surface temperatures are required to reduce adsorption and maintain a high φ even though the kT term works against us. For this reason we choose refractory metals which besides high φ have high melting points.

The Work Function

The electronic work function [183] of tungsten can be determined by several methods: 1) thermionic, from a RICHARDSON plot, 2) photoelectric, from the threshold wave lenght, and 3) the contact potential measurement. The work function changes with the material and also with allotropic modifications of the metal [159, 78]. The work function is anisotropic [206, 156, 148] or it varies for the different crystal faces from 4.25 to 5.26 ev.

Even for a single crystal face different ways of producing a uniform surface have a profound effect on the work function [248].

For polycrystalline tungsten the work function varies between samples depending upon the history of the sample. The accepted average value is 4.52 ev, which also corresponds to the value for the (001) face. The actual measured thermionic φ for clean polycrystalline W wire varies from 4.45 to 4.55 [183].

Initially the DUSHMAN-RICHARDSON equation contained two significant constants—the work function and the emission constant A. Assuming A to be the same for all metals the work function was found to be a linear function of temperature [153, 157, 160]. This temperature dependent part of φ comes out as an exponential constant and can be grouped with A. This makes A vary with the material and the work function is by definition independent of temperature.

The work function is found to be non-uniform over the surface or "patchy". Some areas have values different from others. For a polycrystalline sample this is to be expected.

Linear temperature dependence of the φ as initially defined would not introduce any curvature in the RICHARDSON plot, while nonlinear variation would. Local variations of work function would make the curve concave upward.

In the SAHA-LANGMUIR Theory on which this analysis is based, some authors have introduced a work function linearly dependent upon temperature [215, 137, 151] but this could not account for all measured data. Using weighted summation of terms corresponding to a patchy work function [243] or a work function distribution function [222], the experimental data could be accounted for by this theory. The ionization of Ca, Sr and Ba could reportedly be understood only in terms of a quantum mechanical theory [122].

Due to the sensitiveness of work function to the microscopic structure of the surface and to adsorption of impurities, data is expected to vary for different samples and even for different measurements under the conditions such as would be encountered in ion propulsion. Most of the previous data were accounted for using for tungsten wire $\varphi = 4.51$ to 4.69. Without temperature dependence a better agreement is obtained by using a work function of about 4.62. Because of the probable use of strips or porous tungsten, rather than wire, best φ may be different. For most of this analysis, we choose 4.52 as the work function for clean polycrystalline tungsten. The actual value for the ion rocket emitter geometry may be higher and for most measurements in vacuum chambers we would expect oxidation to increase it further. This latter effect may not occur in space.

Surface Ionization Efficiency

The effect of temperature at constant φ for alkali metals on tungsten is shown in Fig. 1. The reflection coefficients for alkali metals on tungsten are zero. These curves show that at high temperature where adsorption is negligible φ for metals with $V_i < \varphi$ decreases with T while for metals with $V_i > \varphi$ it increases. These curves definitely show that for $\Delta E = \varphi - V_i$ of about .6 ev β's above .9 are realized. This illustrates the reduced efficiency of potassium as a propellant unless the φ can be increased to a value such as that of the W-O (9.2 V).

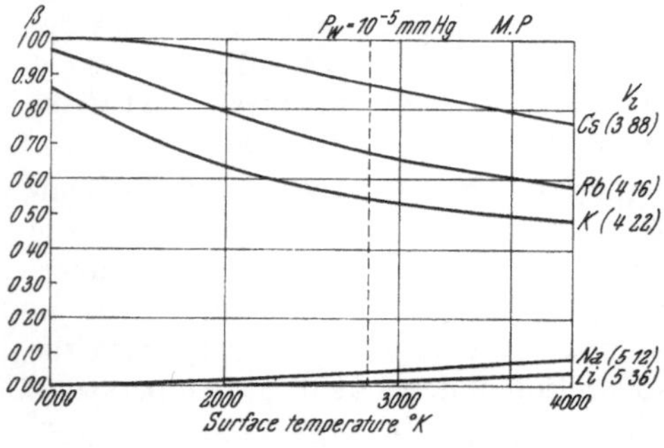

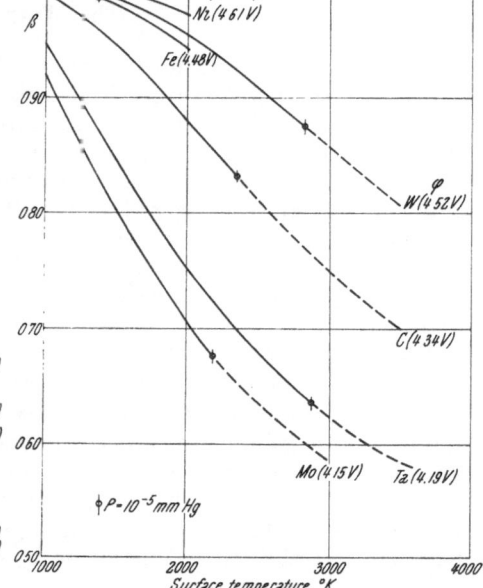

Fig. 1. Ionization of alkali metals on tungsten ($\varrho = 4.52$)

Fig. 2. Ionization of cesium on various metals

The ionization of Cs on different base metals, again assuming constant φ (neglecting adsorption and reflection coefficients) is graphically illustrated in Fig. 2. The useful portion of these curves for ion rocket design extends to temperatures at which the metal vapor pressure increases to the point where an appreciable loss of material occurs during a particular mission. We have selected the pressure of 10^{-5} mm Hg since this is the highest pressure we could probably tolerate in the test tank. For long operation times the rate of loss of surface material is going to decrease the maximum allowable emitter temperature by about 500° K from the $P = 10^{-5}$ mm Hg point. From this standpoint, to get high φ with low evaporation of the base metal either W or Ta and possibly Pt are the only likely materials. Ni and Fe may work below 1200° K if the adsorption in this range is negligible. Other metals would become attractive if coated with a surface layer to increase the work function, oxidation or coating with W being the most likely. Since the oxide layer on W is stable to 3200° K and is formed even at residual air pressure of 10^{-8} mm Hg in about 4 hrs., we can hardly prevent it from forming unless the surface is periodically cleaned. This puts an additional uncertainty in all theoretical calculations since for long mission times a layer may form, depending on the space vacuum conditions. A similar improvement may be expected with Ta and Mo. The oxide layer increases the adsorption of the alkali metal which we have so far neglected.

Adsorption

In the above discussion we assumed a constant work function, thus neglecting adsorption of the alkali metal. These curves give only the limitations imposed by temperature and the properties of the clean surfaces. To evaluate adsorption

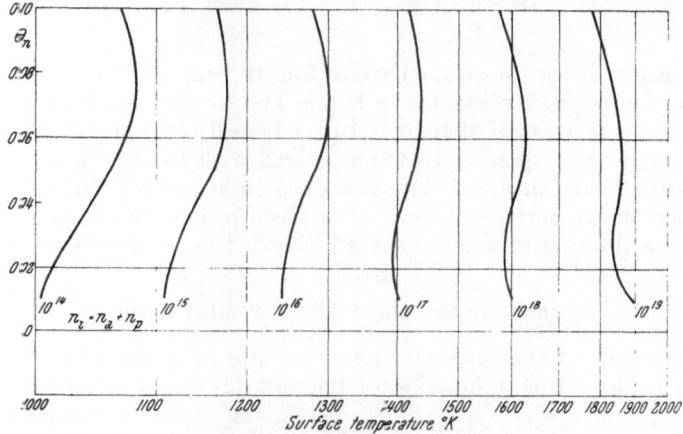

Fig. 3. Cesium coverage of tungsten near critical temperature

effects we need experimental data of which an adequate amount is only available for Cs and other alkali metals on W. For this reason the following discussion will

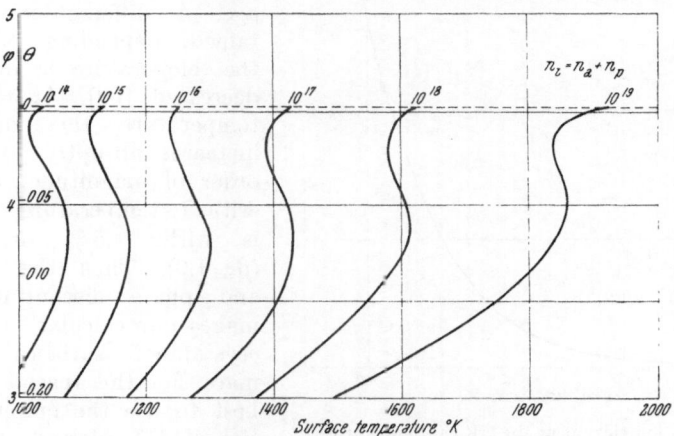

Fig. 4. Work function of tungsten in a cesium atmosphere

deal only with the ionization of Cs on W. The adsorption of Cs on W and its effect on the work function can be deduced from the data of TAYLOR and LANG-MUIR [102].

In order to calculate ion current density j, we must know the work function as a function of T and n_i. In a previous paper [25] this relationship was derived from the contact potential data in Table I [102] and the curves in Fig. 10 [102]. Dr. D. Langmuir has pointed out that curves calculated in this way are shifted some 200—300° K higher than the experimental data represented by Eq. 6 [141]. The data in Fig. 10 [102] actually plot ϵ as a function of T and n_a (Langmuir's ν_a) and only approximate n_i (Langmuir's μ_a) for θ larger than .1. For these curves the assumption was made that $a_i = n_a$ which is true for measurements of atom evaporation for which a retarding field kept any ions from evaporating.

For

$$
\begin{aligned}
.10 < \theta < 1.0 \qquad & n_i \approx n_p \\
.04 < \theta < .10 \qquad & n_i = n_a + n_p \\
\theta < .04 \qquad & n_i \approx n_a .
\end{aligned}
$$

For the range of our interest, curves in Fig. 10 [102] are misleading. A revised set of curves for $\theta < .10$ is presented in Fig. 3. These curves are S shaped. Thus for a given n_i there is a range of temperature where θ can have three values. The work function is correspondingly triple valued with respect to temperature for a fixed n_i as indicated in Fig. 4. This is the region in which Langmuir postulated the existence of two surface phases. At a given n_i two values of θ exist corresponding to a dilute and a concentrated phase. The intermediate value of θ is unstable.

Ion Current and Critical Temperature

The movement of the phase boundary results in a discontinuity in the ion current when plotted as a function of temperature. This temperature is called the critical temperature. Discontinuity in the ion current has been observed by several authors [57, 58, 69, 71]. The critical temperature depends upon n_i and the applied voltage. A different value is obtained, depending on whether the temperature is increased or decreased [69]. At this critical temperature the ion current increases abruptly by about 1 order of magnitude. Above the critical temperature, ionization is within .5% of complete ($\beta \cong 1.0$). Thus the patchiness and phase discontinuity of φ makes our calculations for polycrystalline surfaces approximate. For this reason it appears best to use the empirical equation [141] plotted in Fig. 5

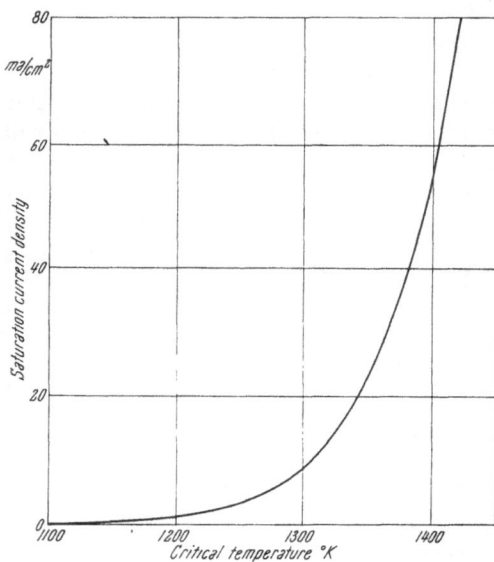

Fig. 5. Cesium ion current at the critical temperature

$$\log_{10} j = 11.993 - 14350/T_g \ \text{ma/cm}^2.$$

This equation represents Langmuir's experimental data up to 2.2 ma/cm².

Ion Emitter Energy Loss

To evaluate the energy loss due to thermal radiation from the incandescent surface, we make some simplifying assumptions. Assume a plane surface radiating to one side only, and into space at absolute zero. Using the STEFAN-BOLTZMANN law

$$P = 5.67 \times 10^{-12} \, \varepsilon_T \, T^4 \text{ in W/cm}^2$$

and the resulting data in Table 1 we calculate the specific power loss for ion generation plotted in Fig. 6.

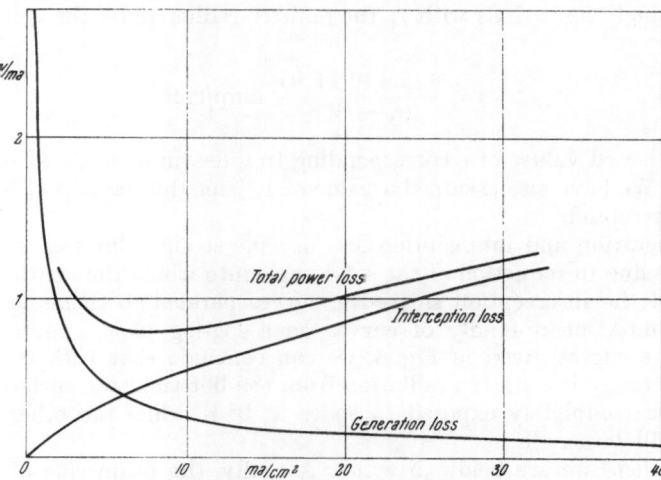

Fig. 6. Optimization of total specific power loss

Table I

K	ε*	W/cm²	$j \, T_c$ ma/cm²	W/ma
1100	.128	1.06	.088	12.0
1150	.135	1.34	.308	4.35
1200	.143	1.68	1.08	1.55
1250	.151	2.09	3.24	.645
1300	.158	2.55	8.95	.285
1350	.167	3.14	23.0	.136
1400	.175	3.81	55.0	.0693
1450	.183	4.59	124	.0370
1500	.192	5.51	272	.020

* From Handbook of Chemistry and Physics, 37th ed., p. 2698. London: Chemical Rubber Publishing Co., 1955.

Design Optimization

For optimal design we require a minimum power loss. Most of the power loss is associated with emission and acceleration. We assume that a certain percentage of the beam is intercepted and the power associated with it is lost. This loss is proportional to the space charge limited current, which follows the 3/2 power law. Since the power per unit current is actually a potential we can plot both emission and acceleration losses as functions of current density. The sum of the two losses represent the total power loss. A linear plot illustrating this is given in Fig. 6. The total power loss per unit of beam current has a minimum which serves as design criterion. This minimum coincides with the point of intersection of the two curves. Having established their importance we proceed to examine the properties of each curve in more detail.

The acceleration loss curve is based on Child's Law which for a cylindrical structure in the range of r_0/r of 4/3 to 5 has this form:

$$j = 5.46 \times 10^{-8} \frac{q}{m} \frac{\frac{-}{2} r V^{3/2}}{(3/4) \, r_0 \, (r_0 - r)^2}$$

which for singly ionized Cs with r_0 the emitter radius and r the collector radius becomes:

$$j = \frac{6.32 \times 10^{-9} \, V^{3/2} r}{(r_0 - r)^2 \, r_0} \text{ amp/cm}^2.$$

We have selected values of r corresponding to a design as outlined in a previous paper [25]. We have also assumed a value of 1 % for the fraction of beam current that is intercepted.

The generation and interception loss become straight lines on a log-log plot. Any change due to reduction of the solid angle into which the emitter radiates or reduction of the interception shifts the curves parallel to themselves. Thus for design optimization a family of curves on a log-log plot is most convenient.

From the curves given in Fig. 6, we can conclude that with the cesium ion rocket the energy loss due to radiation from the hot tungsten surface, even with the one side completely exposed to space at 0° K while the other is perfectly shielded, will be small.

At 5 ma/cm² we are below .5 w/ma. Actually, the beam side of the emitter will be partially radiation shielded, especially in a Pierce geometry rocket, while shielding on the back side will not be perfect.

For operation at higher current densities, the interception fraction will have to be kept below 1 % for maximum efficiency. The problems due to sputtering will probably reduce this limit even more. Some of the intercepted power loss would be recovered by re-radiation to the emitter.

In general, more accurate analysis is practically impossible due to uncertainty in the various parameters. More data are needed in surface ionization for this system and for other materials. Most of the present literature on surface ionization is outlined chronologically in the bibliography.

Acknowledgements

This research was partially supported by the Army Ballistic Missile Agency, Huntsville, Alabama. R. N. Edwards, Supervisor of Ion Propulsion, General Electric Co., suggested this work and the optimization idea and has contributed many valuable criticisms and suggestions. This was also discussed with Dr. Worden. The illustrations were executed by L. Mustain, while the typing was done by A. Breehne.

References

Ion Rocket

1. L. R. Shepherd and A. V. Cleaver, J. Brit. Interplan. Soc. 7, 185, 234 (1948).
2. L. R. Shepherd and A. V. Cleaver, J. Brit. Interplan. Soc. 8, 59 (1949).
3. M. Lunc, Technika Lotnicza 1950.
4. G. F. Forbes, J. Brit. Interplan. Soc. 9, 75 (1950).
5. L. Spitzer, Jr., J. Brit. Interplan. Soc. 10, 249 (1951).
6. H. Preston-Thomas, J. Brit. Interplan. Soc. 11, 173 (1952).
7. L. Spitzer, Jr., J. Amer. Rocket Soc. 22, 92 (1952).
8. H. S. Tsien, J. Amer. Rocket Soc. 13, 87 (1954).

9. E. STUHLINGER, Bericht über den V. Internationalen Astronautischen Kongreß, Innsbruck, 5.—7. August 1954, p. 100. Wien-Innsbruck: Springer, 1955.

10. E. STUHLINGER, IRE Conv. Rec. 1955, Part 10, p. 37—43.

11. E. STUHLINGER, J. Astronautics 3, No. 4. 149 (1955); J. Astronautics 3, No. 1 (1956); J. Astronautics 3, No. 2 (1956).

12. R. W. BUSSARD, J. Brit. Interplan. Soc. 15, 297 (1956).

13. J. J. BARRÉ, Proceedings of the VIIIth International Astronautical Congress, Barcelona, 1957, p. 15. Wien: Springer, 1958.

14. E. STUHLINGER, Proceedings of the VIIIth International Astronautical Congress, Barcelona, 1957, p. 403. Wien: Springer, 1958.

15. H. MICHIELSEN, Astronaut. Acta 3, 130 (1957).

16. M. I. WILLINSKI and E. C. ORR, Amer. Rocket Soc. Repr. 419—57; Jet Propulsion 28, 723 (1958).

17. R. H. BODEN, Rocketdyne R—645; SAE Nat. Aer. Meeting, New York, 1958.

18. R. G. PERELMAN, Nauka i Zhizn 25, No. 7, 60 (1958).

20. D. B. LANGMUIR, ERL—119, 2nd Ann. AFOSR Astronautical Symp., Denver, Colorado, April 28—30, 1958.

21. H. PRESTON-THOMAS, J. Brit. Interplan. Soc. 16, 508 (1958).

22. A. L. HUEBNER and R. H. BODEN, Space Expl. Meeting, San Diego, Calif., Oct. 1958.

23. R. H. BODEN, ASME Aviat. Conf., March 9—12, 1959, Los Angeles, Calif.

24. A. T. FORRESTER, ASME Aviat. Conf., March 9—12, 1959, Los Angeles, Calif., ASME No. 59-AV-35.

25. R. N. EDWARDS and G. KUSKEVICS, ASME Aviat. Conf., March 9—12, 1959, Los Angeles, Calif., ASME No. 59-AV-32.

26. R. N. EDWARDS and H. BROWN, ARS Contr. Sat. Conf., Apr. 30—May 1, 1959, MIT, Cambridge, Mass.

27. S. L. EILENBERG and A. L. C. HUEBNER, ARS Semi-Ann. Meeting, June 8—12, 1959, San Diego, Calif., Paper No. 880—59.

28. Y. C. LEE, ARS Semi-Ann. Meeting, June 8—12, 1959, San Diego, Calif., Paper No. 881—59.

29. H. SHELTON, R. F. WUERKER, and J. M. SELLEN, ARS Semi-Ann. Meeting, June 8—12, 1959, San Diego, Calif., Paper No. 882—59.

30. S. NAIDITCH, ARS Semi-Ann. Meeting, June 8—12, 1959, San Diego, Calif., Paper No. 883—59.

31. R. N. SEITZ and M. J. RAETHER, ARS Semi-Ann. Meeting, June 8—12, 1959, San Diego, Calif., Paper No. 884—59.

Ion Sources

32. M. N. BREDOV, Zh. Tekh. Fiz. 20, 476 (1950).

33. M. G. INGRAM and W. A. CHUPKA, Rev. Sci. Instruments 24, 518 (1953).

34. B. K. KRULIKOVSKII, Naukovi povidomlenna Fizyka, Kiev U. No. 1, 11—12 (1956).

35. H. HINTENBERGER and C. LANG, Z. Naturforsch. 11a, No. 2, 167 (1956).

36. B. G. SAFRANOV, YU. S. AZOVSKII, and G. G. ASEYEV, Prib. Tekh. Eksp. No. 6, 80-2 (1957).

Surface Ionization

37. O. W. RICHARDSON, The Emission of Electricity from Hot Bodies. London: Longmans, Green and Co., 1921.

38. J. J. THOMSON, Rays of Positive Electricity. New York: Longmans, Green and Co., 1921.

39. W. NERNST, Die theoretischen und experimentellen Grundlagen des neuen Wärmesatzes. Halle: W. Knapp, 1924.

40. J. J. THOMSON and G. P. THOMSON, Conduction of Electricity through Gases, p. 382—398. Cambridge: University Press, 1933.

41. R. H. FOWLER, Statistical Mechanics, p. 279. Cambridge, 1929.

42. H. S. W. MASSEY, Negative Ions. Cambridge: University Press, 1950.

43. L. N. DOBRETSOV, Electronic and Ionic Emission. Moscow: Gos. Jzd. Tekh.—Teor. Lit., 1952; also Berlin: VEB Verlag Technik, 1954.

44. COLIN J. SMITHELLS, Tungsten, Its Metallurgy, Properties and Applications. New York: Chemical Publishing Co., 1953.

45. M. N. SAHA, Ionization in the Solar Chromosphere. Philos. Mag. (6) **40**, 472 (1920).

46. M. N. SAHA, Elements in the Sun. Philos. Mag. **40**, 809 (1920).

47. F. W. ASTON, The Mass Spectra of the Alkali Metals. Philos. Mag. (6) **42**, 436 (1921).

48. F. W. ASTON and G. P. THOMSON, The Constitution of Lithium. Nature **106**, 827 (1921).

49. G. P. THOMSON, The Application of Anode Rays to the Investigation of Isotopes. Philos. Mag. (6) **42**, 857 (1921).

50. A. J. DEMPSTER, Positive Ray Analysis of Lithium and Magnesium. Physic. Rev. **18**, 415 (1921).

51. M. N. SAHA, On a Physical Theory of Stellar Spectra. Proc. Roy. Soc. London, Ser. A **99**, 135 (1921).

52. M. N. SAHA, Versuch einer Theorie der physikalischen Erscheinungen bei hohen Temperaturen mit Anwendungen auf die Astrophysik. Z. Physik **6**, 40 (1921).

53. A. J. DEMPSTER, Positive Ray Analysis of Potassium, Calcium and Zinc. Physic. Rev. **20**, 631 (1922).

54. H. N. RUSSEL, The Theory of Ionization and the Sun-spot Spectrum. Astrophysic. J. **55**, 119 (1922).

55. R. H. FOWLER and E. A. MILNE, The Intensities of Absorption Lines in Stellar Spectra and the Temperatures and Pressures in the Reversing Layers of Stars. Monthly Notices Roy. Astronom. Soc. **83**, 403 (1923).

56. R. H. FOWLER, Dissociation-equilibria by the Method of Partitions. Philos. Mag. (6) **45**, 1 (1923).

57. I. LANGMUIR and K. H. KINGDON, Thermionic Phenomena Due to Alkali Vapors. Physic. Rev. **21**, 380 (1923).

58. I. LANGMUIR and K. H. KINGDON, Thermionic Effects Caused by Alkali Vapors in Vacuum Tubes. Science **57**, 58 (1923).

59. M. N. SAHA, On the Physical Properties of Elements at High Temperatures. Philos. Mag. (6) **46**, 534 (1923).

60. H. B. IVES, Positive Rays in Alkali Metal Vapor Thermionic Tubes. Physic. Rev. **21**, 385 (1923).

61. K. H. KINGDON, A Method for Studying the Ionization of the Less Volatile Metals. Physic. Rev. **23**, 778 (1924).

62. F. W. ASTON, The Mass Spectra of Chemical Elements. Part V. Accelerated Anode Rays. Philos. Mag. (6) **47**, 385 (1924); Part VI. Accelerated Anode Rays (Continued). Philos. Mag. (6) **49**, 1191 (1925).

63. W. A. JENKINS, On the Emission of Positive Ions from Hot Tungsten. Philos. Mag. (6) **47**, 1025 (1924).

64. H. E. IVES, Astrophysic. J. **60**, 209 (1924).

65. I. LANGMUIR and K. H. KINGDON, Thermionic Effects Caused by Vapors of Alkali Metals. Proc. Roy. Soc. London, Ser. A **107**, 61 (1925).

66. R. L. KENWORTHY, Emission of Positive Ions from Hot Tungsten Filaments. Physic. Rev. **27**, 112 (1926).

67. C. H. KUNSMAN, A New Source of Positive Ions. Science **62**, 269 (1925).

68. C. H. KUNSMAN, The Thermionic Emission from Substances Containing Iron and Alkali Metal. Physic. Rev. **25**, 892 (1925).

69. T. J. KILLIAN, Thermionic Phenomena Caused by Vapors of Rubidium and Potassium. Physic. Rev. **27**, 578 (1926).

70. C. H. KUNSMAN, The Positive Ion Emission from a Mixture Containing Fe, Al and Cs and the Work Function φ^+ for Cs from this Mixture. Physic. Rev. **27**, 249 (1926).

71. J. A. BECKER, Thermionic and Adsorption Characteristics of Caesium on Tungsten and Oxidized Tungsten. Physic. Rev. **28**, 341 (1926).
72. H. E. IVES, Positive Rays Produced in Thermionic Vacuum Tubes Containing Alkali Metal Vapors. J. Franklin Inst. **201**, 47 (1926).
73. H. A. BARTON, G. P. HARNWELL, and C. H. KUNSMAN, Analysis of Positive Ions Emitted by a New Source. Physic. Rev. **27**, 739 (1926).
74. P. K. MITRA, On the Emission of Positive Electricity from Hot Tungsten in Mullard Radio Valves. Philos. Mag. (7) **5**, 67 (1928).
75. H. B. WAHLIN, The Emission of Positive Ions from Metals. Physic. Rev. **34**, 164 (1929).
76. L. P. SMITH, The Emission of Positive Ions from W at High Temperatures. Physic. Rev. **33**, 279 (1929).
77. J. A. BECKER, J. Amer. Electrochem. Soc. **55**, 153 (1929).
78. A. GOETZ, The Photoelectric Effect of Molten Tin and Two of its Allotropic Modifications. Physic. Rev. **33**, 373 (1929).
79. J. B. TAYLOR, Eine Methode zur direkten Messung der Intensitätsverteilung in Molekularstrahlen. Z. Physik **57**, 242 (1929).
80. E. MEYER, Über die Elektronen- und Positive-Ionen-Emission von Wolfram, Molybdän- und Tantalglühfäden in Kaliumdampf. Ann. Physik (5) **4**, 357 (1930).
81. I. LANGMUIR, S. MACLANE, and K. B. BLODGETT, The Effect of End Losses on the Characteristics of Filaments of Tungsten and Other Materials. Physic. Rev. **35**, 478 (1930).
82. L. P. SMITH, The Emission of Positive Ions from Tungsten and Molybdenum. Physic. Rev. **35**, 381 (1930).
83. H. B. WAHLIN, Emission of Positive Ions from Thoriated Tungsten. Physic. Rev. **35**, 652 (1930).
84. LeRoy L. BARNES, Positive Ions Emitted by Iron and Copper. Physic. Rev. **37**, 218 (1931).
85. H. B. WAHLIN, The Emission of Positive Ions from Metals. Physic. Rev. **37**, 467 (1931).
86. H. MURAWKIN, Massenspektra von Gläsern, Salzen und Metallen nebst Konstruktion eines Kreismassenspektrographen. Ann. Physik **8**, 385 (1931).
87. L. TONES, On the Increase in Surface Area Due to Crystal Faces Developed by Etching. Physic. Rev. **38**, 1030 (1931).
88. H. B. WAHLIN, The Emission of Positive Ions from Cu and Ag. Physic. Rev. **38**, 1074 (1931).
89. P. B. MOON, The Action of Positive Ions of Caesium on a Hot Nickel Surface. Proc. Cambridge Philos. Soc. **27**, 570 (1931).
90. I. LANGMUIR and D. S. VILLARS, Oxygen Films on Tungsten. I. A Study of Stability by Means of Electron Emission in Presence of Caesium Vapor. J. Amer. Chem. Soc. **53**, 486 (1931).
91. A. K. BREWER, The Effect of Adsorbed K^+ Ions on the Photoelectric Threshold of Iron. Physic. Rev. **38**, 401 (1931).
92. I. LANGMUIR, Vapor Pressures, Evaporation, Condensation, and Adsorption. J. Amer. Chem. Soc. **54**, 2798 (1932).
93. P. B. MOON and M. L. OLIPHANT, Surface Ionization of Potassium by Tungsten. Proc. Roy. Soc. London, Ser. A **137**, 463 (1932).
94. I. LANGMUIR and J. B. TAYLOR, The Mobility of Caesium Atoms Adsorbed on Tungsten. Physic. Rev. **40**, 463 (1932).
95. A. M. TYNDALL and C. F. POWELL, Mobility of Positive Alkali Ions in Argon, Neon and Helium. Proc. Roy. Soc. London, Ser. A **136**, 145 (1932).
96. K. T. BAINBRIDGE, The Isotopic Construction of Zinc. Physic. Rev. **39**, 847 (1932).
97. F. S. ASTON, The Isotopic Constitution and Atomic Weights of Caesium, Strontium, Lithium, Rubidium, Barium, Scandium, and Thallium. Proc. Roy. Soc. London, Ser. A **134**, 571 (1932).
98. H. B. WAHLIN, The Emission of Positive Ions from Metals. Physic. Rev. **39**, 183 (1932).

99. LeRoy L. Barnes, The Emission of Positive Ions from Heated Metals. Physic. Rev. **42**, 487 (1932).

100. W. H. Rodebush and W. F. Henry, Molecular Beams of Salt Vapors. Physic. Rev. **39**, 386 (1932).

101. R. C. Evans, The Equilibrium of Atoms and Ions Absorbed on a Metal Surface. Proc. Cambridge Philos. Soc. **29**, 161 (1933).

102. J. B. Taylor and I. Langmuir, The Evaporation of Atoms, Ions, and Electrons from Caesium Film on Tungsten. Physic. Rev. **44**, 423 (1933).

103. I. Langmuir, An Extension of the Phase Rule for Adsorption under Equilibrium and Nonequilibrium Conditions. J. Chem. Physics **1**, 1 (1933).

104. R. C. Evans, The Positive Ion Work Function of Tungsten for the Alkali Metals. Proc. Roy. Soc. London, Ser. A **139**, 604 (1933).

105. I. Langmuir, Advances in Chemistry **2**, 6-9 (1933).

106. M. Nordmeyer, Untersuchungen zur Stoßwirkung langsamer positiver Ionen in Edelgasen. I. Die Ionenquelle. Ann. Physik **16**, 697 (1933).

107. D. Brata, Emission of Metallic Ions from Oxide Surfaces. I. Identification of the Ions by Mobility Measurements. Proc. Roy. Soc. London, Ser. A **141**, 454 (1933).

108. C. F. Powell and D. Brata, Emission of Metallic Ions from Oxide Surfaces, II. Mechanism of the Emission. Proc. Roy. Soc. London, Ser. A **141**, 463 (1933).

109. A. K. Brewer, The Effect of Alkali Ions on the Photoelectric Emissivity of Tungsten. Physic. Rev. **44**, 1016 (1933).

110. N. D. Morgulis, Zh. Eksp. Teor. Fiz. **4**, 684 (1934).

111. L. N. Dobretsov, Zh. Eksp. Teor. Fiz. **4**, 783 (1934).

112. A. I. Anselm, Zh. Eksp. Teor. Fiz. **4**, 678 (1934).

113. N. D. Morgulis, Thermische Ionisation von Natriumdämpfen an einer glühenden Wolframoberfläche. Physik. Z. Sowjetunion **5**, 221 (1934).

114. A. L. Reinmann, The Surface Ionization of Potassium of Tungsten. Physic. Rev. **45**, 898 (1934).

115. H. Altertum, K. Krebs, and R. Rompe, Über die selbständige Ionisation von Natrium- und Cäsiumdampf an glühenden Wolfram- und Rheniumoberflächen. Z. Physik **92**, 1 (1934).

116. A. K. Brewer, The Emission of Alkali Atoms from Various Ammonia Catalysts. J. Chem. Physics **2**, 116 (1934).

117. W. R. Smythe, L. H. Rumbach, and S.S. West, A High Intensity Mass Spectrometer. Physic. Rev. **45**, 724 (1934).

118. G. Gille, Untersuchungen an der Kunsman-Anode. Ann. Physik **21**, 443 (1934).

119. H. B. Wahlin and L. O. Sordahl, Positive and Negative Thermionic Emission from Columbium. Physic. Rev. **45**, 886 (1934).

120. M. J. Copley and T. E. Phipps, The Surface Ionization of Potassium on Tungsten. Physic. Rev. **45**, 344 (1934).

121. M. J. Copley and T. E. Phipps, Reflection Coefficient of Electrons. Physic. Rev. **46**, 144 (1934).

122. R. W. Gurney, Theory of Electrical Double Layers in Adsorbed Films. Physic. Rev. **47**, 479 (1934).

123. M. J. Copley and T. E. Phipps, Surface Ionization of Potassium Iodide on Tungsten. J. Chem. Physics **3**, 594 (1935).

124. M. J. Copley and T. E. Phipps, The Surface Ionization of Potassium on Tungsten. Physic. Rev. **48**, 960 (1935).

125. P. P. Sutton and J. E. Mayer, A Direct Experimental Determination of Electron Affinities, the Electron Affinity of Iodine. J. Chem. Physics **3**, 20 (1935).

126. H. Lüder, Zerstäubung von Metallen durch Aufprall langsamer Ionen und Messung des Schwellenwertes der Zerstäubung. Z. Physik **97**, 158 (1935).

127. L. N. Dobretsov, Zh. Eksp. Teor. Fiz. **6**, 552 (1936).

128. A. N. Guthrie, Surface Ionization of Barium on Tungsten. Physic. Rev. **49**, 868 (1936).

129. L. N. Dobretsov and G. A. Morozow, Zh. Eksp. Teor. Fiz. **6**, 243 (1936).

130. J. P. BLEWETT and E. J. JONES, Filament Sources of Positive Ions. Physic. Rev. 50, 464 (1936).

131. J. H. LEES, Caesium-Oxygen Films on Tungsten. Philos. Mag. (7) 21, 1131 (1936).

132. J. KOCH, Versuche über Erscheinungen beim Auftreten von positiven Caesium-ionen auf einer ausgeglühten Wolframoberfläche. Z. Physik 100, 685 (1936).

133. J. KOCH, Die Herstellung und die nähere Untersuchung einer neuen Alkali-ionenquelle. Z. Physik 100, 669 (1936).

134. K. T. BAINBRIDGE and E. B. JORDON, Mass Spectrum Analysis. Physic. Rev. 50, 282 (1936).

135. J. A. BECKER, Trans. Faraday Soc. 32, 1402 (1936).

136. J. B. MANLEY and S. MILLMAN, The Nuclear Spin and Magnetic Moment of Li^6. Physic. Rev. 51, 19 (1937).

137. J. O. HENDRICKS, T. E. PHIPPS, and M. J. COPLEY, Evidence for Halogen Films on Tungsten in the Surface Ionization of Potassium Halides. J. Chem. Physics 5, 868 (1937).

138. W. R. SMYTHE and A. HEMMENDINGER, The Radioactive Isotope of Potassium. Physic. Rev. 51, 178 (1937).

139. A. HEMMENDINGER and W. R. SMYTHE, The Radioactive Isotope of Rubidium. Physic. Rev. 51, 1052 (1937).

140. R. C. L. BOSWORTH, Studies in Contact Potentials. The Evaporation of Sodium Films. Proc. Roy. Soc. London, Ser. A 162, 32 (1937).

141. J. B. TAYLOR and I. LANGMUIR, Vapor Pressure of Cesium by the Positive Ion Method. Physic. Rev. 51, 753 (1937).

142. R. C. L. BOSWORTH and E. K. RIDEAL, Studies in Contact Potentials. The Condensation of Potassium and Sodium on Tungsten. Proc. Roy. Soc. London, Ser. A 162, 1 (1937).

143. L. N. DOBRETSOV, Electric Field Influence on Surface Ionization. Physik. Z. So-wjetunion 11 (6), 647 (1937).

144. E. M. TSENTER, Zh. Eksp. Teor. Fiz. 8, 632 (1938).

145. W. WELCHER, Über einen Massenspektrographen hoher Intensität und die Trennung der Rubidium-Isotope. Z. Physik 108, 376 (1938).

146. R. H. SLOANE and R. PRESS, The Formation of Negative Ions by Positive Ion Impact on Surfaces. Proc. Roy. Soc. London, Ser. A 168, 284 (1938).

147. J. E. MAYER and I. H. WINTNER, Measurements of Low Vapor Pressures of Alkali Halides. J. Chem. Physics 6, 301 (1938).

148. S. T. MARTIN, On the Thermionic and Adsorptive Properties of the Surfaces of a Tungsten Single Crystal. Physic. Rev. 56, 947 (1939).

149. N. D. MORGULIS and M. P. BERNADINEE, Neutralization and Ionization of Caesium and Potassium of Thoriated Tungsten. Zh. Eksp. Teor. Fiz. 9, 998 (1939). Mem. Fiz. Ukr. SSR 8, No. 1, 35 (1939).

150. I. D. KONOSENKO, Zh. Eksp. Teor. Fiz. 9, 540 (1939).

151. A. A. JOHNSON and T. E. PHIPPS, A Differential Method Applied to the Surface Ionization of Sodium Halides on Tungsten. J. Chem. Physics 7, 1030 (1939).

152. N. MORGULIS, Surface Structure of Thoriated W Examined by Ionic Microscope. Mem. Fiz. Ukr. SSR 8, No. 2, 149 (1940).

153. J. G. POTTER, Temperature Dependence of the Work Function of Tungsten from Measurement of Contact Potentials by the Kelvin Method. Physic. Rev. 58, 623 (1940).

154. N. I. IONOV, C. R. Acad. Sci. USSR 28, 512 (1940).

155. J. J. MITCHELL and J. E. MAYER, An Experimental Determination of the Electron Affinity of Chlorine. J. Chem. Physics 8, 282 (1940).

156. R. SMOLUCHOWSKI, Anisotropy of the Electronic Work Functions of Metals. Physic. Rev. 60, 661 (1941).

157. S. SEELY, Work Function and Temperature. Physic. Rev. 59, 75 (1941).

158. R. H. HAY, The Nuclear Magnetic Moments of C^{13}, Ba^{135} and Ba^{137}. Physic. Rev. 60, 2 (1941).

159. H. B. WAHLIN, Thermionic Properties of the Iron Group. Physic. Rev. 61, 509 (1942).

160. M. D. FISKE, The Temperature Scale, Thermionics and Thermatomics of Tantalum. Physic. Rev. **61**, 513 (1942).
161. K. J. McCALLUM and J. E. MAYER, A Direct Experimental Determination of the Electron Affinity of Chlorine. J. Chem. Physics **11**, 56 (1943).
162. W. WALCHER, Über die Verwendungsmöglichkeiten von Glühanoden zur massenspektroskopischen Isotopentrennung. Z. Physik **121**, 604 (1943).
163. D. T. VIER and J. E. MAYER, A Direct Experimental Determination of the Electron Affinity of Oxygen. J. Chem. Physics **12**, 28 (1944).
164. P. M. DOTY and J. E. MAYER, The Electron Affinity of Bromine and a Study of Its Decomposition on Hot Tungsten. J. Chem. Physics **12**, 323 (1944).
165. S. W. STARODUBTSEV and V. ARIFOV, Bull. Akad. Nauk Uzbek. SSR **4**, 8 (1946).
166. I. ESTERMANN, Molecular Beam Technique Rev. Mod. Physics **18**, 300 (1946).
167. L. N. DOBRETSOV, S. V. STARODUBTSEV, and YU. I. TIMOKHINA, Surface Ionization of Thin Layers of CaO and MgO. Dokl. Akad. Nauk SSSR **55**, No. 4, 303 (1947).
168. G. A. MOROZOV, Surface Ionization of Ba on W. Zh. Tekh. Fiz. **17**, 1143 (1947).
169. G. A. MOROZOV, Zh. Eksp. Teor. Fiz. **17**, 1142 (1947).
170. L. N. DOBRETSOV, Electron and Ion Emission of Thoriated Tungsten in Sodium Vapors. Zh. Exsp. Teor. Fiz. **17**, 301 (1947).
171. I. N. STRANSKI and R. SUHRMANN, Über die Elektronenemission kristalliner Metalloberflächen und ihre Beziehungen zu den Gesetzmäßigkeiten des Kristallbaus. Ann. Physik **436**, 153 (1947).
172. S. V. STARODUBTSEV, Trudy Fiz. Tekh. Inst. Akad. Nauk Uzbek. SSR **2**, 1 (1948).
173. N. I. IONOV, Dissertation. Leningrad Phys.co-Technical Institute, 1948.
174. N. I. IONOV, Ionization of Potassium Iodide, Sodium Iodide, and Cesium Chloride Molecules by Electrons. Dokl. Akad. Nauk SSSR **59**, 467 (1948).
175. U. ARIFOW and G. N. SCHUPPE, Trudy Fiz. Tekh. Inst. Akad. Nauk Uzbek. SSR **2**, Ed. 1 (1948).
176. N. I. IONOV, Zh. Exsp. Teor. Fiz. **18**, 96 (1948).
177. N. I. IONOV, Zh. Exsp. Teor. Fiz. **18**, 174 (1948).
178. I. L. KOFSKY and H. LEVINSTEIN, A Dynamic Method for the Determination of the Velocity Distribution of Thermal Atoms. Physic. Rev. **74**, 500 (1948).
179. G. E. COGIN, The Vapor Pressures of Some Alkali Halides. J. Chem. Physics **16**, 1035 (1948).
180. N. D. MORGULIS, On the Question of Ionization of Atoms and Neutralization of Ions on the Surface of Cathode of Low Conductivity. Zh. Tekh. Fiz. **18**, 567 (1948).
181. F. KNAUER, Die Verweilzeit adsorbierter Alkalien an erhitztem Wolfram. Z. Physik **125**, 278 (1948).
182. M. METLAY and G. E. KIMBALL, Ionization Processes on Tungsten Filaments. J. Chem. Physics **16**, 774 (1948).
183. C. HERRING and M. NICHOLS, Thermionic Emission. Physic. Rev. **21**, 185 (1949).
184. V. M. BUKELSKI, E. YA ZANDBERG, and N. I. IONOV, Negative Ions of Rubidium and Caesium. Dokl. Akad. Nauk SSSR **68**, No. 1, 31 (1949).
185. U. ARIFOV, A. KH. AYUKHANOV, and V. M. LOVTSOV, Determination of the Absolute Ionization Coefficient on the Surface of Heated Tungsten. Dokl. Akad. Nauk SSSR **68** (No. 3), 461 (1949).
186. S. STARODUBTSEV, Zh. Exsp. Theor. Fiz. **19**, 215 (1949).
187. L. DAVIS, B. T. FELD, C. W. ZABEL, and J. R. ZACHARIAS, The Hyperfine Structure and Nuclear Moments of the Stable Chlorine Isotopes. Physic. Rev. **76**, 1076 (1949).
188. R. H. PLUMLEE and L. P. SMITH, Mass Spectrometric Study of Solids: I. Preliminary Study of Sublimation Characteristics of Oxide Cathode Materials. J. Appl. Physics **21**, 811 (1950).
189. S. V. STARODUBTSEV and IU. I. TIMOKHINA, Surface Ionization of Mg on Glowing W. Collection in Honor of the 70th Year of Academician A. I. IOFFE (1950), p. 117. Akad. Nauk USSR.

190. U. ARIFOV, A. KH. AYUKHANOV and V. M. LOVTSOV, Determination of the Absolute Ionization Coefficient of Potassium on the Surface of Heated Metals. Trudy Akad. Nauk Uzbek. SSR **3**, 13 (1950).

191. U. ARIFOV and V. M. LOVTSOV, Migration of Potassium Chloride on the Surface of Incandescent Metals. Trudy Akad. Nauk Uzbek. SSR **3**, 117 (1950).

192. G. COUCHET, Contribution à l'étude de l'émission d'ions positifs par les sels triples du type Li_2O, Ac_2O_3, $4\,SiO_2$. C. R. Acad. Sci. Paris **233**, 1013 (1951).

193. E. W. MÜLLER, Das Feldionenmikroskop. Z. Physik **131**, 136 (1951).

194. M. A. EREMEEV, Emission of Electrons and Reflection of Ions from Metal Surfaces. Dokl. Akad. Nauk SSSR **79** (5), 775 (1951).

195. M. A. EREMEEV, Collision of Alkali Ions with the Surface of a Metal. Trans. Conf. on Cath. Electr. 1952.

196. E. L. BRADY and P. D. ZEMANY, Effect of Various Gases on Potassium Ion Emission from Hot Platinum. J. Chem. Physics **20**, 294 (1952).

197. M. A. EREMEEV and V. V. SHESTUKHINA, Ejection of Electrons and Reflection of Lithium Ions from Tungsten and Tantalum. Zh. Tekh. Fiz. **22**, 1268 (1952).

198. M. A. EREMEEV and V. V. SHESTUKHINA, Ejection of Electrons and the Reflection of Potassium Ions from Tungsten and Tantalum. Zh. Tekh. Fiz. **22**, 1262, 1268 (1952).

199. M. A. EREMEEV and V. G. YUREV, Emission of Electrons and the Reflection of Ions of Potassium and Lithium from Oxidized Tungsten and Tantalum. Zh. Tekh. Fiz. **22**, 1290 (1952).

200. KHEN-BON KIM and I. L. SOKOLSKAYA, Surface Ionization of Sodium on Platinized Tungsten. Vestnik Leningrad Univ. 7, No. 12, 67 (1952).

201. M. A. BREMEEV and T. L. MATSKEVICH, The Emission of Electrons and the Reflection of Potassium Ions from a Surface of Liquid Tin. Zh. Tekh. Fiz. **22**, 1296 (1952).

202. A. M. ROMANOV and S. V. STARODUBTSEV, Surface Ionization of Lithium. Trudy Fiz. Tekh. Inst. Akad. Nauk Uzbek. SSR **4**, 102 (1952).

203. G. COUCHET, Sur la préparation de quelques sources solides ioniques au four à induction et au four solaire. C. R. Acad. Sci. Paris **236**, 1240 (1953).

204. L. N. DOBRETSOV, Surface Ionization in a Strong Electric Field. Zh. Tekh. Fiz. **23**, 417 (1953).

205. M. G. INGHRAM and R. GOMER, Mass Spectrometric Analysis of Ions from the Field Microscope. J. Chem. Physics **22**, 1279 (1954).

206. G. F. SMITH, Thermionic and Surface Properties of Tungsten Crystals. Physic. Rev. 94, 295 (1954).

207. I. N. BAKULINA and N. I. IONOV, Dokl. Akad. Nauk SSSR **99**, 1023 (1954).

208. L. N. DOBRETSOV, Bull. Acad. Sci. Uzbek. SSR No. 3 (1954).

209. G. M. AVAKYANTS, Bull. Acad. Sci. Uzbek. SSR No. 3 (1954).

210. B. LENGYEL and F. TILL, Über die Ionenemission des Glases. Z. physik. Chem. (Leipzig) **203**, 312 (1954).

211. I. N. BAKULINA and N. I. IONOV, Dokl. Akad. Nauk SSSR **105**, 680 (1955).

212. N. A. GORBATYI and G. N. SHUPPE, Zh. Tekh. Fiz. **25**, 1364 (1955).

213. M. A. EREMEEV, Reflection of Alkali Ions from a Metal Surface. Trudy Leningr. Politekh. Inst. im. M. I. Kalinina No. 181, 158 (1955).

214. YU. K. SZHENOV, Surface Ionization of Calcium, Strontium, and Magnesium on Oxidized Tungsten. Zh. Exsp. Teor. Fiz. **29**, 901 (1955); Sov. Phys. J. exper. theor. Physik **2**, 775 (1956).

215. E. H. TAYLOR and S. DATZ, Study of Chemical Reaction Mechanisms with Molecular Beams. The Reaction of K with HBr. J. Chem. Physics **23**, 1711 (1955).

216. R. C. MILLER and P. KUSCH, Velocity Distributions in Potassium and Thallium Atomic Beams. Physic. Rev. 99, 1314 (1955).

217. M. G. INGHRAM and R. GOMER, Massenspektrometrische Untersuchung der Feldemission positiver Ionen. Z. Naturforsch. 10a, No. 11, 863 (1955).

218. F. KIRCHNER and H. KIRCHNER, Über die Ionisation und Desorption durch starke elektrische Felder. Z. Naturforsch. 10a, Nr. 5, 394 (1955).

219. N. D. Morgulis and P. M. Marchuk, Physical Phenomena at the Cathode During Arc Discharge in Caesium Vapor. Ukr. Fiz. Zh. 1, 59 (1956).
220. A. M. Romanov, Ionization of Sodium and Lithium on Hot Tungsten. Diss. Kand. Fiz. Mat. N., Leningr. Fiz. Tech. Inst., Leningrad 1956.
221. P. M. Marchuk, Phenomena Occuring During a Diode Operation with a Tungsten Cathode in Low Pressure Cesium Vapors. Trudy Inst. Fiz. Akad. Nauk Ukr. SSR 7, 3 (1956).
222. D. G. Bulyginsky and L. N. Dobretsov, Investigation of the Surface Work Function Distribution for Oxide Cathodes. Sov. Phys.-Tech. Phys. 1, No. 6, (1957); Zh. Tekh. Fiz. 26, 1141 (1956).
223. E. W. Müller and K. Bahadur, Field Ionization of Gases at a Metal Surface and the Resolution of the Field Ion Microscope. Physic. Rev. 102, 624 (1956).
224. G. N. Shuppe, E. P. Sytaya, and R. U. Kadyrov, Work Function of the Electrons from (110) Face of the Tungsten Single Crystal and Positive Ionization of Sodium on this Face. Izvest. Akad. Nauk SSSR, Ser. Fiz. 20, 1142 (1956). Also Bull. Acad. Nauk SSSR 20, 1035 (1956).
225. G. N. Shuppe, E. P. Sytoi, and R. U. Kadyrov, Positive Surface Ionization of Sodium and Work Function at the Face (110) of a Tungsten Monocrystal. Usp. Fiz. Nauk 59, (No. 2) 363 (1956).
226. G. M. Panchenkov, P. A. Akishin, N. N. Vasilev, O. T. Nikitsin, and S. D. Moiseev, Isotopic Analysis of the Alkali Metals by Means of a Synthetic Aluminosilicate Ion Emitter. Zh. Fiz. Khim. 30, 1330 (1956).
227. P. M. Marchuk, A Study of the Adsorption and Thermionic Properties of Molybdenum in Cesium Vapors under High Pressure. Trudy Inst. Fiz. Akad. Nauk Ukr. SSR 7, 17 (1956).
228. N. I. Ionov, The Surface Ionization of Molecules of Potassium Chloride and Cesium Chloride in an Electric Field. Zh. Tekh. Fiz. 26, 2200 (1956); Sov. Phys. Tech. Phys. 1, 2134 (1957).
229. S. Datz and E. H. Taylor, Ionization on Platinum and Tungsten Surfaces: I—The Alkali Metals. J. Chem. Physics 25, 389 (1956).
230. F. Hughes and H. Levinstein, Interaction of Thermal Atoms with Surfaces, Syracuse Un. Res. Inst. Tech. Report, Sept. 1, 1956.
231. S. Datz and E. H. Taylor, Ionization on Platinum and Tungsten Surfaces: II—The Potassium Halides. J. Chem. Physics 25, 395 (1956).
232. E. W. Müller, Field Desorption. Physic. Rev. 102, 618 (1956).
233. F. Kirchner and H. A. Ritter, Desorption of Positive and Negative Ions through Strong Electric Fields. Z. Naturforsch. 11a, 35 (1956).
234. C. Brunnee, The Ion Reflection and Secondary Electron Emission at the Impact of Alkali Ion on Clean Molybdenum Surfaces. Z. Physik 147 (2), 161 (1956).
235. R. H. V. M. Dawton and K. L. Wilkinson, Plutonium: Evaporation Tests, Ionization Potential and Electron Emission. A.E.R. Establishment (Harwell), Rep. No. 6 P/R 1906, 6 pp. (1956).
236. I. N. Bakulina and N. I. Ionov, Determination of the Energy of Electronic Affinity of Sulphur Atoms by Means of the Method of Surface Ionization. Dokl. Akad. Nauk SSSR 116, 41 (1957). Sov. Phys.-Dokl. 2, 423 (1957).
237. A. M. Romanov and S. V. Starodubtsev, Adsorption and Ionization of Sodium on Hot Tungsten. Zh. Tekh. Fiz. 27, 722 (1957). Sov. Phys.-Tech. Phys. 2, 652 (1958).
238. G. N. Shuppe, E. P. Sytaya and R. U. Kadyrov, Positive Surface Ionization of Sodium and Potassium and the Electron Work Function of Tungsten Single Crystal Faces (110). Trudy SAGU 91, 5 (1957).
239. Oldwig V. Ross, Theorie der Reflexion positiver Ionen auf Metalloberflächen. Z. Physik 147, 184 (1957).
240. E. Ya. Zandberg, Surface Ionization of Potassium Atoms and KCl and CsCl Molecules on Tungsten Filaments in Electric Fields up to 2 MV/Cm. Sov. Phys.-Tech. Phys. 2, 2583 (1958).

241. G. A. GORODETSKII, Reflection of Slow Electrons from a Surface of Pure Tungsten and from Tungsten Covered with Tin Films. II. Zh. Exsp. Teor. Fiz. **34** (1) 13. Sov. Phys.-J. exper. theor. Physik (USSR) **5**, 13 (1958).

242. A. M. ROMANOV, Ionization of Lithium on Tungsten. Zh. Tekh. Fiz. **27**, 1233 (1957). Sov. Phys.-Tech. Phys. **2**, 1125 (1958).

243. J. ZEMEL, Surface Ionization Phenomena on Polycrystalline Tungsten. J. Chem. Physics **28**, 410 (1958).

244. L. N. DOBRETSOV, Theory of Surface Ionization. Trudy Polyt. Inst. Leningr. No. 194 (1958), Radiofizika, p. 143—154.

245. V. N. SHREDNIK, Adsorption of Zirconium and Barium on Tungsten and the Work Function. Izv. Akad. Nauk SSSR, Ser. Fiz. **22**, 594 (1958).

246. C. H. BACHMAN and P. A. SILBERG, Thermionic Ions from Hydrogen Palladium. J. Appl. Physics **29**, 1266 (1958).

247. T. G. FOLANI and R. A. WALLACE, Mass Spectrometric Investigations of the Ionic Emission from Alumina. Amer. Ceram. Soc. Bull. 37 (April 1958), Program 31.

248. F. L. HUGHES, H. LEVINSTEIN, and R. KAPLAN, Surface Properties of Etched Tungsten Single Crystals. Physic. Rev. **113**, 1023 (1959).

249. F. L. HUGHES and H. LEVINSTEIN, Mean Adsorption Lifetime of Rb on Etched Tungsten Single Crystals: Ions. Physic. Rev. **113**, 1029 (1959).

250. F. L. HUGHES, Mean Adsorption Time of Rb on Etched Tungsten Single Crystals: Neutrals. Physic. Rev. **113**, 1036 (1959).

Design Study of an Earth Satellite Evolving from a Four-Step Solid Propellent Rocket Vehicle

S. K. Kumar[1] and B. R. Rau[1]

(With 4 Figures)

(Received July 3, 1959)

Abstract — Zusammenfassung — Résumé

Design Study of an Earth Satellite Evolving from a Four-Step Solid Propellent Rocket Vehicle. As the title itself suggests this paper deals with the use of solid propellents for all the stages of a satellite launching rocket vehicle. The first section deals with the description of the vehicle which will be employed in the launching. The second part deals with the most important portion of the paper, viz. the launching sequence and the control of the trajectory. Various methods which have been employed for the controlled flight of solid propellent rockets are discussed. Details of experiments done in this field are also given. The final chapter gives a description of the 50 Kg. satellite, and its instrumentation. The instrumentation of the satellite is divided into two parts; the first part consists of a number of measuring instruments along with telemetering transmitters. It is the second part that will be of great interest, as it will be a solid propellent rocket placed at the top of the satellite section. The rocket is intended to escape the earth after the satellite has gone into orbit. The experiment will show the possibility of orbital launching.

The paper also contains an appendix where the details of a smaller satellite vehicle, on which development work is on progress, is given.

Konstruktionsstudie für einen Erdsatelliten, der von einer Vierstufen-Festrakete aus gestartet wird. Die vorliegende Arbeit befaßt sich mit der Verwendung fester Brennstoffe für alle Stufen einer Trägerrakete für einen Satelliten. Der erste Abschnitt beschreibt das für den Start benützte Raketenfahrzeug, der zweite Abschnitt den wichtigsten Teil der Arbeit, nämlich die Startfolge und die Bahnkontrolle. Es werden verschiedene Methoden erörtert, die für den kontrollierten Flug von Feststoffraketen benützt worden sind. Es werden ausführliche Angaben über derartige Versuche gemacht. Das Schlußkapitel beschreibt den 50-kg-Satelliten und seine Instrumentierung. Diese wird in zwei Teile gegliedert. Der erste Teil besteht aus einer Anzahl von Meßinstrumenten, zusammen mit Sendegeräten. Von besonderer Wichtigkeit ist jedoch der zweite Teil, da eine Feststoffrakete an der Spitze des Satellitenteiles angebracht werden wird. Die Rakete soll sich aus dem Schwerefeld der Erde befreien, wenn der Satellit seine Umlaufbahn eingeschlagen hat. Das Experiment wird die Möglichkeit des Startes aus der Umlaufbahn zeigen.

Die Arbeit enthält einen Anhang mit den genauen Angaben über ein kleineres Satellitenfahrzeug, an dem die Entwicklungsarbeit fortschreitet.

Etude conceptuelle d'un satellite artificiel lancé à l'aide d'une fusée à quatre étages à propergols solides. La première partie décrit la fusée de lancement. La seconde et la plus importante traite de la séquence des opérations et du contrôle de la tra-

---

[1] Indian Astronautical Society, 2647, Vani Vilas Mohalla, Mysore 2, India.

jectoire. Les diverses méthodes utilisées pour contrôler le vol des fusées à propergol solide sont discutées. Les expériences faites dans ce domaine sont décrites. Le chapitre final décrit le satellite de 50 kg. et son instrumentation. Celle-ci est divisée en deux parties: la première consiste en un certain nombre d'instruments de mesure avec émetteurs télémétriques. La seconde est la plus intéressante; elle consiste en une fusée à propergol solide montée au-dessus de la section satellite. Elle doit se libérer de l'attraction terrestre après la mise en orbite du satellite. L'expérience est destinée à montrer la possibilité du lancement à partir d'une base orbitale.

On trouvera en appendice les détails sur un satellite plus petit en cours de développement.

Introduction

During the last decade liquid propellent rockets have played a great part in the upper atmosphere research and indeed have been responsible for the major achievements in space exploration. However along with the liquid units great strides have been made in the development of solid propellent rockets. Advances made in fuel chemistry have resulted in the discovery of solid fuels which have energy factors equal to that of their liquid counterparts. United States has announced that within the next few years its liquid propellent Intermediate and Intercontinental Missiles will be replaced by solid propellent ones. Thus the solid propellent rockets are regaining their importance that they had lost during the last decade as a potential military weapon and are now being considered as the most reliable form of propulsion. Perhaps it is reliability coupled along with the low cost that singles out solid propellent rockets from other types of propulsion.

One great difficulty that these rockets posed was the lack of directional control. Once the fuel is ignited the reaction is uncontrollable and as the motor is an integral part of the body it was unable to change the thrust axis in order to obtain desired trajectories. But now as these problems are being solved, the use of solid propellents for all the stages of a satellite launching rocket might be considered as a technical feasibility. This paper will attempt to show how this could be practically realised.

Section A

A four step rocket with a total take-off weight of 19.5 tons will be employed to place a 50 Kg. satellite in orbit. The rocket fully assembled will have a length of 7.2 metres with a maximum diameter of 2.3 metres. It is finless and will resemble a modified bullet.

First step is a cluster of twelve rocket tubes each 63.5 cms. in diameter. Of course the use of larger diameter tubings could have increased the total performance of the missile. But as no experience has been gained in handling rockets with diameters exceeding 70 cms., we are forced to rely on smaller diameter tubings. Further the increase in diameter means the increase in wall thickness which in turn adversely effects Mass Ratio. As we have successfully tested rockets with diameters less than 70 cms. we are in a position to predict the exact performance that could be obtained. The material used for the construction of the whole rocket is Duraluminium. Duraluminium has been chosen primarily as it is a light weight metal and as the combustion temperatures that are expected are quite low. As successful tests have been made with Duraluminium nozzles the same metal will be used for the construction of the nozzles of all the tubes. The tubes will be clustered by means of metallic straps while each of the tubes will be welded to the other along their entire length. A thin sheet of metal will be covered over the clustered tubing in order to give it a tubular appearance. The first step is 2.8 me-

Table I. *Technical Details of the Four Step Satellite Vehicle*

	Step 1	Step 2	Step 3	Step 4
Payload (Kg.)	5,204	1,004	258	50
Empty weight (Kg.)	5,892	1,320	301	69
Propellent weight (Kg.)	13,560	3,859	697	187
Final weight (Kg.)	19,452	5,180	998	256
Propellent flow per sec. (Kg./sec.)	263.7	112.5	20.32	13.8
Thrust (Kg.)	44,880	19,145	3,458	1,175
Burning time (secs.)	51.4	34.3	34.3	27.1
Acceleration (*G*'s)	7.4	19.1	27.2	41.6
Length (metres)	2.8	1.2	1.2	0.96
Diameter (cms.)	186	120	84	21
Number of motors..........	12	8	2	1
Mass ratio	3.3	3.9	3.6	3.7

tres long (nozzle included). With a mass ratio of 3.3 it will carry 12,301 Kg. of fuel. The fuel will be consumed in 51.4 seconds and a thrust of 44,880 Kg. will be delivered by the 12 rocket motors.

The second step consists of 8 rocket tubes with a final weight of 4,700 Kg. Each tubing is 1.2 metre long and has a maximum diameter of 50 cms. The tubes

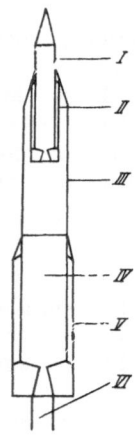

will be clustered together as in the former case. Aluminium cones which exactly fit the nozzle contours of the second stage will be fixed to the first step and the second step is just laid over the cones and firmly fixed by means of compressed air. The second step is expected to ignite two seconds earlier than the first stage power cut-off. This is to ensure that exact angle of ascent is maintained. Two types of fuses will be used to ignite the second step motors. An igniter with a timing devise and another with an electrical circuit sensitive to a definite acceleration (which will be equal or less than that produced in the first step's flight) will be used to trigger the fuse. It is further hoped that a continuous burning charge be placed connecting the fuel charges of the two rockets. Though the latter method might involve some difficulties it has already been tested on a smaller scale. The 8 motors of the second step deliver a total thrust of 19,145 Kg. for 34.3 seconds. 3,501 Kg. of fuel will be carried on this step.

Fig. 1. Details of the final step assembly. *I* The Rocket which will escape the earth; *II* Container of the Escape Rocket; *III* Instrument Section; *IV* Fourth Step Rocket; *V* Container of the Fourth Step Rocket; *VI* Aluminium rod attaching the container to the former step

All the nozzles of the 8 rockets will be fixed so as to ensure a powered ascent in a fixed direction. Two rocket tubes each 1.2 metre long with a diameter of 42 cms. constitute the third step. Carrying 632 Kg. of fuel they deliver a combined thrust of 3,458 Kg. for 34.3 seconds. Having an empty weight of 273 Kg. a Mass Ratio of 3,649 is expected. This step will be attached to its former one as in the first case. Bringing the final step to a horizontal position is done by this step and hence it is the most important step. The nozzles of each rocket tube is inclined at an angle of 4 degrees to the rocket's axis and this angle remains constant. Owing to this angled nozzle a sideward thrust will be obtained. The angle of deflection of the nozzle is so adjusted that by the time the rocket's fuel is exhausted the final step will be moving

in an almost horizontal line with the earth's surface. In order to ensure that a reverse in the trajectory is not resulted (as the nozzles are angled to a definite
direction), small jets are used to check any spin of the rocket. Nitrogen gas stored
at 2000 P.S.I. will be used to run the jets. The control valves of the jets are controlled by a spin detector which is based on gyroscopic principles. Of course this
complication could have been avoided if we had allowed the vehicles to coast.
But as it loses velocity during free flight and as we had to compensate that loss
of velocity by a larger step, we were forced to resort to this method of bringing the
final step to a horizontal position. The final step has a cylinder length of 0.96 metres with a diameter of 35 cms. It carries 169.4 Kg. of fuel and delivers a thrust of
1,175 Kg. for 27 seconds. To this tubular rocket is attached the instrument
section. Unlike the former stages this step is not fixed to its former step by compressed air. A cylinder of a slightly larger diameter is fixed to the third stage. The
fourth step slides into this cylinder through 4 guide rails. In effect the cylinder
attached to the third step acts as a sort of a launching platform. The outer cylinder is fixed to a solid Aluminium rod which is rotated by a motor kept at the top
of the third stage. As the rod rotates the cylinder containing the 4th stage also
rotates and this provides spin stabilization for the final step. The rotation will be
started before launching.

Section B

The Satellite is expected to go into an equatorial orbit and hence will be launched near the equator. The launching place will have its geographical meridians
similar to that of the Atlantic Missile Testing Range of the United States at Florida.

The ignition of the first step will be done by a system of 12 igniters whose ignition time will be synchronised. In order to ensure proper ignition of all the twelve
tubes, pressure diaphragms will be put to the orifice of all the nozzles and this is
supposed to maintain equal combustion pressure in all the tubes. As the vehicle
is finless a hollow tubular ring is fixed surrounding the twelve tubings at the bottom. This ring will sit in a groove
which is a part of the launching
platform. The first step will have
a vertical flight lasting 20 seconds,
during which time it will be at
an altitude of 6.08 Km. At this
stage begins the controlled flight
which will be responsible for
bringing the missile to the required
angular direction. As already
mentioned the control of a solid
propellent rocket and in particular
directional control in flight has been
one of the difficult problems that
one had to face in solid propellent
rocketry. This difficulty was fully
realised by us and hence the complete work of the two solid fuel research projects of the Indian Astro-

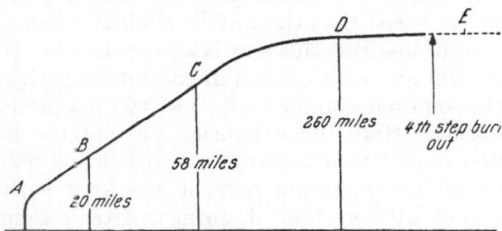

Fig. 2. A preliminary diagram indicating the launching
sequence
A End of the vertical flight, 20 seconds after take off.
B First Step Detachment, 50 seconds after launch;
velocity 1,670 mts./sec., altitude 32 Km. above the earth.
C Second Step power cut, 84 seconds after take off; ve-
locity 4,146 mts./sec., altitude 93.8 Km. above the earth.
D Third Step power cut after curved ascent. Position
almost horizontal; velocity 6476 mts./sec., altitude
416 Km.
E The orbital path begins at 416 Km. after Fourth
Step power cut; velocity 8,260 mts./sec.

nautical Society during 1958 and 1959 was directed towards the final objective of
solving this problem. It can be stated here that countries like India cannot afford
large scale liquid propellent rocket research and hence any progress or contribution from such countries towards the exploration of outer space will have to be

only through the solid propellent rockets. At the initial stages the controlled trajectory was obtained by using clustered single-steps rockets. Usually three tubes were used. Instead of filling all the three tubes with propellent one tube was partially filled with it. At the initial stage all the three motors burned, and a vertical ascent was obtained. As the fuel in one of the tubes was less than the other two, after a definite time interval its motor used to stop, while the other two continued to burn. This change of the thrust in the flight brought about a curved ascent of the rocket. But the main drawback of the system was the rapid change in the ascent angle that was obtained. Angles of the order of 60° from the vertical was obtained with 2 or 3 seconds after the one tube power cut-off. This resulted in some cases in the missile taking a curved path which brought the missile on a downward trajectory while its motors were still functioning. Out of the three experiments conducted two missiles came down with their motors still burning while the other did go at an angle as we had cut down the burning time of the other two tubes to less than 3 seconds after the first tube cut off.

As it was certain that this system was impracticable, effort was concentrated in developing reliable control techniques. The work till now done has resulted in the development of two reliable forms of guiding techniques. In the first method a cluster of three tubes was used. But instead of limiting the propellent supply of one of the tubes, it was filled with a *shaped charge*. As pelletlike fuels were being used, the diameter of the pellet was reduced along a definite length leaving the original diameters at both the ends. As the thrust is directly dependent on the propellent consumption, the reduced area consequently reduced the mass flow per second which in turn reduced the thrust of that particular tube. As a reduced thrust was obtained in one tube, while the thrusts of the other two tubes remained constant a gradual angular ascent was obtained. Since the original diameter of the pellet was maintained at the end after the reduced charge the missile accelerated at full thrust after having attained a definite angle from the vertical. Careful calculations and over a dozen test firings have shown that this method could be depended upon to bring about desired trajectories. However one slight error we noticed was that the missile slightly turned during flight depending on the wind conditions. But this was later checked in other flights by mounting small jets run by compressed nitrogen in conjunction with a gyroscopic instrument. Along with this method similar control of the missile is being effected by the use of tiltable nozzles. Here the expansion part of the nozzle was separated from the motor and was attached to it like a socket and ball arrangement. At the initial stages of flight the expansion portion was kept with its axis coinciding with that of the rocket by a system of springs. After a definite interval of time a timing device brought into action two pairs of springs which tilted the nozzle through a known angle (the angle could be controlled by the tension of the springs employed). The same timer after another time interval brought into action another pair of springs which was responsible in bringing the expansion portion's axis in a line with that of the rocket's axis. Successful experiments have been conducted employing this system. In later experiments the tilting was brought about by the use of a tension cord which is wound and unwounded by an electric motor which works in conjunction with an Integrating Accelerometer. We are also testing the use of electromagnets again with accelerometers for bringing the same results. The satellite rocket employed here is expected to be fitted with the "spring-system" as it is accurate and in particular as we have some experience in designing and developing such systems.

One particular problem that we faced during the initial stages of experimenting with these controlled rocket was determining the angle at which the missile was moving before cut-off. It was possible to get a general idea of the angle through the visual observation. In some cases the rocket was launched in night and it was equipped with an arrangement to leave a white trail and by photographing the trail we were able to obtain some idea of the trajectory that the rocket had obtained. The Department of Defence was approached to obtain auto-following radar equipment, but it was found that their radar sets needed a larger perspective area than our rockets had. Further their range was extremely limited. At this stage the electronic section of our Project was sucessful in designing a transmitter which was capable of varying its signal intensity with the change in the missiles angle from the vertical. But as the design indicated the type of components needed for the construction of the transmitter was not available in India or to be more exact the weight of such an equipment if built with available components was prohibitive. We were fortunate to get the assistance of a reputed foreign electronic firm which untertook the job of developing the transmitter. The transmitter is capable of changing its signal intensity if an angle variation of one second is developed during the flight. The recorded data gave the correct angle of the missile during cut-off and a highly accurate nature of the trajectory[1].

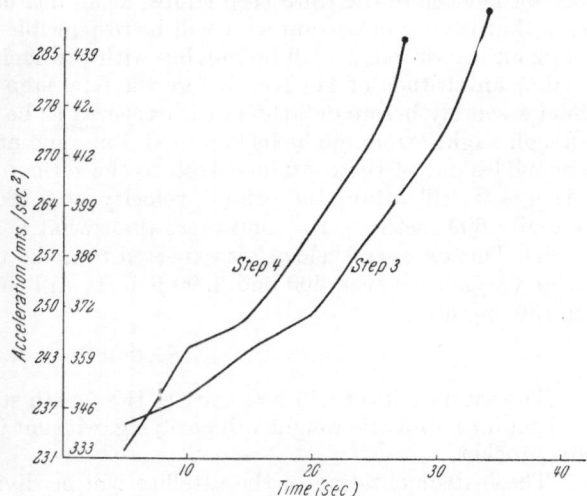

Fig. 3. Step 3 = 231—285 mts./sec.²
Step 4 = 333—439 mts./sec.²

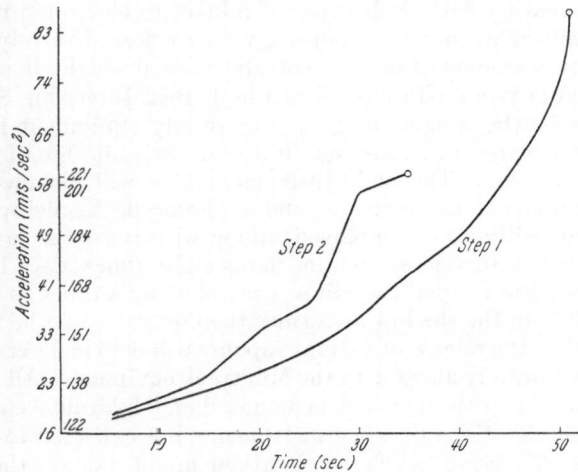

Fig. 4. Step 1 = 16—83 mts./sec.²
Step 2 = 122—221 mts./sec.²

Coming back to the trajectory of the satellite rocket, the first step will detach itself at an altitude of 32 Km. while moving at an angle of 45° to the vertical.

---

[1] When I first wrote to Mr. K. W. GATLAND of the British Interplanetary Society about these experiments and informed him that an angle error of a few seconds was obtained in one test flight, he wrote me back asking how we were able to measure the angles so accurately. Certain security regulations restricted me from telling the work of my electronic colleagues which had resulted in this angle detecting transmitter. It was over four months before I could tell him about it (S. K. KUMAR).

The ignition of the second step is timed 1.5 seconds earlier than the first step cut-off in order to ensure that the second step maintains the same angle of ascent. The second step will accelerate the upper stages to a final velocity of 3,789 metres per second before the third step ignites at an altitude of 93.8 Km. It is the third step that is very important as it will be responsible in bringing the final step to a horizontal position. It will be moving with a velocity of 5,895 metres per second and at an altitude of 418 Km. before the final step is launched. The angle of the final assembly before detachment is expected to be less than 2° to the horizontal though slight errors might be expected. Spinning at a rate of 200 r.p.m. the final step slides out of the container fixed to the former step. Firing its motors for 27 seconds it will attain the orbital velocity of 8,260 metres per second (this is actually 609 metres per second more than what is actually needed to attain the orbit). This excess of velocity is expected to make the satellite go into an elliptic orbit varying between 400 and 3900 Km. It will complete one orbital revolution in 136 minutes.

Section C

The satellite itself will be a part of the fourth step rocket along with which it will get into orbit. Its weight will be 50 Kg. without the empty weight of the fourth step rocket.

The instrumentation of the satellite will be divided into two parts. The first part consists of a number of recording instruments along with telemetering equipment to transmit the information recorded. A continuously transmitting transmitter will be employed. As we have had no experience in either manufacturing or experimenting with such types of miniature electronic instruments (except the transmitter we have developed), we were forced to rely on data published elsewhere. It is assumed that the instruments will weigh about 6 Kg. as indicated by Mr. J. Foley in a paper published in J. Brit. Interplan. Soc. dated August, 1956. However the weight of the power supply equipment for continuous transmission is estimated to be around 10 Kg. as the radio life of the satellite is expected to be one week. The main instrumentation will include two Geiger counter tubes, a resistence thermometer and a photocell. A microphone for detecting meteoritic hit will also be employed. Along with these the satellite will be equipped with a device to release sodium flares eight times after it has entered the orbit. (The sodium evaporator will be controlled by a timer so that it releases the flares when it is in the shadow area since then only it could be visually observed.) It is hoped that the release of sodium vapours will help in determining the orbit of the satellite accurately along with the Minitrack equipment All this equipment will be placed inside a capsule which is mounted on a hydraulic cushion. This is because the final accelerations that the instruments are expected to withstand are of the order of 40 G's. Such terrific accelerations might damage the instruments. But if they are mounted on a hydraulic cushion, as the acceleration increases, the liquid in the cylinder gets compressed owing to the increased weight of the capsule. In effect the 'G' forces acting on the instruments will be much less than what it would have been if the instruments were simply fixed to the rocket. In fact the extension of this principle might be used to send a biological specimen in a solid propellent satellite rocket. The animal will be in a sealed capsule and it will be made to float in a chamber filled with a heavy density liquid. As the acceleration increases the capsule begins to sink in the liquid, which results in the rise in the liquid level. But the capsule will not be experiencing *any* '*G*' *forces at all.*

The second part of the instrumentation will contain a small but powerful rocket. The rocket whose details are given in Table II will be attached to the

Table II. *Technical Details of the Escaping Rocket from the Satellite*

Length	72 cms.
Diameter.......................	9 cms.
Propellent weight	23 Kg.
Total weight	34 Kg.
Mass ratio......................	3.2
Burning time	1.02 sec.
Payload........................	4 Kg.

satellite vehicle in the same manner as the final step was attached to its former step. It will employ a fast burning fuel with a high exhaust velocity. Here we would like to say that such a fuel has already been developed. The fuel while burning at the rate of 1.21 metres per second gives an exhaust velocity of 2,742 metres per second. The rocket is equipped with a transmitter and batteries with life time of 48 hours. The transmitter will be switched on before the launching in order to make sure that it might not fail. The main object of keeping this rocket is to demonstrate the already accepted method of launching a rocket to escape the earth from a satellite orbit. As already explained the rocket will already be possessing an initial velocity of 8,260 per second and it only needs an additional velocity of 2.650 metres per second to escape the earth. The launching of the escaping rocket will be controlled by (i) a timing device which launches it automatically after a definite time interval, (ii) by a device which launches it on receipt of a coded signal from a ground station. We have already developed and tested an instrument to perform the later task. Both the types will be used to ensure that the rocket *is* launched. As the burning time is very small, within seconds after the launch the rocket will be speeding away from the earth having obtained the escape velocity. The escaping trajectory will be almost tangent to the satellite's orbit. The successful operation of this project will demonstrate practically the technique of orbital launching which will definitely play a major part in Interplanetary Flight.

Epilogue

This paper emphasises two main points. Firstly the use of an all Solid Propellent rocket vehicle for launching a satellite. In the long run the Solid Propellent satellite launcher will not only prove to be cheap but also highly reliable. In future tonnes of equipment will have to be hauled into orbit for the construction of Space Stations. If Solid Propellent rocket launchers are employed in Dr. VON BRAUN's Space Station project. we estimate that its cost will be cut down by more than 40%. More than all, almost all the launchers will be reliable. The details of this paper are worked out with a fuel which has been tested in India. We are sure that more powerful solid propellents are in existence in other countries and the use of such fuels will greatly increase the efficiency of the vehicle. Finally the cost of each rocket even if they are manufactured individually will be less than Rs. 300,000 or about $ 60,000.

Appendix

The details of a four-step satellite vehicle employing the "Shaped Charge" principle for the control during flight are given below (Table III). It is expected to deliver a 5 Kg. satellite into an equatorial orbit. This rocket is the final goal of "Project H.A.R.R. (High Altitude Rocket Research)" of the Indian Astronautical Society.

Table III

	Step 1	Step 2	Step 3	Step 4
Payload (Kg.)	342.43	72.04	15.83	5.00
Empty weight (Kg.)	428.30	91.05	21.46	7.07
Propellent weight (Kg.)	1315.40	251.38	50.57	8.75
Final weight (Kg.)	1743.70	342.43	72.04	15.83
Propellent flow per sec. (Kg./sec.)	31.50	9.77	2.44	0.51
Thrust (Kg.)	5915.00	1833.00	458.00	95.8
Burning time (secs.)	42.8	25.7	20.7	17.0
Length (metres)	1.52	0.91	0.73	0.61
Diameter (cms.)	122.0	44.5	22.2	10.0
Number of motors..........	4	1	1	1
Mass ratio	4.07	3.76	3.35	2.24
Final velocity (metres/sec.) ...	1990	4268	6304	8086

Tracking Objects Within the Solar System Using Only Doppler Measurements[1]

By

Robert R. Newton[2]

(With 1 Figure)

(Received August 27, 1959)

Abstract — Zusammenfassung — Résumé

Tracking Objects Within the Solar System Using Only Doppler Measurements.
In this paper, we assume that an artificial planetoid emits radiation, of reasonably
stable frequency, which can be received by a tracking station on Earth. Using no
information except the time dependence of the Doppler shift, we find that the orbital
elements can be completely determined, with a precision of about five significant
figures, during one-half day's tracking from one station. Using realistic values for
power available, noise level, etc., we expect that the effective range for Doppler
tracking should be at least fifty million kilometers.

**Bahnverfolgung von Objekten innerhalb des Sonnensystems unter ausschließlicher
Verwendung von Doppler-Messungen.** In dieser Arbeit wird angenommen, daß ein
künstlicher Planetoid Strahlung von leidlich stabiler Frequenz aussendet, die von
einer Meßstation auf der Erde aufgenommen werden kann. Selbst wenn wir keine
andere Information außer der Zeitabhängigkeit der Doppler-Verschiebung benützen,
finden wir, daß die Bahnelemente vollständig bestimmbar sind. Dabei wird während der
Beobachtungsdauer eines halben Tages von einer Station aus eine Genauigkeit von
etwa 5 Stellen erreicht. Unter Verwendung realistischer Werte für die verfügbare
elektrische Leistung, den Rauschpegel usw. kann man erwarten, daß die wirksame
Reichweite für Dopplermessung mindestens 50 Millionen Kilometer sein wird.

Sur la poursuite par seul effet Doppler de planétoides artificiels du système solaire. Le
planétoide artificiel est supposé émettre un rayonnement de fréquence stable, reçu
par une station terrestre. Sans autre information que la dépendance du temps dans
l'effet Doppler, les éléments de l'orbite peuvent être déterminés avec une précision
de cinq chiffres significatifs, après une demi-journée de réception par la station.
Le rayon d'action efficace de cette méthode doit dépasser cinquante millions de
kilomètres pour des valeurs réalistes de la puissance disponible et du niveau de bruit.

1. Introduction

In this paper, we assume that an object which is executing a trajectory
within the solar system is also a source of radiation which can be detected on
Earth. Because of the combined motions of the object and the detector, there

---

[1] This work was supported by the Department of the Navy, Bureau of Ordnance,
under Contract NOrd 7386.

[2] Applied Physics Laboratory, The Johns Hopkins University, Silver Spring,
Maryland, U.S.A.

will be a Doppler shift in the frequency of the detected radiation, and we assume that this Doppler shift can be measured.

We ignore relativistic effects, and consider only the classical Doppler shift,

$$\Delta\nu = -\,(\dot{\varrho}/c)\,\nu\,, \tag{1.1}$$

where $\Delta\nu$ is the Doppler shift, ϱ is the distance between source and observer, $\dot{\varrho}$ is its time derivative, c is the velocity of light, and ν is the radiated frequency. If ν is known, $\dot{\varrho}$ is directly inferable from $\Delta\nu$ For simplicity in what follows, we shall speak of $\dot{\varrho}$ itself as the primary observable.

Let us further assume that $\dot{\varrho}$ is measured as a function of time[1] over some finite time interval. We wish to investigate the extent to which a complete description of the motion can be inferred from our knowledge of $\dot{\varrho}(t)$.

Suppose that the equations of motion of the source are known, and that the motion of the observer is known. We can then set up a system of spherical polar coordinates, ϱ, Φ, Ψ, say, with the observer as origin, to use as generalized coordinates for describing the motion of the source. By integrating the equations of motion, we obtain ϱ, Φ, Ψ, and their derivatives, as functions of time and of six constants of integration. Some of these constants will occur in, and can be determined from, the known function $\dot{\varrho}$; the number that occur is a measure of the extent to which we can infer the motion from $\dot{\varrho}$ alone. This number in turn depends upon the symmetry of the situation, that is, upon the types of transformation which leave invariant the known force field and the known function $\dot{\varrho}$.

The simplest case is that of no forces, when there is complete spherical symmetry about the origin. The reader can readily convince himself that knowledge of $\dot{\varrho}$ yields three constants, namely, the speed, the time of closest approach, and the range at closest approach (hence the range ϱ at any time). This fact is the basis for a type of "miss-distance" indicator widely used in ballistics, whose origin we do not know, and also for certain work in tracking artificial satellites [1, 3, 5]. Three constants must remain undetermined; these can be taken as two angles which orient the plane containing the motion and the origin, and one angle which orients the trajectory in that plane. Arbitrary changes in these three angles obviously leave $\dot{\varrho}$ unaltered.

The only example we know of in which a trajectory was completely determined by using only Doppler information, leaving aside work in which Doppler information was used only to give range data, is in the work of GUIER and WEIFFEN-BACH [2]. They describe how the six orbit parameters of an artificial earth satellite may be determined using Doppler measurements obtained during a single "pass" from a single observing station. In this case, the force field is the gravitational field of the earth; if the observer were stationary and if the earth had spherical symmetry, there would be cylindrical symmetry about the line joining the observer and the center of the earth, leaving one parameter undetermined. Both the motion of the observer (upon the rotating earth) and the oblateness of the earth remove the degeneracy and allow the determination of the sixth parameter; of these two effects, the motion of the observer is the more important, and is the only one considered by GUIER and WEIFFENBACH.

In the present application, we take the observer to be upon the surface of the earth, and take the source of radiation to be a particle moving in the gravi-

---

[1] The time to be used is the time of emission of the radiation, not the time of reception. We shall discuss in Sect. 4 the procedure to be followed when the time of emission is not directly measurable.

tational field of the sun. The only symmetry operation, in the absence of the earth's rotation, is reflection in the plane of the ecliptic. Thus, the orbit of the particle can be determined completely except for the sign of the orbital inclination. This degeneracy is again removed by the rotation of the earth, because of the inclination of the ecliptic with respect to the equator.

At present, we are primarily interested in the accuracy with which the motion of the source can be determined. For this purpose, it should suffice to take the orbits of both the earth and the source to be Keplerian ellipses about the sun. In what follows, we shall calculate the Doppler frequency (or $\dot{\varrho}$ rather) in such a situation as a function of the ellipse parameters, including specifically the rotation of the earth, and shall then present some calculations upon the accuracy with which this function can be inverted to yield the orbit parameters. Finally, we shall discuss briefly some matters dealing with experimental feasibility and with perturbations upon the function earlier derived.

2. Doppler Frequency as a Function of Orbit Parameters

The simplest way to calculate the Doppler frequency is first to calculate ϱ, the distance between particle and observer, and then to differentiate ϱ. Under the assumption of Keplerian orbits, this is a standard calculation which we shall outline briefly.

We take an xyz coordinate system, with xy being the plane of the ecliptic, assumed invariant, and take x through the vernal equinox, whose precession we ignore. We choose units of length and time such that the semi-major axis of the earth's orbit and the earth's mean motion are both unity, and take the origin of time to be the vernal equinox. The coordinates of the center of the earth are thus given by:

$$\left. \begin{aligned} x_E &= r_E \cos(\theta_E + \varphi_E), \quad y_E = r_E \sin(\theta_E + \varphi_E), \\ z_E &= 0, \qquad\qquad\qquad\quad r_E = (1 - \varepsilon_E^2)(1 + \varepsilon_E \cos\theta_E)^{-1}. \end{aligned} \right\} \tag{2.1}$$

In these, ε_E is the eccentricity, θ_E the true anomaly, and φ_E the longitude of perihelion, of the earth in its orbit.

In describing the orbit of the particle being tracked, we use the semi-major axis a, the eccentricity ε, the orbital inclination i, the longitude of the ascending node $\psi + (\pi/2)$, the argument of perihelion $\varphi - (\pi/2)$ measured from the ascending node, and the time at perihelion t_p. The forms used for the longitude of the node and the argument of perihelion make (ψ, i, φ) a set of EULER's angles which carry z into the normal to the orbit and x into the position of peri-helion (see Fig. 1). Taking X to be an axis through perihelion, Z to be normal to the orbit, and Y so as to complete a right-handed system, the rotational matrix which multiplies a vector (XYZ) to yield a vector (xyz) is

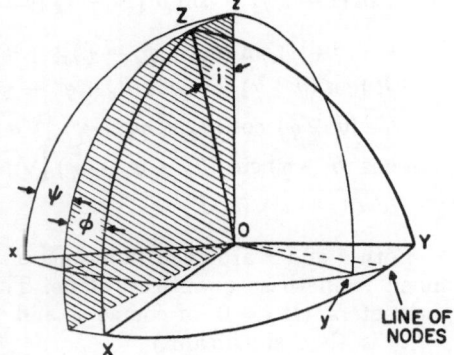

Fig. 1. Heliocentric coordinate systems. xy is the plane of the Earth's orbit, XY is the plane containing the planetoid orbit

$$\begin{pmatrix} \cos\varphi\cos i\cos\psi - \sin\varphi\sin\psi, & -\sin\varphi\cos i\cos\psi - \cos\varphi\sin\psi, & \sin i\cos\psi \\ \cos\varphi\cos i\sin\psi + \sin\varphi\cos\psi, & -\sin\varphi\cos i\sin\psi + \cos\varphi\cos\psi, & \sin i\sin\psi \\ -\cos\varphi\sin i & , & \sin\varphi\sin i & , & \cos i \end{pmatrix} \quad (2.2)$$

The XYZ coordinates of the particle, its true anomaly θ, and its eccentric anomaly u, are given by:

$$X = R\cos\theta, \qquad Y = R\sin\theta, \qquad Z = 0,$$
$$\theta = \cos^{-1}(\cos u - \varepsilon)/(1 - \varepsilon\cos u), \quad u - \varepsilon\sin u = (t - t_p)/a^{3/2}, \quad \Big\} \quad (2.3)$$

where R, the distance of the particle from the sun, is

$$R = a\,(1 - \varepsilon^2)/(1 + \varepsilon\cos\theta). \tag{2.4}$$

Using Eqs. (2.2), (2.3), and (2.4), we have for the xyz coordinates of the particle,

$$x = R\,[\cos(\theta + \varphi)\cos i\cos\psi - \sin(\theta + \varphi)\sin\psi],$$
$$y = R\,[\cos(\theta + \varphi)\cos i\sin\psi + \sin(\theta + \varphi)\cos\psi], \tag{2.5}$$
$$z = -R\cos(\theta + \varphi)\sin i.$$

We first calculate r, the distance from the particle to the center of the earth. r^2 is given by

$$r^2 = (x - x_E)^2 + (y - y_E)^2 + (z - z_E)^2.$$

Substituting Eqs. (2.1) and (2.5) into this,

$$r^2 = r_E{}^2 + R^2 - 2r_E R\,[\cos(\theta + \varphi)\cos i\cos(\theta_E + \varphi_E - \psi) + {} + \sin(\theta + \varphi)\sin(\theta_E + \varphi_E - \psi)]. \tag{2.6}$$

Calculating $\dot r$, the quantity of interest, is easiest done by first calculating $r\dot r$ from differentiation of r^2, and then dividing by $\sqrt{r^2}$. The result for $r\dot r$ is:

$$r\dot r = \varepsilon\,[a\,(1 - \varepsilon^2)]^{-\frac{1}{2}}\sin\theta\,\{R - r_E\,[\cos(\theta + \varphi)\cos i\cos(\theta_E + \varphi_E - \psi) + {}$$
$$+ \sin(\theta + \varphi)\sin(\theta_E + \varphi_E - \psi)]\} + \varepsilon_E\,(1 - \varepsilon_E{}^2)^{-\frac{1}{2}}\sin\theta_E\,\{r_E - {}$$
$$- R\,[\cos(\theta + \varphi)\cos i\cos(\theta_E + \varphi_E - \psi) + \sin(\theta + \varphi)\sin(\theta_E + \varphi_E - \psi)]\} + {}$$
$$+ \sin(\theta + \varphi)\cos(\theta_E + \varphi_E - \psi)\,\{\sqrt{a\,(1 - \varepsilon^2)}\,(r_E/R)\cos i - \sqrt{1 - \varepsilon^2}\,(R/r_E)\} + {}$$
$$+ \cos(\theta + \varphi)\sin(\theta_E + \varphi_E - \psi)\,\{\sqrt{1 - \varepsilon_E{}^2}\,(R/r_E)\cos i - {}$$
$$- \sqrt{a\,(1 - \varepsilon^2)}\,(r_E/R)\}. \tag{2.7}$$

Both r and $\dot r$ are even functions of i, so that the sign of i cannot be determined from these quantities alone. There is no degeneracy in the other orbit parameters. (If $i = 0$, of course, φ and ψ cannot be distinguished; only the sum $\varphi + \psi$ is then significant.)

The component of velocity of the observer as he is carried upon the surface of the rotating earth is not parallel to the ecliptic, and removes this uncertainty in i. In order to calculate this component of velocity, let us take a set of axes $\xi\eta\zeta$, with origin at the center of the earth, parallel to xyz respectively. Since the ξ-axis is, by definition, along the line of the descending node of the earth's orbit, the North Pole must lie in the $\eta\zeta$ plane, and makes an angle β ($\approx 23°$) with the ζ-axis.

Let λ be the co-latitude of the receiving station, and let γ be its longitude measured from the meridian through ξ, which is fixed in space. γ is a function of time:

$$\gamma = \gamma_0 + \omega t . \tag{2.8}$$

ω is the sidereal rate, approximately $366^1/_4$ in our units.

If we let r_0 be the radius from the center of the earth to the observer, we can readily verify that the $\xi\eta\zeta$ coordinates of the observer are

$$\left. \begin{aligned} \xi &= r_0 \sin \lambda \cos \gamma , \\ \eta &= r_0 \left[\sin \lambda \cos \beta \sin \gamma - \cos \lambda \sin \beta\right] , \\ \zeta &= r_0 \left[\sin \lambda \sin \beta \sin \gamma + \cos \lambda \cos \beta\right] . \end{aligned} \right\} \tag{2.9}$$

The distance ϱ from the observer to the source particle is calculated from

$$\varrho^2 = r_0{}^2 + r^2 - 2\xi (x - x_E) - 2\eta (y - y_E) - 2\zeta (z - z_E) . \tag{2.10}$$

We have already calculated r^2 and $r\dot{r}$, and need calculate only the differences $\varrho^2 - r^2$ and $\varrho\dot{\varrho} - r\dot{r}$, from which ϱ and $\dot{\varrho}$ are readily obtained.

The calculations are straight forward, though tedious. The result for $\varrho\dot{\varrho} - r\dot{r}$ can be broken up into two parts, which are, respectively, even and odd functions of i. The even part must, of course, be included in any actual tracking, but can be ignored for present purposes. It contributes small terms rather like those in $r\dot{r}$; its inclusion should not affect our conclusions about the accuracy of the method, whose investigation is our principal object at present. The odd part is

$$(\varrho\dot{\varrho} - r\dot{r})_{odd} = r_0 \sin i \left\{\omega R \sin \lambda \sin \beta \sin \gamma \cos (\theta + \varphi) + \right.$$

$$\left. + \left[a (1 - \varepsilon^2)\right]^{-\frac{1}{2}} \left[\sin \lambda \sin \beta \cos \gamma - \cos \lambda \cos \beta\right] \left[\sin (\theta + \varphi) + \varepsilon \sin \varphi\right]\right\} . \tag{2.11}$$

Of the two terms appearing, the first is the more important for values of a near unity and for ε near zero.

For the present, the difference between ϱ and r can be ignored.

3. Accuracy of Determining the Parameters

The quantity $\dot{\varrho}$ is a function of time and of the six parameters a, ε, φ, ψ, i, and t_p. This function is determined by combining Eqs. (2.11), (2.10), (2.7), and (2.6). It does not seem useful to work with this function other than numerically.

Suppose that we have an experimental measurement of $\dot{\varrho}$, call it $v(t)$, which extends over some time interval (t_0, t_1). Because of experimental limitations in time resolution, not all values of v are in effect independent; let us suppose that values of v at times separated by Δt do represent independent values, so that $v(t)$ is a function defined only at a set of values of t separated by Δt. We then wish to choose the six parameters in $\dot{\varrho}$ which make $\dot{\varrho}$ the best fit to v, in some sense.

The usual criterion of the best fit is that the parameters are to be chosen so as to make

$$f(a, \ldots, t_p) = N^{-1} \sum_{j=1}^{N} \{\dot{\varrho} (t_j, a, \ldots, t_p) - v(t_j)\}^2 \tag{3.1}$$

a minimum, where the t_j are all the values of t for which $v(t)$ is defined. Since $\dot{\varrho}$ is a complicated function of the parameters, finding the minimum is a difficult

calculating task, which we shall not discuss. We wish instead to investigate
the errors in the final values of the parameters which result from experimental
errors in v, assuming no further errors to be introduced by the calculations.
This is a standard problem in statistics, whose solution we shall outline briefly
(see [7], Sects. 108 and 125). Let $\dot{\varrho}_0(t)$ be the value of $\dot{\varrho}$ for the correct set of
parameters $a_0, \ldots t_{po}$, and write

$$v\,(t) = \dot{\varrho}_0\,(t) + \delta\,(t)\,. \tag{3.2}$$

We assume that the experimental errors $\delta\,(t_j)$ are small and random. Further,
let x_a, $a = 1 \ldots .6$, be the quantities $a - a_0, \ldots, t_p - t_{po}$, and expand $\dot{\varrho}$ through
terms linear[1] in the x_a,

$$\dot{\varrho} = \dot{\varrho}_0\,(t) + x_a\,\dot{\varrho}_c\,(t)\,, \tag{3.3}$$

where $\dot{\varrho}_a = \delta\dot{\varrho}/\delta x_a$, and we are using the summation convention on a. One readily
finds, upon substituting from Eqs. (3.2) and (3.3) into (3.1), that

$$f = N^{-1}\,x_a x_\beta \sum_j \dot{\varrho}_a\,(t_j)\,\dot{\varrho}_\beta\,(t_j) + \sigma_\delta{}^2 - (2/N)\,x_a \sum_j \delta\,(t_j)\,\dot{\varrho}_a\,(t_j)\,, \tag{3.4}$$

where σ_δ is the standard deviation of the $\delta(\tfrac{t}{j})$. The values of x_a which result
from minimizing f are

$$x_a = \sum_j \delta\,(t_j)\,A_{a\beta}{}^{-1}\dot{\varrho}_\beta\,(t_j)\,, \tag{3.5}$$

where $A_{a\beta}$, a coefficient in a symmetric matrix, is given by:

$$A_{a\beta} = \sum_j \dot{\varrho}_a\,(t_j)\,\dot{\varrho}_F\,(t_j)\,, \tag{3.6}$$

and $A_{a\beta}{}^{-1}$ is a coefficient in the inverse matrix.

x_a in Eq. (3.5) is the experimental error in the corresponding parameter
which corresponds to a given set of errors $\delta(t_j)$ in the measurement of v. σ_a,
the standard deviation of x_a, is given by

$$\sigma_a{}^2 = \sigma_\delta{}^2\,A_{aa}{}^{-1}\,; \tag{3.7}$$

that is, the σ_a are governed by the diagonal elements in the inverse matrix.

In order to apply Eq. (3.7) to our problem, we first approximated the sum
in Eq. (3.6) by an integral, and let

$$A_{a\beta} = (\varDelta t)^{-1}\int_{t_0}^{t_1}\dot{\varrho}_a\,(t)\,\dot{\varrho}_\beta\,(t)\,dt = B_{a\beta}/\varDelta t\,. \tag{3.8}$$

We then have

$$\sigma_a{}^2 = \sigma_\delta{}^2\,\varDelta t\,B_{ac}{}^{-1}\,. \tag{3.9}$$

$\varDelta t$ and σ_δ are characteristic of the experimental procedure used. The B_{aa} depend
upon the experimental procedure only through the time interval (t_0, t_1) during
which observations are made, and can be calculated theoretically for any particular
interval.

---

[1] If the linear term should happen to vanish for one or more of the x_a, let $x_a{}^n$
be the first power which occurs, and replace $x_a{}^n$ by x_a.

We evaluated numerically the various partial derivatives $\dot{\varrho}_a$ for several time points in an interval (t_0, t_1), for a particular set of orbit parameters, and calculated the matrix elements $B_{\alpha\beta}$ by numerical integration. Finally, we inverted the B matrix to obtain the elements B_{aa}^{-1}. The particular set of orbit parameters[1] used in these calculations was:

$$\left.\begin{array}{llll} \varepsilon = 0.13\,, & a = h^2/(1-\varepsilon^2)\,, & h = 1.063\,, \\ \varphi = 1.5192\,, & \psi = 1.5708\,, & i = .03\,, & t_p = -0.0516\,. \end{array}\right\} \quad (3.10)$$

Also, for simplicity, we used $\varepsilon_E = 0$. This value of ε_E would not be acceptable in actual tracking, but its use should not affect calculations upon the accuracy of the method. We also used the values $\lambda = \gamma_0 = \pi/2$.

These parameters describe a particle with node at the vernal equinox, perihelion a few days before, with a perihelion distance of unity and an aphelion distance about 1.3. At $t = 0$, when tracking is assumed to start, the particle is several hundred thousand kilometers from Earth. Thus, we have a situation which approximates that of initially establishing the orbit of an artificial planetoid launched from Earth. The time interval (t_0, t_1) over which measurements are made was taken to be the half-day following $t = 0$.

Reasonable values for Δt and σ_δ are 1 sec., or 2×10^{-7} in our units, and 1 meter/sec., or 3×10^{-5} in our units. These values will be justified in the next section. Using these values, the resulting standard deviations of the parameters, and the corresponding inverse matrix elements, are given in Table I.

Table I. *Accuracy of the Parameters*

Parameter	Inverse matrix element	Standard deviation
h	2.4×10^5	6×10^{-6}
ε	2.5×10^5	6×10^{-6}
φ	6.6×10^5	11×10^{-6}
ψ	1.3×10^5	5×10^{-6}
i	3.1×10^6	23×10^{-6}
t_p	4.0×10^5	8×10^{-6}
a	—	16×10^{-6}

Two principal numerical processes used in the calculations are differentiation and the inversion of a matrix. In order to increase accuracy, we used the so-called "two-sided" derivative, that is, we used, to approximate a derivative:

$$f'(x) \approx [f(x + \Delta x) - f(x - \Delta x)]/2\Delta x\,.$$

About four significant figures were lost in the differentiation process. The matrix to be inverted had numerous off-diagonal elements comparable with the diagonal ones, so that further loss of significant figures occured during the inversion. The accuracy of the inversion is difficult to estimate, so we tested it by making small alterations in the matrix elements. Our final conclusion is that the standard deviations of the parameters are accurate to about twenty or thirty per cent.

We conclude that Doppler tracking, under the conditions assumed, gives the orbital elements with an accuracy of about 10^{-5}, using a tracking period

---

[1] It was convenient in the calculations to use h, the angular momentum, rather than a.

of $^1/_2$ day. The accuracy should increase fairly rapidly with additional tracking time, as long as we stay within the effective range of the equipment. If we look at the form of the matrix elements $A_{\alpha\beta}$ in Eq. (3.6), we see that the ability to determine the parameters depends upon the extent to which the $\dot\varrho_a(t)$ differ for different a. As we increase the tracking time, we not only gain by having more data, but also by allowing more time for the differences among the $\dot\varrho_a$ to grow.

4. Perturbations and Experimental Feasibility

In this discussion, we shall assume that the particle to be tracked carries a radio transmitter of reasonably stable frequency; such a particle is presumably an artificial planetoid[1]. The first question is the range over which such radio tracking is feasible. The range is a function of power available, of antenna dimensions, of band-width requirements, of noise level, and of signal-to-noise requirements for the desired accuracy.

The band-width required depends upon the uncertainty in our *a priori* knowledge of the received frequency. In initial acquisition, the band-width must be fairly wide, but the range is then short, and requirements are not stringent. As knowledge of the orbit improves, the accuracy of predicting the Doppler shift and hence of the received frequency improves, and the band-width needed decreases. At long ranges, supposing a reasonable amount of previous tracking, the band-width can be decreased to probably one cycle per second; within this band-width we need a signal-to-noise ratio of unity [6] in order to obtain a time resolution of 1 sec. and a frequency error of 1 cycle per sec.

The noise level expected is somewhat uncertain. At 500 megacycles, the dominant sources of noise are apparently man-made ([4], p. 763); the median noise field strength in a band-width of 10 kc/sec received by a half-wave dipole antenna, for urban locations, is 10 μv/m. Thus, within a band-width of 1 cycle/sec, we need a signal S of 0.1 μv/m.

The range is given by

$$R = \sqrt{30 G_r G_t P/S} , \tag{4.1}$$

where R is range in meters, P is transmitted power in watts, G_r is the receiver antenna power gain over a half-wave dipole, and G_t is the transmitter antenna gain over an isotropic antenna. For a parabolic receiving antenna 30 m in diameter ([4], p. 703), G_r is about 7200; assuming no gain in the transmitter antenna, we get

$$R = \sqrt{30 \times 7200\, P}\,/10^{-7} = 4.6 \times 10^9 \sqrt{P} ,$$

about one-thirtieth of the Earth-Sun distance for $P=1$ watt.

It is probably possible to have a much greater effective range. By a judicious choice of the receiving site, it may be possible to decrease the required S by a factor of five or ten; if there is orientation control of the transmitting antenna, it may be possible to have a G_t of perhaps 10; use of a larger receiving antenna is perhaps feasible; and a transmitted power of 4 watts is perhaps feasible. A useful range of at least 50,000,000 km. is probably not unreasonable.

A further experimental problem deals with refraction of the received signal. Most of the refraction probably occurs in the ionosphere, and is highly unpredictable. Fortunately, the effects of ionospheric refraction vary inversely with

---

[1] Under some circumstances, use of Doppler shifts in the visible region might be feasible. The present writer does not feel qualified to discuss the accuracy of this process.

the square of the frequency. If we transmit two frequencies (say an octave apart for simplicity), and compare the received signals, we can in effect measure the refraction, and hence eliminate it.

We cannot do this without cost, however. The power radiated in each frequency is only half of the available power, so that the noise level achieved will increase. Also, the propagation paths for the two frequencies will not be identical, so that the refraction effects will not be accurately measured in regions where the ionospheric properties are changing rapidly. The net effect is probably to double the error in calculated velocity over what we estimated above neglecting refraction.

There is also the problem of frequency drift of the transmitted radiation. Here, there are several possibilities. For one, it may be possible to stabilize the frequency to at least one part in 10^9, particularly in the stable environment of a planetoid. This makes the frequency drift negligible at 500 megacycles. For another, it is conceivable to radiate the original frequency on Earth, and to have a transponder on the vehicle which adds a constant frequency and re-radiates the altered frequency. Frequency drift could be made negligible by this procedure.

Even if frequency drift cannot be eliminated, it is still not a serious problem provided the drift is reasonably steady. For example, if the radiated frequency drifts linearly over the period of observation, the received Doppler frequency contains eight unknown parameters, which are the central frequency, its drift rate, and the six orbit parameters. Addition of two parameters increases the complexity of the calculations, but does not sensibly affect the accuracy of the final results.

In summary, a frequency error of 2 cycles/sec. standard deviation at 500 megacycles, with a time resolution of 1 sec., seems feasible, leading to a standard deviation of about 1 meter/sec. in velocity. In any one set of observations, the errors will depend upon ionospheric conditions at the time.

Let us turn now to various effects omitted in the treatment of Sect. 2. We assumed there that the orbits of both Earth and the planetoid are Keplerian ellipses with one focus at the sun. Perturbations which cause deviations from such ellipses cause no trouble in principle, so long as they are known. The fundamental assumption is that the equations of motion are known; complexity in these equations causes complexity in the calculations, but need not affect the accuracy of the results.

Unknown perturbations will continue to be handled as they have always been. Their initial effect will be to cause larger residual errors than we expect; ultimately, systematic effects in the residuals will be discovered, and finally removed by refinements in the equations of motion.

One unknown quantity is the ratio between the laboratory and the astronomical units of length. The astronomical unit is used in the orbital description, while the laboratory unit is used in the velocity of light. By leaving this ratio (or equivalently, the velocity of light in astronomical units) as a parameter to be determined, we can determine this ratio to the same accuracy as the rest of the tracking. This accuracy, for a reasonable period of tracking, probably exceeds the present accuracy with which this ratio is known.

A final point arises in connection with the time to be used in solving the equations of motion. As we mentioned in footnote 1, p. 436, the time to be used is local time at the planetoid, but the time directly measured is Earth time. These times differ by the transit time of the radiation. If initial acquisition in the tracking occurs at such a distance that transit time is appreciable, we can proceed in two ways. One is to replace t where it earlier occurred (i.e., local time on the planetoid) by $t — (\varrho/c)$, where t is now Earth time. Since ϱ, the unknown range, now occurs here as well as in all previous places, the complexity of the calculations may become unbearable.

Alternatively, we may at first either ignore or guess at the transit time, determine a preliminary ephemeris, use this ephemeris to compute approximately the transit time, and continue to iterate through this cycle until satisfactory accuracy is obtained. This process puts the transit time on the same basis, with regard to the calculations, as perturbations in the equations of motion.

References

1. R. R. BROWN et al., Proc. IRE 45, 1552 (1957).
2. W. H. GUIER and G. C. WEIFFENBACH, Theoretical Analysis of Doppler Radio Signals from Earth Satellites. Johns Hopkins Univ. Applied Physics Lab. Bumblebee Series Report 276, April, 1958.
3. A. M. PETERSEN and Staff of Stanford Research Institute, Proc. IRE 45, 1553 (1957).
4. *Reference Data for Radio Engineers*, 4th Ed. New York: International Telephone and Telegraph Corp., 1956.
5. Royal Aircraft Establishment Staff. Nature 180, 937 (1957).
6. E. SHOTLAND, Accuracy of Zero Counters and Their Use in High-Precision Doppler Measuring Devices. Johns Hopkins Univ. Applied Physics Lab. Internal Memorandum BBD-569, Feb. 9, 1959.
7. E. WHITTAKER and G. ROBINSON, The Calculus of Observations, 4th Ed. London: Blackie and Son, 1944.

Nuclear Rocket Missions and Associated Power Plants

By

John J. Newgard[1] and Myron M. Levoy[1]

(With 7 Figures)

(Received July 13, 1959)

Abstract — Zusammenfassung — Résumé

Nuclear Rocket Missions and Associated Power Plants. The power plant requirements for a number of terrestrial escape to orbit and space missions, utilizing open
cycle nuclear power plants are presented. Among the particular missions analyzed
are single-stage, large payload, nuclear boosted, escape vehicles; chemically boosted,
secondstage nuclear rockets escaping with large payloads, and terrestrial orbiting
nuclear vehicles capable of moving into far space to perform such missions as orbiting
Mars.

A "family" of hydrogen cooled, solid-fuel graphite element, graphite core moderated, BeO reflector moderated reactor power plants are analyzed. The general reactor
system structure under diverse size and power conditions but similar for all "family"
members is presented. Such power plant considerations as heat transfer and fluid
flow, nuclear statics and power plant dynamics utilizing reflector control are considered for the missions in question. Principal problem areas, such as core corrosion-
erosion, reactor startup, and possible power plant perturbations leading to accidents
are discussed.

In conclusion, it is shown that a basic consistent philosophy of nuclear rocket
flight, utilizing the power plants discussed, may be developed.

Aufgaben und Kraftanlagen für Kernenergie-Raketen. Man kann Kernraketen
sowohl für irdische Aufgaben mit hohem Impuls als auch für den Antrieb von Raum-
schiffen verwenden. Für diese Aufgaben kann man eine Familie von Reaktoren
verwenden, die sich hauptsächlich durch ihre Größe und Leistung unterscheiden.

Eine typische Aufgabe der ersten Art wäre die Sendung einer Nutzlast von 50 000 kg
in den Weltraum mit einer einstufigen Kernrakete von 800 t Gewicht. Wenn der
Strukturkoeffizient (Trockengewicht ohne Reaktor/Totalgewicht) als 0,10 angenommen wird, würden der benötigte Reaktor und Reflektor 34 000 kg wiegen. Der
Reaktorkern in Form eines 2,9-m-gleichseitigen Zylinders mit 40% Hohlraum für
das Kühlmaterial würde ungefähr 120 000 MW entwickeln.

Für dieselbe Aufgabe könnte man die Kernrakete als zweite Stufe einer zweistufigen
Rakete verwenden, in der der Kernstufe durch die erste Stufe (chemisch mit F_2
und N_2H_4 als Treibstoffen) eine Geschwindigkeit von 3900 m/sec gegeben wird.
Dafür ist ein Reaktor mit einem 1,65-m-zylindrischen Kern nötig, der 8000 kg wiegt
und 22 000 MW entwickelt. (Trotz dieser Verringerung in der Leistung könnte man
nur wenig an Gesamtgewicht der zweistufigen Rakete gegenüber der einstufigen
sparen, es sei denn, daß beide Stufen mit Kernantrieb versehen werden.)

---

[1] Reaction Motors Division, Thiokol Chemical Corporation, Denville, New Jersey,
U.S.A.

Eine typische Raumfahrtsaufgabe würde eine Kernrakete mit 15 t Gesamtgewicht sein, um 2600 kg Nutzlast von einer Bahn in die Höhe von 500 km über die Erde zum Mars als „Sonde" mit Minimal-Energie zu schicken. In diesem Fall würde die Kraftanlage (einschließlich Pumpen, Düsen, Struktur und Schutzschild) rund 7000 kg wiegen und 1250 MW in ihrem 1,5-m-zylindrischen Kern entwickeln.

Für jede Aufgabe wählt man den passenden Reaktor aus einer Familie von Kernreaktoren mit Uran-Graphit-Elementen, gekühlt mit Wasserstoff als Treibmittel. Die Reaktoren verwenden auch Graphit als Baumaterial, das — zusammen mit den obigen Elementen — teilweise als Moderator dient. Weitere Moderation wird durch einen 15 cm dicken, zylindrischen BeO-Reflektor erhalten, der den Kern umschließt und mit Wasserstoff (Treibmittel) gekühlt wird. Die Höchsttemperatur der Gase aus dem Kern ist ungefähr 2200° C für eine Grenztemperatur der Graphitelemente zwischen 2500 und 2800° C. Der spezifische Impuls ist daher 750 bis 800 sec für ein irdisches „Boost"-Fahrzeug und fast 900 sec für ein Raumschiff. Das Gas aus dem Reflektor hat etwa 830° C Temperatur und wird zum Antrieb der Treibstoff-Turbopumpe verwendet und dann als Impuls verwertet (es ist nur ein kleiner Bruchteil des Gesamtgases).

Der Reaktorkern besteht aus einer großen Anzahl von Graphitplatten, die mit Uran geladen und in Elementbündel zusammengruppiert sind. Separatoren halten die Platten auseinander und schaffen so die Hohlräume für den Kühlstoff: 40% Hohlraum für die „Boost"-Rakete bis hinunter zu 5% in einem Raumfahrzeug-Reaktor. Metalle werden aus kernphysikalischen Gründen so wenig wie möglich verwendet, da sogar 0,2% Metall den Reaktorkern stark vergiften. Die Elementbündel werden daher durch Graphittraversen und Lateralstützen am Platz gehalten. Die Traversen werden ihrerseits von der Wand der Raketendruckkammer gehalten. Der BeO-Reflektor besteht aus bearbeiteten Blöcken mit Kühllöchern. In manchen Blöcken befinden sich kreisförmige Öffnungen zur Aufnahme der unten beschriebenen Kontrollplatten. Die Reflektorblöcke sind mit den Graphittraversen oder mit der Wand der Raketenkammer verbunden.

Zur Kontrolle der Reaktorleitung dienen zwölf Bor enthaltende Platten aus nicht rostendem Stahl in der Form von Zylinderteilen, deren Drehung bis 180° die wirksame Reflektordicke vergrößert oder verkleinert. Die kritische Masse kann so verdoppelt werden, wie dies für manche Aufgaben nötig sein mag, z. B. bei Korrosion und Erosion der Graphitelemente oder Xenonvergiftung. Da im Reflektor eine tiefere Temperatur herrscht als im Kern, arbeitet dieses Kontrollsystem unter weniger unangenehmen Bedingungen als eines mit Kontrollstangen durch den Kern.

Die wichtigen dynamischen Probleme der Kernrakete werden erwogen. Das „Boost"-Fahrzeug soll schnell beschleunigen, um Treibmittel zu sparen, so lange nicht Erwägungen der Kernphysik oder der Festigkeit dagegen sprechen. Für Übergänge zwischen Bahnen um die Erde oder für die Raumschiffahrt wird das Problem der Wärmeübertragung nach Schließung des Reaktors akut, ebenso wie die Vergiftung durch Xenon bei manchen Aufgaben. Bei Operationen unter Spitzenleistung können arge Störungen erwartet werden: durch Änderung der Dichte des Wasserstoffs, Zerstörung und Auswurf eines Graphitelements oder Elementbündels und Erosion und Korrosion des Reaktorkernes. Eine plötzliche Vergrößerung der Dichte des Wasserstoffs von 25% kann dazu führen, daß der Reaktor prompt kritisch wird.

Trotz dieser Probleme stellt die beschriebene Kernkraftanlage einen guten Ausgangspunkt für die Entwicklung der ersten Generation von Kernraketen dar. Sie kann verschiedenen Aufgaben angepaßt werden, und zwar durch Änderung der Kerngröße, der Anzahl und Größe der Elemente und Bündel, deren Urangehalt und Leistung.

Quelques missions assignables aux fusées nucléaires et systèmes propulsifs correspondants. Pour un certain nombre de missions hors du champ de gravitation terrestre les systèmes propulsifs nucléaires à cycle ouvert doivent présenter des caractéristiques qui font l'objet de cette étude. Les fusées nucléaires à un seul étage et charge utile élevée, capables de se libérer de l'attraction terrestre; les fusées à premier étage chimique et second étage nucléaire à grande charge utile et les fusées orbitales nucléaires capables de pénétrer l'espace en profondeur pour accomplir des missions telles qu'orbiter Mars sont parmi celles prises en considération.

Une famille de réacteurs à combustible solide, refroidis à l'hydrogène, avec modérateur à noyaux de graphite ou réflecteur en BeO est analysée. La structure générale du réacteur est similaire pour tous les membres de la famille pour des dimensions et puissances différentes. Dans le cadre des missions en question, des facteurs tels que l'échange de chaleur et l'écoulement du fluide échangeur, la statique des réactions nucléaires et la dynamique du réacteur à modération par réflecteur sont pris en considération. On discute les principaux problèmes qui se posent: tels que la corrosion et l'érosion du noyau, le démarrage de la réaction nucléaire et les perturbations qui pourraient dégénérer en accidents.

En conclusion on montre qu'une philosophie de base de la propulsion par réaction nucléaire, basée sur l'utilisation de ce type de réacteur, peut être élaborée.

I. Introduction

Chemical rockets, though still in a state of technical improvement and growth, have a marked upper limit on possible performance. The limit results from the fact that chemical rockets rely upon the heat of chemical reaction between an oxidizer and fuel for the expansion of the combustion products which constitute the propellant.

The heats of combustion for suitable propellant combinations rarely give temperatures in excess of 8000° F or average molecular weights of combustion products, $\bar{M}$ below 9. For example, F_2—N_2H_4, one of the best high temperature oxidizer-fuel combinations, has a combustion temperature, T_c of 7740° F, and F_2—H_2 one of the best low molecular weight combinations has an $\bar{M}$ of 8.9 at maximum specific impulse.

The specific impulse, I_{sp}, defined as rocket thrust per unit weight flow of propellant, is an important figure of merit in evaluating rocket performance. It is proportional to $\sqrt{\dfrac{T_c}{\bar{M}}}$. For the best combinations of T_c and $\bar{M}$, I_{sp} for chemical rockets is limited to about 400 sec.

Another important parameter in rocket performance calculations is the mass ratio, defined as total propellant weight divided by gross vehicle weight. If the rocket contains a nuclear reactor power plant, the weight of this plant must be included in the gross vehicle weight. Performance is improved if the mass ratio is high. Hence, a nuclear rocket power plant must have low weight and must be capable of creating and expelling high temperature low molecular weight exhaust gases.

Although many types of nuclear rocket power plants can be considered for nuclear rocket applications, this paper is limited to the open-cycle solid reactor core type [1-8]. This is the most likely "first generation" reactor system. The considerations of low weight reactor cores capable of operating at the very highest temperatures compatible with sound structure and at reasonable fissionable material loadings leads to the choice of low molecular weight ceramics or ceramic-like materials such as graphite [6]. Low molecular weight propellant gas (which serves as reactor coolant) makes the choice of H_2 inevitable if any appreciable specific impulse increase over the best chemical systems is to be attained. For example, use of CH_4, NH_3, or He would give approximately half the specific impulse of a H_2 nuclear rocket operating at the same temperature.

II. Performance—First and Second Stage Nuclear Rocket Boosters

The reactor core design is governed by the mission to be performed, at maximum heat transfer and fluid flow consistent with a structurally stable system. Performance calculations indicate that single and multi-stage chemical rockets compare favorably with nuclear rockets until thrust levels exceed some hundreds of thousands of pounds. For a solid core nuclear rocket reactor with definite temperature limitations (e.g., graphite limited to about 5000° F) maximum specific impulses of the order of 800 can be obtained. This is an improvement of about 100 % over the best chemical systems. Fig. 1 illustrates the results of a performance study in which a single-stage nuclear rocket (Point A on the Figure) is compared with high performance liquid propellant rockets for an escape mission with a 100,000 lb. payload. The weight of this graphite reactor power plant is 41,600 lb. including a 6 in. BeO reflector with 10 % void. The core is a 7.2 ft. by 7.2 ft. right cylinder with 40 % void, generating 58,600 MW.

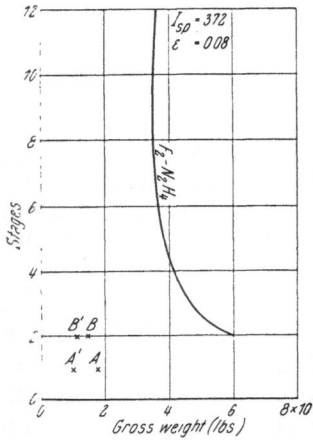

Fig. 1. Required stages for nuclear and chemical escape rockets carrying payloads of 100,000 lbs. Point A: 1-stage H_2 nuclear rocket, $I_{sp} = 750$, $\varepsilon = 0.15$ (excluding reactor). Point B: 2-stage rocket, 1st stage: F_2—N_2H_4, $I_{sp} = 372$, $\varepsilon = 0.08$; 2nd stage: H_2 nuclear, $I_{sp} = 750$, $\varepsilon = 0.15$. Points A' and B' are same as above except $\varepsilon = 0.10$ for nuclear stages

The structure factor ε (vehicle dry weight/total weight) includes tankage, pumps, combustion chamber, nozzles, and associated hardware, but excludes the reactor system. (Reactor weight is added into the gross weight.) ε's for both the chemical and nuclear rockets have been very conservatively estimated. Reductions in ε will considerably improve the already superior nuclear rocket performance, as seen in Fig. 1. A reduction in the structure factor from 15 to 10 % reduces the weight of the reactor power plant to 26,700 lb. The core is now a 6.4 ft. by 6.4 ft. right circular cylinder with 40 % void generating 35,000 MW. The corresponding vehicle is indicated by Point A' of Fig. 1.

Point B on Fig. 1 represents an optimized, minimum weight two-stage escape vehicle (nuclear-powered second stage) with a 100,000 lb. payload. The booster first stage is a F_2-N_2H_4 chemical rocket. The graphite nuclear reactor power plant required for the second stage weighs approximately 21,000 lb. including a 6 in. BeO reflector with 10 % void. The reactor core is a cylinder approximately 5.8 ft. in diameter, 5.8 ft long, with 40 % void, generating 27,500 MW. These results indicate the weight and stage superiority of nuclear powered rockets for a high payload escape (or satellite) mission.

III. Performance—Nuclear Space Vehicle

It is necessary, for a meaningful, reliable space mission, that certain realistic boundary conditions be placed on the nuclear vehicle. First a minimum feasible reactor power plant weight (including structures, pumps, and nozzles) must be established. At the present time, for an essentially thermal neutron-energy machine, a minimum weight of about 12,000 to 14,000 lb. has been calculated for a solid-core reactor (one member of a family of reactors which are discussed throughout this report). It is quite important that the reactor be "thermal"

for a space vehicle because of the unusually difficult remote control problems
in the system which should not be compounded with fast reactor control problems.
This problem is discussed briefly, below. The heavy reactor weight penalty
immediately suggests the need for going to the larger gross weight space vehicles.

A second boundary condition for the first nuclear space flights is establishment of a useful mission in terms of minimum ΔV and restart operation. A
minimum-energy Mars probe with $\Delta V = 11{,}600$ ft./sec. is chosen. This excludes

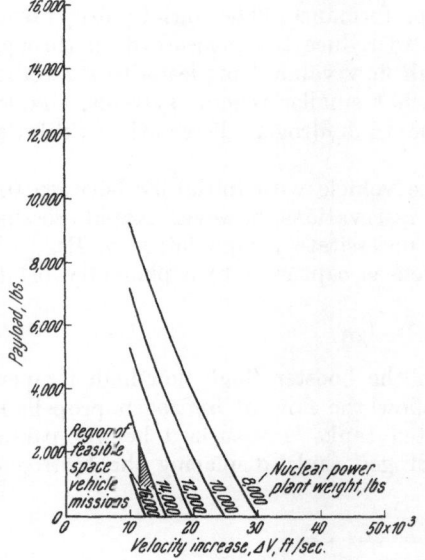

Fig. 2. Payload vs. mission velocity increase, Δv,
for a range of nuclear power plants $\sim$. Specific
impulse = 900 sec. Vehicle gross weight = 25,000 lb.
Tankage weight = 0.05 propellant weight

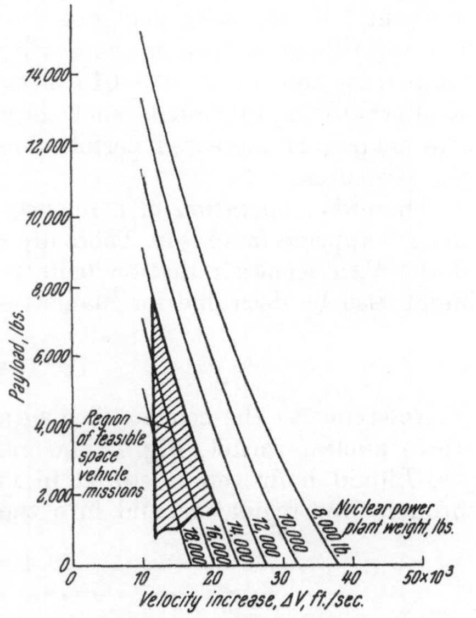

Fig. 3. Payload vs. mission velocity increase, Δv,
for a range of nuclear power plants $\sim$. Specific
impulse = 900 sec. Vehicle gross weight = 33,500 lb.
Tankage weight = 0.05 propellant weight

guidance requirements. For this mission, restart is unnecessary. Midcourse
guidance corrections ($\Delta V \sim 1000$ ft./sec.) may be performed with a chemical
rocket system weighing about 800 lb.

A third boundary condition is the minimum useful payload per mission.
Such a curve of minimum useful payload vs. ΔV per mission is difficult to
establish at present. From literature sources which tend to conflict, a rather
flat curve at 750 to 2000 lb. payload was established, increasing with mission ΔV.

These three boundary conditions are plotted in Figs. 2 and 3 for nuclear
space vehicles with specific impulses of 900 sec. (about 4000° F hydrogen temperature) and gross vehicle weights of 25,000 and 33,000 lb. The increased area
of feasible operation with increased vehicle gross weight is clear.

A typical vehicle for a minimum energy Mars probe from a 300 mile orbit
would have a gross weight of 33,000 lb., a payload of $\sim$ 5800 lb., a 16,000 lb.
reactor (including nozzle, pumps, structures) and a tankage weight of about
700 lb. (5 % of the propellant weight). These correspond to a specific impulse
of 900 sec. for a reactor exit hydrogen temperature of about 4000° F. The thrust
would be about 63,000 lb. for a flow rate of hydrogen propellant of 70 lb. sec.
leading to an initial acceleration of $\sim$ 2 g.

Reactor weight need not increase in going to heavier space vehicles. It is
possible to operate the reactor at a fixed power and propellant flow rate and

permit the initial vehicle acceleration g to vary with gross weight. For the typical 33,000 lb. space vehicle mentioned above, reduction in g down to a value of 0.2 does not substantially effect performance. (This cut-off value of g varies for different missions and gross weights from 0.1 to about 0.4 g.) Therefore, the same basic reactor system may be used for a range of space vehicle gross weights from below 20,000 lb. to about 300,000 lb.

On the other hand, instead of increasing vehicle gross weight, it is possible to reduce the flow rate in the typical vehicle described above from 70 lb./sec. to about 7 lb./sec. with negligible loss in performance. (The initial g drops from 2 to 0.2.) Such a drop in flow will tend to reduce the pressure drop through the reactor core to about 0.015 of the high flow value. This leads to a number of alternate improvements such as somewhat smaller reactor systems, simpler core shapes, or increased performance due to hydrogen dissociation at lower gas pressures.

Therefore, operation of a nuclear space vehicle with initial g's between 0.2 and 1.0 appears favorable. There are some reservations, however. Rapid crossing of the Van Allen radiation belts might necessitate a high initial g. High g's might also be desirable for maneuvers such as capture into a planetary orbit.

IV. Overall Design

Referring to the schematic diagram of the booster (high flow-high temperature) nuclear rocket, Fig. 4, one may follow the flow of hydrogen propellant gas. Liquid hydrogen is stored in insulated tanks. Pressurized helium passes through flow regulators and into the hydrogen tank, displacing the hydrogen

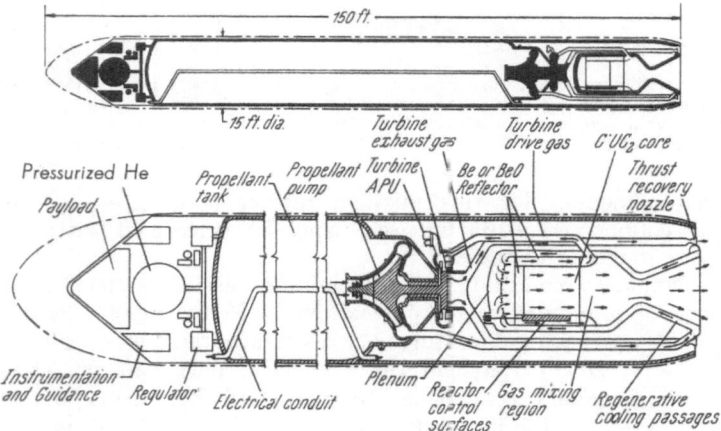

Fig. 4. Nuclear rocket schematic

which is forced through the propellant pump. From the turbopump, the hydrogen at high pressure, flows down to the exit end of the nozzle system through ducts.

At the end of the nozzle, the hydrogen enters a torus shaped distribution plenum. Here, it enters the regenerative cooling passages which surround the hot walls of the nozzle and reactor chamber. The cryogenic hydrogen coolant flows up the outside of nozzle and chamber walls and into the entry plenum of the reactor. Most of the hydrogen then flows down through the end reflector and reactor core where its temperature is raised to about 4000° F while its pressure drops considerably. The hot gas is then ejected through a convergent-divergent nozzle (DeLaval nozzle) where the sensible heat of the gas is converted into kinetic energy which provides thrust.

A small percentage of the hydrogen at the entry plenum flows into cooling passages in the neutron reflector which surrounds the core. Orificing of the reflector passages determines the amount of flow. Typical reflector exit gas conditions would be 1500°F and 400 psia. This gas is then ducted back to the drive turbine of the turbopump. By-pass valves permit control variations. After passing through the turbine, the gas exhaust is used for thrust recovery in small auxiliary nozzles.

A very small amount of hydrogen gas from the entry plenum passes down through reactor structural columns, for cooling purposes.

A series of core elements types were studied. It was found that parallel flat plates arranged in bundles gave lowest pressure drops under the severe operating conditions of the core. Flat plates were found to be superior to rods because of the simpler spacer arrangement and less flow blockage due to the spacers.

Table I gives a summary of reactor parameters for typical nuclear powered rockets useful as first and second stage boosters and space vehicles.

Table I. *A Range of Parameters for a Nuclear Reactor for Rocket Propulsion*[1]

Thrust	4×10^3 to 4×10^6 lbs.
Specific Impulse, I_{sp}	750 – 900 sec.
Propellant Flow Rate	7.0 to 3000 lbs./sec.
Reactor Nominal Core Size	4×4 min.; to 7.5 ft. diameter $\times$ 7.5 ft. length right cylinder.
Power Density in Core Elements (average)	.15 to 15 kw/cc max.
Void Fraction in Core	0.05 to 0.40
Core Element Materials	Uranium loaded graphite (density 1.9 gm/cc)
Gas Temperature at Regenerative-coolant passage inlet (main stream H_2 used as regenerative coolant, then passed through reactor)	37° R (H_2 boiling point)
Gas Temperature, Reactor Outlet	about 3000°F—4200° F
Element Shape	Parallel plates or rods
Plate Thickness	30 mils minimum
Reactor Weight (including reflector and core structure)	14,000 to 50,000 lb.

V. Reactor Core Design

A prototype 5 ft. diameter, 5 ft. long cylindrical solid fuel element reactor, designed for maximum feasible core power density (10 kw/cc in the core elements), temperature, and propellant flow, illustrates all the basic features of a nuclear power plant for rocket propulsion. This reactor is capable of performing the mission designated as B' in Fig. 1, if ε for the second stage nuclear rocket is reduced from 15 % to 10 %. (It should be noted that for the second stage nuclear rocket, large tankage reductions are not as crucial for overall system weight savings as in single stage nuclear rockets.) The reactor core might consist of a large number of parallel plate fuel elements which may be made of uranium loaded graphite. Plates may be unclad or may utilize a coating, depending

---

[1] Low weights, specific impulse, power densities, thrust, temperatures, correspond to small reactor sizes and vice versa.

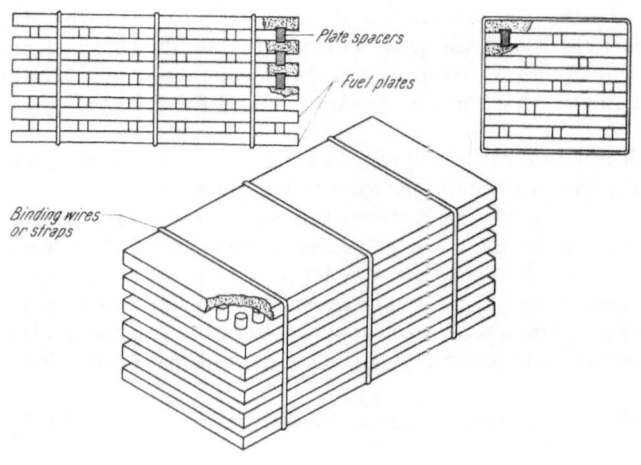

Plate spacers

fuel plates

Binding wires
or straps

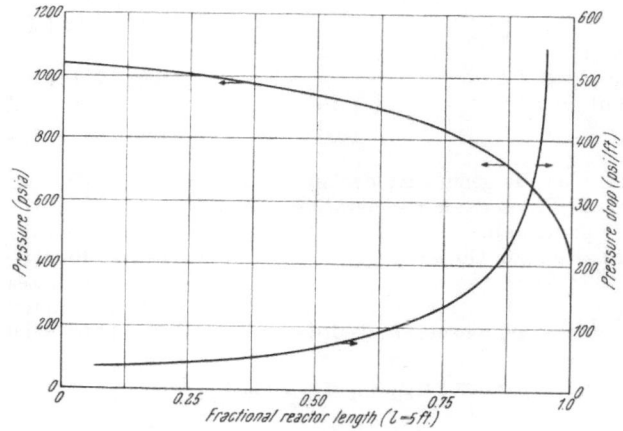

(Void fraction = flow area/total area.) This results in a rapidly increasing pressure drop as the gas approaches the hot end of the reactor, particularly at low values of void fraction. The increase in pressure drop is almost exponential. Fig. 6, for a typical core, illustrates the rapidly increasing pressure drop.

A more complex design might utilize a variable void fraction through the core (higher void fraction at the hot end, lower at the cold end). Such variations would reduce the overall pressure drop across the core without introducing much additional fabrication difficulty. Also, a saving in core structural weight might be effected due to lower coolant drag forces on the structures.

The design at each point through a typical graphite reactor is limited by three considerations: 1. Maximum allowable graphite plate temperature must be below about 5000°F. 2. The thermal stress in the plate may not exceed the allowable stress. 3. The minimum plate thickness must be held to some reasonable level, perhaps 30 mils, from fabrication and rigidity considerations.

From the nuclear point of view, we have considered reactor systems consisting of cylindrical reactor cores varying from 3.5 ft. diameter, 3.5 ft. length to 10 ft. diameter, 10 ft. length, with void-volumes varying from 30 to 50%. A 6 in. reflector with 10% coolant void surrounds the core in all cases. (Axial reflection is not considered.)

Some interesting conclusions can be drawn. A large peak to average power ratio exists in a homogeneously loaded system. This will require an asymmetric power loading radially, that is, a delta loading (more fuel at core-reflector interface) for optimum power generation and highest possible temperature from the core. The critical mass is extremely sensitive to void volume, the smallest variation being associated with the largest reflector. In addition, large reactivity is controlled by reflector changes at 30—50% core void volume. For example, critical mass varies more than a factor of two between the fully reflected reactor and the unreflected reactor in the case of a 5 ft. diameter, 5 ft. length nuclear core. This permits numerous operational contingencies requiring large reactivity insertions or removals. Such contingencies are anticipated in the nuclear rocket (as discussed below) and are one of the factors leading to reflector control. Although the reactor may be intermediate, a large thermal power is present for those systems greater than 4 ft. in diameter, particularly when somewhat reflected. This simplifies control considerations.

The difficulties in utilizing core inserted poison control rods also leads to the adoption of reflector control. Also, core rods lead to more difficult cooling problems than reflector drums. They must be axially removed and inserted tending to make the overall system less compact. They must operate in a high temperature region subject to some destruction. This may lead to control rod channel blockage. In addition, core rods occupy valuable volume which might otherwise be occupied by fuel elements.

VI. Reflector Design and Control Plate Design

The reactor core is surrounded by a BeO reflector which slows down fast neutrons and reflects a large fraction of them back into the core. The reflector is nominally 6 in. thick and extends almost the entire length of the reactor core. Coolant holes run axially through the blocks and permit cooling of the blocks with hydrogen gas to tolerable temperatures and temperature gradients.

Gamma heating is the principal mode of heat production in the reflector. Approximately 10% of the 6 in. thick reflector lateral cross sectional area consists of 1/8 in. diameter hydrogen coolant holes.

As previously indicated, the hydrogen gas leaves the reflector at about 1500°F and is then used to drive the turbopump.

The control of the nuclear reactor is accomplished by means of stainless steel control plates containing boron, stationed around the reactor core. The plates are curved portions of right cylinders. The plates must be cooled due to the heating accompanying capture of neutrons by the boron and consequent α-particle emission. Each plate is completely jacketed by coolant passages.

The BeO reflector blocks are machined to take the control plate assemblies. The plates fit into curved slots in the assembled BeO blocks. When the plate faces in toward the reactor core, the BeO reflector is cut off from the core neutrons in part by the borated surface. As the plate is rotated out, more fast and slow neutrons can enter the reflector. Some slowing down occurs there and most of these neutrons are returned or reflected to the core. Hence, the power level of the reactor may be controlled by the movement of these plates.

To achieve maximum reactor control, movement of the control plate through 180° can be provided. This method of varying the effective reflector thickness can achieve $1/2$ to 2 % reactivity control for each of 12 control plates.

An analysis of deflections of the composite shroud and control plate under maximum load indicates that enough stiffness may be built into the plate to avoid binding of the plate in the BeO slot, for reasonable slot widths. The width of the slot must take into account plate deflection; machining, fabrication, and placement tolerances; and thermal creep and expansions of structural support, BeO blocks, and control plate during operation.

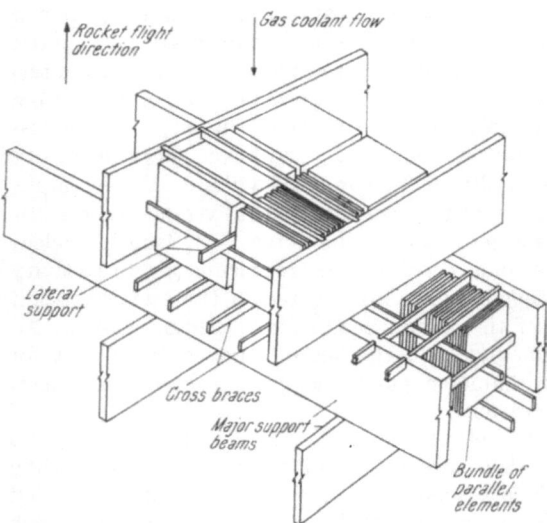

Fig. 7. Graphite structural support for a nuclear reactor with parallel plate fuel elements

VII. Core Structural Support Design

The amount of structural metal or other material with a high capture cross section introduced into the reactor core must be severely limited to prevent poisoning of the core and consequent increase in fissionable material inventory. Structural material must also retain strength up to about 4500°F. Graphite is a likely structural material as well as fuel element material since it has the above features and also has excellent thermal shock properties.

A structural design which seems quite promising consists of a matrix of lateral graphite beams and cross braces (Fig. 7). The fuel element bundles are positioned and held in place by means of the cross braces. Tubular tension members for reinforcement are supported from the cold end of the reactor and run the length of the reactor. These columns are cooled with internally flowing hydrogen. Structural support beams are attached to these tension members and to the rocket outer wall, while lateral support beams are attached to the major beams and the rocket wall. Each layer of beams is oriented 90° to the ones just above and below. Cross braces support the elements and hold them against lateral loads. The reflector blocks are attached to the cross beams by means of keys.

VIII. Some Dynamic Operating Considerations

A suitable safe reactor period must be chosen by an optimum manipulation of the control plates. Fluid flow rates will be held to a relatively low value as temperatures of the core increase consistent with structural stability, until the proper temperature profile through the core is established. (Criticallity can occur at some point in this heat-up phase.) Power increase to required operating level can now occur by increasing the fluid flow rate. This must be done rapidly to avoid propellant waste. The reactor will have to be carefully coupled with the control of turbopump and valves.

Space missions requiring the retaining and restarting of the reactor pose additional difficulties. The reactor must be cooled on shutdown to remove the fission product decay afterheat. The slow bleed of propellant required may be used for extremely low thrust—low g propulsion similar to an ion rocket. For coasts between Earth and Mars, this may be reasonably satisfactory since the coast occurs in gravityfree space, or nearly so. In this case, the overall mission would have 3 phases: a relatively high thrust escape spiral from earth, followed by a very low thrust coast, with a final high thrust maneuver (reactor restart) into Mars orbit. Since the coast time is quite long (260 days) no xenon override is required.

For a mission from a 300 mile orbit to a stationary orbit around the earth, the problem is more serious. The low g coast occurs in a relatively high gravitational field leading to considerable performance losses. In addition, xenon override (because of the large thermal neutron component designed into these systems) becomes a problem in such orbital transfers where reactor restart is required during the first day. Decay heat during transfer must be safely removed and poses a weight penalty upon the entire system unless the open cycle system can be closed under these circumstances.

During power operation (perhaps sooner) various reactivity changes will occur due to changes in the material composition of the core. Reactivity calculations were carried out for a 5 ft. diameter, 5 ft. long cylindrical reactor used as a second stage containing 40% void volume and reflected by a 6 in. BeO reflector, 10% void.

The following reactor core changes were considered:

(1) Reactivity change due to change in hydrogen density in the core void volume.

(2) Uniform decrease in the solid volume of the core (including both fuel and moderator) due to corrosion or erosion of the system.

(3) Non-uniform decrease due to destruction of a plate or bundle in the core matrix and its consequent discharge out of the core by the propellant gases.

Table II. *Reactivity Changes* $\left(\dfrac{\delta k}{k}\right)^1$

Core Disturbance	3 in. Reflector	6 in. Reflector
Addition of H_2	1.58×10^{-2}	2.90×10^{-2}
1% decrease in core solid volume	$-.566 \times 10^{-2}$	$-.550 \times 10^{-2}$
1% decrease in fuel volume	$-.419 \times 10^{-2}$	$-.414 \times 10^{-2}$

[1] In a 5 ft. long, 5 ft. diameter cylindrical reactor with 40% core void, 10% reflector void.

Table II summarizes some results. Each of these changes in the reactor was assumed to be distributed uniformly throughout the reactor core volume. It is important to note the large positive reactivity change associated with even small changes in the hydrogen density in a core of this size. Control of small density fluctuations will pose severe problems since this control is also important in burnout considerations.

IX. Hydrogen Pump Design

It is desirable to have a hydrogen pump with the simplest possible design following conventional hydraulic practice.

Pump bearings and seals are among the critical items. Hydrostatic-type bearings of hydrogen bled from the high pressure source can float the rotating shaft. Other methods may place the shaft outside the cold region so that bearings operate in an ambient environment. However, there are indications that simple stainless steel roller bearings or ball bearings will run satisfactorily immersed in liquid hydrogen. This type of bearing may be the most desirable immediate development.

Maximum impeller tip velocities up to 1300 ft./sec. may not be sufficient to generate required heads (800 to 1400 psi.) with low density hydrogen at high flows. Therefore, multi-stage pumps, preferably of the axial flow type, may be required in order to get higher pressures without higher tip velocity.

X. Nozzle Design and Regenerative Cooling

A typical nozzle design for the nuclear boost rocket is described in Table III.

Table III. *Nozzle Parameters for a Nuclear Rocket*[1]

Chamber diameter	60 in.
Throat diameter	34 in.
Exit diameter	96 in.
Exit area/throat area[2]	8.0
Throat area	909 sq. in.
Pressure at nozzle outlet	14.7 psi to about 0
Length of calming section between reactor core and convergent section	18"
Length of convergent section	10"
Convergent angle	45°
Total length of nozzle plus calming section	140 in.

The hydrogen propellant appears to be satisfactory as a coolant for the reactor chamber and nozzle wall. An important consideration in any regenerative cooling design is vapor formation in bulk boiling which might produce burnout due to reduction in coolant flow. Since the hydrogen in the regenerative cooling portion of the cycle of necessity must be well over the critical pressure (188 psi), boiling as such will not occur.

The regenerative coolant for the nozzle and chamber will flow through tubular passages which constitute the nozzle and chamber wall. These tubular passages might be made of nickel, a nickel alloy such as Inconel X, or a stainless steel.

---

[1] Coupled to a 5 ft. length, 5 ft. diameter cylindrical (15,000 MW) nuclear power plant.

[2] This value may range from 8 to 40 depending on the flight profile of the rocket mission.

The hot gas side of these tubes may require a ceramic insulating layer. The tubular passage wall thickness may run as low as 40 mils. The tubes will be wire-wrapped on the outside and will be brazed or welded together.

An important consideration in the chamber vessel design is the fact that in addition to the usual hoop-stresses due to gas pressure, a very considerable axial force from g-load and drag in the boost vehicle is transmitted to the shell from the reactor core via the structural support members. A relatively heavy wall or thick reinforcing ribs may therefore be necessary in the chamber. Such added metal not only decreases overall rocket performance due to greater dry weight, but also leads to higher temperatures and thermal stresses since considerable gamma absorption from the core is unavoidable without resort to heavy shields.

XI. Conclusion

This paper has shown that reasonable missions can be accomplished with a graphite type nuclear reactor open cycle power plant. The weight variation of such plants can be between 14,000 to over 50,000 pounds. All can be controlled similarly, utilizing a thermalizing reflector control. All can be constructed similarly. The principal problems in the heavier systems are power—structural ones since these boost-type systems will be subjected to high internal drag and acceleration forces at high power output and also subject to large scale perturbations. Such systems can be developed within present technology and a prototype development of one of these power plants, for example the 5 ft. length, 5 ft. diameter system at its maximum capability, will show the potentialities of this concept in nuclear propulsion.

Acknowledgment

The authors would like to thank R. BARROW, J. GIOVANNUCCI, W. T. HARRISON, and H. PEDERSEN, for their analytical and design contributions to this paper.

References

1. H. S. SEIFERT, M. M. MILLS and M. SUMMERFIELD, Physics of Rockets: Dynamics of Long Range Rockets. Amer. J. Physics 15, 266 (1947).
2. L. R. SHEPHERD and A. V. CLEAVER, The Atomic Rocket. Brit. Interplan. Soc. 8, 23 (1949).
3. S. A. COLGATE and R. L. AAMODT, Plasma Reactor Promises Direct Electric Power. Nucleonics 15, No. 8, 50 (1957).
4. R. W. BUSSARD, Concepts for Future Nuclear Rocket Propulsion. Jet Propulsion 28, 223 (1958).
5. R. W. BUSSARD and R. D. DeLAUER, Beyond Tomorrow. Nucleonics 16, No. 7, 73 (1958).
6. M. M. LEVOY and J. J. NEWGARD, Rocket-Reactor Design. Nucleonics 16, No. 7, 66 (1958).
7. J. J. NEWGARD and M. M. LEVOY, Nuclear Rocket Propulsion. Scient. Amer. 200, No. 5, 46 (1959).
8. F. E. ROM and P. G. JOHNSON, Nuclear Rockets for Interplanetary Propulsion. Paper 63R, SAE National Aeronautic Meeting, New York, March 1959.

On Fast Corpuscles of the Upper Atmosphere

By

V. I. Krassovsky, I. S. Shklovsky, G. I. Galperin and E. M. Svetlitsky[1]

(With 2 Figures)

(Received August 2?, 1959)

Abstract — Zusammenfassung — Résumé

On Fast Corpuscles of the Upper Atmosphere. A direct discovery of powerful electron fluxes of low energy at great heights in the upper atmosphere made with the aid of the Third Soviet Sputnik is reported. The observed electrons have a typical energy of about 10 keV and show anisotropy with respect to magnetic force lines. Some implications of the observed phenomena are briefly discussed.

Über schnelle Teilchen in der Hohen Atmosphäre. Es wird über die direkte Auffindung intensiver Elektronenströme von niedriger Energie in großen Höhen der Hohen Atmosphäre berichtet, die mit Hilfe des dritten sowjetischen Sputnik entdeckt wurden. Die beobachteten Elektronen weisen eine charakteristische Energie von ungefähr 10 keV und eine Anisotropie bezüglich der magnetischen Kraftlinien auf. Einige Folgerungen aus den beobachteten Erscheinungen werden kurz erörtert.

Sur les particules rapides de la haute atmosphère. Le troisième satellite soviétique a permis la découverte directe de flux puissants d'électrons de faible énergie dans la haute atmosphère. Les électrons observés ont une énergie typique d'environ 10 keV et présentent une anisotropie par rapport aux lignes de force magnétiques. Quelques conséquences des phénomènes observés sont brièvement discutées.

An experiment for direct discovery of soft electrons in the upper atmosphere has been carried out at the third Soviet sputnik launched in May 15, 1958 [1, 2, 3]. Measurements of their slightly harder component were carried out afterwards by other authors [4, 5]. Similar experiments were carried out by means of rockets in the auroral region [6].

The study of cosmic rays was not the aim of our investigation. We were interested in the processes which are essentially effective in the upper atmosphere. Although a series of indications were in favour of the existence in the upper atmosphere of the soft electrons mentioned above, many of our colleagues, in particular those studying cosmic rays, felt sceptic about this, especially concerning the region beyond the auroral zone.

Fluorescent screens of ZnS activated with Ag of 2 mg/cm² were used as indicators. Luminescence of each of these two screens was measured by a photoelectron multiplier. There were thin aluminium absorbing foils of 0.4 and 0.8 mg/cm² in front of the fluorescent screens. The choice of the minimum thickness of the foils used (0.4 mg/cm²) was dependent on the appearance of microscopic pores,

---

[1] Institute for Atmospheric Physics of the Academy of Sciences of the U.S.S.R., Moscow, U.S.S.R.

through which the light is passing when the instrument is directed to the sun. The block-scheme of the instrument is shown in Fig. 1. The characteristics of the amplifiers have been made logarithmic with an accuracy of some ten per cent.

The indicators were calibrated in a parallel monoenergetic flux of electrons. Within the instrument scale and due to some inhomogeneities in this thickness of the foil the instrument recorded, by both channels, the electrons from 10 keV. When the electrons were of more than 50 keV the results of measurements by two channels were practically the same. As a result of calibration we obtained the dependence of the intensity ratio of the photofluxes of two photomultipliers upon the energy of electrons. The sensitivity of the indicators used for the electrons of 10 keV was 30 times lower as compared to one for the electrons of 40 keV with a thin foil and 3.10^3 lower with a thick foil. This made it possible to estimate the average energy of observed nonmonoenergetic fluxes of electrons in the upper atmosphere. They also caused different signals in two channels. The energy of electrons of monoenergetic flux corresponding to the observed ratio of signals will be called "equivalent" energy of nonmonoenergetic electron flux. The value of the "equivalent" energy is obviously dependent on the shape of the energy spectrum. It should be stated, however, that due to the strong decrease of sensitivity with the decrease of energy, the "equivalent" value shows that the majority of the registered electrons are of lower energy than the "equivalent" one.

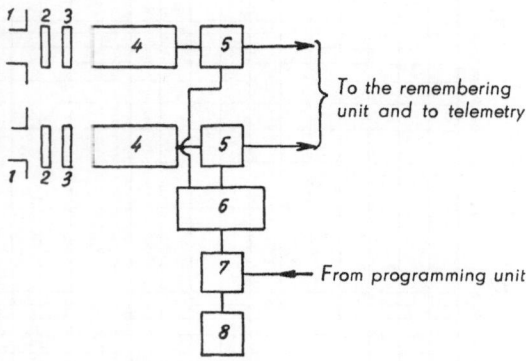

Fig. 1. Block diagram of the instrument for measurements of solar corpuscular radiation. *1* Collimating blend (solid angle $1/_4$ steradian); *2* Absorbing foil (0.4 and 0.8 mg/cm²); *3* Fluorescent screen; *4* Photomultiplier; *5* Cathode follower; *6* High voltage converter; *7* Switching relay; *8* Batteries

The experiment in question is characterised by the fact that it practically recorded only the electrons of tens of keV. The indicators used did not respond to the X-radiation created by such electrons in the atmosphere and in the sputnik's body, as rather thin fluorescent screens were used. These screens entirely absorbed the electrons of the above indicated energy and only negligibly small portions of X-radiation passing through them. Protons of some tens of keV did not affect the indicators because they were absorbed by the aluminium foils. If the X-radiation had also been registered the data obtained would distort the information about the latitudinal and height distribution of corpuscles. As was reported earlier, the upper atmosphere especially in the auroral zone becomes a source of X-rays under the effect of fast electrons. Approximately one half of this radiation inevitably gets into outer space. Therefore, the instruments on a high altitude rocket or sputnik will be irradiated by the X-radiation arising not only in their bodies but also in the entire upper atmosphere. Since the X-rays are most intensively generated near the auroral zone and this zone becomes visible above the equator at a distance of approx. 2.5 of the Earth radius from its centre, the maximum flux of this radiation above the equator occurs near the indicated height. In the experiments of Van Allen and others the penetrating X-radiation was recorded along with the hard electrons and protons. It makes interpretation of the obtained results indirect and not quite definite.

The telemetering data obtained were processed as follows. First of all, the difference between the logarithms of fluxes of two indicators was determined and then according to it by using the calibrating curve the "equivalent" energy

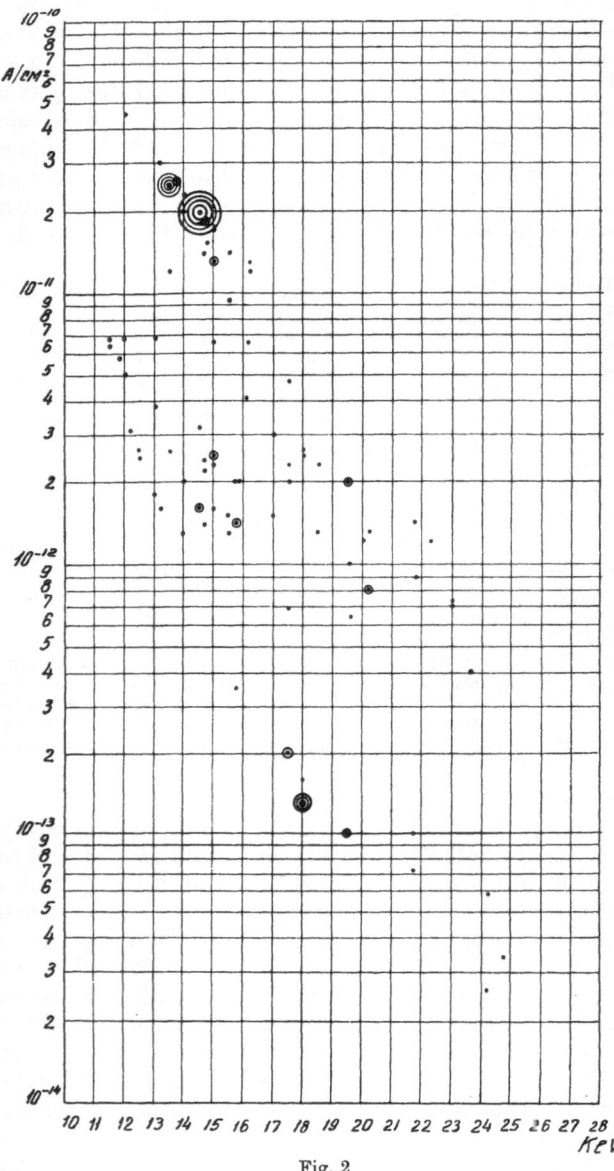

of electrons. Since the intensity of the screen's luminescence was calibrated for the wide range of the electron energies, the "equivalent" current of the monoenergetic flux of electrons and then its "equivalent" power, which is the product of the "equivalent" energy and current was determined. Thick aluminium diaphragms, which provided registration of corpuscles within the solid angle $1/4$ steradians were placed in front of the indicators. To obtain the values of the "equivalent" current and power per steradian the measured values were multiplied by 4. It should be noted, however, that the electron fluxes may be essentially anisotropic.

The radiotelemetering data available allow a number of new conclusions to be made, very important from the point of view of geophysics. It is for the first time that fluxes of soft electrons of 10 keV have been discovered directly while exploring the upper atmosphere. They were recorded at the heights of 470 up to 1880 km above sea level. The minimum intensity was recorded at the height of about 1300 km above the geomagnetic equator. The "equivalent" energy being about 20 keV their minimum current was approximately estimated as 10^{-14} A cm^{-2} steradian^{-1} at night. (The current of 10^{-10} A cm^{-2} corresponds to the electron flux of $6.2 \cdot 10^8$ cm^{-2} sec^{-1} i.e. for the particle energy of 10 keV it corresponds to the energy flux of 10 erg cm^{-2} sec^{-1}.) At the middle and polar latitudes up to 60°

Fig. 2

of geomagnetic latitude for the electrons of the "equivalent" energy of 12 keV, the current of 5.10^{-11} A cm^{-2} steradian^{-1} is ordinary at night and sometimes more than 10^{-10} A cm^{-2} steradian^{-1}. Higher values of the recorded current sometimes were even observed at the geomagnetic latitude of —4°. It is interesting to note that the occurrence of the powerful fluxes recorded is observed in the nonequatorial zone of higher ionization in the F-region, described in [7].

Before the launching the intensity of the electron fluxes was not expected to be so great, and because of it the instruments in many cases were off scale and it was impossible to estimate the intensity and the "equivalent" energy of the recorded electrons. The simultaneous recording of the two indicators, when they were not off scale, were few. Fig. 2 illustrates the dependence of the intensity of the electron fluxes upon their "equivalent" energy at the heights of 1720 up to 1880 km above the area of —42° to —54° of geomagnetic latitude at night (May 15, 1958 in the South of Pacific Ocean). The values of the "equivalent" energy in keV are plotted along the abscissa axis in the linear scale and the densities of their "equivalent" currents in A cm^{-2} are plotted in logarithmic scale along the ordinate axis, if they correspond to parallel flux of the electrons (i.e. without multiplying the measured values by 4 steradians^{-1}). Reoccurred values of equal intensities with equal "equivalent" energies of particles are marked with concentric circles. With the increase of the "equivalent" energy of electrons their number rapidly decreases.

Taking into account the sense of the "equivalent" energy introduced one cannot exactly determine the energy spectrum according to this graph, nevertheless, if one assumes its form as $N\,(>E) \sim E^{-\gamma}$, then $\gamma \geq 8$. After the measurements by S. N. VERNOV and his collaborators at the first cosmic rocket [5] in the region of harder energies, the energy spectrum goes on decreasing with the growth of energy; for the region of energies of 20—100 keV $\gamma \sim 5$ and for even harder energies at the edges of the zone of high intensity $\gamma \sim 3$.

While the sputnik was rotating around its axes essential changes took place in the intensities of the electron fluxes. Sometimes, however, variations of intensity and the "equivalent" electron energy could be observed for much shorter periods of time than of its rotation. The shortest variations, as was recorded, took place for a period of $1/2$ sec. More rapid ones than these were not recorded. The rotation of the sputnik changed the position of the indicator's windows with respect to the direction of the magnetic force lines. Comparison with the known orientation of the sputnik shows that in the majority of cases the fluxes of maximum intensity lie near the plane perpendicular to the magnetic vector. But in equatorial and in polar regions some inverse cases were recorded. The values of the "equivalent" energy of electrons at the lower latitude were higher. The maximum value recorded was 40 keV. In the polar regions only the minimum values of about 10 keV were recorded.

The solid angle can be estimated around a magnetic force line; moving within it at the height of the measurements, the particle will reach the given height in the ionosphere [8]. At the geomagnetic latitude of 50°, for example, the critical angle between the magnetic and velocity vectors is to be of 50° in order to provide the particle penetration lower than F-region from the heights of 1500—2000 km; such a solid angle is larger than the angle of the indicator. In view of this it seems possible to estimate the total flux of fast electrons, penetrating into the lower layers of the atmosphere. Thus, for example, for the cases of minimum currents at the high latitudes this energy flux is of the order of 1 erg cm^{-2} sec^{-1} and this is quite sufficient for additional ionization and heating of the upper atmosphere. By value it is close to one, which is necessary after BATES [9] and CHAPMAN [10] to support the temperature gradient of 5° per km in the upper atmosphere and which cannot be provided by hard electromagnetic solar radiation.

Since the electron fluxes recorded are most intensive at the higher geomagnetic latitudes they may be assumed to be sources of heating and expansion of the upper atmosphere, discovered by retardation of the sputniks [11, 12]. The variations of the intensity of these electron fluxes may be dependent on the solar activity and the correlation between the sputnik's retardation and the integral effect of the chromospheric flares may be due to them [13]. A great power of electron fluxes of small energy is very important for the explanation of the equatorial ring current and a number of other important geophysical effects. An essentially less powerful hard component can be apparently not determinative for these problems although detailed data about it are very important for astronautics as the penetrating radiation is very dangerous for living beings in the cosmos.

Many authors put forth the hypothesis [14, 15, 16, 17] that hard charged particles kept by the magnetic field of the Earth are the charged products of the β-decay of the albedo neutrons—the secondary products of the interaction between the cosmic rays and the Earth's atmosphere.

The relative role of some or other mechanisms of generation of soft and hard components is being discussed at present. However, the energy flux of electrons of 10 keV penetrating into lower layers of the atmosphere and being absorbed there, is so great that it can't be due to cosmic rays. The number of trapped hard electrons resulting from albedo neutrons' β-decay increases essentially, if slowing down is taken into account for upward diffusion of neutrons, originated from stars. In this case the number of β-decays of neutrons in the inner belt amounts to $3 \cdot 10^{-3} \sec^{-1}$ above 1 cm² of the Earth's surface.

Besides that, there are apprehensions that the measurements of a hard component of the corpuscular fluxes can be complicated by the registration of the products of nuclear explosions.

References

1. V. I. Krassovsky, Yu. M. Kushnir, and G. A. Bordovsky, Usp. Fiz. Nauk 64, No. 3, 425 (1958).
2. V. I. Krassovsky, Yu. M. Kushnir, G. A. Bordovsky, and E. M. Svetlitsky, Report to V CSAGI, Moscow, August 1958.
3. V. I. Krassovsky, Yu. M. Kushnir, G. A. Bordovsky, and E. M. Svetlitsky, Iskusstv. Sputniki Zemli, No. 2 (1959).
4. J. A. Van Allen, C. McIlwain, and G. Ludwig, Report to V CSAGI, Moscow, August 1958.
5. S. N. Vernov, A. E. Chudakov, P. V. Vakulov, and Yu. I. Logachev, Doklady Akad. Nauk SSSR 125, 304 (1959).
6. L. N. Meredith, L. R. Davis, S. P. Heppner, and O. E. Beur, Report to V CSAGI, Moscow, August 1958.
7. E. Appleton, Nature 157, 691 (1946).
8. V. D. Pletnev, Izv. Akad. Nauk SSSR, Ser. Geofiz. No. 8 (1959).
9. D. R. Bates, Proc. Physic. Soc. (London) B 54, 805 (1951).
10. S. Chapman, Smithson. Contrib. Astrophysics 2, No. 1 (1957).
11. M. L. Lidov, Iskusstv. Sputniki Zemli, No. 1 (1958).
12. P. E. Eliasberg, Iskusstv. Sputniki Zemli, No. 1 (1958).
13. T. Nonweiler, Nature 122, 468 (1958).
14. H. Griem and S. F. Singer, Physic. Rev. 99, 608 (1955).
15. S. F. Singer, Bull. Amer. Physic. Soc. 1, 229 (1956).
16. S. F. Singer, Physic. Rev. Lett. 1, 171 (1958).
17. S. N. Vernov, P. V. Vakulov, E. V. Gorchakov, Yu. I. Logachev, and A. E. Chudakov, Iskusstv. Sputniki Zemli, No. 2, 61 (1959).

Recent Developments and Designs of the Ion Rocket Engine[1]

By

Robert H. Boden[2]

(With 13 Figures)

(Received August 27, 1959)

Abstract — Zusammenfassung — Résumé

Recent Developments and Designs of the Ion Rocket Engine. Recent developments of ion rocket engines have proceeded rapidly in the last two years under research programs sponsored by the Air Force Office of Scientific Research. In order to understand the trend of this progress a brief summary of the basic relationships governing the operation of electrical propulsion systems are summarized. Fundamental design parameters of ion rocket engines are tabulated for both English and metric systems.

The cross-sectional area of the engine is estimated by considering a flat-plate electrode configuration and the modifications imposed on this idealized situation by the presence of an aperture through which ions are emitted.

The relationship of engine design, vehicle design, and the mission results from the characteristic velocity required to accomplish the objective. Methods of mission analysis are referenced, but not discussed in detail. These methods result in characteristic velocity requirements which in turn establish the specific impulse or exhaust velocity the engine must develop.

The specific impulse cannot be arbitrarily chosen if the maximum payload is to be transported in the vehicle. The variation of payload with specific impulse is demonstrated with some typical examples. The overall variation depends upon the weight of the vehicle per unit power developed and the efficiency of conversion of primary energy into directed kinetic energy in the exhaust jet. The analytical relationship among the characteristic velocity, weight per unit power of the vehicle, efficiency and the specific impulse to obtain maximum payload is graphically presented. The maximum payload for a mission accomplished with an electrically powered vehicle is estimated from these data.

The characteristic velocity of the mission also determines the operating time of the electrical rocket engine. In general it is desirable to use as low a specific impulse powerplant as possible in order to reduce the operating time; however, the operating time is also dependent upon the thrust-to-weight ratio of the vehicle. This ratio depends on the capability of the vehicle to generate adequate amounts of power. The limiting thrust-to-weight ratio has been summarized graphically for several types of vehicles

---

[1] The research programs described here were performed by the Advanced Design and Research Sections of the Engineering Department of Rocketdyne, a Division of North American Aviation, Inc. The programs were sponsored by the Air Force Office of Scientific Research and the Aeronautical Research Laboratories, Wright Air Development Center, Air Research and Development Command, United States Air Force.

[2] Staff Scientist, Rocketdyne, A Division of North American Aviation, Inc., Canoga Park, California, U.S.A.

including ion, plasma, and chemical rockets and modern aircraft. Comparative maximum payloads of chemical and ion rockets are presented in terms of the mission characteristic velocity.

The experimental program and techniques for verifying these analyses on a specific type of ion rocket engine conceived at Rocketdyne are presented. The experiments are conducted in high vacuum environments. The ion thrust chamber discussed in this paper uses electrostatic acceleration of ions generated from surface contact ion sources. Preliminary evaluation of the results of the program indicate that a flyable prototype engine can be produced within a few years under the impetus of a progressive development program.

Jüngste Entwicklungen und Entwürfe eines Ionenraketenmotors. Entwicklungen des Ionenraketenmotors haben in den letzten zwei Jahren unter dem vom Air Force Office of Scientific Research geförderten Programm große Fortschritte gemacht. Um die Richtung dieses Fortschrittes zu verstehen, wird eine kurze Zusammenfassung der grundlegenden Beziehungen gegeben, die die Operation des elektrischen Antriebssystems beherrschen. Fundamentale Entwurfsparameter des Ionenraketenmotors werden in englischen und metrischen Systemen tabuliert.

Die Querschnittfläche des Motors wird mit Hilfe einer Flachplatten-Elektroden-Anordnung geschätzt, aber auch der Modifikationen, die durch die Gegenwart einer Öffnung, durch die Ionen ausgesendet werden, auf diese idealisierte Situation einwirken.

Die Beziehung zwischen dem Maschinenentwurf, Fahrzeugentwurf und dem Auftrag folgt aus der charakteristischen Geschwindigkeit, die benötigt wird, um das Ziel zu erreichen. Methoden einer Auftrags-Analyse sind angegeben, werden aber nicht im einzelnen besprochen. Diese Methoden führen zu charakteristischen Geschwindigkeitsbedürfnissen, die ihrerseits wiederum den spezifischen Impuls aufstellen oder die Auspuffgeschwindigkeit, die der Motor entwickeln muß.

Der spezifische Impuls kann nicht willkürlich gewählt werden, wenn eine maximale Nutzlast im Fahrzeug transportiert werden soll. Die Veränderung der Nutzlast mit dem spezifischen Impuls wird an Hand einiger typischer Beispiele demonstriert. Die Gesamtveränderung hängt ab von dem Gewicht des Fahrzeuges pro Einheit erzeugter Kraft und der Leistungsfähigkeit der Umwandlung von ursprünglicher Energie in gerichtete, kinetische Energie in der Auspuffdüse. Die analytische Beziehung zwischen der charakteristischen Geschwindigkeit, Gewicht pro Krafteinheit des Fahrzeuges, Leistungsfähigkeit und dem spezifischen Impuls, um eine maximale Nutzlast zu erhalten, wird graphisch dargestellt. Die maximale Nutzlast für eine Mission (Auftrag), die mit einem elektrisch angetriebenen Fahrzeug unternommen wird, wird mit Hilfe dieser Daten geschätzt.

Die charakteristische Geschwindigkeit der Mission bestimmt ebenfalls die Arbeitszeit des elektrischen Raketenmotors. Im allgemeinen ist es wünschenswert, eine möglichst wenig spezifische Impuls-Kraftanlage zu benutzen, um die Arbeitszeit zu verringern; doch hängt die Arbeitszeit auch von dem Schub-Gewichtsverhältnis des Fahrzeuges ab. Dieses Verhältnis ist abhängig von der Fähigkeit des Fahrzeuges, angemessene Mengen von Kraft zu erzeugen. Das begrenzende Schub-Gewichtsverhältnis wurde für verschiedene Typen von Fahrzeugen einschließlich der Ionen-, Plasma-, chemischen Raketen und modernen Flugzeuge graphisch zusammengefaßt. Vergleichbare maximale Nutzlasten werden als Ausdruck der charakteristischen Geschwindigkeit der Mission angegeben.

Das experimentelle Programm und die Techniken zur Bestätigung dieser Analysen, die an einer spezifischen Type einer Ionenrakete bei Rocketdyne erhalten worden sind, werden wiedergegeben. Die Experimente wurden im Hochvakuum ausgeführt. Die Ionenstoßkammer, die in dieser Abhandlung diskutiert wird, benutzt elektrostatische Beschleunigung der Ionen, die von einer Oberflächenkontakt-Ionenquelle erzeugt werden. Vorläufige Berechnungen der Ergebnisse des Programms zeigen, daß eine fliegbare Prototyp-Maschine in einigen Jahren, unter dem Antrieb eines fortschrittlichen Programmes, erzeugt werden kann.

Progrès récents et études dans le domaine des propulseurs ioniques. Une progression rapide des études de propulseurs ioniques a été faite pendant les deux dernières années sous l'égide du Bureau des Projets Scientifiques des Forces Aériennes. Pour mieux comprendre la nature de ce progrès, l'auteur donne un résumé des relations fondamentales qui gouvernent le fonctionnement des systèmes à propulsion électrique. Les paramètres d'études de base des propulseurs ioniques sont présentés en unités anglaises et métriques.

Le maître-couple du moteur est calculé à partir d'une électrode-plaque en tenant compte des corrections nécessaires, la configuration idéalisée étant affectée par la présence d'un orifice prévu pour l'émission ionique.

Les relations entre le dessin du moteur et du véhicule et sa mission dépendent de la vitesse caractéristique imposée par l'objectif à atteindre. Des références pour les méthodes d'analyse de mission sont données, sans discussion détaillée. Ces méthodes permettent de définir la vitesse caractéristique qui, à son tour, établit l'impulsion spécifique ou la vitesse d'éjection du propulseur.

L'impulsion spécifique ne peut être choisie d'une façon arbitraire si le véhicule doit transporter une charge payante maximum. Quelques exemples typiques illustrent les variations de charge payante en fonction de l'impulsion spécifique. La variation totale dépend du poids du véhicule par unité de puissance développée et du rendement de la conversion de l'énergie primaire en énergie cinétique dirigée à l'éjection. Les rapports analytiques entre la vitesse caractéristique, la puissance massique, le rendement et l'impulsion spécifique nécessaires pour obtenir une charge payante maximum sont illustrés graphiquement. La charge payante maximum pour une mission accomplie par un véhicule à propulsion électrique est calculée à partir de ces données.

La vitesse caractéristique de la mission détermine également la durée de fonctionnement du propulseur électrique. En général il est désirable d'utiliser un propulseur à impulsion spécifique aussi basse que possible afin de réduire la durée de fonctionnement; cependant, la durée de fonctionnement dépend également de la poussée massique du véhicule. Ce rapport est gouverné par la capacité du véhicule de produire une puissance suffisante. Les valeurs limite de la poussée massique ont été résumées graphiquement pour plusieurs types de véhicules comprenant les propulseurs ioniques, plasmas, fusées à propergols chimiques ainsi que les avions modernes. Des comparaisons de charges payantes maxima de fusées chimiques et propulseurs ioniques sont données en termes de vitesses caractéristiques de mission.

Sont aussi présentés: le programme expérimental et les techniques de vérification d'analyses d'un propulseur ionique d'un modèle établi par Rocketdyne. Ces expériences ont lieu dans un vide très poussé. La chambre de poussée décrite dans la présente conférence fonctionne par l'accélération électrostatique d'ions produits par un système de surfaces de contact. Sous réserve d'un programme de travail énergique, l'évaluation préliminaire des résultats acquis démontre qu'il serait possible de construire en quelques années un prototype de propulseur capable de fonctionner en vol.

Nomenclature

a	Acceleration	h	Number of electrons stripped from particle
A	Equivalent weight of particle with charge of single electron	I	Current
A_i	Atomic or molecular weight of particular species of particle	I_s	Specific impulse
		J	Current density
A_t	Cross sectional area of particle beam or thrust chamber	m_i	Mass of particle, i'th species
		M_g	Gross weight of vehicle
C	Average exhaust velocity	M_o	Gross mass of vehicle
C_i	Velocity of particle, i'th species	$\dot{M}$	Propellant flowrate
d	Separation of acceleration electrodes	M_e	Weight of powerplant section
e	Charge of electron	M_s	Weight of structure
F	Thrust of rocket engine	M_x	Weight of payload
G	Thrust-to-weight ratio	M_p	Weight of propellant
		$\dot{N}$	Particle flowrate

P Power in exhaust jet
P_g Gross power of vehicle
t Engine operating time
V Accelerating potential
$\varDelta v$ Characteristic velocity
α Power per unit weight

η Ratio of exhaust jet power to gross
     vehicle power

Subscripts

i refers to particular particle spe-
     cies

Premises

The operation of the reaction-type electrical propulsion systems discussed
in this paper is based on the premise that thrust is obtained by the expulsion
of material at high velocity from the engine. The analyses of the system are
obtained from consideration of NEWTON's laws of motion. The significant variables
of interest to the engineer and scientist involved in space projects are the thrust
generated by the engine system, F, and the gross power, P_g that must be generated in the vehicle. An alternative power variable is the power in the jet exhaust,
P. The latter two variables are conveniently related by the engine efficiency, η.

The independent variables of the analysis are the current, I, accelerating
potential, V, the particle weight, m, the degree of ionization of the ejected
particles, h, the gross weight of the vehicle, M_g, and the thrust-to-weight ratio,
G. Other variables are the particle flowrate, $\dot{N}$, and the exhaust velocity, C.

The analyses of this paper are generalized to include the case of many kinds
of particles being ejected from the electrical rocket engine. The thrust for this
general case is given by NEWTON's laws:

$$F = M_g G = \sum \dot{N}_i m_i C_i \tag{1}$$

in which the subscript i refers to the particular species of particle, ion, electron,
or charged dust particles.

The power that must be converted into directed energy flow is expressed by
the relation:

$$P = 1/2 \sum \dot{N}_i m_i C_i^2 = IV. \tag{2}$$

The exhaust velocity, C_i, of a particular particle species is expressed in
terms of the potential drop, V, of the accelerating fields; the degree of ionization,
h_i; the particle mass, m_i; and the charge on the electron. The degree of ionization,
h_i, is the number of electron charges carried on the particle. In the case of an
ion rocket, it is the number of electrons stripped from the atom in the ionization
process and is an integer.

$$C_i = \left(\frac{h_i e V}{150 m_i} \right)^{1/2}. \tag{3}$$

The total current in the exhaust jet is the sum of the currents of each species
of charged particle. The total current must vanish if a stable thrust is generated
by an electrical rocket. In the case of an ion rocket engine, the current of the
ion jet must be balanced by the current of an electron emitter.

$$I = \sum I_i = \frac{10^{-9} e}{2,99} \sum h_i \dot{N}_i \text{ amperes.} \tag{4}$$

The propellant flowrate is closely related to the total current. It is the sum
of the products of the particle rates and the particle masses.

$$\dot{M} = \sum \dot{N}_i m_i \tag{5}$$

These five relations are sufficient to describe the over-all operation of the electrical rockets discussed in this paper.

Design Parameters

Two useful design parameters for electrical rocket engines result from consideration of the thrust current and velocity relationships, eqs. (1), (3) and (4). Since all the particles will be accelerated through the same potential drop, V, of the accelerating electrical field, the substitution of the velocity eq. (3) into eq. (1) is the first step in the development of the parameters. The thrust on the vehicle becomes:

$$F = \sum N_i (h_i m_i)^{1/2} \left(\frac{e V}{150} \right)^{1/2}.$$ (6)

Use of the current equation reduces the thrust relationship to terms which can be measured by straightforward physical techniques.

$$F = \sum I_i \left(\frac{m_i}{h_i} \right)^{1/2} \left(\frac{e V}{150} \right)^{1/2}.$$ (7)

When this equation is modified by expressing the current in terms of the ratio of the current of each particle species to the total current, the thrust can be expressed in terms of the equivalent mass of a particle carrying a single electron charge. The thrust on the vehicle becomes

$$F = I m^{1/2} \left(\frac{e V}{150} \right)^{1/2}$$ (8)

in which the definition of the weighted, equivalent particle mass is:

$$m^{1/2} = \sum \frac{I_i}{I} \left(\frac{m_i}{h_i} \right)^{1/2}.$$ (9)

Singly ionized particle weight is then defined by:

$$A^{1/2} = \frac{I_i}{I} \left(\frac{A_i}{h_i} \right)^{1/2}.$$ (10)

A_i is the atomic, or molecular weight if ions are used to generate thrust. If charged particles are used, A_i is the weight of the particle in grams.

The parametric design equation for the thrust results after taking into account the mechanical and electrical units which the investigator wishes to use. In the cgs system of units, the thrust becomes:

$$F = 14.39 \, I V \left(\frac{V}{A} \right)^{1/2} \text{ dynes,}$$ (11)

and in the foot-pound-second system,

$$F = 3.235 \times 10^{-5} \, I V \left(\frac{V}{A} \right)^{1/2} \text{ lb .}$$ (12)

The parameter V/A expressed in volts per gram occurs in all of the engineering design parameters:
Specific Power
Power per Unit Thrust
Average Exhaust Velocity
Exhaust Current per Unit Thrust
Propellant Flowrate per Unit Thrust

Specific Impulse
Minimum Throat Area per Unit Thrust
Engine Operating Time

The detailed derivations of these parameters are developed in [1]. They are summarized in Table I for the English and metric unit systems.

Dimensions of Ion Thrust Chamber

From the particle current and the limiting current density, an estimate of the minimum dimensions of ion thrust chambers can be obtained. The current density is assumed to be limited by space charge and not by emission from the ion source. The density of this current is described by CHILD's law [6].

A quantitative analysis of the problem is not practicable except when the thrust chamber can be described in terms of a single dimension, i.e., when the particle paths follow the lines of electric force as they do when the thrust chambers are of rectangular, cylindrical, and spherical symmetry. In the case of the ion rocket thrust chamber, the following conditions are assumed:

(a) The electrodes are infinite, plane equipotential surfaces;

(b) The number of charged particles emitted at the repelling electrode exceeds the demand;

(c) The particles start from rest at the repelling electrode;

(d) Particles have simple positive or negative charge, only one species being present in the thrust chamber; and

(e) The accelerating voltage is constant and has been constant sufficiently long for the current to reach its steady-state value.

Under these conditions the current density of the particles is given by the relation:

$$2.997 \times 10^9 \, J = \frac{1}{9 \, \pi \, d^2} \left(\frac{V}{299.7}\right)^{3/2} \sum \left(\frac{e h_i}{2 \, m_i}\right)^{1/2}. \tag{13}$$

Where

J = current density, amp/unit area
e = charge on the particle, esu
m_i = mass of the charged particle, gm
V = accelerating voltage, volts
d = separation of accelerating electrodes, cm or ft.

The dimensions of J are amperes/cm^2 or amperes/sq ft depending upon whether d is expressed in centimeters or feet.

Evaluated in terms of the parameter V/d, the current density becomes:

$$J = 0.5468 \times 10^7 \, \frac{A}{d^2} \left(\frac{V}{A}\right)^{3/2} \text{amp/unit area}. \tag{14}$$

The cross-sectional area per unit thrust of the ion thrust chamber is then, from eqs. (14) and Table I:

$$\frac{A_t}{F} = \frac{I}{J} = 0.1271 \times 10^7 \left(\frac{V}{d}\right)^{-2} \text{cm}^2/\text{dyne} \tag{15}$$

in cgs units,

$$\frac{A_t}{F} = 5.651 \times 10^{11} \left(\frac{V}{d}\right)^{-2} \text{sq ft/lb.} \tag{16}$$

in fps units,

A_t = cross-sectional area of ion thrust chamber.

Table I. Summary of Performance Parameters, Electrical Rocket Engines

	Definition	
	English System	Metric System
Specific Power	$\dfrac{P}{M} = 3.089 \times 10^4 \, G \left(\dfrac{V}{A}\right)^{1/2}$ watts/lb $\dfrac{P}{M_g} = 41.42 \, G \left(\dfrac{V}{A}\right)^{1/2}$ hp/lb	$\dfrac{P}{M} = 0.0695\, a \left(\dfrac{V}{A}\right)^{1/2}$ watts/gm
Power per Unit Thrust	$\dfrac{P}{F} = 3.089 \times 10^4 \left(\dfrac{V}{A}\right)^{1/2}$ watts/lb $\dfrac{P}{F} = 41.42 \left(\dfrac{V}{A}\right)^{1/2}$ hp/lb	$\dfrac{P}{F} = 0.0695 \left(\dfrac{V}{A}\right)^{1/2}$ watts/dyne
Average Exhaust Velocity	$C = 4.557 \times 10^4 \left(\dfrac{V}{A}\right)^{1/2}$ ft/sec	$C = 13890 \left(\dfrac{V}{A}\right)^{1/2}$ meters/sec
Exhaust Current per Unit Thrust	$\dfrac{I}{F} = \dfrac{3.089 \times 10^4}{A \left(\dfrac{V}{A}\right)^{1/2}}$ amp	$\dfrac{I}{F} = \dfrac{0.0695}{A \left(\dfrac{V}{A}\right)^{1/2}}$ amp/dyne
Propellant Flowrate per Unit Thrust	$\dfrac{\dot{M}}{F} = 7.060 \times 10^{-4} \left(\dfrac{V}{A}\right)^{-1/2}$ sec^{-1}	$\dfrac{\dot{M}}{F} = 7.199 \times 10^{-7} \left(\dfrac{V}{A}\right)^{-1/2}$ sec^{-1}
Specific Impulse	$I_s = 1417 \left(\dfrac{V}{A}\right)^{1/2}$ sec	$I_s = 1417 \left(\dfrac{V}{A}\right)^{1/2}$ sec
Minimum Throat Area per Unit Thrust (Ion Rocket Only)	$A_t = 5.651 \times 10^{11} \left(\dfrac{V}{d}\right)^{-2}$ sq ft	$A_t = 127.1 \left(\dfrac{V}{d}\right)^{-2}$ (m)^2/lb
Engine Operating Time	$t = \left[1 - e^{-\dfrac{\Delta v}{4.557 \times 10^4 \left(\frac{V}{A}\right)^{1/2}}}\right] \dfrac{\left(\dfrac{V}{A}\right)^{1/2}}{7.060 \times 10^{-4}\, G}$ sec	$t = \left[1 - e^{-\dfrac{\Delta v}{1.368 \times 10^6 \left(\frac{V}{A}\right)^{1/2}}}\right] \dfrac{\left(\dfrac{V}{A}\right)^{1/2}}{7.199 \times 10^{-7}\, a}$ sec
Equivalent Particle Weight	$A = \left[\sum \dfrac{I_i}{I}\left(\dfrac{A_i^{1/2}}{h_{ii}}\right)\right]^2$ gm	$A = \left[\sum \dfrac{I_i}{I}\left(\dfrac{A_i^{1/2}}{h_{ii}}\right)\right]^2$ gm

Conclusions about the dimensions of the ion rocket thrust chambers are limited by the assumptions upon which the analyses are based. In practice, parallel flat plates assumed here are not feasible. An accelerating electrode array such as the PIERCE system [10] in which particles are discharged through an aperture, or an arbitrary array of open electrodes within which a uniform field is generated must be used. The area relationship developed here holds for these electrode arrays.

The area of the thrust ion chamber will be proportional to the thrust and inversely proportional to the square of the field strength, V/d. The decrease of the ion thrust chamber cross section per unit thrust with increasing field strength, V/d, is shown in Fig. 1. It is believed that the proportionality constant will be greater than the values given in eqs. (15) and (16). The area is independent of the propellant used.

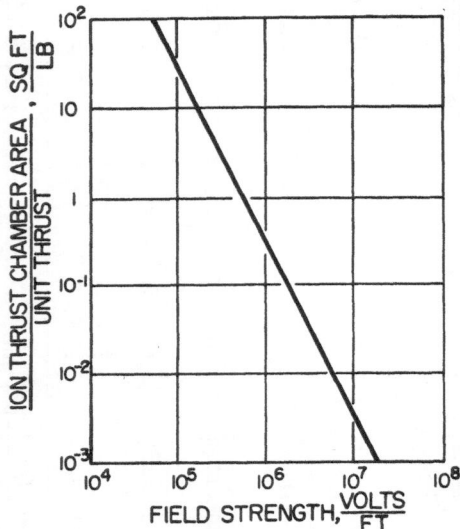

Fig. 1. Influence of field strength on ion thrust chamber cross section

Unfortunately, PIERIE's method cannot be exactly applied to the design of systems of high perveance for which it was originally designed. For high perveance, the ratio of $I/(V)^{3/2}$, the cathode aperture must be made comparable in size with the cathode-to-anode spacing. This results in some non-uniformity and in a decrease in the fields at the anode surface. In practice it is estimated that the area may be about three times the computed values.

The design problem is to adjust the accelerating voltages and the electrode spacings to minimize the cross section of the thrust chambers, yet not to force the vehicle power-to-weight ratio (specific power) to values which are impractical to achieve. Because the minimum area depends upon the field strength, V/d, of the accelerating electrode array, the ion aperture must be made comparable to the cathode to anode spacing, and the relation V/A must be kept to a value within the limits of the specific power capabilities of the vehicle. Therefore, the configuration of the ion engine will probably be a long narrow strip. If compactness is a requirement, an array of narrow apertures formed in a rectangular pattern, or possibly a spiral configuration, can be used. These arrangements are necessary to achieve a low specific impulse, desirable for short missions.

Characteristic Velocity

Application of the performance parameters (summarized in Table I) in the design of an ion rocket engine requires a knowledge of the mission. Mission analyses for vehicles using rocket engines have been popular topics of investigation for many scientists and engineers in recent months. [3] through [9] cover a wide variety of methods for investigating low-thrust space travel.

The critical variable to be determined from mission analyses, and which affects the application of the propulsion system, is the characteristic velocity,

Δv. This variable is defined as the change in velocity of the vehicle in free space from the engine thrust program. After the characteristic velocity has been determined from a mission analysis, the specific impulse, or jet exhaust velocity, and operating time of the rocket engine can be estimated.

Eq. (1) is integrated to determine the characteristic velocity:

$$\Delta v = \int_{t_0}^{t} a \, dt = \int_{t_0}^{t} \frac{\Sigma \dot{N}_i m_i C_i}{M_o - \Sigma \dot{N}_i m_i t} \, dt. \tag{17}$$

When this operation is performed, the characteristic velocity is given in the form long familiar to the engineer working in the rocket field:

$$\Delta v = C \ln \frac{M_g}{M} \tag{18}$$

in which M_g is the gross weight of the vehicle, and M is the weight at termination of the engine thrust program.

If a controlled flight is visualized as a series of changes under constant thrust from one orbit to another, the total characteristic velocity required is the sum of the characteristic velocities for each maneuver.

$$\Delta v \text{ total} = C \ln \frac{M_g}{M_1} + C \ln \frac{M_1}{M_2} + C \ln \frac{M_2}{M_3} + \cdots - C \ln \frac{M_{n-1}}{M_n}$$

$$= C \ln \frac{M_o}{M_g}. \tag{19}$$

In this equation M_1, and $M_2 - - - M$ are the vehicle weights at the end of the first second $- - -$ and n'th maneuvers. M_n is the empty weight of the vehicle, all the propellant being expended.

The exhaust velocity, C, is defined by the ratio

$$C = \frac{\Sigma \dot{N}_i m_i C_i}{\Sigma \dot{N}_i m_i}. \tag{20}$$

The average exhaust velocity can be expressed in terms of the parameter V/A, and is included in Table I. When C is divided by the gravitational constant, g, the specific impulse is obtained. The specific impulse is a convenient parameter for discussion of electrical power-plants because it has the same numerical value in

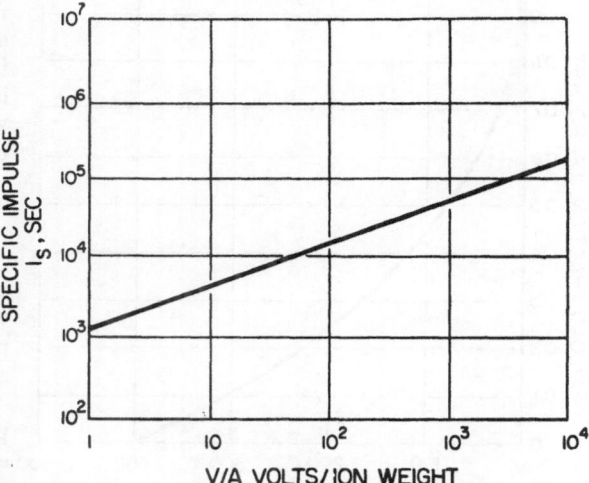

Fig. 2. Specific impulse, ion rocket

both English and metric units. For this reason much of the discussion of this paper will relate the engine and vehicle parameters to the specific impulse and Fig. 2 is presented to give a graphical relation between it and the design parameter, V/A.

Once the characteristic velocity for a mission has been specified, the power-plant specialist can predict the maximum payload which the vehicle can carry

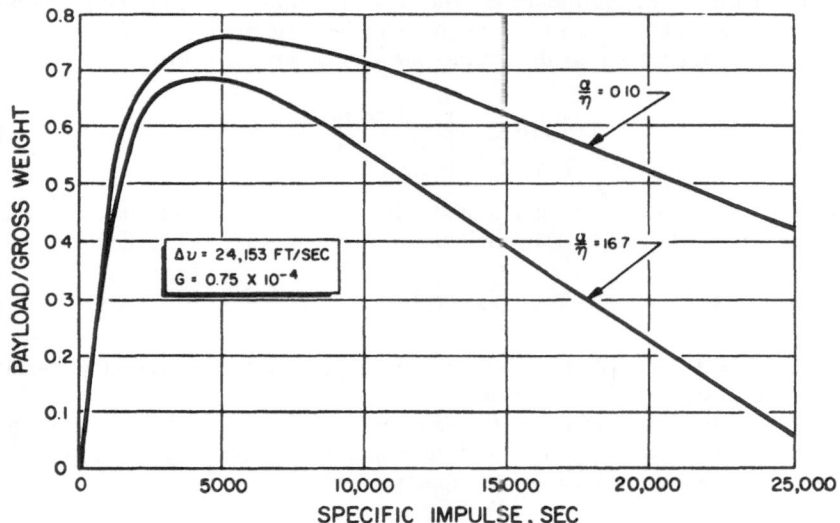

Fig. 3. Typical variation of payload vs specific impulse, ion rocket

on the mission. The gross weight of the vehicle includes the payload weight, M_z, the propellant weight, M_p, and the powerplant section weight, M_e. The powerplant section weight includes the weight of the powerplant, vehicle structure and the tanks for the propellant. In practice this weight is found to be proportional within close limits to the total power of the vehicle, P_g.

$$M_e = a P_g . \qquad (21)$$

The proportionality factor a is expressed in pound weight of vehicle per horse-power, the reciprocal of the specific power.

The powerplant generates power for guidance, controls, communications and other demands of the vehicle in-cluding power losses attend-ant to the generation of thrust in addition to that poured into the exhaust jet. In terms of the characteristic velocity of the mission, Δv, and the specific impulse of the engine, the gross power require-ments are:

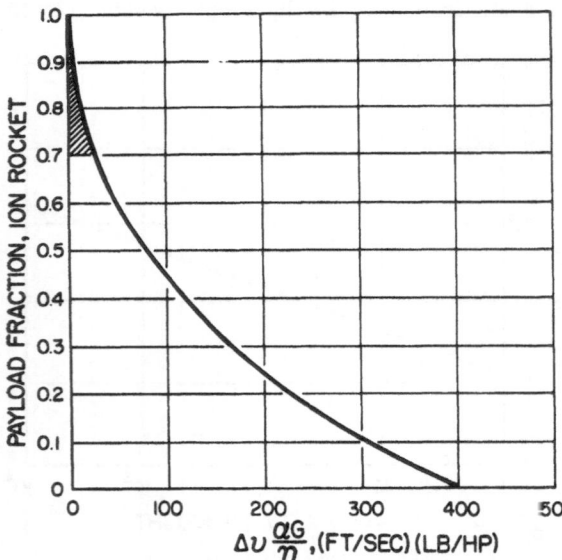

Fig. 4. Reduction of payload with mission requirements

$$P_g = \frac{\dot{M} C^2}{1100\eta} = \frac{\varDelta v\, F}{1100\,\eta \ln\left(1 - \dfrac{Gt}{I_s}\right)} \tag{22}$$

in which η is the function of the total electrical power transmitted to the jet.

From these relations and the characteristic velocity, the ratio of the payload weight to the gross weight of the vehicle,

$$\frac{M_x}{M_g} = 1 - \frac{M_e + M_p}{M_g} \tag{23}$$

becomes:

$$\frac{W_x}{W_g} = \frac{M_x}{M_g} = 1 + \frac{9.10 \times 10^{-4}\, G a \varDelta v}{\eta \ln\left(1 - \dfrac{Gt}{I_s}\right)} - \frac{Gt}{I_s} \tag{24}$$

in which G is the thrust-to-weight ratio of the vehicle. The payload ratio increases rapidly with specific impulse, exhibits a relatively flat maximum, then decreases more slowly with specific impulse greater than the value for maximum payload,

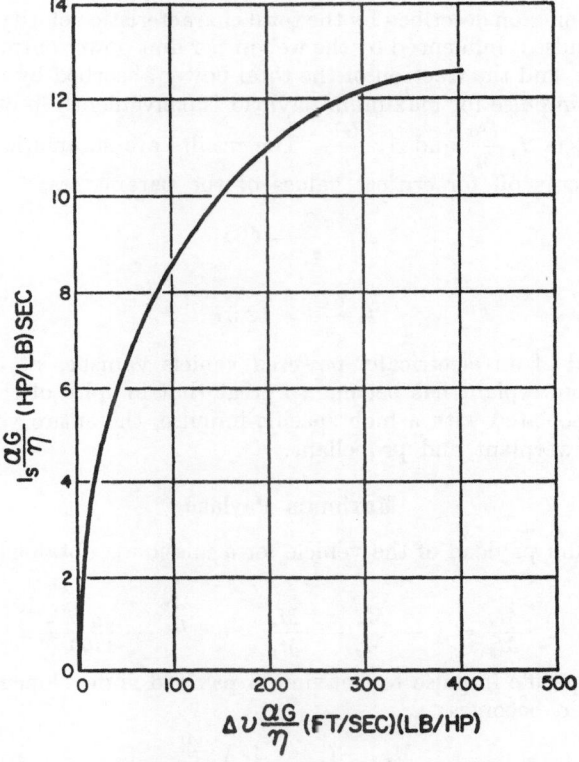

Fig. 5. Specific impulse for maximum payload of rocket engine

Fig. 3. The data of Fig. 4 are for traveling from a 100-mile satellite orbit to the moon's orbit. In selecting a specific impulse for a particular design, a slightly high value is desirable to prevent a possible drop in efficiency of the powerplant over a long operating period.

To find the specific impulse for which the ratio of payload weight to gross

weight is a maximum for a given mission, eq. (24) is differentiated with respect to I_s and set equal to zero. This operation yields the following relationship:

$$\left(1 - \frac{Gt}{I_s}\right) \ln^2 \frac{1}{\left(1 - \left(\frac{Gt}{I_s}\right)\right)} = 9.10 \times 10^{-4} \frac{\Delta v \, Ga}{\eta}. \tag{25}$$

Expressions for $\left(1 - \frac{Gt}{I_s}\right)$ are obtained by taking anti-logs of eq. (25) and from the expression for the characteristic velocity, Δv. These are equated, yielding the transcendental relations:

$$e^{-\frac{\Delta v}{g I_s}} = e^{-\left[\frac{0.91 \times 10^{-3} \, \Delta v \, G a}{\eta_s - \frac{\Delta v}{I_s^2}}\right]^{1/2}}. \tag{26}$$

When the exponents of both sides of this equation are equated, the equation describing the specific impulse for maximum payload weight is obtained:

$$\left(\frac{\Delta v}{g I_s}\right)^2 e^{-\frac{\Delta v}{I_s g}} - \frac{\Delta v \, Ga}{1100 \eta} = 0. \tag{27}$$

For a given mission described by the total characteristic velocity ratio required, the specific impulse is influenced by the weight per unit power output of the powerplant section, α, and the fraction of the total power absorbed by the exhaust jet, η. The specific impulse for maximum payload is conveniently presented in terms of the parameters $I_s \frac{Ga}{\eta}$ and $\Delta v \frac{Ga}{\eta}$. The results are summarized in Fig. 5.

The curve cuts off for critical values of the parameters,

$$\Delta v \frac{a G}{\eta} = 404.7 \tag{28}$$

and

$$I_s \frac{a G}{\eta} = 12.57 . \tag{29}$$

The payload of an electrically powered vehicle vanishes at this point. The weight of the powerplant has become so great that in spite of the reduced propellant flow associated with a high specific impulse, the entire vehicle weight is absorbed in powerplant and propellant.

Maximum Payload

The maximum payload of the vehicle for a mission is obtained from eqs. (19) and (27).

$$\frac{M_z}{M_g} = 1 - \frac{M_p}{M_g} - \frac{M_e}{M_g} = e^{-\frac{\Delta v}{I_s g}} - \frac{a g G}{1100} I_s. \tag{30}$$

When the specific impulse for maximum payload is developed by the rocket engine, the ratio becomes:

$$\frac{M_z}{M_g} = \left(1 - \frac{\Delta v}{I_s g}\right) e^{\frac{\Delta v}{I_s g}} \tag{31}$$

in which $I_s g$ is the specific impulse obtained from Fig. 5. The variation of this ratio can be expressed as a function of the specific impulse, or the characteristic velocity, the specific mass and efficiency of the powerplant and the thrust-to-weight ratio. The variation of the payload-to-gross weight ratio with the mission parameter, $\Delta v \frac{a G}{\eta}$, is summarised in Fig. 6.

The shaded area under the curve has the following significance: If an ion rocket vehicle is powered by a nuclear reactor, the mimimum mass of the powerplant section is controlled by the critical mass of the reactor, the structural mass, and the basic component weights. As the total energy demand increases above the minimum value, the weight of the powerplant section of the vehicle increases approximately proportional to the power output. The accurate value of the limiting maximum payload must be determined from the particular design and size of the vehicle.

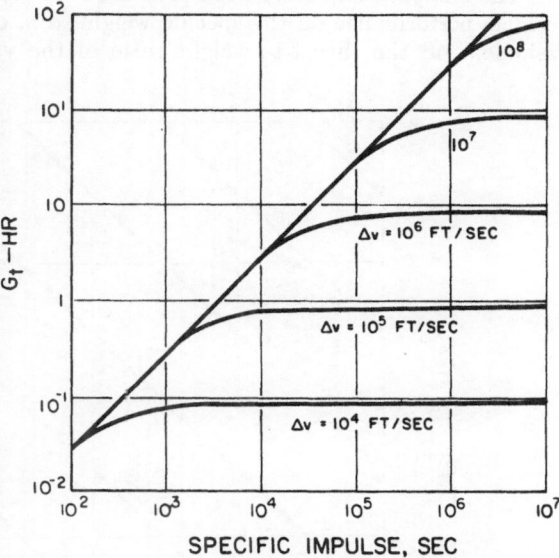

Fig. 6. Operating time of low-thrust vehicle

After the mission requirements have been determined by establishing the payload fractions and the characteristic velocity, Δv, the rocket powerplant requirements are established from the mission

parameter, $\Delta v \dfrac{a\,G}{\eta}$. The specific mass of the rocket engine section, a, the fraction of the total power into the exhaust jet, η, and the thrust-to-weight ratio G, must be adjusted to meet achievable engineering values.

Operating Time

The time required to perform a maneuver is determined from the characteristic velocity:

$$t = \frac{M_o - M}{M} = \left(1 - e^{-\frac{\Delta v}{C}}\right) \frac{M_o C}{F} \ \text{sec} \ . \tag{32}$$

It is convenient to express the operating time in terms of the sea-level specific impulse, I_s, and thrust-to-weight ratio, G:

$$Gt = I_s \left(1 - e^{-\frac{\Delta v}{I_s g}}\right) \text{sec} \ . \tag{33}$$

This product is shown as a function of specific impulse in Fig. 6. For a given characteristic velocity, Δv, and thrust-to-weight ratio, G, the operating time rapidly approaches a limiting value. For example, to achieve a characteristic velocity of 10^5 ft/sec, the same operating time is required for a vehicle propelled by an engine developing a specific impulse of 10^{-4} seconds or higher. The operating time is materially decreased by reducing the specific impulse to as low a value as possible. However this decreased time is obtained at the expense of increased propellant flow. The operating time for a maximum payload rocket is obtained by using the specific impulse for maximum payload in eq. (33).

Specific Power

The analyses that have been presented clearly indicate the dependence of the vehicle performance on the specific weight, a/η, of the powerplant section of the vehicle, and the thrust-to-weight ratio of the vehicle. As shown in Fig. 3, the

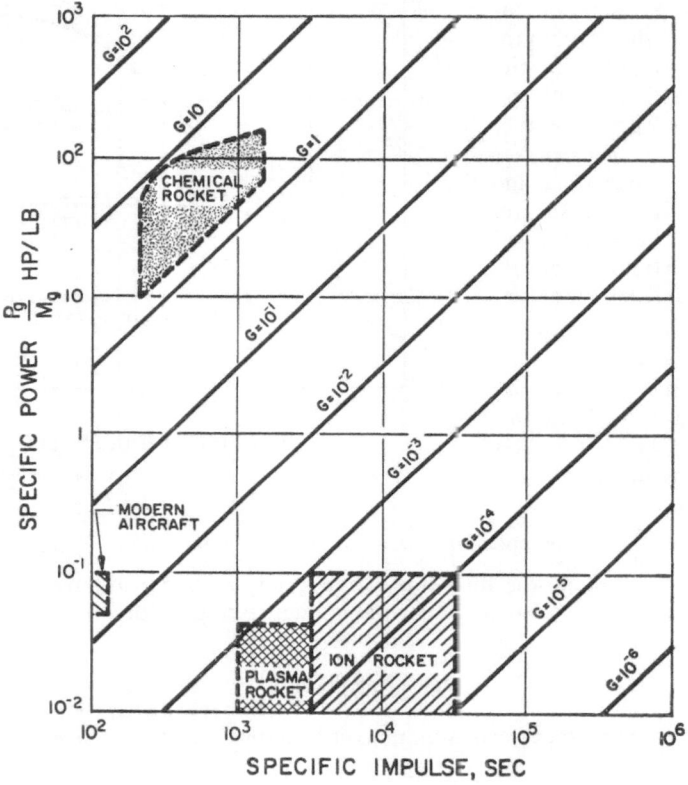

Fig. 7. Specific power requirements of vehicles

payload weight decreases with specific weight, shifts the optimum specific impulse and makes the payload become increasingly sensitive to engine operation at specific impulses above the optimum values. Fig. 4 indicates that the payload rapidly decreases with increasing specific weight of the powerplant section and thrust-to-weight ratio and the specific impulse of the engine. The relation for the specific power is developed from the basic power relation, eq. (22):

$$P_{\theta} = \frac{1}{1100} \frac{\dot{M} C^2}{\eta g} \text{ hp} \tag{22}$$

and from the specific impulse:

$$I_{\theta} = \frac{F}{\dot{M}} = \frac{C}{g} \text{ sec} \tag{34}$$

and thrust:

$$F = M_{\theta} G \text{ lb} . \tag{35}$$

The specific power of the vehicle is then:

$$\frac{P_g}{M_g} = \frac{1}{a}\left(\frac{M_e}{M_g}\right) = \frac{0.0292\,G\,I_s}{\eta}\ \text{hp/lb}\ . \tag{36}$$

The specific power of the vehicle, thrust-to-weight ratio and specific impulse of modern aircraft and various chemical rockets have been computed from performance data. The same performance parameters for the plasma rocket and the ion rocket have been estimated. These data are summarized in Fig. 7. If the powerplant manufacturer is capable of building a nuclear-powered electrical propulsion system of a specific power comparable to that of modern aircraft, an electrical rocket engine which will develop a thrust-to-weight ratio between 10^{-4} and 10^{-3} appears assured.

Comparative Performance

Comparative performance of a chemical rocket vehicle and an ion rocket vehicle is most fairly made from consideration of an impulse type chemical rocket. If the rocket engine action is short, the thrust-to-weight ratio can be approximately unity. The characteristic velocity to accomplish a given mission is then 41 percent of that required for a low-thrust rocket [6, 9]. The payload ratio for a low thrust-to-weight chemical rocket is then:

$$\frac{M_x}{M_g} = \mathrm{e}^{-\frac{\varDelta v}{I_s g}} - \frac{M_s}{M_g} \tag{37}$$

in which M_s is the structural mass.

There is no reduction in the required characteristic velocity by using a thrust-to-weight ratio greater than unity. The mission time is shortened for a high-thrust chemical rocket.

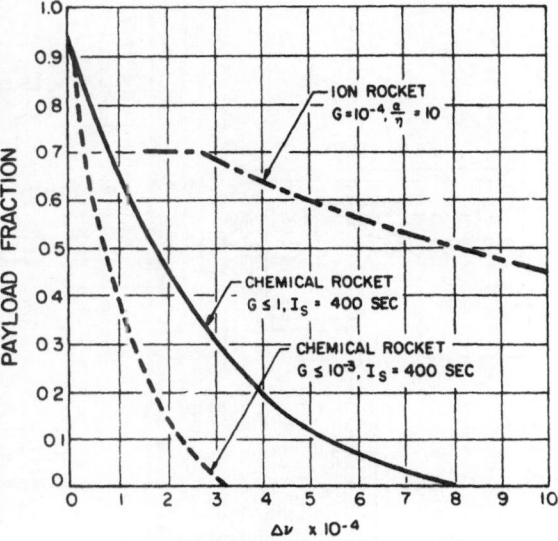

Fig. 8. Comparative performance of ion and chemical rocket

$$\frac{M_x}{M_g} = \mathrm{e}^{-\frac{0.41\,\varDelta v}{I_s g}} - \frac{M_s}{M_g}\ . \tag{38}$$

A good high-energy chemical propellant has a specific impulse of approximately 400 sec at altitude. This value is used for presenting a picture of the variation of payload ratio, Fig. 8. The ratio of structural weight to gross weight, M_s/M_0, can be between 0.07 and 0.10, or higher. The value 0.07 is used here.

Fig. 8 summarizes the payload performance of a low thrust-to-weight ratio chemical rocket, $G < 10^{-3}$; a high-thrust chemical rocket, $G \geq 1$, and an ion rocket for which the thrust-to-weight ratio is 10^{-4} and the ratio of the specific mass of the powerplant section to the fraction of the total power into the jet, α/η, is 0.1. These values of G and α/η are believed to be achievable with the current technology in a nuclear-powered ion vehicle [1]. Significant factors to be noted in the data of Fig. 8 are that an ion rocket vehicle can carry a large payload long after the

payload of a single-stage chemical rocket under optimum operating conditions
has vanished; a chemical rocket is superior to the ion rocket for missions requiring
a characteristic velocity not exceeding five thousand feet per second. A lower
limit for the specific impulse of an ion rocket engine is anticipated because of the
tendency of ions to strike the accelerating electrode system. This lower limit of
characteristic velocity is expected to be between 6000 and 8000 ft/sec [1]. The
plasma jet may be satisfactory in the narrow intermediate range between the
chemical and ion rockets, depending upon the efficiency of the conversion of
vehicle power to power in the jet. This is presently an uncertain factor and
requires experimental proof.

The Ion Rocket Engine

The Rocketdyne concept of the ion rocket engine evolved from these analyses
is a system including three major subsystems: the energy source, the power
converter, and the ion thrust chamber.

The heart of the thrust-producing system is the thrust chamber (see Fig. 9).
For the purpose of the schematic illustriation, the thrust chamber lies within the

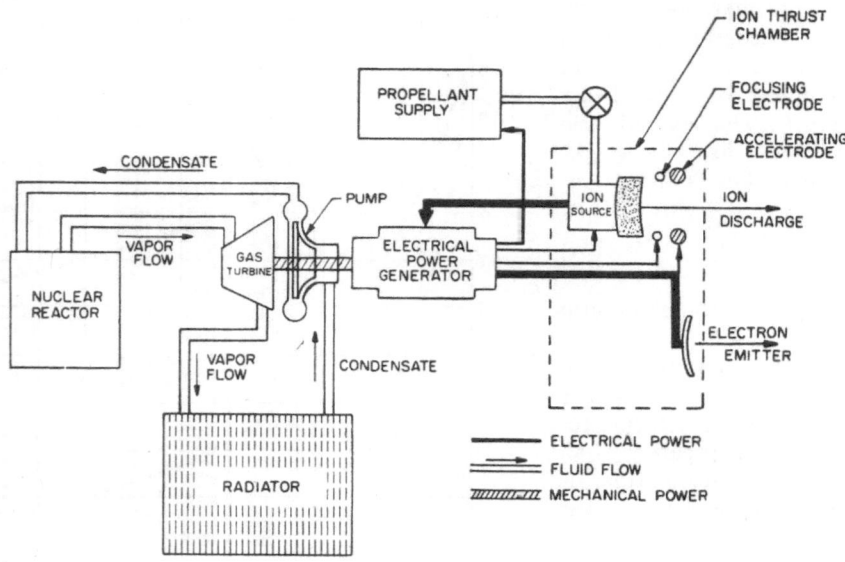

Fig. 9. Typical ion propulsion system

rectangle enclosed by the dashed lines. This thrust chamber includes an ion source,
an array of electrodes to focus and accelerate them, and an electron emitter.

Vaporized propellants are fed to the ion source, where they are ionized either
by contact with heated metals of high work function, or by being subjected to
electron bombardment within electromagnetic fields. The latter method of ionization is commonly described as the arc source.

The fully ionized propellant is conducted from the ion source through the
focusing electrodes. The focusing electrodes direct the ion current through the
accelerating electrode, which imparts kinetic energy to the ions. This results in
the required changes of momentum of the particles to develop the major portion
of the thrust for the vehicle. The accelerating electrode has been set at the po-

tential of the vehicle skin for the purpose of safety and for minimizing the electrical field outside of the vehicle.

The ion discharge is balanced by an equal flow of oppositely charged particles, otherwise the surface of the vehicle develops a high induced charge, causing the ion flow to slow down and stop. An electron emitter is shown in Fig. 9. The electron discharge can be achieved by thermionic emission from a heated tungsten plate or by field emission of electrons from discharge points incorporated in the accelerating electrodes.

Power to drive the ion thrust chamber is obtained from high-speed electrical power generators. Three types may be applied to the ion rocket engine: a permanent magnet generator, an a—c generator, or an electrostatic generator. The distribution of power to the various components of the system is shown by the solid lines in Fig. 9. Most of the power is absorbed in pumping the electrons, which are collected at the ion source, up to the potential of the accelerating electrode.

Experimental Program

A well-balanced electrical propulsion research program is sponsored by the Air Research and Development Command of the United States Air Force at Rocketdyne. Comprehensive analyses of electrical rocket engines, and parallel

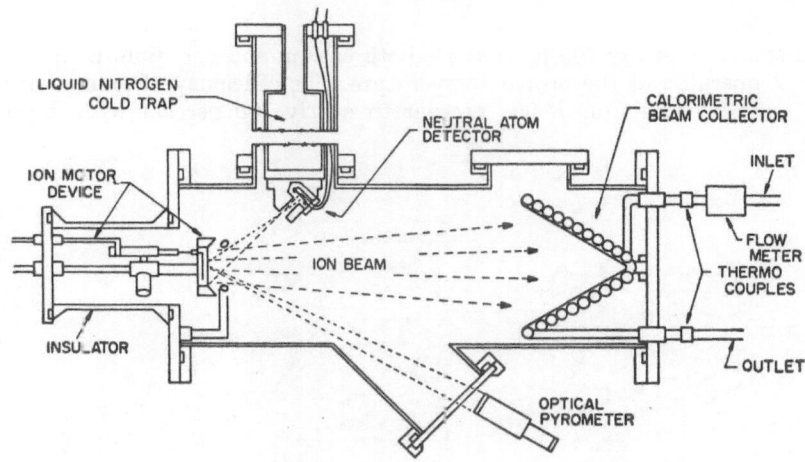

Fig. 10. Schematic of ion thrust chamber test system

experimental programs to verify these analyses, are being pursued. Some typical experimental devices which have been under test recently are described later in this paper.

The current experimental program is based on an electrostatic ion thrust chamber with cesium metal as the propellant. Cesium was selected because its vapor can be singly ionized and the ionization process is readily controlled. Preliminary analyses of arc sources presaged a difficult control problem. In an arc source, many species of ions are generated. The quantity of each is dependent upon the accelerating voltage of the ionizing electrons, ambient pressure and temperature of the ionizing chamber, and the strength of the magnetic field used to stabilize the arc. Analysis of the ionization products of uranium tetrachloride from an arc source was taken from data published by the Oak Ridge National

Laboratories [11]. The equivalent singly ionized particle weight was calculated from Eq. 10. The equivalent weight from one set of data was 99. The weight estimated from a second set was 177. The efficiency of ionization of the surface-

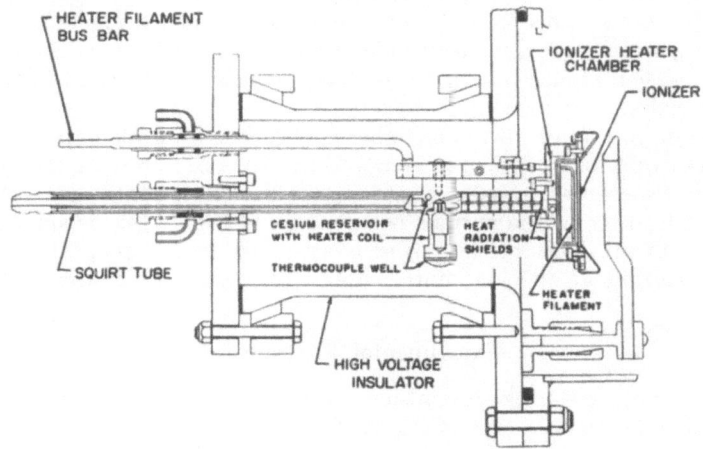

Fig. 11. Schematic of typical ion thrust chamber

contact source is nearly 100 percent for both cesium and rubidium if the ionizer surface is operated at the proper temperature. The efficiency of ionization from an arc source varies from a few percent to nearly 100 percent with improved

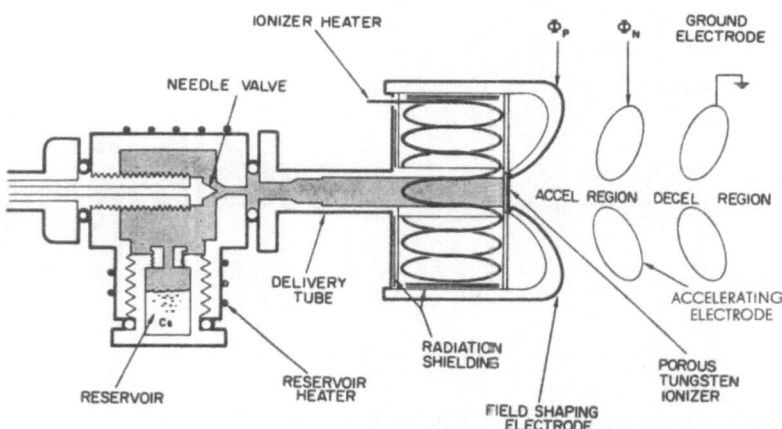

Fig. 12. Schematic of modified ion thrust chamber

designs. From the preliminary analysis it was concluded that, although the arc type source afforded a wider range of charged particle weights, the control problem associated with obtaining high efficiency introduced problems that made the surface-contact ion source superior for the initial research program. The experimental research thrust chambers illustrated in the following figures all incorporate the surface contact ion source.

The vacuum tank of the test apparatus is an 8-in. diameter, stainless steel unit. Heat exchange coils are wrapped around it for heating and cooling as needed.

The ion thrust chamber is mounted at the left hand end of the tank on an interchangeable cover plate through a suitable high-voltage insulator. This high-
voltage section is surrounded by a grounded protective cage.

Either a thrust-measuring device or a beam-power-measuring ion collector
is mounted on an interchangeable plate which closes the right hand end of the
tank. Shuttered observation windows are installed at the top. A similar window
is mounted at an angle at the bottom center of the tank. Pyrometric measurements
of the ion source are made through this observation port. The large cylinder at the
upper left encloses a cold trap and neutral particle detector.

The pumping system is beneath the work bench. Fig. 10 is a schematic of
the vacuum tank.

A typical research thrust chamber is shown schematically in Fig. 11. The ion
source is a surface-contact unit with an opening slit capable of delivering approxi-

Fig. 13. Thrust measurement device

mately 0.1 ampere of cesium ions. An accelerating electrode with a slit matching
the one of the source is mounted on a post attached to the interchangeable tank
end plate. The accelerating electrode is at ground potential.

A modification of the research thrust chamber with an array of electrodes is
shown in Fig. 12. The ion beam may be focused or decelerated by adjusting the
electrodes at the front of the source.

Thrust measurement is accomplished with the instrument shown in Fig. 13.
The ion beam is collected on a serrated carbon plate. This plate is mounted on a
central rod to which is attached a flat strip suspension system, designed to eliminate side sway. The central rod carries a deflection measuring transducer and
a damping system, shown at the center of the mounting plate. Calibration of the
thrust measurement system is accomplished with a micrometer positioner mounted
in the rear of the mounting plate and not visible.

The devices discussed in this section are currently in use at the Ion Research
Laboratory at Rocketdyne's Propulsion Field Laboratory. Analytical results

have been verified and many questions for which comprehensive theoretical analyses could not be obtained have been answered.

The results of both the completed analytical and experimental programs and those currently under way indicate that the ion rocket engine is a promising powerplant to supplement the high-thrust chemical and nuclear rocket engines now in the development or production stage. When the complete results of the program are released (these are beyond the scope of this paper) it is believed that the current performance estimates of the ion rocket engine will prove conservative. A flyable prototype ion engine can be achieved within a few years under the impetus of a progressive development program.

References

1. R. H. Boden, The Ion Rocket Engine. (Rocketdyne, A Division of North American Aviation, Inc., Canoga Park, California.) AFOSR TN 57-573 (R-645), 3 July 1958.
2. M.I.T. Staff of E. E. Dept., Applied Electronics. New York: J. Wiley, Inc., 1943.
3. E. Levin, Low Thrust Transfer Between Circular Orbits. (Rand Corporation, Santa Monica, California.) ASME Paper No. 58-AV-2, American Society of Mechanical Engineers, New York, March 1958.
4. Robert H. Fox, Powered Trajectory Studies for Low Thrust Space Vehicles. (University of California Radiation Laboratory, Berkeley, California.) ARS Paper 879-59, American Rocket Society, New York, June 1959.
5. R. H. Boden, Recent Developments in Ion Propulsion Systems for Space Travel. (Rocketdyne, A Division of North American Aviation, Inc., Canoga Park, California.) ASME Paper 59-AV-45, American Society of Mechanical Engineers, New York, March 1959.
6. J. H. Irving and E. K. Blum, Comparative Performance of Ballistic and Low Thrust Vehicles for Flight to Mars. Second Annual AFOSR Symposium, Denver, Colorado, April 1958.
7. E. Rodriguez, A Method of Determining Steering Programs for Low Thrust Planetary Vehicles. (Autonetics, A Division of North American Aviation, Inc., Downey, California.) ARS Paper 645-58, American Rocket Society Semi-Annual Meeting, Los Angeles, California, June 9—12, 1958.
8. M. Camac, Reduction of Flight Time and Propellant Requirements of Satellites with Electric Propulsion by the Use of Stored Electrical Energy. (Avco Research Laboratory, Avco Mfg. Corp., Everett, Mass.) AFOSR TN 58-1013, October 1958.
9. H. S. Tsien, Take-off From a Satellite Orbit. J Amer. Rocket Soc. 11, 233 (1953).
10. J. R. Pierce, Rectilinear Electron Flow in Beams. J. Applied Physics 23, 548 (1940).
11. A. Guthrie and R. K. Wakerling, Sources and Collectors for Use in Calutrons. TID-5218, University of California Radiation Laboratory, Berkeley, California, June 1949.

Experiments in Space Cabin Simulators[1]

By

George R. Steinkamp[2, 5], Willard R. Hawkins[3, 5] and George T. Hauty[4, 5]

(With 6 Figures)

(Received August 27, 1959)

Abstract — Zusammenfassung — Résumé

Experiments in Space Cabin Simulators. Following five days of physical, psychiatric, and psychological examination and two days of indoctrination and training, volunteers were subjected to a simulated flight of seven days duration in a one-man hermetically sealed chamber.

Chamber pressure and the partial pressure of oxygen were maintained at 380 mm Hg. and 159 mm Hg. respectively. Partial pressure of CO_2, temperature, and humidity were maintained at optimal levels. Within the 50 cubic feet of allowable space, the subject could assume two positions, seated and prone. He could not stand erect or walk. During flight, radio silence was maintained and visual contact denied; accordingly, gross behavior was monitored by closed circuit television and physiological behavior by recording respiratory and cardiovascular functions. Throughout the flight, the subject was committed to alternate four hour periods of work and rest. Work was provided by an operator system which permitted an objective and quantified appraisal of spatial discrimination vigilance, perceptual judgment and problem-solving.

The results reveal the effects of such a physical environment; of the impoverishment of the sensory environment; of the drastic change imposed on accustomed day-night cycling; and, the efficacy of diets.

Versuche unter raumfahrtäquivalenten Bedingungen in Unterdruckkammern. Nach fünftägiger ärztlicher, psychiatrischer und psychologischer Untersuchung und zweitägiger theoretischer und praktischer Belehrung wurden freiwillige Versuchspersonen sieben Tage lang in einer hermetisch geschlossenen Ein-Mann-Kammer Bedingungen, wie sie in der Kabine eines Raumfahrzeuges herrschen würden, ausgesetzt.

Kammerdruck und Sauerstoff-Teildruck betrugen 380 mm Hg beziehungsweise 159 mm Hg. Kohlensäure-Teildruck, Temperatur und Feuchtigkeit wurden auf physiologisch günstigstem Niveau gehalten. Der Nutzraum der Kabine betrug 50 Kubikfuß. Die Versuchsperson konnte sitzen oder auf dem Bauch liegen. Aufrecht stehen oder gehen war nicht möglich. Weder Funkverbindung noch Sicht nach außen

---

[1] The contents of this manuscript reflect the personal views of the authors and are not to be construed as a statement of official Air Force policy.

[2] Lt. Colonel, USAF (MC), Chief, Astroecology.

[3] Major, USAF (MC), Chief, Biogravics.

[4] Ph. D., Associate Professor of Experimental Psychology.

[5] Space Medicine Division, School of Aviation Medicine USAF, Brooks Air Force Base, Texas, U.S.A.

waren während des „Fluges" vorgesehen. Das Verhalten der Versuchsperson wurde
jedoch durch Fernsehen beobachtet. Kreislauf und Atmungstätigkeit wurden fort-
laufend registriert. Die Versuchsperson war angehalten, während des Versuches
abwechselnd vier Stunden zu arbeiten und vier Stunden zu ruhen. Die Arbeit bestand
in der Bedienung einer psychologischen Test-Apparatur, die die objektive und quantitative Beurteilung der Versuchsperson in Beziehung auf Raumgefühl, Aufmerksamkeit, Wahrnehmung und Unterscheidung verschiedener Sinneseindrücke und ihre
Fähigkeit, Probleme zu lösen, erlaubte.

Die Versuchsergebnisse zeigten die physiologischen Auswirkungen des Mikro-
klimas einer geschlossenen Kabine, der Reduktion von Sinneseindrücken und des
drastisch veränderten Arbeit-Ruhe-Zyklus. Außerdem wurde die Zweckdienlichkeit
der ausgewählten Diät geprüft.

Expériences avec des simulateurs de cabine spatiale. Après cinq jours d'examens
physique, psychiatrique et psychologique et deux jours d'entraînement et d'indoctrination des volontaires ont été soumis dans une cabine monoplace hermétiquement
close à un vol spatial simulé d'une durée de sept jours.

La pression de cabine et la pression partielle d'oxygène ont été maintenues respectivement aux niveaux de 380 et 159 millimètres de mercure. Les pressions partielles
de CO_2, la température et l'humidité ont été maintenues à des niveaux optima.
Dans les 50 pieds cubiques d'espace disponible seules les positions assise et couchée
étaient possibles. Le sujet ne pouvait ni se mettre debout ni marcher. Durant le vol
il ne pouvait y avoir ni contact visuel ni auditif par radio. Le comportement général
était enregistré par télévision et le comportement physiologique par enregistrement
des fonctions respiratoires et cardio-vasculaires. Des périodes alternées de quatre
heures de travail et quatre heures de repos étaient imposées. Le travail était fourni
par un dispositif opératoire permettant d'apprécier la discrimination spatiale, le
degré de vigilance, les capacités de perception et de résoudre des problèmes.

Les résultats révèlent les effets des conditions physiques ambiantes; l'apauvrisse-
ment de l'environnement sensoriel; modifications drastiques du cycle habituel diurne-
nocturne. Ils montrent aussi l'efficacité de diverses diètes.

A series of seven-day simulated space flights have been conducted for the
purpose of determining biomedical requirements for manned space flight. In
these flights, the subjects' role has been more than serving as a passive biological
specimen. They have been committed as an integral component in a man-machine
system; and, as such, they have been systematically exposed to several relevant
conditions of the closed ecological environment of a space capsule. The duration
of seven days was selected because it allows time for depletion of initial reserves
and subsequent biological adaptation to the imposed conditions.

As a result of the integrated problems undertaken by this series of studies,
and the critical importance of determining their joint effects, an extensive inter-
disciplinary effort is an undeniable requirement. Accordingly, the result to be
reported do represent such an effort; and, as a consequence, are considered
to have implications for many of the bioastronautic problems and areas of
interest.

SAM Space Cabin Simulator

The developmental history and the present characteristics of the SAM Space
Cabin Simulator (Fig. 1) which was delivered to the School in 1954 have been
described in great detail in a previous report[1].

---

[1] Human Experiments in the Space Cabin Simulator: I, School of Aviation
Medicine Report 59-101, July 1959.

The essential characteristics of this chamber were as follows:

1. *Carbon dioxide control:* The absorption apparatus consisted of three two-pound cannisters of Baralyme through which the cabin air was continuously

Fig. 1

drawn. It had been found by previous study that this amount would absorb approximately 450 grams of CO_2 per hour. Therefore, in order to maintain

Fig. 2

minimal CO_2 levels and allow a relatively-wide margin of safety, the subject was required to renew the six pounds of absorbent every eight hours. As a further safeguard, the CO_2 level was constantly monitored by means of a Liston-Beckman non-disperse, infra-red CO_2 detector (Model 16) and constantly recorded on a strip chart.

2. *Oxygen Control:* A Pauling-Beckman detector constantly sensed the metabolically induced drop in O_2 within the cabin. This information activated a Brown potentiometer which recorded the data on a chart. The controlling recorder simultaneously activated a mercury switch which operated a solenoid valve in the O_2 source line and allowed a sufficient amount to enter the chamber to maintain the desirable partial pressure level. To calibrate the O_2 analyzer, sample gases of known concentration were passed through the analyzers into a previously evacuated receiving vessel at the existing chamber pressure. With this technique, the chamber atmosphere was not diluted by the sample gases; the chamber pressure was not altered; and the subject was not able to sense the time or duration of the calibration procedure.

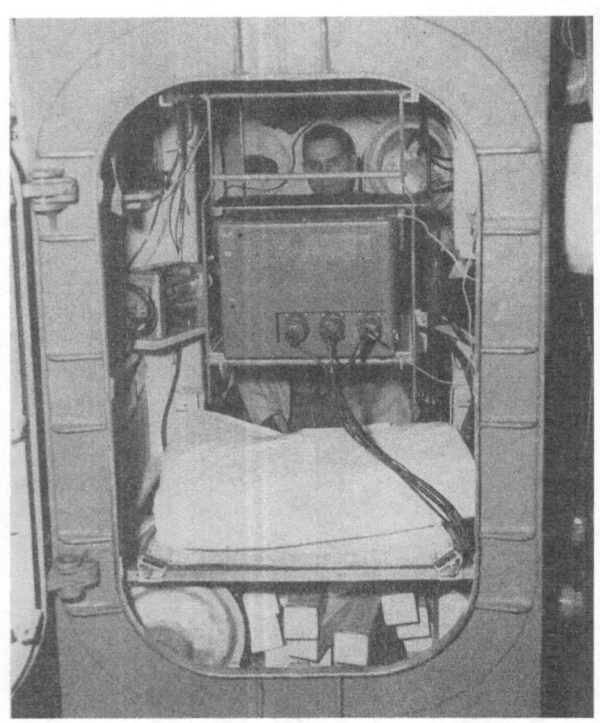

Fig. 3

3. *Temperature Control:* This was accomplished by attaching to the chamber a heat sink which consisted of a sealed pipe (seven inches in diameter and 40 inches long) wrapped with a layer of refrigerating coils covered with insulating material. This provided a temperature gradient across the skin of the chamber rather than direct cooling of the ambient atmosphere. The cabin air was circulated through this heat sink by a master blower, cooled, and the condensed water vapor was collected in a reservoir and supplemented the water supply for personal hygiene.

4. *Humidity Control:* In addition to the negative control provided by the heat sink, a positive control was achieved by an automatic vaporizer.

5. *Odor Control:* An activated charcoal unit, demonstrating a high degree of efficiency, was placed within the air flow circuit of the master blower.

6. *Closed Circuit Television:* Two systems provided for a constant monitoring of the subject and his activities and for the presentation of his operator task. With respect to the latter, it was necessary to locate the complete operator system outside of the cabin, due to limitation of space and to the desirability of immediate repair and/or replacement of parts during flight. Consequently, a fixed camera transmitted the display console of the operator system to a monitor within the chamber. This monitor and its associated control console

are shown in Fig. 2. The task required of the subject involved the functions of spatial discrimination, perceptual judgment, vigilance, and problem solving. The proficiency with which these functions were executed was electrically recorded.

7. *Command Panel:* To permit the exchange of information and instructions, a panel containing 24 switches and signal lights was installed within the chamber with a duplicate panel at the external monitoring station. Provision was made for two-way operation so that any signal light could be activated by the subject or the ground observers.

8. *Habitability:* The small amount of space (50 cubic feet) available to the subject was sufficient nevertheless for a collapsible hammock arrangement and an aircraft cockpit seat (Fig. 3). A feces receptacle built into the seat provided for storage. A metal tank beneath the floor served as a urine receptacle. A standard hot-cup, in which water could be heated to boiling temperature, was the only available food preparation unit.

Procedure

Pre-Flight Indoctrination: Following a week of extensive and definitive medical evaluation, two days were devoted to intensive pre-flight indoctrination. This began with a period of initial familiarization with the School of Aviation Medicine Space Cabin Simulator and all allied equipment. The subject was then given a briefing for a two-hour indoctrination flight involving the following:

Ascent to 18,000 feet simulated altitude; O_2 maintained at 159 mm Hg.; CO_2 maintained below 5 mm Hg.; temperature and relative humidity at 80° F and 46 percent respectively. First hour—cabin familiarization. Second hour—practice at the operator work system and termination with 15 minute descent to ground level. The flight was conducted accordingly and the subject instructed to respond to all command signals presented.

On the second day, a briefing was given for a six hour indoctrination flight designed to provide final preparation:

Explanation of each event of the seven day flight schedule.

Application of physiological sensors, explanation of reapplication by subject, and

Recording of all data regularly obtained during actual flight.

Four hour cycle of practice at the operator work system.

Presentation of every command signal with subject making appropriate responses.

Preparation and consumption of a hot meal.

Preparation and testing of sleeping equipment.

The flight was then conducted.

Flight: At 0730 hours of the third day, final preparations for the flight were begun. The subject was weighed and given a briefing designed to clarify all points of doubt or uncertainty. Physiological sensors were applied and he then was dressed in a loose fitting scrub suit. The subject entered the cabin and was given a detailed briefing on the location and storage of all items. Following this, the recordings of his ECG and respiration and the amplification of his audible heart sounds were tested for adequacy. Finally, when everything was checked out to the satisfaction of the principal investigators, the hatch was closed and simulated ascent began.

The daily schedule of principal activities to which the subjects were committed during their flights is given in Table I.

Table I. *Daily Flight Schedule*

0810 Alarm	1700 Begin work
0845 15 minute alert	2100 Stop work
0846 Baralyme change	2101 Off duty
0855 Report	0010 Alarm
0900 Begin work	0045 15 minute alert
1300 Stop work	0046 Baralyme change
1301 Emergency check	0055 Report
1305 Off duty	0100 Begin work
1610 Alarm	0500 Stop work
1645 15 minute alert	0501 Off duty
1646 Baralyme change	0810 Alarm
1655 Report	

Following the alarm, which is intended to awaken the subject, there is a 30-minute period for personal hygiene, meal preparation, eating, etc. These, at least, are the duties recommended for this period. However, the subject still has complete freedom of choice. At the end of this period, a signal was given alerting him to the fact that his next work period would begin in 15 minutes. One minute later, a signal was given to change the Baralyme in the air flow system. Next, the subject was signalled to make a report over the communication system of cabin temperature, oral temperature, and relative humidity. At 0900 hours, he was given the signal to begin work at the simultaneously activated operator system, and at 1300 hours, the signal to stop work. One minute later, the emergency check was signalled. This required the subject to apply the oxygen mask and use the walk-around bottle until he received the all clear signal. During this time, he was also required, of course, to check oxygen pressure and if this had fallen below 100 psi, he must plug into the refiller line and signal the ground crew for refill. At 1305 hours, he received the signal for the beginning of his uninterrupted three hour rest or ad lib period.

It will be noted that throughout each day of the seven days of flight, the subject was committed to a 4:4 ratio of work and rest. This represented a drastic revision of the subjects' accustomed day-night cycle, particularly, since the third work period (0100-0500) occupied that portion of the day which the subjects typically devoted to sleep.

During the flight, the ground crews follow a schedule of required activities. These included calibration of the O_2 and CO_2 analyzers at 0820, 1620 and 0020 hours; hourly recordings of cabin temperatures, humidity, pressure, and partial pressures of O_2; and CO_2 and hourly recording of heart and respiratory rates. In addition to these scheduled activities was the constant monitoring of the subject by means of the closed circuit television system which permitted the observers to maintain logs of subjects' activities and significant events. At the onset of the latter, a kinescope could be activated so that the event could be filmed for subsequent study.

Throughout the entire flight, the attempt was made to maintain environmental conditions at constant levels: cabin pressure at 380 mm Hg., partial pressure of O_2 at 159 mm Hg., partial pressure of CO_2 below 5 mm Hg., temperature at 86° F, and relative humidity at 45 percent. Certain other conditions to which the subject was exposed are unique and, therefore, must be mentioned.

The air flow blowers produced a constant background of noise of approximately 84 db. The requirements for periodically photographing and for monitoring the subject resulted in a constant high level of illumination. Finally, the observation ports were closed so that the subject never had visual access to the outside and radio silence was constantly maintained by the ground personnel.

Two and a half hours prior to the termination of flight, descent began and continued at an extremely slow rate until ground level was reached. The subject was taken from the cabin, immediately weighed and hospitalized for his postflight medical evaluation. At this time, the cabin contents were carefully analyzed. The respective amounts of urine, condensate, and unused potable water were measured. Determinations of amount and type of food and juices consumed were made. Fecal material was weighed. All material written by the subject was collected.

Post-Flight Evaluation: Immediately following termination of the flight, the subjects were again given a complete physical evaluation. Special attention was given to circulatory and pulmonary function studies. A comprehensive neuropsychiatric interview was also a part of the procedure.

The circulatory studies on the tilt table, with repetition of the stresses, such as breath-holding, hyperventilation and carotid pressure, revealed a common finding. Heart rate accelerated much more rapidly than on their original examinations, and in each case, recovery was noticeably slower. In general, their immediate physical condition resembled the pattern found among convalescent patients during prolonged bed rest. However, within 24 hours these functions returned to normal.

Subjects

Without exception, all subjects were athletically inclined and possessed vigorous physiques but varied widely in general bodily contour. Subject I was the largest of the group, weighing 190 pounds and 71 inches in height. Subject II was the smallest, being 65 inches tall and weighing 140 pounds. Their ages were 22, 29, 34, 39, and 35 years respectively.

The first subject was chosen on the basis of general superior physical condition and a superior adult level of intelligence. He did not have college background; however, he had successfully met the mental and physical requirements of the United States Air Force for pilot training. For general purposes, then, he can be considered to be comparable to the other four except in background experience and training.

The remaining four subjects were volunteers from a highly select group of the pilot population of the United States Air Force. As such, these shared similar backgrounds in training, duty situations and career goals. A general evaluation would be that all five subjects were highly comparable in basic ability. In the case of experience, however, Subject I represents a dissimilarity by virtue of his age and lack of exposure to the high degree of training and professional discipline required of pilots of high performance aircraft.

Results and Discussion

Physiological: An examination of the heart and respiratory rate data reveals no indication that a physiologic adaptation occurred in these five subjects as a result of being subjected to an altered work-sleep cycle. In fact, it was a little surprising to see how little variation there was in a particular subject's heart rate and respiration from day to day. In spite of the altered living cycle and the accumulative effect of the combined stresses imposed on the subjects, each one has his own characteristic pattern of high and low values that was consistently followed throughout the flight.

Psychophysiological: The principal indication of adjustment to the 4:4 ratio of work and sleep is given by the proficiency with which the subjects executed the tasks that constitute the operator system. Accordingly, the mean proficiency of the four different tasks was computed for each hour of work for each subject. For brevity, the obtained proficiency curves for only three of the seven days of flight are presented (Fig. 4, 5, and 6).

In Fig. 4, we see the levels of proficiency attained by each subject during each of the three work periods of day 1. For the first subject these curves ex-

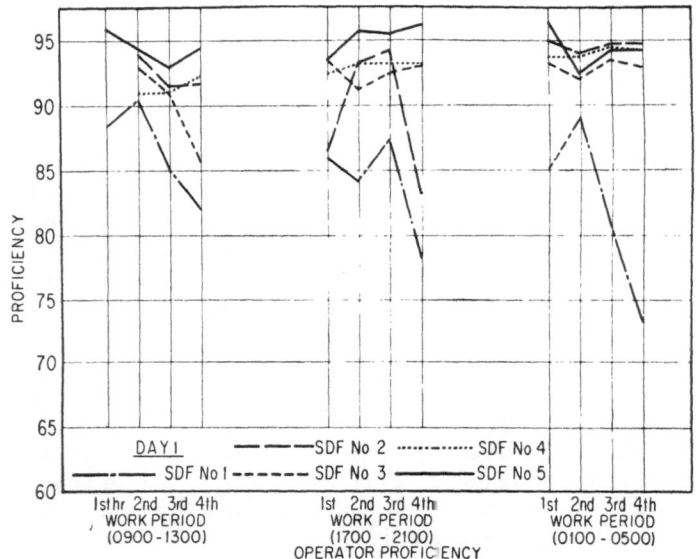

Fig. 4. Simulated seven day space flights

hibit a completely predicable form. That is, decrement occurs within each work period and, further, for each successive period of work the magnitude of decrement increases. In comparison, the proficiency curves of the other four subjects are seen to be different. As would be expected, the proficiency of these subjects is at a higher initial level but, more pertinent to the problem being explored, is the extent to which these initial levels are sustained. Decrement does occur for three of these subjects during the first work period and for one during the second work period. However, during the third period of work which may be considered as being the most critical, suprisingly little decrement in proficiency occurs.

During the third day (Fig. 5), and in the case of the first subject, a trend which began during the previous day has now become clearly established. On the first day of flight, this subject was most proficient during the first period of work (0900-1300). Now, his highest level is achieved during the second work period (1700-2100). In fact, it is during this particular period of day 3 that the least difference between the subjects is evidenced. But, during the subsequent work period, the differences in proficiency decrement previously noted, appear again.

During the fourth, fifth and sixth days of flight, the first subject continued to exhibit profound decline in proficiency while the other subjects continued

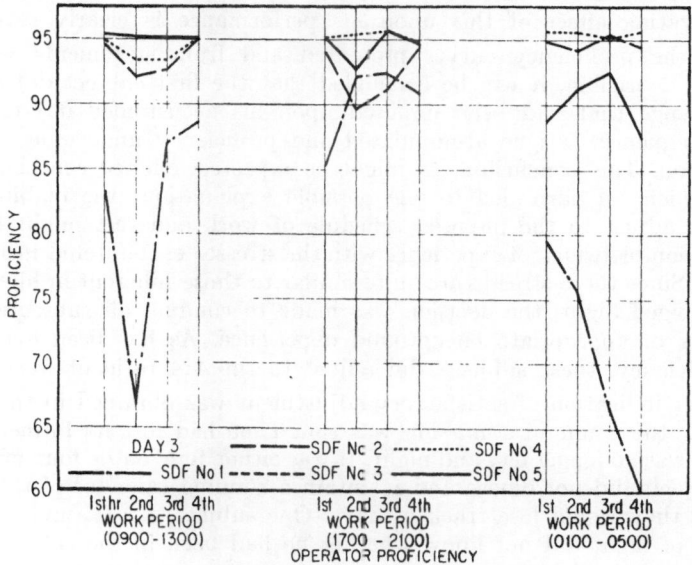

Fig. 5. Simulated seven day space flights

to maintain their respective high initial levels of proficiency with little or no decrement. On the seventh and final day (Fig. 6) radio silence was broken prior

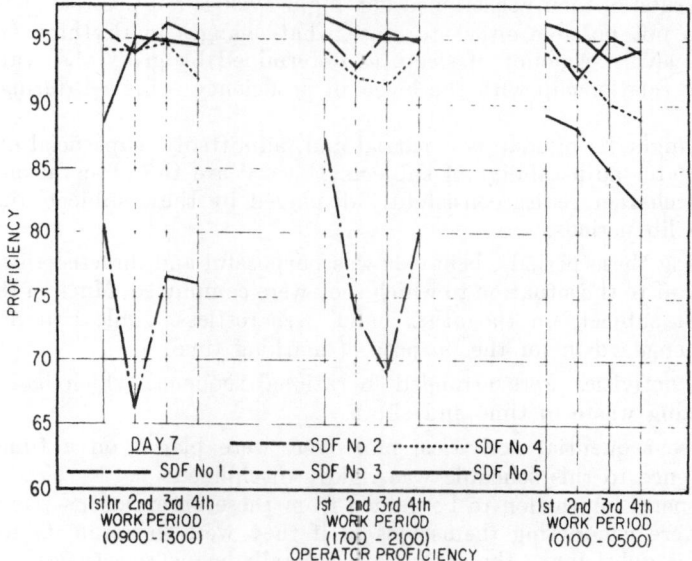

Fig. 6. Simulated seven day space flights

to his last work period and verbal contact established with the first subject for the purpose of allaying his hostility which by this time had approached a level thought to constitute a threat to his well-being. The subject was congratulated for a job well done, questioned on his physical condition, and then told of the group of individuals that would be present at the termination of his flight. The motivating effect of this upon his performance is clearly revealed.

From the proficiency curves presented and from statements volunteered during his debriefing, it can be concluded that the first subject did not adjust to the change that had been imposed upon his accustomed day-night cycle. As a consequence, fatigue accumulated and proficiency underwent progressive deterioration. This conclusion, as might be expected, elicited considerable speculation which, in turn, led to one possible explanation: the inability of this subject to adjust to the imposed schedule of work and rest might have been due to his complete lack of experience with the stresses encountered in his simulated flight. Since these stresses are quite similar to those inherent in high performance prolonged flight, the decision was made to conduct all subsequent flights with pilots of appropriate background experience. As has been evidenced by their proficiency, these subjects did adjust to the 4:4 ratio of work and rest.

Another indication of satisfactory adjustment was obtained in their debriefings. Here, the common admission was that time had lost its former meaning. That is, it was no longer day and night. It was either four on or four off. Further, if the subjects did not devise and maintain a running calendar of work periods and days, they would lose track of days. One subject, for example, during his third day of flight did not know whether he had been in the cabin two or six days. This produced his only moment of panic. By counting up his empty food cartons, he was able to reestablish the correct day of flight. He then constructed a calendar so he wouldn't "lose track of time again." During his debriefing, he was insistent on the importance of being time oriented.

To explain specifically why these subjects did adjust to what was an equally drastic revision of their accustomed day-night cycle is not possible at the present time. It is not possible either to state what changes in rhythmicity occurred at cellular levels. Amount of sleep, as determined by gross observation, bears little or no relationship with the levels of proficiency achieved during the work periods.

What might be offered as a partial and, admittedly, superficial explanation for the greater adjustability of subjects II to V are the observations made of behavioral characteristics consistently displayed by these subjects during their rest or ad lib periods.

1. During these periods, behavior was purposeful and directed toward objectives relevant to the situation to which they were committed. Much of the activity of the first subject, on the other hand, was restless, aimless in nature, and seemingly engaged in for the purpose of marking time.

2. The activities were arranged in rational sequence which had the effect of minimizing waste in time and effort.

3. These sequential activities, moreover, were placed on a time schedule and adherence to this schedule was rigidly disciplined.

The general impression to be gained from these observations was that these subjects were conducting themselves as if they were in actual flight. In fact, during their debriefings, the statement generally made was that the simulated seven-day space flight was conceived simply as another difficult job to be done and that they were going to get it done to the best of their ability. There can be little doubt that such experience was the differentiating factor in the case of initial levels of proficiency, appropriateness of behavior, and realistic task-oriented attitudes toward a stressful situation. But experience of this nature, however, must not be taken as the principal factor determined modifiability of day-night cycling. What must be considered is the fact that these particular subjects represented the end-products of a long-term, continuous, and valid process of selection. Quite likely, their greater adaptability is attributable to certain common, as yet unidentified, psychophysiological characteristics and the conditioning to which these characteristics have been exposed.

Equal in importance to the problem of day-night cycling and its modifiability is the problem of sensory deprivation. While the sensory environment to which the subjects were committed for seven days was markedly impoverished, it can be definitely stated that aberrancy of behavior was never observed nor reported. This may be at variance with a priori expectation but actually is not too surprising. The reason is twofold: Firstly, scheduling work and rest in the manner described had the predictable effect of nullifying the condition essential to extreme deprivation of sensory input. This condition, in brief, is that of requiring highly motivated individuals to attend to a small perceptual field of work so intensively and for such a prolonged period of time that the individual, in effect, deprives himself of the ambient stimuli which otherwise would have been sufficient for a normal level of sensory input. Such a level, perhaps, was occasioned by the periodic command signals, by the regularly occurring work and rest activities, and, probably by the close physical proximity of the monitoring ground crews. With respect to the latter, only one subject expressed the impression that he felt isolated at any time. The others stated that they were separated from the ground crews by only the thickness of the cabin shell.

Nutrition: Since the study of the cumulative effects of fatigue on work performance was one of the main objectives of the experimental design, it was determined that the diet for each subject would be optimal, insofar as the limitations of facilities within the cabin allowed.

Foods were selected from stock USAF Commissary items and a few portions were taken from the standard USAF In-Flight Ration Number 7 to expand or individualize the diet. Inquiry was made of each subject prior to the experiment to determine individual preferences and each diet was modified accordingly. The group offered no striking food likes or dislikes, and only very minor variation of basic menus was necessary.

The objectives of the dietary regime for this series were:

1. Maintenance of good nutritional status for the seven-day period.

2. Avoidance of monotony of diet, by pre-selection of varied and interesting items.

3. Selection of items which allowed ease of preparation and consumption.

Items selected for the diet were packed in 21 boxes (three per day) and stored in the cabin in such a manner that the subject could have easy access to consecutive meals throughout the seven-day period. The total weight of all food and containers was 50 pounds and occupied 1.1 cubic feet of space within the cabin.

Each box contained a "main meal" and a "snack meal." The subject was given a "meal period" signal on the command panel every eight hours during the flight. This signal came at the end of the rest period, 40 minutes prior to the beginning of the next work period. The subject opened the lunch box corresponding to the eight hour period and prepared and ate as much of the main meal as he desired. Each box contained, in addition, "snacks" which he was able to eat during the work period (e.g., fruit juice, gum, hard candy, nuts, coffee, tea).

An average of 2400 calories per day was pre-packed for each subject. The daily average of protein was 102 grams, of carbohydrates 336 grams, and of fat 72 grams. Fluid requirements were met by the provision of two liters of water per day.

The concept and construction of the diet, as outlined, were universally accepted by the subjects participating in this series of experiments. Each subject reported favorably on the diet during the post-flight debriefing and indicated that the optimal diet given had been a factor in the maintenance of good morale during the seven-day test period.

The Sterilization of Space Vehicles to Prevent Extraterrestrial Biological Contamination[1]

By

Richard W. Davies[2] and Marcus G. Comuntzis[2]

(*Received August 27, 1959*)

Abstract — Zusammenfassung — Résumé

The Sterilization of Space Vehicles to Prevent Extraterrestrial Biological Contamination. Present knowledge of the planets Mars and Venus is compatible with the possibility both of an indigenous life and of the maintenance and rapid spread of terrestrial microorganisms. The introduction of terrestrial organisms as contaminants on planetary probes might so distort the biology of either planet as to constitute a scientific catastrophe.

The investigation of life on the planets is among the most significant scientific objectives of our time, and is at the same time the most sensitive to irremediable harm. Space probes which have any likelihood of a landing, intentional or accidental, should be subject to careful sterilization.

Extraterrestrial contamination has not received sufficient international attention. Two reports have been put out by a scientific committee (CETEX) representing the International Scientific Unions. These reports attempted to set a tone for developing a code of conduct in space research, but did not offer any specific suggestions as to how probes should be sterilized.

Sterilizing space probes is an engineering nuisance, however, the same has been said of the sterilization preparations for surgery. In both instances, the techniques cannot be left to the last minute.

Four phases of payload sterilization are recommended. They are:
1. sterile assembly,
2. built in or intrinsic sterilization,
3. terminal sterilization,
4. maintaining sterilization.

The third phase, terminal sterilization is the most important. The entire payload would be sterilized if it were placed in dry steam at 160° C. for twenty minutes. However, twenty percent of the components, such as transistors and antennas, cannot endure this temperature. Some parts which cannot be subjected to 160° C. can be sterilized with chemicals such as ethylene oxide gas or beta propiolactone. These gases can penetrate the interstices of some components.

The interiors of some instruments cannot be sterilized by gases and will not function properly after being subjected to heat sterilization. (Dielectric materials in transponders are examples.) The interiors of these components should be fabricated utilizing oils, waxes and potting compounds which are already sterile. Substances of biological origin such as casein and shellac should be avoided. This is phase one.

---

[1] This paper presents the results of one phase of research carried out at the Jet Propulsion Laboratory, California Institute of Technology, under Contract No. NASw-6, sponsored by the National Aeronautics and Space Administration.

[2] Jet Propulsion Laboratory, California Institute of Technology, Pasadena, California, U.S.A.

In some instances, sterilization can be built into components (phase 2) by coating parts with organo-metallic compounds that kill bacteria on contact.

The fourth and final stage of sterilization is keeping clean what has already been sterilized. Contact with the air and last minute adjustments will cause local surface contamination. Ultraviolet solar radiation would kill the least radiosensitive dormant anaerobic bacteria in a few hours provided, of course, the bacteria are not shaded from the sun. It is most probable, however, that certain parts of the payload surface will never see the sun during its space journey. The payload shroud can be employed, not only to protect the payload during the launch, but also to contain sterilizing gases at normal pressure, thereby maintaining surface sterility.

In order to minimize chemical contamination of a kind that might confuse later investigators, it is recommended that a careful molecular inventory be made of each mission together with a replica of each package. It is also recommended that a micro-biological survey be made of the launching site area.

Owing to the delicate nature of many instruments in the payload, penetrating radiation sterilization has limited usefulness.

Die Sterilisierung von Raumfahrzeugen zur Vermeidung außerirdischer biologischer Verseuchung. Unser gegenwärtiges Wissen über die Planeten Mars und Venus schließt die Möglichkeit sowohl eingeborenen Lebens wie auch einer Erhaltung und schnellen Ausbreitung irdischer Mikroorganismen nicht aus. Die Einbringung irdischer Organismen auf planetarischen Sondenfahrzeugen als Verseuchungsursachen könnte die Biologie beider Planeten so sehr stören, daß eine wissenschaftliche Katastrophe die Folge wäre.

Die Erforschung des Lebens auf den Planeten gehört zu den bedeutendsten wissenschaftlichen Aufgaben unserer Zeit; gleichzeitig ist sie diejenige Aufgabe, die am ehesten einen nicht wieder gutzumachenden Schaden bewirken kann. Infolgedessen sollten Raumsonden, für die irgendeine Wahrscheinlichkeit einer — beabsichtigten oder zufälligen — Landung besteht, einer sorgfältigen Sterilisierung unterzogen werden.

Außerirdische Verseuchung hat bis jetzt keine ausreichende internationale Aufmerksamkeit erfahren. Zwei Berichte wurden von einem wissenschaftlichen Komitee (CETEX) herausgegeben, das dem International Council of Scientific Unions (ICSU) angehört. Diese Berichte versuchten, Richtlinien zur Entwicklung eines Code des Verhaltens bei der Raumforschung aufzustellen, brachten jedoch keine eingehenden Vorschläge über das Vorgehen zur Sterilisierung von Sondenfahrzeugen.

Die Sterilisierung von Raumsonden ist eine technische Unannehmlichkeit, doch ist das gleiche über die Vorbereitungen zur Sterilisierung in der Chirurgie gesagt worden. In beiden Fällen kann man die Verfahren nicht der letzten Minute überlassen.

Es werden vier Phasen der Sterilisierung der Nutzlast empfohlen, und zwar:

1. Sterile Montage;
2. eingebaute oder innere Sterilisierungsanlagen;
3. Sterilisierung im Endstadium;
4. Sterilisierungserhaltung.

Die dritte Phase (Endsterilisierung) ist die wichtigste. Die ganze Nutzlast würde sterilisiert werden, wenn sie 20 Minuten lang bei 160° C einem trockenen Dampf ausgesetzt würde, doch könnten 20 Prozent der Bestandteile, wie zum Beispiel Transistoren und Antennen, dieser Temperatur nicht standhalten. Einige Teile, die 160° C nicht vertragen, können mit Chemikalien, wie Äthylenoxyd oder β-Propiolakton sterilisiert werden. Diese gasförmigen Stoffe können die Zwischenräume mancher Komponenten durchdringen.

Das Innere mancher Instrumente kann aber nicht mit Gasen sterilisiert werden; sie würden nach einer Hitzesterilisierung nicht richtig funktionieren. Daher sollte die Innenausstattung dieser Bestandteile unter Benützung von Ölen, Wachsen und Bindestoffen hergestellt werden, die bereits steril sind. Substanzen von biologischem Ursprung, wie Casein und Schellack, sollte man vermeiden. Dies ist Phase eins.

In manchen Fällen kann die Sterilisierung in die Komponenten eingebaut werden (Phase zwei), indem die Teile mit organometallischen Verbindungen überzogen werden, die Bakterien bei Berührung töten.

Das vierte und Schlußstadium der Sterilisierung hält rein, was bereits sterilisiert worden ist. Berührung mit der Luft und Justierungen in letzter Minute können lokale Oberflächenverseuchung verursachen. Ultraviolette Sonnenstrahlung würde die wenigst strahlungsempfindlichen, latenten, anaerobischen Bakterien in wenigen Stunden töten, vorausgesetzt natürlich, daß die Bakterien nicht gegen die Sonne abgeschirmt sind. Es ist aber sehr wahrscheinlich, daß bestimmte Teile der Oberfläche der Nutzlast während der ganzen Reise durch den Raum niemals vom Sonnenlicht getroffen werden dürften. Eine Umhüllung der Nutzlast kann verwendet werden, nicht nur, um sie während des Starts zu schützen, sondern auch, um sterilisierende Gase bei Normaldruck zu halten, wodurch die Oberflächensterilität aufrecht erhalten würde.

Um chemische Verunreinigung einer Art, welche spätere Forscher verwirren könnte, auf ein Mindestmaß zu senken, wird empfohlen, sorgfältige molekulare Bestandsaufnahmen für jede Fahrt zu machen, zusammen mit einer Nachbildung jedes Stückes. Auch eine mikrobiologische Überprüfung des Geländes der Startanlage wird vorgeschlagen.

Wegen der empfindlichen Natur vieler Instrumente in der Nutzlast hat eine Sterilisierung mittels durchdringender Strahlung nur beschränkten Nutzen.

La stérilisation des véhicules spatiaux contre la contamination biologique des milieux extraterrestres. La connaissance actuelle des planètes Mars et Vénus est en accord avec les possibilités de vie propre et de prolifération des micro-organismes terrestres. Une catastrophe scientifique pourrait résulter des bouleversements biologiques dus à l'apport d'organismes terrestres par les sondes planétaires.

L'étude de la vie sur les planètes est un objectif significatif de nos temps; elle est aussi sensible à un dommage irrémédiable. Les sondes planétaires, qui ont quelque chance d'entrer en contact avec une planète, devraient être soumises à une stérilisation soignée.

Ce problème de contamination n'a pas reçu assez d'attention de la part des milieux internationaux. Deux rapports présentés par le comité CETEX de l'I.C.S.U. se bornent à développer un code de conduite pour la recherche de l'espace; ils n'ont fait aucune proposition spécifique quant à la stérilisation des sondes.

La stérilisation des sondes est un travail ennuyeux; on a dit la même chose des préparatifs de stérilisation en chirurgie. Dans les deux cas les techniques ne peuvent être développées à la dernière minute.

Quatre phases de la stérilisation de l'instrumentation sont recommandées. Ce sont:

1) la stérilisation en cours de montage;
2) la stérilisation interne ou intrinsèque;
3) la stérilisation finale;
4) le maintien de la stérilisation.

La troisième phase est la plus importante. L'instrumentation complète serait stérilisée par un bain de vapeur sèche à 160° C durant vingt minutes. Cependant 20% des éléments, comme les semi-conducteurs et les antennes, ne peuvent tolérer cette température. Ces éléments peuvent être stérilisés par produits chimiques tels l'oxyde d'éthylène ou le propiolactone beta. Ces gaz peuvent pénétrer les interstices de certains éléments.

Certains instruments ne peuvent être stérilisés intérieurement par des gaz ni fonctionner convenablement après stérilisation à la chaleur. Ils doivent alors être fabriqués à partir d'huiles, cires et composés déjà stériles en évitant les matériaux d'origine biologique comme la caséine et le shellac. C'est la phase un.

Dans certains cas la stérilisation est réalisée par un revêtement organo-métallique qui tue les bactéries par contact. Phase 2.

La phase finale consiste à maintenir la stérilité. Le contact avec l'air et les manipulations finales peuvent causer des contaminations locales. Les rayons ultra-violets du soleil peuvent tuer les bactéries dormantes anaérobies les moins sensibles en quelques heures, à condition qu'elles ne soient pas à l'ombre. Il est suggéré d'utiliser le capot protecteur de l'instrumentation non seulement pour la phase de lancement mais aussi comme enveloppe d'un gaz stérile à pression normale assurant la stérilité de surface.

Pour diminuer le danger de contamination chimique qui menacerait les enquêteurs futurs il est recommandé de faire un inventaire moléculaire soigneux lors de chaque mission et de prévoir un double de chaque ensemble.

On recommande aussi un examen biologique du site de lancement. La stérilisation par rayonnement pénétrant ne se prête pas à la nature délicate de beaucoup d'instruments.

Introduction

Speculation on the existence of extraterrestrial life is sufficiently commonplace to suggest that the concept is subtly imbedded in our social culture. However, it is difficult to verify the origin of the extraterrestrial life concept because of a tendency to ascribe original authorship of many ideas to antiquity[1].

The discovery of life on any of the planets would be one of the most exciting events in human history. Satisfying society's general curiosity would, however, be only one facet of the discovery. The event would also have tremendous scientific interest because, next to the synthesis of living matter in the laboratory, it would be the most important step that could be made toward understanding of the problem of the origin of life. Systems based on nucleic acid and proteins as bearers of life may or may not be unique. This is one of the fundamental questions that the discovery of extraterrestrial life might answer.

The biological importance of the planets is not limited to detecting and studying life on them. Even if no life is found to exist, the opportunity to sample organic compounds on the planets might give some valuable clues to the origin of life. Sterile worlds may provide the information necessary to the understanding of the organic chemical processes that preceded the development of life on the earth.

Present knowledge (as well as the lack of it) of the planets Mars and Venus is compatible with the possibility both of an indigenous life and of the support and rapid proliferation of terrestrial micro-organisms. The introduction of terrestrial organisms and contaminants might so distort the biology of either planet as to constitute a scientific catastrophe. The processes are irreversible and they make the search for life on other planets most sensitive to irremediable harm. If the earth were sterile, it would require only months or years to universally populate it with the descendants of a single cell. A common bacterium, $E. Coli$, has a mass of 10^{-12} grams and a fission interval of 30 minutes. Ideally, it would take 66 hours for the progeny to attain the mass of the earth. The progeny never reach this magnitude principally because the food supply is insufficient. Nevertheless, this extrapolation illustrates that the exponential growth rate of bacteria is truly explosive and, therefore, the timescale of planetary biological distortion need not be long. Indeed, it could be considerably less than the time interval of earth-planet oppositions. Space probes which have any likelihood of a landing, intentional or accidental, should be subject to careful sterilization.

Definition of Biological Contamination

It is convenient to separate biological contamination into two kinds, pollution and infection. Biological pollution is meant to be a deposit of a large enough number of micro-organisms to be scientifically significant, as such,

---

[1] According to P. DUHEM, Le Système du Monde (Vol. 1, pp. 17-18, 1913) the Pythagorean PHILOLAUS (5th century B.C.) postulated the existence of an inhabited anti-earth which was always in opposition to the earth in relation to the sun.

without further growth. Infection is meant to describe the growth of one or more viable organisms. Likewise, pollution can be divided into two categories; viable pollution, which does not grow by nature of its environment, and non-viable pollution.

Pollution

Pollution is a type of contamination that applies to the Moon, Mars, and Venus. Pollution would be most likely if a mammal were splattered on any one of these three bodies. For example, the Moon's area is 4×10^{13} square meters, and the intestines of a mammal can contain 10^{12} micro-organisms per kilogram. If the mammal died in flight, the putrified tract could contain 10^{13} micro-organisms per kilogram [1]. Present techniques are capable of detecting one micro-organism per square centimeter. These techniques could be immediately extended to detect one micro-organism per square meter. Future improvements in technique may increase the detecting sensitivity by a few orders of magnitude; therefore, a single probe leaving a residue of from 10^9 to 10^{10} dead bacteria could provide a misleading background noise for future investigators[1].

Infection

Infection appears least likely on the Moon because water is the lowest common denominator of all known terrestrial organisms, and all present evidence of solvents on the Moon's surface is highly controversial [2, 3]. The hypothesis that beneath the lunar surface material one would find both water traces and relics of primitive organisms is not so unreasonable as to warrant the immediate dismissal of the matter of infection.

Mars is arid by terrestrial standards; its polar caps consist of thin hoar frost, and dense terrestrial type water clouds have never been observed. However, the polar caps retreat and the equatorial dark areas advance with the onset of the Martian spring. The pressure (85 mb. or less) and temperature (200-300° K) are so low it is frequently supposed that the presence of liquid water on the surface is very rare. This point has been refuted (4, 5]. If salts are present on the Martian surface, an anti-freeze mechanism can occur. It is reasonable to suggest that the dark areas of the planet contain salts, perhaps in the form of deposits left behind by dried-up seas.

SINTON [6, 7] has found three small absorption dips, at 3.43, 3.57, and 3.67 microns, associated with the dark areas of Mars. This suggests the presence of organic matter on Mars, but the question of its origin is an open question.

These few facts indicate Mars may be a promising subject, both for basic biological research and *infection*.

Theories of Venus are so varied, and the facts so few, it is imperative to be very cautious, at least in the early stages of exploration.

Space-Flight Environment

On cursory examination, probe sterilization may appear to be unnecessary because the space-flight environment is so hostile to terrestrial organisms. Several self-sterilizing mechanisms which immediately suggest themselves are:

---

[1] These comments would be unnecessary but for the fact that many people (not all laymen) presuppose that no significant discoveries will be made until a man is landed on the planets.

1. Ultraviolet radiation from the sun
2. Space vacuum
3. High temperatures on the Moon's surface
4. Heat of impact, or impact explosion on the Moon
5. Heat of entry into a planetary atmosphere.

Parts of the probe will never be exposed to sunlight. Ultraviolet radiation will destroy organisms which are nakedly exposed, but its penetrating power is so low that organisms can survive if surrounded by only a small group of dead ones.

Laboratory vacuum is employed to help preserve micro-organisms. There is, as yet, no knowledge on space vacuum being bactericidal. Perhaps this question can be answered in the near future by means of a satellite experiment.

The Moon and the Planets most likely have cracks and fissures on their surfaces which would protect organisms from exposure to high temperatures and ultraviolet radiation [5, p. 306].

A probe hard-landed on the Moon would have an impact velocity of approximately 3 kilometers per second. This is not sufficient kinetic energy to melt or vaporize the probe on impact, but it is sufficient to scatter parts of the payload all over the Moon's surface if the initial impact were on a hard surface, such as a mountain. The orbital velocity of a satellite in the Moon's gravitational field is roughly 2 kilometers a second. It is contended that the probe would bury itself in the Moon's surface. We do not believe that any of the supporting arguments presented thus far are sufficiently convincing to be dogmatic.

A probe that unintentionally enters an atmosphere has a high probability of coming in at a shallow angle, which is the ideal approach for a successful landing [8, 9]. The probe may shed a few parts during the planet fall, but the bulk of it would strike the surface. A successful descent on Mars would be comparatively simple because the tenuous atmosphere of Mars extends so far out from its surface. Furthermore, meteors have been found whose interiors show no evidence of having been heated appreciably [10]. It is evident that the rigors of a space journey are not a reliable means of preventing biological contamination.

International Discussion

CETEX (Committee on Contamination by Extraterrestrial Exploration), representing the International Scientific Unions, has published two reports [11, 12] in an attempt to set a tone for developing a code of conduct in space research. These reports imply that, particularly as regards biological exploration, a purely national program does not have much chance of being fruitful.

The CETEX reports recommend the sterilization of space probes, but they do not suggest a procedure for sterilizing probes nor do they suggest what tolerances would be acceptable. In this paper, we discuss both an operational approach to sterilization and the value judgements that will have to be faced by the operational agencies responsible for launching space vehicles.

Operational Tactics

Sterilizing space probes is an engineering nuisance; however, the same ordeal has confronted surgical crews for quite some time. In both instances, anticipation of the task is necessary.

At this time, it is possible to anticipate and recommend four phases of payload sterilization for all deep space missions. They are, in sequence:

1. Sterile assembly of components, particularly heat sensitive ones,
2. Built-in sterilization of parts, particularly where traces of water are admissible,
3. Terminal sterilization,
4. Maintaining sterilization.

A microbiological testing procedure must also be integrated into the sterilization operations.

Terminal Sterilization

Phase three, terminal sterilization, is the most important operation and shall be discussed first.

All known micro-organisms perish when subjected to dry steam at 160° C for twenty minutes [13]. There is a time-temperature effect. Micro-organisms can survive much higher temperatures over a shorter period of time; such as, the flash temperatures in explosions. However, approximately 20% of the components that go into payloads with which we are now familiar cannot endure 160° C. A more general disinfectant for this purpose is ethylene oxide gas [14].

Ethylene oxide (C_2H_4O) is the simplest of the ethers. It is a very small molecule and therefore dissolves in many substances, such as rubber, plastic, and oil. As a result of these properties, under slight pressure ethylene oxide is quite penetrating, working its way into the small interstices of most components. It is non-corrosive, and its human toxicity is low.

Ethylene oxide is a few thousand times more effective as a sporicide than other powerful disinfectants [15]. Viruses are more sensitive to ethylene oxide than many other organisms, whereas, they are much more resistant to radiation.

Ethylene oxide is inflammable in air in concentrations as low as 3%. However, a mixture of 10% ethylene oxide and 90% carbon dioxide (sometimes called carboxide) is not inflammable even when infinitely diluted with air. This mixture at 2 atmospheres pressure and 25° C would sterilize most parts of the probe in four hours. The sterilization could take place in a polyethylene tent and left there to retain its sterility for quite some time.

This part of the sterilization technique is well established. The U.S. Army Chemical Corps has sterilized many pieces of delicate laboratory apparatus without damage. They have also sterilized Air Force bombers and a commercial aircraft, in which a vial of live polio virus was accidentally broken.

Gaseous sterilization will not prove effective on certain impenetrable components. For these parts (paper capacitors for example) heat sterilization or radiation can usually be employed.

It is impractical to sterilize an entire payload with radiation. It is useful for certain small, sealed heat-sensitive components such as mylar capacitors.

The radiation dose required for some specified degree of sterilization is proportional to the natural logarithm of the number of bacteria. For 10^5 bacteria per gram of material, a dosage of 10^6 to 10^7 rem is required for good sterilization depending upon the organism. Actually 10^5 bacteria is a very high bacteria loading for most payload materials.

The Jet Propulsion Laboratory selected some sealed heat-sensitive components for radiation treatment by the General Electric Corporation. Two packages of identical parts were exposed to 10^6 and 10^7 rem from a Co^{60} source of gamma rays. A majority of these components withstood 10^7 rem. The most important exceptions were transistors and mercury cell batteries.

We estimate that, between gas, heat, and radiation (terminal sterilization), 95% of the payload parts can be readily sterilized without fear of degrading their performance characteristics.

Sterile Assembly

The removal of dust and foreign particles from the space probe eliminates a major source of biological pollution and it is, at the same time, an engineering virtue. (Most atmospheric pollution is borne by dust particles, except perhaps in closed rooms crowded with human beings.)

The washing and scrubbing of parts of the payload with water and detergents (or other more acceptable solvents) can reduce the number of microbes on the probe by three orders of magnitude.

Other aspects of sterile assembly include using compounds that are made sterile. Parts such as screws and bolts can be dipped in any of a number of sterilizing solutions. If screws and fitting holes are made to fit exactly, then care must be taken to sterilize before joining. Such fittings will remain sterile. If the fittings are not perfectly joined, the ethylene oxide gas will penetrate and sterilize these interstices.

Built-In Sterilization

Wherever possible, substances which are inimical to the well being of microorganisms should be employed. Certainly, substances of biological origin, such as casein glue or shellac, should be avoided.

Recently, germicides that contain organo-metallic compounds as active ingredients have been used to disinfect hospitals. These substances might prove valuable during the fabricating of sealed components with parts that get slightly contaminated with handling. This reduction of the contamination load during the initial stages provides an opportunity to attempt terminal sterilization radiation at a considerably reduced dosage, something of the order of 10^4 rem.

Built-in sterilization is not so much a specific technique as it is a philosophy of preparation for terminal sterilization.

Maintaining Sterilization

Once the space probe is sterilized, it will be necessary to mount it on the rocket boosters. The technical problem is then one of keeping microbes from coming into contact with the probe.

The probe is encased in a protective metal shroud during the launch phase of the space flight. The shroud can be employed to house a disinfectant atmosphere throughout the count down and flight through the atmosphere. The disinfectant can be either carboxide, employed in the terminal phase, or a faster acting but less penetrating gas, such as beta-propiolactone or ethylene imine.

Testing Procedure

In the past, several identical payloads were made for each mission. If this policy can be continued, it will not be difficult to produce convincing statistical arguments as to whether or not the payload meets the desired sterilization standard. Difficulties may arise, however, when the payloads become larger and more expensive. In this respect, it would be most practical to turn terminal sterilization and the sterilization certification over to an organization outside the space-flight groups. It would still be the space agency's responsibility to integrate this independent statistical estimate of sterilization with the other probabilities involved. This brings us to the problem of determining acceptable contamination tolerances.

Biological Contamination Tolerances

Now we get to the heart of the matter as it is not practical to pursue codes of conduct and to employ testing techniques unless the community places a subjective value upon what the biologists want to protect. Discussions of the ethics of contamination are made confusing by people who persist in believing that sterility is an absolute, to which only a yes or no answer applies.

The answer to the question of probe sterility can be given only in terms of probabilities. When large numbers of micro-organisms are subjected to lethal treatment, the live count drops off exponentially with time, or approximately so. The process is mathematically similar to the radioactive decay of an unstable nucleus. The death of a micro-organism has no clear-cut definition.

A group of biologists in the United States, including some of the nation's most eminent microbiologists, biochemists, and biophysicists, who are also sensitive to the engineering areas in space research, have given this problem some intensive thought.

For Mars and Venus. the consensus is that the probability of landing one viable organism should be less than one in a million. This means that, if the probability of successfully impacting a probe were judged *a priori* to be one in a hundred, it would be necessary to sterilize the payload to a tolerance of one chance in ten thousand that it have a live organism. We are investigating what degrees of sterilization can be expected as the space program evolves.

As previously indicated, the status of the Moon as a biologically interesting target is considerably more doubtful than that of the planets; therefore, it is more difficult to get an intuitive grasp of what tolerances are acceptable. We tentatively suggest that one chance in ten (perhaps one hundred) of a viable organism remaining on the probe be an acceptable infection tolerance. We also suggest that pollution be kept less than 10^8 dead organisms per probe for Moon and planetary shots.

These tolerance levels are submitted here for general evaluation, with the understanding that, as more information on the celestial bodies becomes available, the levels should be revised.

Recommendations and Conclusions

Planetary biology is one of the most exciting areas of space exploration. The unnecessary destruction of potential information in this research field by contamination would be an uncultural event. It is feasible to sterilize probes in such a manner that the loss of information to future investigators is minimized. This can be accomplished utilizing ethylene oxide, heat and radiation, accompanied by the sterile assembly of special components, as sterilizing agents.

Pollution tolerances should be kept to 10^8 dead bacteria per missile. Infection tolerances should be kept to less than 10^{-6} per missile for the planets and 10^{-1} for the Moon.

A molecular inventory, preferably in the form of payload duplicates, should be kept for each space flight. More information on the chemical composition of space-probe materials should be acquired.

An agency specially qualified to handle sterilization should perform the terminal disinfection and ascertain the degree of sterilization.

Acknowledgments

American scientists have been very patient while rocket technicians have picked their brains for information of value to space research. It has been precisely by this technique that we accumulated the facts contained in this paper. We hope that, by recognizing the gravity of the problem, we have partially compensated for our lack of originality.

Numerous people working with the National Academy of Sciences have assisted us in formulating our ideas and we thank them all. In particular we want to mention Joshua Lederberg for his characteristic insight, and Charles Phillips for making his work known to us.

References

1. J. Lederberg and D. B. Cowie, Moondust. Science 127, 1473 (1958).
2. D. Alter, The Alphonsus Story, Proceedings of Lunar and Planetary Exploration Colloquium. North American Aviation, Inc., Los Angeles, California, January 12, 1959.
3. N. A. Kozyrev, Volcanic Activity on the Moon? Sky and Telescope 18 (3), 123, 131 (1959); also, Observations of a Volcanic Process on the Moon. Ibid. 18 (4), 184 (1959).
4. C. Sederholm, H. Weaver, and C. Sagan, Private Communication, 1959.
5. G. De Vaucouleurs, Physics of the Planet Mars, p. 269. London: Faber and Faber, Ltd., 1954.
6. W. M. Sinton, Further Evidence of Vegetation on Mars. Science 130 (3384), 1234 (1959).
7. G. A. Tikhov, Is Life Possible on Other Planets? J. Brit. Astronom. Assoc. 65 (3), 193 (1955).
8. C. Gazley, The Penetration of Planetary Atmospheres. Rand Report, P-1322. The Rand Corporation, Santa Monica, California, February 24, 1958; also, Private Communication.
9. D. R. Chapman, An Approximate Analytical Method for Studying Entry into Planetary Atmospheres. TN 4276, National Advisory Committee on Aeronautics, Washington, D.C., May 1958.
10. F. Whipple, Private Conversation.
11. Contamination by Extraterrestrial Exploration. Science 128, 887 (1958).
12. Contamination by Extraterrestrial Exploration. Nature 183 (4666), 925 (1959).
13. H. O. Halvorsen, Spores. Washington, D.C. American Institute of Biological Sciences, 1957.
14. C. R. Phillips and S. Kaye, The Sterilizing Action of Gaseous Ethylene Oxide. Amer. J. Hygiene 50, 270 (1949).
15. C. R. Phillips and B. Warshowsky, Chemical Disinfectants. Ann. Rev. Microbiol. 12, 525 (1958).